CONTINUOUS ISSUES IN NUMERICAL COGNITION

CONTINUOUS ISSUES IN NUMERICAL COGNITION

HOW MANY OR HOW MUCH

Edited by

AVISHAI HENIK

Department of Psychology and Zlotowski Center for Neuroscience, Ben-Gurion University of the Negev, Beer Sheva, Israel

AMSTERDAM • BOSTON • HEIDELBERG • LONDON
NEW YORK • OXFORD • PARIS • SAN DIEGO
SAN FRANCISCO • SINGAPORE • SYDNEY • TOKYO
Academic Press is an imprint of Elsevier

Academic Press is an imprint of Elsevier
125 London Wall, London EC2Y 5AS, UK
525 B Street, Suite 1800, San Diego, CA 92101-4495, USA
50 Hampshire Street, 5th Floor, Cambridge, MA 02139, USA
The Boulevard, Langford Lane, Kidlington, Oxford OX5 1GB, UK

British Library Cataloguing-in-Publication Data
A catalogue record for this book is available from the British Library

Library of Congress Cataloging-in-Publication Data
A catalog record for this book is available from the Library of Congress

ISBN: 978-0-12-801637-4

For information on all Academic Press publications
visit our website at https://www.elsevier.com/

Publisher: Mara Conner
Acquisition Editor: Natalie Farra
Editorial Project Manager: Kristi Anderson
Production Project Manager: Chris Wortley
Designer: Matthew Limbert

Typeset by Thomson Digital

Dedication

To Rachel

Contents

11. Magnitudes in the Coding of Visual Multitudes: Evidence From Adaptation

FRANK H. DURGIN

12. Ordinal Instinct: A Neurocognitive Perspective and Methodological Issues

ORLY RUBINSTEN

13. Discrete and Continuous Presentation of Quantities in Science and Mathematics Education

RUTH STAVY, REUVEN BABAI

14. Interaction of Numerical and Nonnumerical Parameters in Magnitude Comparison Tasks With Children and Their Relation to Arithmetic Performance

SWIYA NATH, DÉNES SZŰCS

List of Contributors

Christian Agrillo Department of General Psychology, University of Padova, Padova; Cognitive Neuroscience Center, Padova, Italy

Daniel Ansari Numerical Cognition Laboratory, Department of Psychology; Brain & Mind Institute, University of Western Ontario, Ontario, Canada

Reuven Babai Department of Mathematics, Science and Technology Education, The Constantiner School of Education and The Sagol School of Neuroscience, Tel Aviv University, Tel Aviv, Israel

Amit Benbassat Department of Computer Science, Ben-Gurion University of the Negev, Beer-Sheva, Israel

Michael J. Beran Department of Psychology and Language Research Center, Georgia State University, Atlanta, GA, United States of America

Angelo Bisazza Department of General Psychology, University of Padova, Padova; Cognitive Neuroscience Center, Padova, Italy

Marinella Cappelletti Department of Psychology, Goldsmiths College University of London, London; University College London, Institute of Cognitive Neuroscience, London, United Kingdom

Jenny Yun-Chen Chan Institute of Child Development, University of Minnesota, Minneapolis, MN, United States of America

Roi Cohen Kadosh Department of Experimental Psychology, Oxford, United Kingdom

Alain Content Laboratoire Cognition, Langage & Développement, Université Libre de Bruxelles, Brussels, Belgium

Maria Dolores de Hevia University Paris-Descartes, Sorbonne Paris Cité, Paris; Laboratory of Psychology of Perception, Paris, France

Frank H. Durgin Swarthmore College, Department of Psychology, Swarthmore, PA, United States of America

Nancy Dyson University of Delaware, Newark, DE, United States of America

Titia Gebuis Department of Molecular and Cellular Neurobiology, Center for Neurogenomics and Cognitive Research, Neuroscience Campus Amsterdam, VU Amsterdam, Amsterdam, The Netherlands

Wim Gevers Center for Research in Cognition and Neurosciences (CRCN), Université Libre de Bruxelles and UNI—ULB Neurosciences Institute, Brussels, Belgium

Shai Itamar Department of Psychology, Ben-Gurion University of the Negev, Beer-Sheva; The Zlotowski Center for Neuroscience, Ben-Gurion University of the Negev, Beer-Sheva, Israel

Nancy C. Jordan University of Delaware, Newark, DE, United States of America

Arava Y. Kallai Department of Psychology, Ben-Gurion University of the Negev, Beer-Sheva; The Zlotowski Center for Neuroscience, Ben-Gurion University of the Negev, Beer-Sheva; Department of Psychology, Max Stern Yezreel Valley College, Emek Yezreel, Israel

Gali Barabash Katz Department of Cognitive Sciences, Ben-Gurion University of the Negev, Beer-Sheva, Israel

Tali Leibovich Department of Cognitive and Brain Sciences, Ben-Gurion University of the Negev, Beer-Sheva; The Zlotowski Center for Neuroscience, Ben-Gurion University of the Negev, Beer-Sheva, Israel; Department of Psychology, The University of Western Ontario, Canada

Susan C. Levine University of Chicago, Chicago, IL, United States of America

Stella F. Lourenco Emory University, Atlanta, Georgia, United States of America

Michèle Mazzocco Institute of Child Development, University of Minnesota, Minneapolis, MN, United States of America

Maria Elena Miletto Petrazzini Department of General Psychology, University of Padova, Padova, Italy

Kelly S. Mix Michigan State University, East Lansing, MI, United States of America

Swiya Nath Department of Psychology, University of Cambridge, Cambridge, United Kingdom

Nora S. Newcombe Temple University, Philadelphia, PA, United States of America

Julie Nys Laboratoire Cognition, Langage & Développement, Université Libre de Bruxelles, Brussels, Belgium

Audrey E. Parrish Department of Psychology and Language Research Center, Georgia State University, Atlanta, GA, United States of America

Bert Reynvoet Brain & Cognition, Faculty of Psychology and Educational Sciences, KU Leuven, Leuven; Faculty of Psychology and Educational Sciences, KU Leuven, Kulak, Belgium

Orly Rubinsten Edmond J. Safra Brain Research Center for the Study of Learning Disabilities, Department of Learning Disabilities, University of Haifa, Haifa; Center for the Neurocognitive Basis of Numerical Cognition, Israel Science Foundation, Israel

Delphine Sasanguie Brain & Cognition, Faculty of Psychology and Educational Sciences, KU Leuven, Leuven; Faculty of Psychology and Educational Sciences, KU Leuven, Kulak, Belgium

Maria Sera Institute of Child Development, University of Minnesota, Minneapolis, MN, United States of America

Moshe Sipper Department of Computer Science, Ben-Gurion University of the Negev, Beer-Sheva, Israel

Karolien Smets Brain & Cognition, Faculty of Psychology and Educational Sciences, KU Leuven, Leuven, Belgium

H. Moriah Sokolowski Numerical Cognition Laboratory, Department of Psychology; Brain & Mind Institute, University of Western Ontario, Ontario, Canada

Ruth Stavy Department of Mathematics, Science and Technology Education, The Constantiner School of Education and The Sagol School of Neuroscience, Tel Aviv University, Tel Aviv, Israel

Dénes Szücs Department of Psychology, University of Cambridge, Cambridge, United Kingdom

Acknowledgments

Over 2 years ago, in September 2013, I organized a symposium titled "Continuous Issues in Numerical Cognition" during the 18th Conference of the European Society for Cognitive Psychology (ESCoP) in Budapest, Hungary. I asked speakers in the planned symposium to examine the accepted views in the field of numerical cognition and, in particular, the possibility that perception and evaluation of noncountable dimensions (eg, sizes or amounts) might contribute to the development of numerical cognition. Several contributors to the current volume, namely, Christian Agrillo, Marinella Cappelletti, Tali Leibovich, Gali Katz, Stella Lourenco, and Bert Reynvoet presented at the symposium. This is an occasion to thank them for their participation in that symposium and to thank ESCoP for the opportunity.

The symposium was financed in part by the European Research Council (ERC), which is currently supporting my research on this theme (advanced researcher grant 295644). I would like to extend my thanks to the ERC for supporting my research, the symposium, and the current venture.

In the last 8 years, my research in the field of numerical cognition was supported by the Israel Science Foundation (ISF) and I would like to thank the ISF for their continuous support.

Last but not least, I would like to thank my students and colleagues each of whom contributed his or her own way to develop ideas, thoughts, and plans of research throughout the years.

Avishai Henik

Introduction

Avishai Henik

Department of Psychology and Zlotowski Center for Neuroscience, Ben-Gurion University of the Negev, Beer Sheva, Israel

Numerical cognition is important for the development of solid knowledge in arithmetic and the latter is a good predictor of success in life. Reynvoet, Smets, and Sasanguie (Chapter 9) devote the first part of their contribution to discuss the importance of the number sense (a set of abilities and intuitions regarding the representation of numbers and underlying mathematical achievement) in various activities. They note differences in the definition of the number sense but stress that regardless of these differences, there is a wide agreement that number sense is crucial for higher mathematical abilities and success in life (see also Jordan & Dyson, Chapter 3).

In recent years, research in numerical cognition has expanded and we have witnessed a flourishing field that produced major advances toward unraveling the building blocks of numerical cognition and its development. Research in the area of numerical cognition led to a widely accepted view of the existence of innate, domain-specific, core numerical knowledge based on the ability to perceive and manipulate discrete quantities (eg, enumeration of dots). This core knowledge involves the intraparietal sulcus (IPS) in the brain. Moreover, a deficiency in this core knowledge is thought to be the basis for arithmetic disability. An influential model (Feigenson, Dehaene, & Spelke, 2004) suggests the existence of two systems. One system represents large, approximate numerical magnitudes, that is, the approximate number system (ANS), and a second system devoted to small sets of objects, is based on the ability to track several objects separately, and enables the precise knowledge of the number of objects. However, several findings suggest that this wide agreement needs to be examined carefully. In particular, similar to sensitivity to discrete quantities (eg, enumeration of objects), infants as well as adults show sensitivity to noncountable continuous dimensions like perception of area and size. Moreover, numbers are intimately associated with noncountable dimensions (eg, area, brightness, time). As is the case in many other research areas, questions and doubts regarding accepted views continuously fuel research and bring up new developments and insights.

In order to study enumeration or evaluation of quantities in adults and children, researchers have used various tasks. One example is a comparative judgment task. Participants are presented with two arrays of dots and asked to decide which array has more dots. As has been suggested by many authors, the two dot arrays involve not only numerosity but also other visual properties; average diameter, total surface (the sum of the surface area for the dots in each array), area extended (the smallest contour that includes all of the dots, as if an elastic band were wrapped around the dots), density (area extended divided by total surface), and total circumference (the sum of the circumferences of all dots in an array). Because researchers were interested in numerosity and its development, they attempted to control the visual confounding variables. Interestingly, it has been suggested that such control might be either almost impossible or very complicated (Gevers, Cohen Kadosh, & Gebuis, Chapter 18; Leibovich, Kallai, & Itamar, Chapter 16; Lourenco, Chapter 17). The reason is apparently the natural correlation between average diameter, total surface and other visual properties and numerosity, which is very difficult to eliminate.

Several chapters in this volume discuss the relationship between countable and noncountable (possibly confounding) variables. De Hevia (Chapter 2) presents evidence that numerical information is closely related to space; both to spatial extent, reflecting magnitude, and to spatial positions, reflecting order. She suggests that there are "fundamental, privileged mappings between number and spatial extent" which are "functional at birth, before the acquisition of language, counting, symbolic knowledge, or before extensive experience with correlations between these dimensions, and remain constant throughout childhood and adulthood after the onset of education." Mix, Hutenlocher, and Levine (2002a, 2002b) were one of the first groups that stressed the difficulty in disentangling the effects of number and visuo-spatial perception in studies of early development. In a current chapter herein, Mix, Levine, and Newcombe (Chapter 1) review early infant research on this subject and suggest that in spite of the difficulty in isolating effects of discrete numbers using looking time methods, infants gain a sense of number due to their sensitivity to other quantitative percepts. They discuss the contribution of visuo-spatial processes to the development of a general quantification system. Similarly, Leibovich and coworkers; (Leibovich & Henik, 2013; Leibovich, Kallai, & Itamar, Chapter 16) suggest that access to number sense may be gained through the natural correlation between numerosity and spatial extent or other visual properties of environmental stimulation. Interestingly, Mix et al. (2002a, 2002b, Chapter 1) discuss perceptual learning processes (eg, cue redundancy, chunking or unitization, development of selective attention toward specific dimensions that were previously fused in perception) by which infants might gain access to the number sense from correlated percepts.

Another relevant possibility is described by Gevers, Cohen Kadosh, and Gebuis (Chapter 18). These authors propose that various sensory properties of stimuli are taken into account when estimation of numerosity is required. The various (confounding) properties are weighed and modulate evaluation of quantities.

Importantly, Content and Nys (Chapter 10) and Durgin (Chapter 11) present what seem to be two opposing views regarding the ability to extract pure numerosity from nonsymbolic displays. Content and Nys present two experiments with 3–6-year-old children. They report that children were able to extract numerosity in visual arrays without counting. Moreover, they provide evidence that children's estimates were not based on continuous dimensions of stimuli. Durgin examined the relationship between texture density and perceived number in adaptation experiments with adults. Durgin summarizes that adaptation to dense or numerous displays affects both numerosity comparison and numerosity estimation. As a reminder of Mix and colleagues' (Mix et al., 2002a, 2002b) early suggestions, Durgin points out that density rather than numerosity was adapted in his adaptation experiments.

Two contributions, one by Beran and Parrish (Chapter 8) on nonhuman primates and one by Agrillo, Petrazzini, and Bisazza (Chapter 7) on basal vertebrates, present reviews of the debate regarding animals' ability to use discrete numerical information. These two contributions suggest that animals use visual and other nonnumerical information in order to extract magnitude. This suggestion notwithstanding, there seem to be indications that vertebrates can also represent and use discrete quantities. However, as Beran and Parrish suggest, " … whether animals are capable of exhibiting the boundless, cardinality-based system for exact representation of number concepts is also unclear, and perhaps unlikely." (p. 187).

The need to control or to eliminate effects of the confounding visual properties might have additional effects and in particular, it might introduce additional mental operations. Reynvoet, Smets, and Sasanguie (Chapter 9) suggest that various tasks designed to measure the number sense (eg, habituation in infants and comparative judgment in older children) might assess additional processes rather than purely reflect numerical processing. A similar suggestion is discussed by Nath and Szucs (Chapter 14). Let us go back to the comparative judgment task mentioned earlier. In order to control the effects of irrelevant visual properties, researchers attempt to reduce the correlation between numerosity and various visual properties. Commonly, this is achieved by having both congruent and incongruent stimuli. In congruent stimuli all properties are compatible so that the array with more dots is also characterized by a larger density. In contrast, incongruent stimuli are characterized by incompatibility between the various variables so that the array with more dots has a smaller density. Hence, across stimuli there is no correlation between numerosity and visual properties. Gevers and colleagues suggest that in spite of this fact, participants might still use visual properties in comparison of numerosities, and Nath and Szucs suggest that such a design introduces the need to inhibit the irrelevant dimension so that the task involves both numerosity comparison and inhibitory processes. Note that Nath and Szucs discuss another task that might require inhibitory processes—the numerical Stroop or the size congruity task. In this task (Henik & Tzelgov, 1982), participants are presented with two digits that differ both in their numerical values and in their physical sizes. In separate blocks, participants are asked to decide which digit is numerically larger and ignore the physical sizes or to decide which digit is physically larger and ignore the numerical values. Commonly, participants are slower to respond to incongruent trials (eg, large 3 and small 5) than to congruent trials (eg, small 3 and large 5) regardless of task. The size congruity effect in physical judgments (when the numerical values are irrelevant) has been considered as an indication for automatic processing of numerical values. Nath and Szucs suggest that this task, rather than measuring the effects of numerical values, is a good predictor of inhibitory control. It is clear that researchers' attempts to control irrelevant variables might introduce mental operations that were not intended to be measured in the first place. The potential effect of inhibitory processes is discussed by Cappelletti (Chapter 5) also. Cappelletti's contribution revolves around effects of aging on numerical abilities. She proposes that numerical abilities in the aged population might be compromised. One example is related to the ability to estimate or enumerate large sets of data (Cappelletti reports a larger Weber fraction in aged than in young populations). However, she suggests that such decline in numerical abilities might be related to some domain-general factors, like deterioration in inhibitory processes, rather than deficiencies in numerical functions per se.

Stavy and Babai (Chapter 13) examine continuous and discrete modes of presentations in comparisons of perimeters, dots, and areas. They suggest that for a given stimulus and task, there is a compatible (and preferred) mode of presentation. Changing modes of presentation seem to change the saliency of the relevant and irrelevant aspects of the stimulus. Accordingly, such a change might improve or be detrimental to performance. For example, in their third study, discrete presentation of area (presentation of area as a scattered array of rectangles) decreased comparative judgment of areas. It seems that an incompatible mode of presentation might increase interference of irrelevant aspects of the stimuli.

Mazzocco, Chan, and Sera (Chapter 4) draw attention to the role of language in the development of the number sense. They discuss the fact that number words may relate to discrete or continuous quantities and their meaning is influenced by context (eg, six short walks are more walks than two long walks but involve less walking than two long walks). Mazzocco and colleagues suggest number words may be confusing (eg, two pound apples) and that the ability to extract meaning from number words may be an important aspect of developing a solid number sense. It is possible that sensitivity to contextual changes in number word meanings affects learning trajectories during the earliest years of schooling and may predict development of numerical cognition.

Rubinsten (Chapter 12) points out that humans are prewired to perceive order and that the ability to extract order might be a major building block in the development of numerical abilities. This is reminiscent of Gelman's early work (Gelman, 1978, 1980) on counting. Gelman suggested that the ability of a young child to count is based on five principles—the counting principles. The first two principles are: "(a) The one-one principle—each item in an array must be tagged with one and only one unique tag; (b) The stable-order principle—the tags used must be drawn from a stably-ordered list (p. 161)." (Gelman, 1980). Hence, the ability to extract and use order seems to be essential for the development of counting. Katz, Benbassat, and Sipper (Chapter 6) examine whether counting might have evolved from a more primitive size perception system. They used evolutionary computation techniques to generate artificial neural networks (ANNs) that excelled in size perception. They suggest that these systems presented an advantage in evolving the ability to count compared with ANNs that evolved to excel in counting only. Moreover, ANNs that trained to perceive size with continuous stimuli presented better counting skills than those trained with discrete stimuli. It might be the case that the ability to compare sizes or similar properties might be essential for the ability to extract order.

Several authors suggested the existence of a general magnitude system. Sokolowski and Ansari (Chapter 15) review the literature related to the neural underpinning of numerical cognition. They conclude that numerical information is processed by a general magnitude system that involves the right parietal cortex whereas the left parietal cortex seems to support the processing of symbolic information. The general magnitude system that involves the right parietal lobe processes both numerical and nonnumerical continuous magnitudes such as space, time, and luminance. Lourenco (Chapter 17) presents evidence for the existence of a general magnitude system that represents magnitude across numerical and nonnumerical, and discrete and continuous magnitudes. The idea is that integration rather than differentiation is the rule and exchange among magnitudes does not necessitate mediation by other systems such as language. Moreover, Lourenco argues that this general system of magnitude representation is operational from early life. What neural system might support such direct connections between magnitude? Lourenco suggests that one possibility is the existence of population neurons that share coding for different magnitudes (eg, in terms of more vs. less relations). Leibovich and colleagues

(Chapter 16) suggest that both discrete and continuous magnitudes are processed together. Accordingly, they suggest the existence of an approximate magnitude system (AMS) rather than the ANS.

Jordan and Dyson (Chapter 3) draw attention to the need to monitor indications for proper development of number sense as early as possible (eg, kindergarten) because children with poor number sense are likely to encounter difficulties in acquiring arithmetic abilities later on. They present studies of interventions aimed at improving understanding of number, number relations, and number operations in high-risk children. They suggest that number-fact practice seems to produce more gains than other aspects of intervention (eg, number-list counting strategies). Interestingly, quite a few contributions in this volume suggest that the development of number sense is related to various aspects of the environment (eg, visual-spatial perception, and perception and evaluation of noncountable dimensions). One question is whether diagnosis and intervention of such nonnumerical dimensions might also help in the development of the number sense.

The current volume encompasses contributions regarding various aspects of numerical cognition and its development. Contributors discuss central issues in the field, they raise doubts regarding accepted views, bring up new developments, provide interesting insights, and propose important theoretical solutions that will fuel future research and discussions.

References

Feigenson, L., Dehaene, S., & Spelke, E. (2004). Core systems of number. *Trends in Cognitive Sciences, 8*, 307–314.

Gelman, R. (1978). Counting in the preschooler: What does and does not develop. In R. S. Siegler (Ed.), *Children's thinking: What develops?* (pp. 213–240). Hillsdale, N.J: Erlbaum.

Gelman, R. (1980). What young children know about numbers. *Educational Psychologist, 15*, 54–68.

Henik, A., & Tzelgov, J. (1982). Is three greater than five: the relation between physical and semantic size in comparison tasks. *Memory & Cognition, 10*, 389–395.

Leibovich, T., & Henik, A. (2013). Magnitude processing in non-symbolic stimuli. *Frontiers in Psychology, 4*, 375.

Mix, K. S., Huttenlocher, J., & Levine, S. C. (2002a). Multiple cues for quantification in infancy: is number one of them? *Psychological Bulletin, 128*, 278–294.

Mix, K. S., Huttenlocher, J., & Levine, S. C. (2002b). *Quantitative development in infancy and early childhood*. Oxford: Oxford University Press.

SECTION I

DEVELOPMENT

CHAPTER

1

Development of Quantitative Thinking Across Correlated Dimensions

Kelly S. Mix, Susan C. Levine**, Nora S. Newcombe*[†]

*Michigan State University, East Lansing, MI, United States of America; **University of Chicago, Chicago, IL, United States of America; [†]Temple University, Philadelphia, PA, United States of America

OUTLINE

Continuous Issues in Numerical Cognition. http://dx.doi.org/10.1016/B978-0-12-801637-4.00001-9

The origins of human quantification have been the subject of intense study for decades. Methodological advances in the 1960s allowed researchers to ask what perceptual information infants notice, and this led to a proliferation of research on numerical sensitivity. Based on initial evidence, there were strong claims about innate capacities for number and these claims defined the research agenda for many years. However, the basis for these claims has not gone unchallenged. Some investigators have questioned whether infants' quantitative responses are based on number at all, or rather, a host of correlated, nonnumerical percepts, such as contour length (eg, Mix, Huttenlocher, & Levine, 2002a).

Indeed, recent findings have led to a novel proposal—that human quantification is rooted in a generalized magnitude system that integrates these various perceptual streams into a unified representation and only later becomes differentiated into separate signals (Cantrell & Smith, 2013; Newcombe, Levine, & Mix, 2015). This starting point is a departure from previous accounts and raises many new questions. For example, in what sense are quantitative representations generalized? One possibility is infants do not notice all the different dimensions, but rather, respond to a lower level stimulus, such as intensity. Another possibility is that infants somehow combine the dimensions to arrive at a summative evaluation. Also, by what mechanisms do quantitative percepts become differentiated? Object exploration, language, and measurement could all play important roles (Mix, Huttenlocher, & Levine, 2002b; Newcombe et al., 2015).

Although the extant literature does not provide answers to these questions, we can seek clues in the historical roots of infant research and the literature on visuospatial perception and perceptual learning. From these sources, we may begin to frame alternative hypotheses that can advance research in this area. Toward that end, this chapter will begin with a brief review of the looking time methods used to study infant quantification and the challenges involved in isolating number using them. Second, we will consider the visuospatial processes that might contribute to quantification in a generalized magnitude system and the various ways these processes might be integrated. Third, we will identify learning mechanisms by which a generalized magnitude system could become differentiated into separate quantitative dimensions and eventually be reassembled into a coherent system in which discrete and continuous quantification are coordinated.

1.1 THE USE OF LOOKING TIME TO MEASURE INFANT QUANTIFICATION

1.1.1 Early Work on Infant Perception: The Advent of Looking Time Measures

Contemporary research on infant cognition is rooted in the pioneering efforts of psychologists in the 1960s, who developed methods to ask whether preverbal infants notice a change. Originally, these methods were used to ask about low-level sensitivity to form, orientation, color, and so forth (eg, Bower, 1967; McGurk, 1970; Ruff & Birch, 1974; Spears, 1964; Watson, 1966). But depending on how experimenters construct the contrasting stimuli, rather sophisticated concepts can be tested. To illustrate, researchers have used these perceptual discrimination methods to test whether 11-month-olds understand symbolic metaphors, such as arrows pointing up to indicate ascendancy and arrows pointing down to indicate descendancy (Wagner, Winner, Cicchetti, & Gardner, 1981).

The details of particular methods vary across studies, but they are all based on simple approach/withdrawal responses, such as looking toward or away from a display. Preferential looking is one way to capitalize on these responses. Two stimuli are presented side-by-side and infants' looking time is measured. If infants notice a difference, and prefer one stimulus over the other, they should look longer toward one display or the other. Habituation is another approach. Infants see a series of displays until they lose interest, as indicated by decreasing looking time, and then they are shown a display that differs in some way. If infants notice the change, they should look longer at the novel display because it is more interesting than the familiar, now boring, displays (though in some cases, preference for familiarity has been obtained). Interestingly, these methods were adapted from those used to test visual preferences in chicks, where the dependent measure was pecks instead of looks (Fantz, 1961).

The initial application of these methods was straightforward. Psychologists simply wanted to know what infants see or hear. For example, Fantz (1961) showed 2-month-olds various geometric forms and observed that they looked longer toward forms with greater complexity, such as a bull's eye (Fig. 1.1). There was a proliferation of such findings as theorists began mapping out the dynamic interplay among infant attention, stimulus characteristics, innate orienting biases, and fluctuations in arousal (Banks & Ginsburg, 1985; Cohen, 1972; Gardner & Karmel, 1984; Haith, 1980; Karmel, 1969a,b; Ruff & Birch, 1974). The notion that infants react to various stimulus dimensions based on low-level attentional biases and arousal was subsequently abandoned when these methods were applied to abstract constructs, such as number or shape, and innate representational systems were posited. Yet, the "rules babies look by" (eg, Haith, 1980) are engaged in any looking time task, even when the aim is to test higher-level concepts (Oakes, 2010).

Recent examinations of the attentional processes underlying looking time measures have continued to emphasize the contribution of low-level processing (Schoner & Thelen, 2006; Shultz & Bale, 2001; Sirois & Mareschal, 2002), as well as the artificiality of these measures relative to perceiving live scenes (Wass, 2014). In a particularly well-articulated model, Schoner and Thelen (2006) demonstrated that what seem to be dishabituation responses based on infant knowledge and higher-level concepts can actually be fully explained in terms of infant arousal, preferences

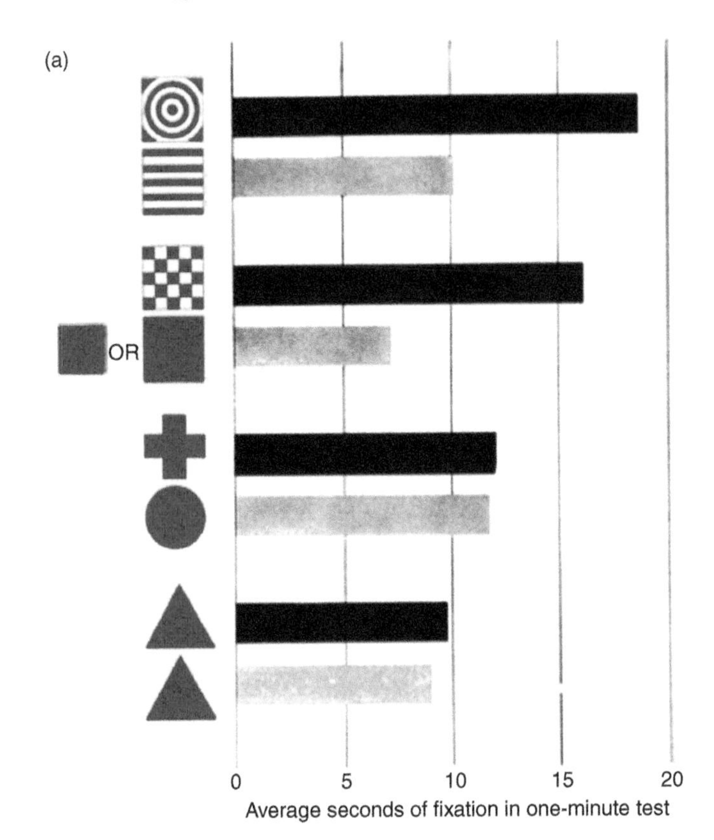

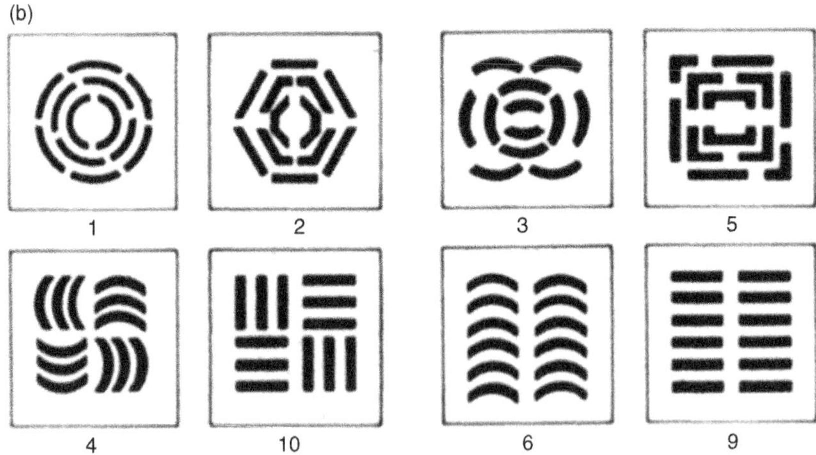

FIGURE 1.1 (a) Infants' looking toward various displays in paired comparisons. (b) Stimulus patterns used to test visual preferences in 13-week-old infants. The numerals under each pattern indicate its order of preference, with 1 being the pattern with the longest looking times and 10 being the pattern with the shortest looking times. *Part a: from Fantz (1961). Part b: from Ruff & Birch (1974).*

for novelty and familiarity, and theoretically irrelevant stimulus shifts (eg, similarity along irrelevant dimensions, depth of habituation, timing of shifts, etc.). These parameters are frequently unreported in research on higher-level concepts, such as number. Sometimes they have been controlled by counterbalancing for order, but order effects may be obscured if the sample sizes are not large enough to detect interactions or only a subset of the data is analyzed (eg, Starkey, Spelke, & Gelman, 1990; Xu & Spelke, 2000). As a result, it may be difficult to isolate higher-level processing, such as categorization, comparison, and so forth, using looking time methods because these methods cast a broad net—one that includes low-level perception and shifting attention weights based on superfluous inputs and changes in arousal.

The possible effects of low-level processing in looking time measures also help explain why, in many domains, these methods indicate an ability is present early, whereas active choice methods indicate the same ability is lacking in older children. Such discontinuities raise questions about whether looking time measures and active choice measures tap different underlying mechanisms, or they tap the same mechanisms but the "strength" of the knowledge needed to demonstrate competence on looking time tasks is lower (see Frick, Mohring, & Newcombe, 2014, for a recent review).

1.1.2 Studies of Infant Number

Initial investigations into the origins of number adapted looking time measures to numerical stimuli. For example, infants were shown displays of two objects until they reached habituation, and then they were tested with a novel set of two or a new number, like three (Starkey, Spelke, & Gelman, 1983, 1990; Strauss & Curtis, 1981). It was immediately apparent that infants could discriminate small sets (<4) under a variety of conditions (heterogeneous sets, homogeneous sets, sequential presentation, etc.). Habituation, preferential looking, and violation of expectation experiments subsequently demonstrated discrimination of sets (eg, Xu & Spelke, 2000; Xu, Spelke, & Goddard, 2005), intermodal equivalence (Starkey et al., 1983, 1990), and calculation (Wynn, 1992). (For a thorough review of this literature, see Cantrell & Smith, 2013).

The dominant interpretation has been that infants are endowed with two systems for discrete quantification (eg, Feigenson, Dehaene, & Spelke, 2004). One system, based on object-tracking representations studied in adults (eg, Kahneman, Treisman, & Gibbs, 1992; Trick & Pylyshyn, 1994), applies to small sets and uses spatial pointers to represent and track the precise number of items. The other system (ie, the approximate number system or ANS) is used to represent the numerosities of large sets and, though inexact, is based on discrete numbers. The notion of two distinct systems is bolstered by evidence that small and large number representations differ at both the behavioral and the neural levels in infants as well as adults (eg, Hyde & Spelke, 2009, 2011). However, these interpretations may be optimistic about the central role of discrete number because it is nearly impossible to isolate number in these looking time procedures (Soltesz, Szucz, & Szucs, 2010). As noted previously, looking time tasks measure only an approach response to one stimulus display or another. Although it is straightforward to vary the dimension of interest, it is not obvious how to control all the other dimensions that might drive such a response.

Indeed, recent evidence indicates that infants attend to a range of quantitative cues and use them interchangeably. For example, they form expectations about number based on length and temporal duration, as well as vice versa (de Hevia & Spelke, 2010; Hyde, Porter, Flom, & Ahlander-Stone, 2013; Lourenco & Longo, 2010; Srinivasan & Carey, 2010). Moving beyond visuotemporal dimensions to purely auditory stimuli, Srinivasan and Carey found no transfer between length and loudness, but there may be transfer from length to pitch (Dolscheid, Hunnius, Casasanto, & Majid, 2014). These findings suggest that the system underlying quantitative discriminations is not number specific, but rather may reflect the engagement of a generalized magnitude system (Cantrell & Smith, 2013; Newcombe et al., 2015).

Further support for this view comes from studies showing that number and spatial extent are linked in adults, preschool children, and infants (Cantrell, Boyer, Cordes, & Smith, 2015; Hurewitz, Gelman, & Schnitzer, 2006; Soltesz et al., 2010). For example, preschoolers' sensitivity to both number and area is based on the same Weber fractions with similar growth patterns for each, and children accurately use the word "more" in both number and area contexts, beginning at the same ages (Odic, Libertus, Feigenson, & Halberda, 2013; for slightly different results, see Odic, Pietroski, Hunter, Lidz, & Halberda, 2012). Neuroimaging studies have also revealed that topographic field maps for numerosity are based on nonnumerical sensory information and discrete number cannot be disentangled from these other cues (Gebuis, Gevers, & Cohen Kadosh, 2014). Finally, a recent study demonstrates that 9-month-olds make finer grained numerical discriminations when area and contour length covary with number (Cantrell et al., 2015).

1.2 GENERALIZED OVER WHAT?

The proposition of a generalized magnitude system brings us back to the difficulty of controlling quantitative dimensions because it is unclear what is being generalized or how. Anyone who has ever tried to design an experiment that isolates number can attest to the variety of perceptual streams that present potential confounds. However, these do not constitute a simple laundry list of variables. There is an implicit hierarchy of perceptual correlates of number that vary in their generality. Depending upon which of these correlates infants perceive, and which ones drive infant attention, there are different ways a magnitude system could be generalized.

One class of percepts includes dimensions that are clearly quantitative and related to discrete items, but are not numerical. This class would include cumulative contour length, surface area, and volume—dimensions that are continuous yet are inherent to the individual objects in a set (Fig. 1.2a). For example, if the task is to compare two apples to three apples, one way to do it is to detect the volume of each apple and sum over them. When infant number researchers design controls for nonnumerical variables, they often do so with this summing in mind. That is, they calculate the contour length of each item in a set and then control the total contour length summed over items in one display versus another (eg, Clearfield & Mix, 1999). Indeed, when Feigenson, Carey, and Spelke (2002) found that infants responded to contour length instead of number in their displays, they interpreted the effect precisely this way—that infants were representing each object as a discrete member of the set, and then extracting the contour lengths of each object and summing over them to generate a response. Thus, one sense of generalization could be that even continuous dimensions of a numerical set are derived from discrete number—they are essentially all signals pointing in the same direction for the same reason (ie, numerosity).

A second class of correlated percepts includes continuous dimensions that are quantitative but are inherent to the set as a whole rather than the individual objects, such as density and overall area. Because these variables are inversely related, they are impossible to control in a single trial. If two sets cover the same surface area, the more numerous set is necessarily more dense. If two sets are equally dense, the more numerous set covers a larger surface area (Fig. 1.2b). Because both dimensions constitute highly reliable cues to quantity, it is conceivable that infants rapidly learn both and switch between them. Interestingly, though, children have trouble coordinating set–set level correspondences, such as density, with object–object level correspondences, as Piaget (1952) demonstrated many years ago. So, if infants and young children use such cues, it is unlikely they do so with an awareness of the coordinated relations between them—an important point to which we will return.

Complexity is another set-level quantitative variable that is based on the notion that some visual scenes have more information to process than others (Fig. 1.3). Across studies using humans from infancy to preschool and other species, such as rats and chicks, complex stimuli have elicited

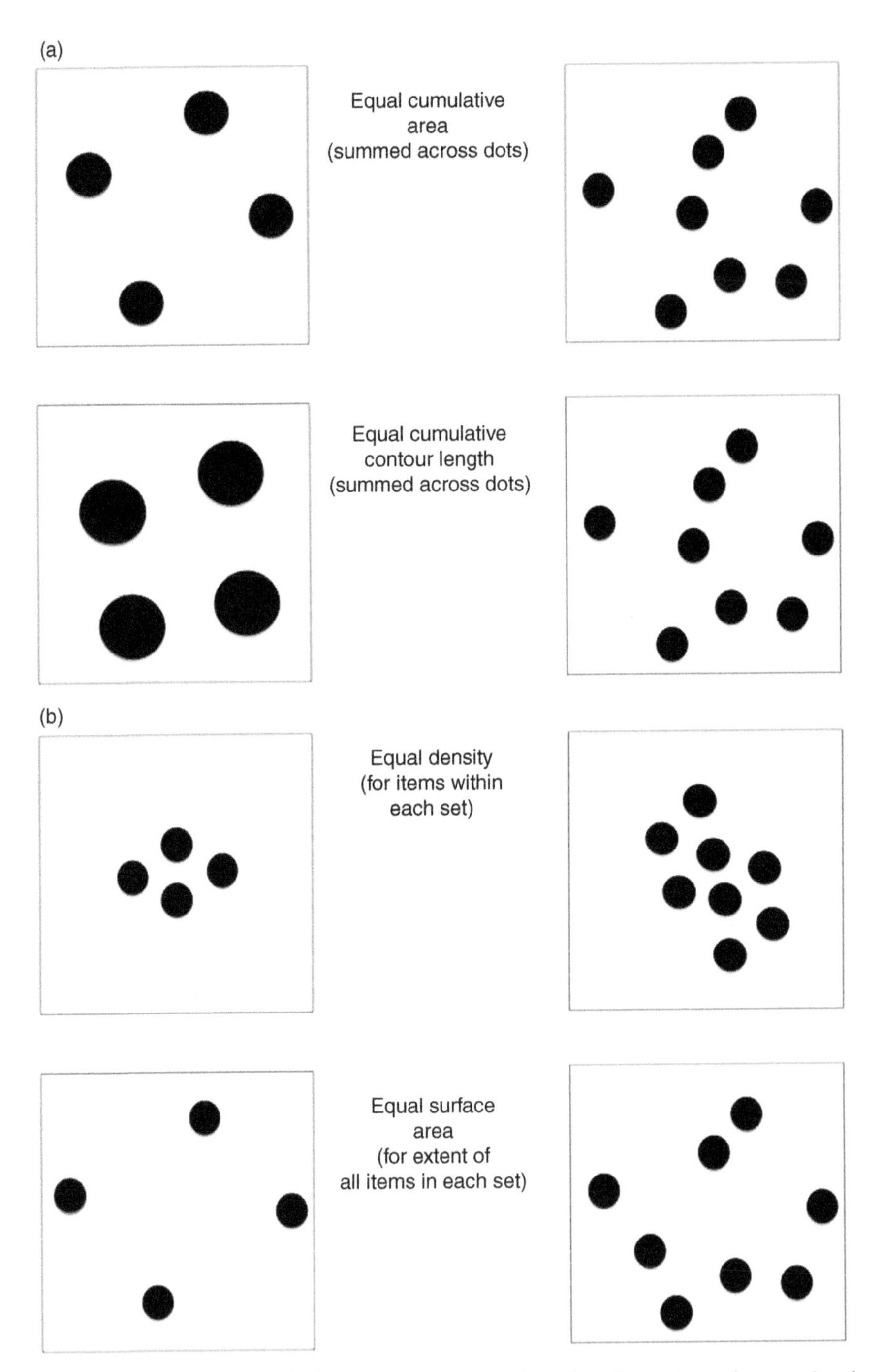

FIGURE 1.2 **The interrelations among area, contour length, and number in visual displays for both item level (a) and set level (b) measures.** The diameters of the dots are small = 0.32 in., medium = 0.45 in., and large = 0.65 in.

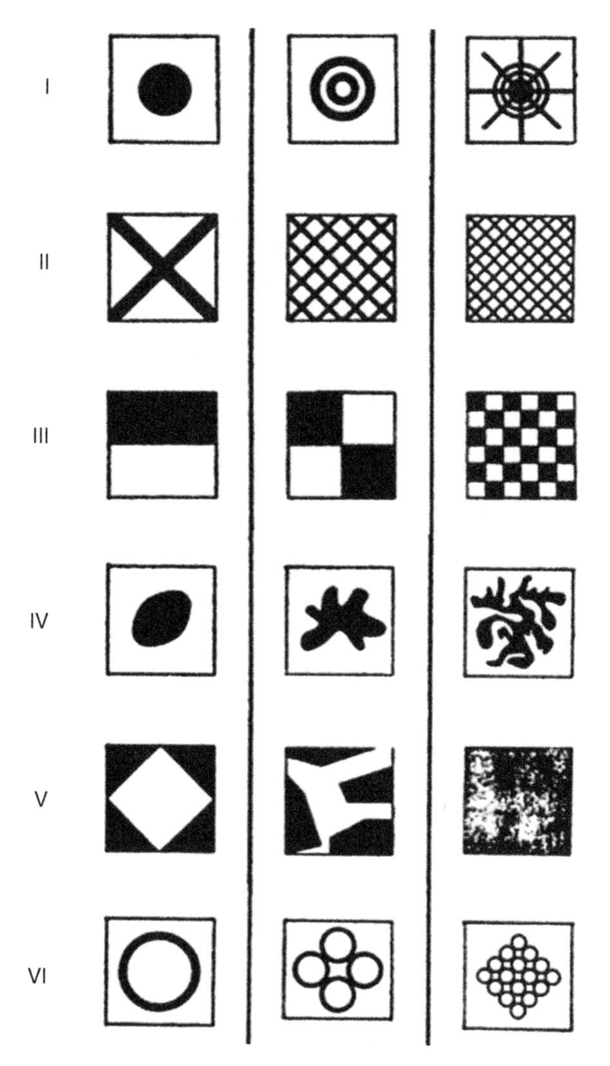

FIGURE 1.3 **Sample stimuli that vary in complexity.** For each stimulus set (I–VI), the stimuli vary from low to high complexity moving left to right. *From Cantor et al. (1963).*

preferential responses (eg, Cantor, Cantor, & Ditrichs, 1963; Karmel, 1969a,b; Spears, 1964). Complexity was not seen as interchangeable with number, but rather, on a par with other attentional attractors, like novelty and incongruity (Berlyne, 1958). Still, many metrics one might use to operationalize "more information" are likely to correlate with number. For example, in studies showing that infants preferred checkerboards with more squares (eg, 128 vs. 32), researchers noted that the displays with more elements also had more angles, more intersections, more contour, more redundancy, smaller elements, and a higher density of black-white contour per unit of area (Hershenson, Munsinger, & Kessen, 1965; Karmel, 1969a,b; Munsinger & Weir, 1967; Spears, 1964). These dimensions of complexity are worth careful consideration because research on visual perception has indicated that contours, along with direction changes, are the most informative parts of a visual scene (Attneave, 1954).

Interestingly, the preference for greater complexity only holds up to some limit—a limit that shifts depending upon age and context—after which too much complexity leads to orientation toward the less complex display (Brennan, Ames, & Moore, 1966; Hershenson, 1964; Hershenson et al., 1965; Maisel & Karmel, 1978; Karmel, 1969b). Infants also are slower to habituate to relatively complex stimuli (Caron & Caron, 1969). These effects are unlikely to be based on number per se because the stimuli were not always comprised of discrete sets of individuals and even when they were, different configurations of the same number of identical elements (eg, stacks of bars vs. a bull's eye made of the same bars) yielded strong complexity preferences (cf. Fantz & Nevis, 1967) (Fig. 1.2). Moreover, with studies reporting an optimal complexity of nearly 600 elements for 14-week-olds (Brennan et al., 1966), the likelihood of engaging even approximate enumeration seems low. Instead, perhaps infants look longer because there is simply more to see (ie, more intersections, more edges, etc.).

A third class consists of variables that are less obviously quantitative, yet highly correlated with numerosity under normal circumstances. It includes basic visual percepts like brightness, texture, and spatial frequency. Even though these information streams are widely recognized as central drivers of visual attention (eg, Balas & Woods, 2014; Norcia et al., 2005), they have been virtually ignored in the literature on quantitative development (as noted by Cantrell & Smith, 2013). For example, it is well known that infants and adults use texture gradients to make figure-ground discriminations (Lamme, Van Dijk, & Spekreijse, 1992; Romani, Caputo, Callieco, Schintone, & Cosi, 1999). In these studies, the same numbers of elements are used in both target and control displays—only the direction of the elements is varied—so these studies are not testing number discrimination. Yet it is plausible that the ability to detect such differences in texture contributes to the discrimination of large set sizes. Many studies have shown that adults' numerosity judgments are affected by variations in the spacing of elements, amount of surface area surrounding a set, and the set's texture density (Birnbaum & Veit, 1973; Durgin, 1995; Sophian & Chu, 2008). Indeed, there is evidence suggesting that only by coordinating texture density with surface area can adults make accurate numerosity judgments for large sets (eg, from 20 to 320 elements) (Durgin, 1995). It seems plausible that such processes underlie infants' sensitivity to differences in large sets rather than an enumeration process, such as ANS.

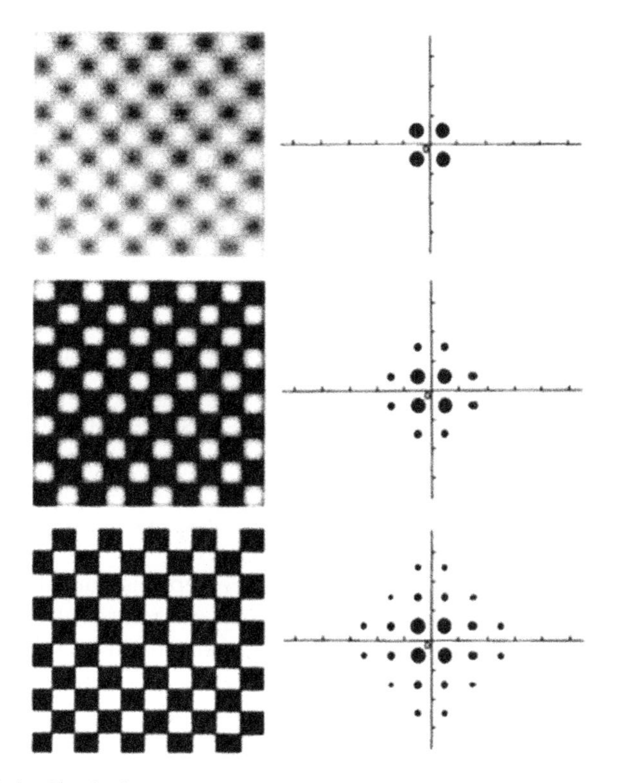

FIGURE 1.4 **Checkerboards increasing in spatial frequency from top to bottom.** The plots to the right indicate the number and density of spatial frequency components. *From Banks & Ginsburg (1985).*

Spatial frequency refers to the ratio of dark–light contrast in a scene and it underlies nearly all of visual perception (Fig. 1.4). In short, a scene can be apprehended in a global way as an array of coarse blobs based on low spatial frequencies, or it can be perceived in terms of fine-grained local differences using high spatial frequencies (Gwiazda, Bauer, Thorn, & Held, 1997; Hegdé, 2008). Low and high spatial frequencies appear to be processed in different parts of the brain (Kauffmann, Ramanoel, & Peyrin, 2014). Importantly, these separate activations occur on a differential time course that facilitates the well-known shift from coarse-to-fine processing when people are perceiving visual scenes (Hegdé, 2008; Navon, 1977; Schyns & Oliva, 1994). Research has clearly demonstrated that infants are sensitive to differences in spatial frequency (eg, Ellemberg, Lewis, Liu, & Maurer, 1999; Hainline & Abramov, 1997) and this sensitivity supports a range of visual processing, including decidedly nonquantitative tasks, such as assessing environmental threats (Stein, Seymour, Hebart, & Sterzer, 2014). Importantly, relative attention to different spatial frequencies changes over infant development and within presentations based on depth of processing (Otsuka, Ichikawa, Kanazawa, Yamaguchi, & Spehar, 2014). Such findings have important implications for research on infant number concepts—particularly processing of large visual sets—because it is likely the stimuli vary in terms of spatial frequency. In a review of infant complexity research published at the time, Banks and Ginsburg (1985) argued that all the reported effects could be explained in terms of responses to spatial frequency. Spatial frequency has not been reported in current infant number studies but it may mediate infants' responses in the same way Banks and Ginsburg argued it did for experiments using checkerboards. Moreover, if depth of processing determines whether infants will detect high spatial frequencies, as one would need to detect numerical differences between large sets, then stimulus duration would be a critical variable in this work. That is, infants may need longer exposures to detect numerical differences, so successful discrimination for one set size or another, or at one age level or another, could be due to differences in exposure duration across experiments.

The fourth and most general class of percepts is stimulus intensity. Humans can sense overall stimulus intensity and reflexively direct attention so as to regulate their internal levels of stimulation (Sokolov, 1963). Intensity can be varied in different ways, by changing the loudness and length of auditory tones, for example, or the brightness of visual displays. Differences in intensity elicit changes in autonomic nervous system responses, such as heart rate deceleration (see Graham & Keen-Clifton, 1966, for a review). These responses have been demonstrated in all modalities including visual, auditory, tactile, and thermal across many species and age groups. Although these responses may not be related to number except in the most raw quantitative sense, the operation of such mechanisms could affect the results of number experiments. Because stimulus intensity is amodal, responses to intensity could explain, at least in part, why adults and infants respond equally to quantitative stimuli presented in different modalities (eg, Barth, Kanwisher, & Spelke, 2003; Lipton & Spelke, 1997) and may explain why infants seem sensitive to cross-modal stimulation (eg, Starkey et al., 1990), as Lewkowicz and Turkewitz (1981) argued. Responses to stimulus intensity have also been implicated in habituation experiments (see Turpin, 1986, for a review) and thus could play an important role in infants' responses across a range of experiments. The failure to report psychophysical measures in infant number experiments makes it impossible to judge to what extent intensity is varying within and across experiments, but this variation could be at least partially driving many of the reported results.

This brief review illustrates that there are many perceptual correlates of number and they are not all strictly quantitative. More importantly, though, they are not interchangeable. Instead, they form a hierarchy in which responses to specific dimensions, like number, can be reduced to responses to more general dimensions, like spatial frequency. This distinction is crucial to our examination of what might constitute a generalized magnitude system. It is possible infants are aware of separate, specific dimensions like area or number and switch among them somehow, but it is equally possible that responses to all of these specific dimensions are driven by reaction to a more general percept, like spatial frequency or intensity, and it is in that sense that the whole system is a general one. In light of this, it is critically important that research begin to disentangle the relative contributions of these percepts, at least to the extent that looking time measures allow, as others have argued (Cantrell & Smith, 2013).

1.3 MECHANISMS OF DIFFERENTIATION

A central challenge for generalized magnitude accounts is explaining how children eventually differentiate and coordinate the quantitative dimensions reviewed earlier in this chapter. Much of the infant number research was designed with the idea that number and perhaps other quantitative dimensions are perceived as separate because there are innate systems that are tuned to these streams, much as the Language Acquisition Device was believed to seek grammatical information (Chomsky, 1965). On this view, the difficult hurdle is designing stimuli that provide clear evidence of these systems. However, if we assume a rich, but less number-specific starting point, how might this differentiation be achieved?

We can look to the literature on perceptual learning for suggestions about the likely catalysts of quantitative differentiation. Perceptual learning is the interplay of bottom up and top down processes that allow us to make sense of the environment. It is thought to begin very early in life, as infants focus their attention based on salience or signal strength and the information gained then influences attention in the next encounter. In the following sections we outline the ways that three well-known drivers of perceptual learning might come into play in quantitative differentiation.

1.3.1 Redundant Cues

One of the main determinants of attention, particularly early in learning, is cue redundancy. When multiple cues are associated with the same outcome or point in the same perceptual direction, it draws learners' attention and helps them detect the relatively strong signal (vs signals for which there is only a single perceptual cue). For example, Bahrick and Lickliter (2000) found that 5-month-olds were better able to learn a complex rhythm when it was simultaneously presented in two modalities rather than one modality or the other (see also Bahrick & Lickliter, 2002).

Cue redundancy has been used to explain a range of developmental achievements, including speech perception, object perception, and categorization (Goldstone & Byrge, 2014).

Cue redundancy is one of the key components of a recent account of infant quantification (Cantrell & Smith, 2013). On this view (the Signal Clarity Hypothesis), numerical representations are strongest and most likely to be noticed when quantitative cues are correlated and pointing in the same direction (eg, one set is more than another based on number, contour, area, brightness, complexity, and spatial frequency). When correlated dimensions vary randomly or are pitted against number, the number representations themselves become more fragile and less likely to sustain attention. Therefore, situations with maximal redundancy are most likely to direct infants' attention toward quantity in general and number in particular.

To illustrate how this might work, Cantrell and Smith (2013) pointed out that physical quantification may be inherently multidimensional in the same way that color perception is multidimensional. That is, people perceive differences in color without necessarily isolating the dimensions that underlie these differences. If this is true for number, then we might imagine that very young infants respond reflexively to intensity and then begin to notice that high intensity correlates with other percepts, like spatial frequency, and one-by-one, notice more signals all moving in the same direction.

This is one possible scenario that corresponds to what is already known about infants' responses to quantitative dimensions, but Cantrell and Smith made no specific claims about which quantitative dimensions were noticed and when. Indeed, this view is not in opposition to the idea that number is noticed and represented early in life, and further experimentation is needed to determine which of many possible scenarios is true. However, the basic mechanism they propose is certainly plausible, addresses a number of oddities in the infant literature, and is consistent with recent neuroimaging work that suggests neurons become tuned to number via statistical learning and perceptual narrowing (eg, Stoianav & Zorzi, 2012).

1.3.2 Unitization

Another perceptual learning mechanism that could be implicated in number learning is chunking or unitization (Goldstone & Byrge, 2014). With this process, learners begin to notice that certain features hang together, and this unit becomes a powerful organizer of future perception. An example would be face perception, for which infants rapidly learn that faces are a reliable configuration and subsequently this unit—the face—holds their attention more than the parts of a face or the same face in different configurations (eg, Morton & Johnson, 1991).

The same process could unfold for quantitative cues. For example, it is possible for a sense of "more" to hold together as a unit and for a time, obscure the separate dimensions that contribute to this sense. This is essentially what was proposed by Cantrell and Smith (2013) and may be what underlies evidence of cross-dimensional transfer in infants (eg, de Hevia & Spelke, 2010; Lourenco & Longo, 2011; Starkey et al., 1990). That is, through redundant cues, infants may detect a signal for quantity that supports rough ordinal judgments, but one that is based on a chunking of all the perceptual streams. Responses in cross-modal tasks, thus, might appear to be based on some abstraction of quantitative equivalence (eg, as argued by Starkey et al., 1990) but could actually arise from fused percepts. Indeed, the mechanism of unitization suggests that individual perceptual streams actually become less distinguishable over time if they are highly correlated, as in the earlier face perception example.

1.3.3 Selective Attention

A third perceptual learning mechanism is the development of selective attention (ie, attention toward specific dimensions that were previously fused in perception) (Goldstone & Byrge, 2014). Although stimulus differentiation and selective attention unfold naturally over development, based on bottom-up perceptual cues (eg, Kemler & Smith, 1978), it has also been manipulated in perceptual learning experiments for which people are trained to notice various dimensions they previously ignored when percepts were chunked. For example, Goldstone and Steyvers (2001) created arbitrary categories by morphing one face into another and then randomly selected a dimension to serve as the basis of the novel, to-be-learned category. Although these arbitrary dimensions were not available to the adults in their study before training, they were readily acquired with exposure and feedback. In the following sections, we will examine two sources of information that might direct children's attention toward number—one source that is bottom up (object segregation) and one that is top down (language).

1.3.3.1 Object Segregation

The ability to tell one object from another is arguably discrete number in its purest sense because the awareness that one object is separate from another necessarily implies there are two of them (eg, Xu & Carey, 1996). Infants learn to distinguish objects in the environment using a variety of perceptual cues, but these are not all available from the earliest stages. One of the earliest and most reliable cues for object segregation is movement (Haith, 1966). Infants' sensitivity to this cue was demonstrated in Kellman's and Spelke's (1983) rod-and-box experiments, in which 4-month-olds perceived either one or two rods emerging

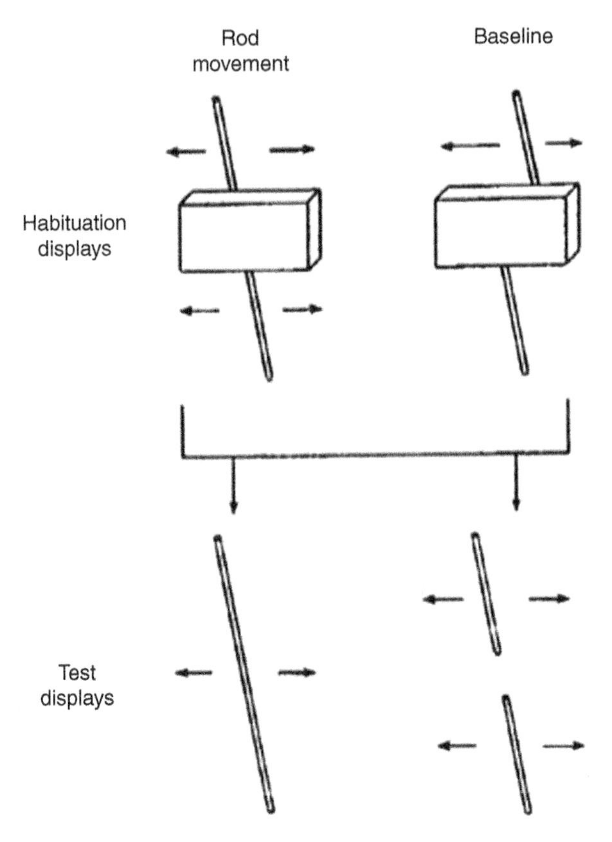

FIGURE 1.5 **Rod and box stimuli used to test whether infants use movement cues to segregate objects.** *From Kellman & Spelke (1983).*

from behind a box depending upon whether they saw the two visible rod segments moving together during familiarization. Subsequent studies have confirmed that very young infants are sensitive to movement cues and use them to tell one object from another (Kestenbaum, Termine, & Spelke, 1987) (Fig. 1.5).

Despite an awareness of object features, like color, size, texture, and shape, and the ability to bind cooccurring features via associative learning (Slater, Mattock, Brown, Burnham, & Young, 1991), infants do not immediately connect feature clusters to particular object identities or locations. Instead, the associations between features and objects become evident and develop between 4 and 13 months of age (Oakes, Ross-Sheehy, & Luck, 2006; Wilcox & Baillargeon, 1998; Wilcox & Chapa, 2002, 2004). It is thought that featural information is gradually incorporated into object concepts as infants accumulate encounters with stable configurations of features in individuals and categories (Needham, 2001a). Binding features to locations may aid in this process, and the whole system feeds itself as the expanding catalog of known objects reduces the degrees of freedom when infants encounter new scenes, thus freeing up cognitive resources to learn more object correspondences (Needham, 2001b).

Interestingly, there is a set size limit for tracking features and locations, and it increases with age (Ross-Sheehy, Oakes, & Luck, 2003). Specifically, 4-month-olds recognize these bindings for only one object at a time, whereas 13-month-olds can do so for up to four objects. This implies that infants may lack the working memory capacity to track color in at least two objects before 4 months of age, so they are unlikely to use features to segregate objects and detect numerosity before that time. It is worth noting that the upper limit of four objects is similar to the upper limit proposed for the object file representation of number in infancy (Feigenson et al., 2004), multiple object tracking in adults (Trick & Pylyshyn, 1994), and the subitizing range in from age 5 years on (Klahr & Wallace, 1976). This correspondence in set size lends support to the notion of shared process that might underlie nascent representations of number, but the evidence that this tracking system develops and is sensitive to context indicates that it is the product of learning and perhaps brain maturation (Oakes et al., 2006). Indeed, even in adults, multiple object tracking is not automatic and immutable. Instead, the object tracking process appears to be quite dynamic and sensitive to shifts in tracking speed, cognitive load, and spacing between objects (Spencer, Barich, Goldberg, & Perone, 2012).

In a classic demonstration, Xu and Carey (1996) showed that this progression from spatial location to object features played a critical role in infants' awareness of numerical identity. Infants saw two distinct objects (eg, a duck and a truck) emerging from either side of an occluder and then returning. In one condition, the objects were revealed in sequence. To adults who understand that object features tell us how many objects are hidden (ie, a duck cannot just turn into a truck), this would be enough information to indicate two objects are hidden behind the screen instead of one. However, infants were 12 months old before they showed surprise when the occluder was removed to reveal only one object hiding. Not so when spatial cues were offered. When 10-month-olds saw the objects emerge simultaneously on either side of the occluder, they were surprised (and looked longer) when only one object was revealed at test. They appeared to know that one object cannot be in two locations at the same time. This finding suggests that spatial location and movement are important cues to object segregation (see also Newcombe, Huttenlocher, & Learmonth, 1999) and thus, constitute important inputs for quantitative differentiation. Indeed, if cardinal number is eventually represented via spatial location, as a way to decouple it from correlated continuous variables, this might explain why people readily associate canonical patterns and numerosity (eg, two = line, three = triangle, etc.) (Ashkenazi, Mark-Zigdon, & Henik, 2013; Mandler & Shebo, 1982).

1.3.3.2 *Language*

Top-down processing could also contribute to selective attention to number through the impact of language. First, word learning is closely related to object individuation and spatial location. When children learn nouns, their attention is directed toward objects because that is what nouns name. Research demonstrates that these mappings are supported by consistent spatial locations, such that 16- to 18-month-olds are unable to learn novel words for objects that shift locations. Indeed, consistent spatial location may be more important to early word learning than referential intent (Samuelson, Smith, Perry, & Spencer, 2011). Thus, one can think of the entire system that includes object recognition, naming, spatial location, and number, as conceptually and developmentally interconnected.

Second, the count words might orient attention toward the dimension of number. Count words first emerge in productive speech during toddlerhood (Fuson, 1988; Mix, 2009), so it is reasonable to assume infants have been exposed to them during the same period they are learning to distinguish objects from one another. Because count words name the property of number, they could be potent attention-directing cues. Whenever adults count objects or name quantities, they invite children to notice that the same words apply to a variety of items that differ in area, contour, and complexity. Count words signal that number is a distinct property, independent of these other quantitative dimensions. It may not be coincidental that counting ability helps children recognize numerical matches in the face of perceptual variability (Mix, Huttenlocher, & Levine, 1996; Posid & Cordes, 2015). Just as top–down training helped adults discover new dimensions along which faces can be sorted (Goldstone & Steyvers, 2001), count words may help children isolate the dimension of discrete number.

Indeed, for people in cultural groups without a well-developed count list, as well as for nonhuman animal species, representations of number beyond the subitizing range remain approximate, in accord with Weber's law (Agrillo, 2014; Gallistel, 1990; Gordon, 2004; Meck & Church, 1983; Pica, Lemer, Izard, & Dehaene, 2004 Spaepen, Coppola, Spelke, Carey, & Goldin-Meadow, 2011). This pattern may be explained, at least partially, because these large set sizes remain undifferentiated magnitudes. Even adults in a numerate culture who do not have access to a linguistic count list, as is the case for deaf homesigners in Nicaragua, show this pattern (Spaepen et al., 2011), whereas nonhuman animals that have been taught the count words up eight can make precise discriminations of these larger quantities (eg, Pepperberg & Carey, 2012).

1.4 MECHANISMS OF REINTEGRATION

Our focus thus far has been the nature of a generalized magnitude system and the means by which specific perceptual streams might become differentiated, but a mature understanding of quantification includes both number and its correlates, integrated into a coherent conceptual framework. Adults know, for example, that whether you count three elephants or three mice, you still have three. They also know that a pile of sand may be equivalent to a pile of lead in terms of volume, but it will likely differ in weight. In the parlance of perceptual learning, adults no longer treat quantitative dimensions as fused percepts, but rather move fluidly among them, depending on the task at hand. The ability to coordinate continuous and discrete dimensions is apparent in childhood (eg, on active choice tasks such as Piaget's number conservation task) but it is not immediate or easy. By what mechanism is this coordination achieved?

We have proposed that children reintegrate quantitative dimensions when they begin unitizing spatial extent and counting these units (Mix et al., 2002b; Newcombe et al., 2015), rather than counting discrete physical objects exclusively (eg, Shipley & Shepperson, 1990). This unitizing is evident in advanced mathematical skills such as measurement and fractions, but its origins can be traced in the development of spatial location and proportional scaling—skills that emerge concurrently with quantification, as we discuss in the next section. The proposal is that the experience of dividing continuous amounts, such as spatial extent, provides the foundation for coordinating discrete and continuous quantification, even if the two are not explicitly connected until somewhat later.

1.4.1 Proportional Scaling and Spatial Division

The early emergence of spatial division has been demonstrated in studies where infants and children must find a hidden object in a rectangular sandbox. Even 5-month-olds seem surprised when an object is hidden in one location and then revealed in another, suggesting at least limited ability to remember locations in metric space (Newcombe, Sluzenski, & Huttenlocher, 2005). By 18 to 24 months of age, children can accurately remember the hiding location and retrieve the object, but only when a single object is hidden (Huttenlocher, Newcombe, & Sandberg, 1994). The ability to find two hidden objects, or remember the location of one object over a substantial delay (eg, 2 min), develops only gradually over the preschool years (Sluzenski, Newcombe, & Satlow, 2004).

Interestingly, toddlers' bias patterns for search in the sandbox suggest the early availability of Bayesian combination of categorical and fine-grained metric information, similar to patterns observed in adults (Huttenlocher, Hedges, & Duncan, 1991). That is, one way to remember location is to use the distance from one end or the other. Another way is to divide the space into regions, such as halves or quarters, and remember the location by associating the target with its spatial category. The latter process is evident when errors are distributed such that they move toward the center of these categorical units and away from their boundaries. Consistent with this, Huttenlocher et al. (1994) noted that toddlers' searches were biased toward the center of the box, suggesting that the sandbox was used as a category.

The use of such categories undergoes further development, however. The spatial categories children use become smaller with age, and hence more informative; adjustment by a smaller category draws estimates to a prototype value closer to the actual location. Also, children become capable of subdividing a space into more than one category between 4 and 8 years, as reflected in a distinctive bias pattern best described as a quintic function (Huttenlocher et al., 1994). Finally, children fare worse when there are two dimensions to consider (eg, radial distance and angle) until about 9 years of age because they cannot coordinate categorical and metric coding (Sandberg, Huttenlocher, & Newcombe, 1996). These developmental patterns may reflect the same processes of perceptual learning we have discussed with respect to the development of quantification (ie, attention tuning, perceptual unitizing, and attribute differentiation). The similar difficulty children exhibit coordinating multiple dimensions within each domain is also noteworthy as both difficulties may be overcome at about the same age due to improved capacity in domain general processes, such as working memory and attention (Newcombe et al., 2015).

Another quantitatively relevant change in spatial processing is the shift from intensive to extensive coding. Initially, children seem to use the sandbox as a reference and remember the hidden object's location in relation to its extent, in the same way one might estimate the volume of juice by comparing its level to the height of the entire glass (eg, half full) (Huttenlocher, Duffy, & Levine, 2002; Huttenlocher, Levine, & Ratliff, 2011; Vasilyeva, Duffy, & Huttenlocher, 2007). Indeed, children only begin to represent length or volume in terms of absolute amount starting at age 4 years and eventually reaching mature performance by 8 years (Duffy, Huttenlocher, & Levine, 2005; Vasilyeva et al., 2007). This pattern of findings raises an interesting question about perceptual quantification, as measured in the infant looking time tasks reviewed earlier in this chapter—namely whether infants in ANS and other number experiments are noticing a change in absolute quantity or a change in proportion, relative to the boundaries of the visual display (eg, screen).

1.4.2 Symbolic Representation

Children readily transfer their intuitive sense of spatial scaling to symbolic representations (eg, maps) starting at 3 years of age (Huttenlocher,

Newcombe, & Vasilyeva, 1999). Although finding hidden objects using a map or model is initially harder than placing visible objects in accord with a map or model, this problem is quickly overcome (Huttenlocher, Vasilyeva, Newcombe, & Duffy, 2008). Between 3 and 5 years of age, children's precision in scaling tasks improves (Frick & Newcombe, 2012), as does their ability to perceive equivalent proportions despite differences in scale (eg, up to 1:19.2) (Vasilyeva & Huttenlocher, 2004), though larger scaling factors continue to predict lower accuracy at least until age 10 years (Boyer & Levine, 2012).

Performance on these tasks seems to be supported by perceptual estimation—recognizing that roughly a quarter of the way across on a map implies roughly a quarter of the way across on a referent space (Möhring, Newcombe, & Frick, 2014)—and as such, has important links to mathematics. Indeed, perceptual estimation is the precursor to conventional measurement and several mathematical skills that involve translation between continuous and discrete quantification by way of unit measures, including proportional reasoning, fractions, and number line estimation. Perhaps not coincidentally, these are also among the most challenging stumbling blocks of elementary school mathematics, suggesting that achieving this coordination is not easy. In the following analysis, we will argue that the same problem complicates learning in each of these mathematical topics, namely, the problem of identifying the relevant countable units onto which number symbols are mapped.

1.4.2.1 *Measurement*

American children struggle with unit-based measurement problems across many grade levels (National Assessment of Educational Progress (NAEP) 1996, 2000, 2003). When the ruler is not aligned with the object to be measured (Fig. 1.6), it is common for children to either read off the number at one end (ie, 5 in Fig. 1.6), or count the hash marks (ie, 4 in Fig. 1.6), rather than to count the units (Levine, Kwon, Huttenlocher, Ratliff, & Dietz, 2009; Solomon, Vasilyeva, Huttenlocher, & Levine, 2015). The problem is at least partially attributable to the difficulty of imposing

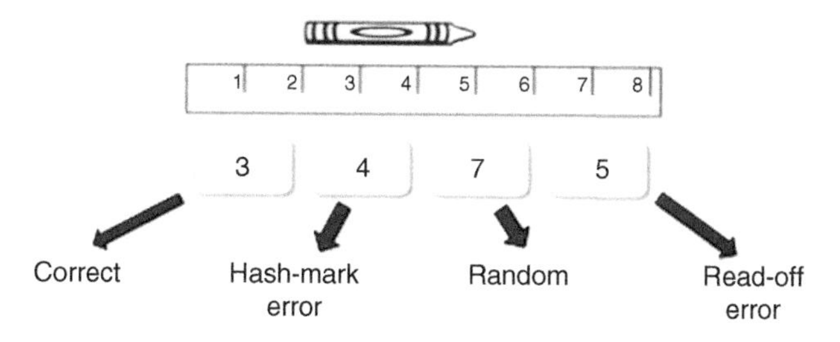

FIGURE 1.6 **Common responses on a measurement task.** *Based on Solomon et al. (2015).*

discrete units onto a continuous measurement instrument because kindergarten and second grade children fare better on misaligned problems when the units are discrete, adjacent pennies than when the units are represented as on a conventional ruler (Solomon et al., 2015). Recent training experiments have found that emphasizing the relevant countable units and presenting children with disconfirming evidence on misaligned ruler problems are effective ways to help young children avoid these errors (Kwon, Congdon, Ping, & Levine, in preparation; Congdon, Ping & Levine, in preparation).

1.4.2.2 *Proportional Reasoning and Probabilities*

In proportional reasoning, there is a similar problem: when countable units are salient, children, at least through fifth grade, have difficulty concentrating on spatial extent when they should. For example, in the problem shown in Fig. 1.7, adults would not say that the correct answer is the alternative that shows 2 units because the bar on the left has 2 units, yet children are seduced by this error until they are 8 or 9 years of age (Boyer, Levine, & Huttenlocher, 2008). However, performance is significantly improved when children can use an intuitive perceptual strategy based on intensive coding (ie, the continuous condition), where number does not provide an alternative interpretation (Boyer et al., 2008; Boyer & Levine, 2012).

Additional evidence comes from tests of probabilistic reasoning. Children were shown two donut-shaped forms that were divided into

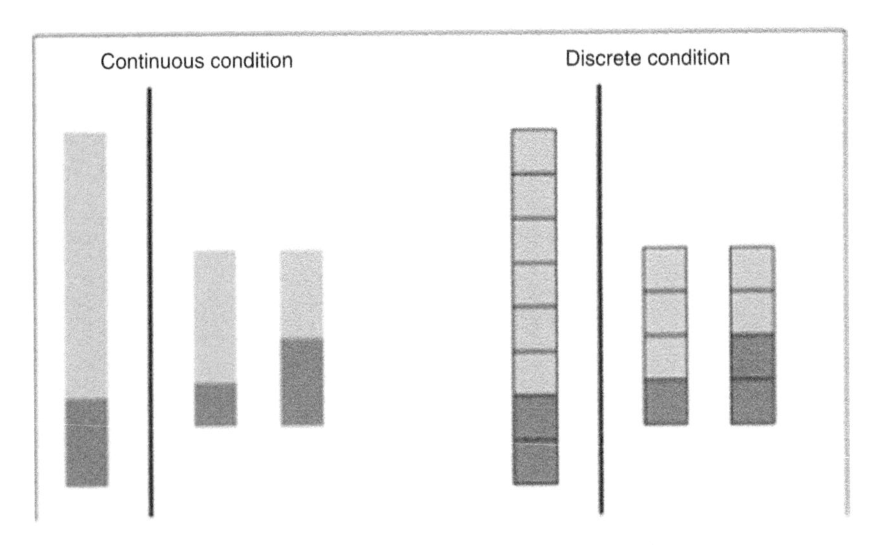

FIGURE 1.7 **Examples of continuous and discrete proportional reasoning tasks.** *From Boyer & Levine (2012).*

red and blue regions, each with a spinner in the center. Their task was to decide for which donut the spinner was most likely to land on one color or the other. Performance on this task was above chance by age 6 years, but only when the different colored regions were presented as continuous amounts. When the red and blue regions were divided into several equal-sized, bounded units, children performed worse and did not begin to succeed until 10 years of age (Jeong, Levine, & Huttenlocher, 2007). Again, the continuous task may be easier because these quantities can be mapped onto magnitude representations more readily.

Across all these tasks, the developmental challenge is to impose discrete, countable, equal-size units onto continuous amounts and to know which system to use, when, and how. Research indicates formal instruction in proportional and probabilistic reasoning should build on children's intuitions about continuous amounts and intuitive proportional reasoning and then provide strong analogies to these same amounts with discrete units imposed (Boyer & Levine, 2015). In other words, for proportional reasoning and probabilities, as for measurement, children need support to see how spatially defined units and counting go together.

1.4.2.3 *Fractions*

The same challenge likely underlies children's notorious difficulties learning fractions (eg, Nunes, Light, & Mason, 1993; Resnick & Ford, 1981). Fractions are essentially the integration of spatial extent and counting, by way of unit measures, where the size of the unit is based on the number of divisions of a whole into parts (eg, quarters are smaller than halves because more divisions mean smaller pieces). The tricky thing about fractions is that the units and counts are buried in the notation. Whereas we explicitly name the count and the unit in measurement (eg, 4 in.), these notions are merely implied when we say, for example, three-quarters. In this case, we mean we have a count of three pieces where each piece (unit) is the size of the whole divided into four. Not surprisingly, children struggle to learn formal fractions in school settings, despite the emergence of nonverbal fraction concepts in preschool (eg, Mix, Levine, & Huttenlocher, 1999), most likely because fraction notation is so opaque (Behr, Harel, Post, & Lesh, 1992).

Consistent with this view, children's errors on fraction tasks usually reflect a failure to coordinate the continuous and discrete dimensions. For example, they ignore portion size in sharing tasks, and tend to give recipients equal numbers without regard for the total amount (eg, Frydman & Bryant, 1988; Miller, 1984; Sophian et al., 1997). Children also interpret fraction names as ratios of cardinal numbers (eg, 3/5 = 3:5) rather than parts and wholes (Paik & Mix, 2003; Mix & Paik, 2008) and perform intuitive fraction tasks better when the quantities are continuous (ie, spatially contiguous), rather than discrete (ie, unitized) (Hunting & Sharpley, 1988).

This lack of coordination might explain why children readily acquire the meanings of common fractions, such as one-half, but are limited to demonstrating this understanding on tasks that require an approximate sense of ratio, such as matching equivalent fraction pictures (Spinillo & Bryant, 1991), completing pictorial analogies (Singer-Freeman & Goswami, 2001), or making equal shares of continuous amounts (Hunting & Sharpley, 1988; Miller, 1984). Perhaps competence on these nonverbal tasks reflects children's ability to recruit magnitude representations as referents without a precise understanding of the referents for fraction symbols (ie, unit counts). Consistent with this, adults and children can represent the meanings of fraction symbols as magnitudes on a mental number line, but this mapping is based on a rough estimate of the absolute quantity represented by a fraction, rather than precise, part-whole relations based on unit measures (deWolf, Grounds, Bassok, & Holyoak, 2014 Siegler, Thompson, & Schneider, 2011).

1.4.2.4 Number Line Estimation

A final manifestation of this translation difficulty is children's performance on number line tasks. When numbers are placed correctly, they are evenly spaced, resulting in a perfectly linear relation between number and position. However, children initially tend to spread small numbers farther apart than they should be, and bunch together the larger numbers, a pattern of responses best fit by a logarithmic function (Opfer & Siegler, 2007; Siegler & Opfer, 2003; Thompson & Opfer, 2008). Responses eventually shift to a more mature, linear function, but the logarithmic-to-linear shift repeats itself several times, depending upon the range of numbers being placed. That is, although 7-year-olds respond linearly on the 0–100 number line, they respond logarithmically on the 0–1000 number line. By 9 years of age, children respond linearly on the 0–1,000 number line, but respond logarithmically on the 0–10,000 number line, and so on. There are competing explanations for these shifts, but they tend to agree on some degree of disconnect between approximate magnitude representations and the more precise representations provided by counting and multidigit numeral interpretation (Ebersbach, Luwel, Frick, Onghena, & Verschaffel, 2008; Landy, Charlesworth, & Ottmar, 2014; Moeller, Pixner, Kaufmann, & Nuerk, 2009; Slusser, Santiago, & Barth, 2013). Interestingly, just as for measurement and fractions, number line performance is associated with better subsequent learning and performance in mathematics more generally (eg, Booth & Siegler, 2006; Ramani & Siegler, 2008; Schneider, Grabner, & Paetsch, 2009), and children with poor number line performance are more likely to have mathematics learning disabilities (Geary, Hoard, Byrd-Craven, Nugent, & Numtee, 2007). Perhaps the reason all of these tasks are predictive is that they tap the same ability to divide into equal units and enumerate these units.

1.5 CONCLUSIONS

In this chapter, we explored what it could mean for quantitative development to be rooted in a generalized magnitude system. We first considered the ways in which magnitudes might be generalized based on a review of the correlated dimensions of quantity. We concluded that although discrete number may be difficult to isolate using looking time methods, a sense of number could be embedded in infants' sensitivity to other quantitative percepts. Indeed, the strength of this correlated signal could be what helps infants notice and represent number (Cantrell & Smith, 2013). We identified several perceptual learning processes by which infants and toddlers might distinguish number from this correlated perceptual stream, as well as specific cues to discrete number, including object segregation and counting. Finally, we proposed that these differentiated dimensions ultimately become coordinated by way of spatial division and unit measures. The challenges children face in this coordination process were illustrated in four examples: measurement, probabilistic reasoning, fractions, and number line placement. It is hoped that this conceptual framework will help spur research that explicates more precisely how the spectrum of quantitative dimensions becomes coordinated over development.

References

Adams, R. J., & Courage, M. L. (1996). Monocular contrast sensitivity in 3- to 36-month-old human infants. *Optometry and Vision Science*, *73*, 546–551.

Agrillo, C. (2014). Numerical and arithmetic abilities in non-primate species. In R. Cohen Kadosh, A. Dowker (Eds.), *Oxford handbook of numerical cognition*. New York: Oxford University Press.

Ashkenazi, S., Mark-Zigdon, N., & Henik, A. (2013). Do subitizing deficits in developmental dyscalculia involve pattern recognition weakness? *Developmental Science*, *16*(1), 35–46.

Attneave, R. (1954). Some informational aspects of visual perception. *Psychological Review*, *61*(3), 183–194.

Bach, M., & Meigen, T. (1998). Electrophysiological correlates of human texture segregation, an overview. *Documenta Ophthalmologica*, *95*, 335–347.

Bahrick, L. E., & Lickliter, R. (2000). Intersensory redundancy guides attentional selectivity and perceptual learning in infancy. *Developmental Psychology*, *36*, 190–201.

Bahrick, L. E., & Lickliter, R. (2002). Intersensory redundancy guides early cognitive and perceptual development. In R. V. Kail (Ed.), *Advances in child development and behavior* (pp. 153–187). (Vol. 30). New York, NY: Academic Press.

Balas, B., & Woods, R. (2014). Infant preference for natural texture statistics is modulated by contrast polarity. *Infancy*, *19*(3), 262–280.

Banks, M. S., & Ginsburg, A. P. (1985). Infant visual preferences: a review and new theoretical treatment. *Advances in Child Development and Behavior*, *19*, 207–246.

Banks, M. S., & Salapatek, P. (1976). Contrast sensitivity function of the infant visual system. *Vision Research*, *16*(8), 867–869.

Baroody, A. J. (1998). *Fostering children's mathematical power: an investigative approach to K–8 mathematics instruction*. Mahwah, NJ: Erlbaum.

Baroody, A. J., & Hume, J. (1991). Meaningful mathematics instruction: the case of fractions. *Remedial and Special Education, 12*(3), 54–68.
Barth, H. C., & Paladino, A. M. (2011). The development of numerical estimation: evidence against a representational shift. *Developmental Science, 14*(1), 125–135.
Barth, H. C., Kanwisher, N., & Spelke, E. (2003). The construction of large number representations in adults. *Cognition, 86*, 201–221.
Behr, M. J., Harel, G., Post, T. R., & Lesh, R. (1992). Rational number, ratio and proportion. In D. Grouws (Ed.), *Handbook of research on mathematics teaching and learning* (pp. 296–333). New York, NY: Macmillan Publishing.
Berlyne, D. E. (1958). The influence of the albedo and complexity of stimuli on visual fixation time in the human infant. *British Journal of Psychology, 49*, 315–328.
Birnbaum, M. H., & Veit, C. T. (1973). Judgmental illusion produced by contrast with expectancy. *Perception and Psychophysics, 1*, 149–152.
Booth, J. L., & Siegler, R. S. (2006). Developmental and individual differences in pure numerical estimation. *Developmental Psychology, 41*, 189–201.
Bower, T. G. R. (1967). Phenomenal identity and form perception in an infant. *Perception and Psychophysics, 2*, 74–76.
Boyer, T. W., & Levine, S. C. (2012). Child proportional scaling: Is 1/3 = 2/6 = 3/9 = 4/12? *Journal of Experimental Child Psychology, 111*(3), 516–533.
Boyer, T. W., & Levine, S. C. (2015). Prompting children to reason proportionally: processing discrete units and continuous amounts. *Developmental Psychology, 51*, 615–620.
Boyer, T. W., Levine, S. C., & Huttenlocher, J. (2008). Development of proportional reasoning: where young children go wrong. *Developmental Psychology, 44*(5), 1478–1490.
Brennan, W. M., Ames, E. W., & Moore, R. W. (1966). Age differences in infants' attention to patterns of different complexities. *Science, 150*, 354–356.
Cantor, G. N., Cantor, J. H., & Ditrichs, R. (1963). Observing behavior in preschool children as a function of stimulus complexity. *Child Development, 34*, 683–689.
Cantrell, L., Boyer, T. W., Cordes, S., & Smith, L. B. (2015). Signal clarity: An account of variability in infant quantity discrimination tasks. *Developmental Science, 18*, 877–893.
Cantrell, L., & Smith, L. B. (2013). Open questions and a proposal: a critical review of the evidence on infant numerical abilities. *Cognition, 128*(3), 331–352.
Caron, R. F., & Caron, A. J. (1969). The effects of repeated exposure and stimulus complexity on visual fixation in infants. *Psychonomic Science, 10*(6), 207–208.
Chomsky, N. (1965). *Aspects of the theory of syntax*. Cambridge, MA: MIT Press.
Clearfield, M. W., & Mix, K. S. (1999). Number versus contour length in infants' discrimination of small visual sets. *Psychological Science, 10*(5), 408–411.
Cohen, L. B. (1972). Attention-getting and attention-holding processes of infant visual preferences. *Child Development, 43*, 869–879.
Cohen Kadosh, R., & Walsh, V. (2009). Numerical cognition: reading numbers from the brain. *Current Biology, 19*(19), R898–R899.
Congdon, E. L., Ping, R., Kwon, M., & Levine, S. C. (in prep) Learning to measure through action and gesture: children's starting state matters.
de Hevia, M. D., & Spelke, E. S. (2009). Spontaneous mapping of number and space in adults and young children. *Cognition, 110*, 198–207.
de Hevia, M. D., & Spelke, E. S. (2010). Number-space mapping in human infants. *Psychological Science, 21*(5), 653–660.
de Wolf, M., Grounds, M. A., Bassok, M., & Holyoak, K. J. (2014). Representation and comparison of magnitudes for different types of rational numbers. *Journal of Experimental Psychology: Human Perception and Performance, 40*(1), 71–82.
Dolscheid, S., Hunnius, S., Casasanto, D., & Majid, A. (2014). Prelinguistic infantsare sensitive to space-pitch associations found across cultures. *Psychological Science, 25*, 1159–1168.

Duffy, S., Huttenlocher, J., & Levine, S. (2005). It is all relative: how young children encode extent. *Journal of Cognition and Development*, *6*(1), 51–63.

Durgin, F. H. (1995). Texture density adaptation and the perceived numerosity and distribution of texture. *Journal of Experimental Psychology: Human Perception and Performance*, *21*, 149–169.

Ebersbach, M., Luwel, K., Frick, A., Onghena, P., & Verschaffel, L. (2008). The relationship between the shape of the mental number line and familiarity with numbers in 5- to 9-year-old children: evidence for a segmented linear model. *Journal of Experimental Child Psychology*, *99*(1), 1–17.

Ellemberg, D., Lewis, T. L., Liu, C. H., & Maurer, D. (1999). Development of spatial and temporal vision during childhood. *Vision Research*, *39*, 2325–2333.

Fantz, R. L. (1961). The origin of form perception. *Scientific American*, *204*, 66–72.

Fantz, R. L., & Fagan, J. F. (1975). Visual attention to size and number of pattern details by term and preterm infants during the first six months. *Child Development*, *46*(1), 3–18.

Fantz, R. L., & Nevis, S. (1967). Pattern preferences and perceptual-cognitive development in early infancy. *Merrill Palmer Quarterly*, *13*(1), 77–108.

Feigenson, L., Carey, S., & Spelke, E. S. (2002). Infants' discrimination of number vs. continuous extent. *Cognitive Psychology*, *44*, 33–66.

Feigenson, L., Dehaene, S., & Spelke, E. (2004). Core systems of number. *Trends in Cognitive Science*, *8*(7), 307–314.

Frick, A., & Newcombe, N. S. (2012). Getting the big picture: development of spatial scaling abilities. *Cognitive Development*, *27*, 270–282.

Frick, A., Mohring, W., & Newcombe, N. S. (2014). Development of mental transformation abilities. *Trends in Cognitive Sciences*, *18*(10), 536–542.

Frydman, O., & Bryant, P. (1988). Sharing and the understanding of number equivalence by young children. *Cognitive Development*, *3*(4), 323–339.

Fuson, K. (1988). *Children's counting and concepts of number*. New York, NY: Springer-Verlag.

Gallistel, C. R. (1990). Representations in animal cognition: an introduction. *Cognition*, *37*, 1–22.

Gardner, J. M., & Karmel, B. Z. (1984). Arousal effects on visual preferences in neonates. *Developmental Psychology*, *20*(3), 374–377.

Geary, D. C., Hoard, M. K., Byrd-Craven, J., Nugent, L., & Numtee, C. (2007). Cognitive mechanisms underlying achievement deficits in children with mathematical learning disability. *Child Development*, *78*, 1343–1359.

Gebuis, T., Gevers, W., & Cohen Kadosh, R. (2014). Topographic representation of high-level cognition: numerosity or sensory processing? *Trends in Cognitive Sciences*, *18*, 1–3.

Goldstone, R. L., & Byrge, L. A. (2014). Perceptual learning. In M. Matthen (Ed.), *The Oxford handbook of philosophy of perception* (pp. 1–16). New York, NY: Oxford University Press.

Goldstone, R. L., & Steyvers, M. (2001). The sensitization and differentiation of dimensions during category learning. *Journal of Experimental Psychology: General*, *130*, 116–139.

Gordon, P. (2004). Numerical cognition without words. *Science*, *306*, 496–499 (5695).

Graham, F. K., & Keen-Clifton, R. (1966). Heart-rate change as a component of the orienting response. *Psychological Bulletin*, *65*(5), 305–320.

Greenberg, D. J., & O'Donnell, W. J. (1972). Infancy and the optimal level of stimulation. *Child Development*, *43*, 639–645.

Gwiazda, J., Bauer, J., Thorn, F., & Held, R. (1997). Development of spatial contrast sensitivity from infancy to adulthood: psychophysical data. *Optometry and Vision Science*, *74*, 785–789.

Hainline, L., & Abramov, I. (1997). Eye movement-based measures of development of spatial contrast sensitivity in infants. *Optometry and Vision Science*, *74*, 790–799.

Haith, M. M. (1966). The response of the human newborn to visual movement. *Journal of Experimental Child Psychology*, *3*, 235–243.

Haith, M. M. (1980). *Rules that babies look by: the organization of newborn visual activity*. Potomac, Maryland: L Erlbaum Associates.

Hegdé, J. (2008). Time course of visual perception: coarse-to-fine processing and beyond. *Progress in Neurobiology, 84*, 405–439.
Hershenson, M. J. (1964). Visual discrimination in the human newborn. *Journal of Comparative and Physiological Psychology, 58*, 270–276.
Hershenson, M. J., Munsinger, H., & Kessen, W. (1965). Preference for shapes of intermediate variability in the newborn human. *Science, 147*, 630–631.
Hunting, R. P., & Sharpley, C. F. (1988). Fraction knowledge in preschool children. *Journal for Research in Mathematics Education, 19*(2), 175–180.
Hurewitz, F., Gelman, R., & Schnitzer, B. (2006). Sometimes area counts more than number. *Proceedings of the National Academy of Sciences of the United States of America, 106*, 10382–10385.
Huttenlocher, J., Duffy, S., & Levine, S. C. (2002). Infants and toddlers discriminate amount: Are they measuring? *Psychological Science, 13*, 244–249.
Huttenlocher, J., Hedges, L. V., & Duncan, S. (1991). Categories and particulars: prototype effects in estimating spatial location. *Psychological Review, 98*(3), 352–376.
Huttenlocher, J., Newcombe, N., & Sandberg, E. (1994). The coding of spatial location in young children. *Cognitive Psychology, 27*, 115–147.
Huttenlocher, J., Newcombe, N., & Vasilyeva, M. (1999). Spatial scaling in young children. *Psychological Science, 10*, 393–398.
Huttenlocher, J., Vasilyeva, M., Newcombe, N., & Duffy, S. (2008). Developing symbolic capacity one step at a time. *Cognition, 106*, 1–12.
Huttenlocher, J., Levine, S. C., & Ratliff, K. (2011). The development of measurement: from holistic perceptual comparison to unit understanding. In N. L. Stein, & S. Raudenbush (Eds.), *Developmental science goes to school: implications for education and public policy research* (pp. 175–188). New York: Taylor and Francis.
Hyde, D. C., & Spelke, E. S. (2009). All numbers are not equal: an electrophysiological investigation of small and large number representations. *Journal of Cognitive Neuroscience, 21*, 1039–1053.
Hyde, D. C., & Spelke, E. S. (2011). Neural signatures of number processing in human infants: evidence for two core systems underlying non-verbal numerical cognition. *Developmental Science, 14*(2), 360–371.
Hyde, D. C., Porter, C. L., Flom, R. A., & Ahlander-Stone, S. (2013). Relational congruence facilitates neural mapping of spatial and temporal magnitudes in preverbal infant. *Developmental Cognitive Neuroscience, 6*, 102–112.
Jeong, Y., Levine, S. C., & Huttenlocher, J. (2007). The development of proportional reasoning: effect of continuous vs. discrete quantities. *Journal of Cognition and Development, 8*, 237–256.
Johnson, S. P., & Aslin, R. N. (1995). Perception of object unity in 2-month-old infants. *Developmental Psychology, 31*(5), 739–745.
Kahneman, D., Treisman, A., & Gibbs, B. J. (1992). The reviewing of object files: object-specific integration of information. *Cognitive Psychology, 24*, 174–219.
Kaldy, Z., & Leslie, A. M. (2003). Identification of objects in 9-month-old infants: integrating "what" and "where" information. *Developmental Science, 6*, 360–373.
Karmel, B. Z. (1969a). Complexity, amounts of contour, and visually dependent behavior in hooded rats, domestic chicks, and human infants. *Journal of Comparative and Physiological Psychology, 69*, 649–657.
Karmel, B. Z. (1969b). The effect of age, complexity, and amount of contour on pattern preferences in human infants. *Journal of Experimental Child Psychology, 7*, 339–354.
Kauffmann, L., Ramanoel, S., & Peyrin, C. (2014). The neural basis of spatial frequency processing during scene perception. *Frontiers in Integrative Neuroscience, 8*, 1–14.
Kellman, P., & Spelke, E. (1983). Perception of partially occluded objects in infancy. *Cognitive Psychology, 15*, 483–524.
Kemler, D. G., & Smith, L. B. (1978). Is there a developmental trend from integrality to separability in perception? *Journal of Experimental Child Psychology, 26*(3), 498–507.

Kerslake, D. (1986). *Fractions: Children's strategies and errors*. London: NFER-Nelson.

Kestenbaum, R., Termine, N., & Spelke, E. S. (1987). Perception of objects and object boundaries by 3-month-old infants. *British Journal of Developmental Psychology, 5*, 367–383.

Kibbe, M. M., & Leslie, A. M. (2011). What do infants remember when they forget? Location and identity in 6-month-olds' memory for objects. *Psychological Science, 22*, 1500–1505.

Klahr, D., & Wallace, J. G. (1976). *Cognitive development: an information processing view*. Oxford, UK: Erlbaum.

Kwon, M., Ping, R., Congdon, E. L., & Levine, S. C. (manuscript under review) The power of disconfirming evidence: Overturning children's misconceptions about ruler measurement.

Lamme, V. A., Van Dijk, B. W., & Spekreijse, H. (1992). Texture segregation is processed by primary visual cortex in man and monkey. Evidence from VEP experiments. *Vision Research, 32*, 797–807.

Landy, D., Charlesworth, A., & Ottmar, E. (2014). Cutting in line: discontinuities in the use of large numbers by adults. *Proceedings of the thirty-sixth annual conference of the Cognitive Science Society*. Quebec City, Quebec: Cognitive Science Society.

Levine, S. C., Kwon, M., Huttenlocher, J., Ratliff, K., & Dietz, K. (2009). Children's understanding of ruler measurement: a training study. In N. A. Taatgen, H. van Rijn (Eds.), *Proceedings of the thirty-first annual conference of the Cognitive Science Society* (pp. 2391–2395). Austin, TX: Cognitive Science Society.

Lewkowicz, D., & Turkewitz, G. (1981). Intersensory interaction in newborns: modification of visual preferences following exposure to sound. *Child Development, 52*, 827–832.

Lipton, J. S., & Spelke, E. S. (1997). Discrimination of large and small numerosities by human infants. *Infancy, 5*(3), 271–290.

Lourenco, S. F., & Longo, M. R. (2009). The plasticity of near space: evidence for contraction. *Cognition, 112*(3), 451–456.

Lourenco, S. F., & Longo, M. R. (2010). General magnitude representation in human infants. *Psychological Science, 21*, 873–881.

Lourenco, S. F., & Longo, M. R. (2011). Origins and the development of generalized magnitude representation. In S. Dehaene, & E. Brannon (Eds.), *Space, time, and number in the brain: searching for the foundations of mathematical thought* (pp. 225–244). Cambridge, MA: Academic Press.

Maisel, E. B., & Karmel, B. Z. (1978). Contour density and pattern configuration in visual preferences of infants. *Infant Development and Behavior, 1*, 127–140.

Mandler, G., & Shebo, B. J. (1982). Subitizing: an analysis of its component processes. *Journal of Experimental Psychology: General, 11*(1), 1–22.

McGurk, H. (1970). The role of object orientation in infant perception. *Journal of Experimental Child Psychology, 9*, 363–373.

Meck, W. H., & Church, R. M. (1983). A mode control model of counting and timing processes. *Journal of Experimental Psychology: Animal Behavior Processes, 9*, 320–334.

Miller, K. (1984). Child as the measurer of all things: measurement procedures and the development of quantitative concepts. In C. Sophian (Ed.), *Origins of cognitive skills* (pp. 193–228). Hillsdale, NJ: Erlbaum.

Mix, K. S. (1999). Similarity and numerical equivalence: appearances count. *Cognitive Development, 14*, 269–297.

Mix, K. S. (2008a). Children's equivalence judgments: cross-mapping effects. *Cognitive Development, 23*, 191–203.

Mix, K. S. (2008b). Surface similarity and label knowledge impact early numerical comparisons. *British Journal of Developmental Psychology, 26*, 13–32.

Mix, K. S. (2009). How Spencer made number: first uses of the number words. *Journal of Experimental Child Psychology, 102*, 427–444.

Mix, K. S., & Paik, J. H. (2008). Do Korean fraction names promote part-whole reasoning? *Journal of Cognition and Development, 9*(2), 145–170.

Mix, K. S., Huttenlocher, J., & Levine, S. C. (1996). Do preschool children recognize auditory-visual correspondences? *Child Development, 67*, 1592–1608.
Mix, K. S., Levine, S. C., & Huttenlocher, J. (1999). Early fraction calculation ability. *Developmental Psychology, 35*(1), 164–174.
Mix, K. S., Huttenlocher, J., & Levine, S. C. (2002a). Multiple cues for quantification in infancy: Is number one of them? *Psychological Bulletin, 128*, 278–294.
Mix, K. S., Huttenlocher, J., & Levine, S. C. (2002b). *Quantitative development in infancy and early childhood*. New York, NY: Oxford University Press.
Moeller, K., Pixner, S., Kaufmann, L., & Nuerk, H. -C. (2009). Children's early mental number line: logarithmic or decomposed linear? *Journal of Experimental Child Psychology, 103*, 503–515.
Möhring, W., Newcombe, N. S., & Frick, A. (2014). Zooming in on spatial scaling: preschool children and adults use mental transformations to scale spaces. *Developmental Psychology, 50*, 1614–1619.
Morton, J., & Johnson, M. (1991). CONSPEC and CONLEARN: a two-process theory of infant face recognition. *Psychological Review, 98*(2), 164–181.
Munsinger, H., & Weir, M. W. (1967). Infants' and young children's preference for complexity. *Journal of Experimental Child Psychology, 5*, 69–73.
Navon, D. (1977). Forest before trees: precedence of global features in visual-perception. *Cognitive Psychology, 9*, 353–383.
Needham, A. (2001a). Perceptual, conceptual and representational processes in infancy. *Journal of Experimental Child Psychology, 78*, 98–106.
Needham, A. (2001b). Object recognition and object segregation in 4.5-month-old infants. *Journal of Experimental Child Psychology, 78*, 3–24.
Needham, A., & Baillargeon, R. (1998). Effects of prior experience in 4.5-month-olds' object segregation. *Infant Behavior and Development, 21*, 1–24.
Newcombe, N., Huttenlocher, J., & Learmonth, A. (1999). Infants' coding of location in continuous space. *Infant Behavior and Development, 22*, 483–510.
Newcombe, N. S., Sluzenski, J., & Huttenlocher, J. (2005). Pre-existing knowledge versus on-line learning: what do young infants really know about spatial location? *Psychological Science, 16*, 222–227.
Newcombe, N., Levine, S.C., & Mix, K.S. (2015). Thinking about quantity: the intertwined development of spatial and numerical cognition. *WIREs Cognitive Science, 6*(6), 491–505.
Norcia, A. M., Pei, F., Bonneh, Y., Hou, C., Sampath, V., & Pettet, M. W. (2005). Development of sensitivity to texture and contour information in the human infant. *Journal of Cognitive Neuroscience, 17*(4), 569–579.
Nunes, T., Light, P., & Mason, J. (1993). Tools for thought: the measurement of length and area. *Learning and Instruction, 3*, 39–54.
Oakes, L. M. (2010). Using habituation of looking time to assess mental processes in infancy. *Journal of Cognition and Development, 11*(3), 255–268.
Oakes, L. M., Ross-Sheehy, S., & Luck, S. J. (2006). Rapid development of feature binding in visual short-term memory. *Psychological Science, 17*(9), 781–787.
Odic, D., Pietroski, P., Hunter, T., Lidz, J., & Halberda, J. (2012). Young children's understanding of "more" and discrimination of number and surface area. *Journal of Experimental Psychology: Learning, Memory, and Cognition, 39*(2), 451–461.
Odic, D., Libertus, M. E., Feigenson, L., & Halberda, J. (2013). Developmental change in the acuity of approximate number and area representations. *Developmental Psychology, 49*, 1103–1112.
Opfer, J., & Siegler, R. (2007). Representational change and children's numerical estimation. *Cognitive Psychology, 55*, 169–195.
Otsuka, Y., Ichikawa, H., Kanazawa, S., Yamaguchi, M. K., & Spehar, B. (2014). Temporal dynamics of spatial frequency processing in infants. *Journal of Experimental Psychology: Human Perception & Performance, 40*(3), 995–1008.
Paik, J. H., & Mix, K. S. (2003). U.S. and Korean children's comprehension of fraction names: a re-examination of cross-national differences. *Child Development, 74*, 144–154.

Pepperberg, I. M., & Carey, S. (2012). Grey parrot number acquisition: the inference of cardinal value from ordinal position on the number list. *Cognition, 125*, 219–232.

Piaget, J. (1952). *The child's conception of number*. New York: Basic Books (Original in French, 1941).

Pica, P., Lemer, C., Izard, V., & Dehaene, S. (2004). Exact and approximate arithmetic in an Amazonian indigene group. *Science, 306*(5695), 499–503.

Posid, T., & Cordes, S. (2015). Verbal counting moderates perceptual biases found in children's cardinality judgments. *Journal of Cognition and Development, 16*(4), 621–637.

Ramani, G. B., & Siegler, R. S. (2008). Promoting broad and stable improvements in low-income children's numerical knowledge through playing number board games. *Child Development, 79*, 375–394.

Resnick, L. B., & Ford, W. W. (1981). *The psychology of mathematics for instruction*. Hillsdale, NJ: Erlbaum.

Rieth, C., & Sireteanu, R. (1994). Texture segmentation and "pop-out" in infants and children: the effect of test field size. *Spatial Vision, 8*, 173–191.

Romani, A., Caputo, A., Callieco, R., Schintone, E., & Cosi, V. (1999). Edge detection and surface filling as shown by texture visual evoked potentials. *Clinical Neurophysiology, 110*, 86–91.

Ross-Sheehy, S., Oakes, L. M., & Luck, S. J. (2003). The development of visual short-term memory capacity in infants. *Child Development, 74*(6), 1807–1822.

Ruff, H. A., & Birch, H. G. (1974). Infant visual fixation: the effect of concentricity, curvilinearity, and number of directions. *Journal of Experimental Child Psychology, 17*, 460–473.

Samuelson, L. K., Smith, L. B., Perry, L. K., & Spencer, J. P. (2011). Grounding word learning in space. *PloS One, 6*(12), e28095.

Sandberg, E. H., Huttenlocher, J., & Newcombe, N. (1996). The development of hierarchical representation of two-dimensional space. *Child Development, 67*, 721–739.

Schneider, M., & Siegler, R. S. (2010). Representations of the magnitudes of fractions. *Journal of Experimental Psychology: Human Perception and Performance, 36*(5), 1227–1238.

Schneider, M., Grabner, R. H., & Paetsch, J. (2009). Mental number line, number line estimation, and mathematical school achievement: their interrelations in grades 5 and 6. *Journal of Educational Psychology, 101*, 359–372.

Schoner, G., & Thelen, E. (2006). Using dynamic field theory to rethink infant habituation. *Psychological Review, 113*(2), 273–299.

Schyns, P. G., & Oliva, A. (1994). From blobs to boundary edges: evidence for time- and spatial-scale-dependent scene recognition. *Psychological Science, 5*(4), 195–200.

Shipley, E. F., & Shepperson, B. (1990). Countable entities: developmental changes. *Cognition, 34*(2), 109–136.

Shultz, T. R., & Bale, A. C. (2001). Neural network simulation of infant familiarization to artificial sentences: rule-like behavior without explicit rules and variables. *Infancy, 2*, 501–536.

Siegler, R. S., & Opfer, J. E. (2003). The development of numerical estimation: evidence for multiple representations of numerical quantity. *Psychological Science, 14*, 237–243.

Siegler, R. S., Thompson, C. A., & Schneider, M. (2011). An integrated theory of whole number and fractions development. *Cognitive Psychology, 62*(4), 273–296.

Singer-Freeman, K. E., & Goswami, U. (2001). Does half a pizza equal half a box of chocolates? Proportional matching in an analogy task. *Cognitive Development, 16*(3), 811–829.

Sirois, S., & Mareschal, D. (2002). Models of habituation in infancy. *Trends in Cognitive Sciences, 6*(7), 293–298.

Slater, A., Mattock, A., Brown, E., Burnham, D., & Young, A. (1991). Visual processing of stimulus compounds in newborn infants. *Perception, 20*, 29–33.

Slusser, E. B., Santiago, R. T., & Barth, H. C. (2013). Developmental change in numerical estimation. *Journal of Experimental Psychology: General, 142*(1), 193–208.

Sluzenski, J., Newcombe, N. S., & Satlow, E. (2004). Knowing where things are in the second year of life: implications for hippocampal development. *Journal of Cognitive Neuroscience, 16*, 1443–1451.

Sokolov, E. N. (1963). Higher nervous functions: the orienting reflex. *Annual Review of Physiology, 25*, 545–580.

Solomon, T., Vasilyeva, M., Huttenlocher, J., & Levine, S.C. (2015). The long and the short of it: children's difficulty conceptualizing spatial intervals on a ruler as linear units. *Developmental Psychology, 51*, 1564–1573.

Soltesz, F., Szucz, D., & Szucs, L. (2010). Relationships between magnitude representation, counting, and memory in 4- to 7-year-old children: a developmental study. *Behavioral and Brain Functions, 6*, 1–14.

Sophian, C., & Chu, Y. (2008). How do people apprehend large numerosities? *Cognition, 107*(2), 460–478.

Sophian, C., Garyantes, D., & Chang, C. (1997). When three is less than two: early developments in children's understanding of fractional quantities. *Developmental Psychology, 33*, 731–744.

Spaepen, E., Coppola, M., Spelke, E. S., Carey, S. E., & Goldin-Meadow, S. (2011). Number without a language model. *Proceedings of the National Academy of Sciences, 108*(8), 3163–3168.

Spears, W. C. (1964). Assessment of visual preference and discrimination in the four-month-old infant. *Journal of Comparative and Physiological Psychology, 57*(3), 381–386.

Spencer, J. P., Barich, K., Goldberg, J., & Perone, S. (2012). Behavioral dynamics and neural grounding of a dynamic field theory of multi-object tracking. *Journal of Integrative Neuroscience, 11*(3), 339–362.

Spinillo, A., & Bryant, P. (1991). Children's proportional judgments: the importance of "half". *Child Development, 62*(3), 427–440.

Srinivasan, M., & Carey, S. (2010). The long and the short of it: on the nature and origin of functional overlap between representations of space and time. *Cognition, 116*(2), 217–241.

Starkey, P., & Cooper, R. G. (1980). Perception of numbers by human infants. *Science, 210*, 1033–1035.

Starkey, P., Spelke, E. S., & Gelman, R. (1983). Detection of intermodal numerical correspondences by human infants. *Science, 222*, 179–181.

Starkey, P., Spelke, E. S., & Gelman, R. (1990). Numerical abstraction by human infants. *Cognition, 36*, 97–128.

Stein, T., Seymour, K., Hebart, M. N., & Sterzer, P. (2014). Rapid fear detection relies on high spatial frequencies. *Psychological Science, 25*(2), 566–574.

Stoianav, I., & Zorzi, M. (2012). Emergence of a "visual number sense" in hierarchical generative models. *Nature Neuroscience, 15*(2), 194–196.

Strauss, M. S., & Curtis, L. E. (1981). Infant perception of numerosity. *Child Development, 52*, 1146–1152.

Thompson, C. A., & Opfer, J. E. (2008). Costs and benefits of representational change: effect of context on age and sex differences in magnitude estimation. *Journal of Experimental Child Psychology, 101*, 20–51.

Trick, L., & Pylyshyn, Z. W. (1994). Why are small and large numbers enumerated differently? A limited-capacity preattentive stage in vision. *Psychological Review, 101*, 80–102.

Turpin, G. (1986). Effects of stimulus intensity on autonomic responding: the problem of differentiating orienting and defense reflexes. *Psychophysiology, 23*(1), 1–14.

Vasilyeva, M., & Huttenlocher, J. (2004). Early development of scaling ability. *Developmental Psychology, 40*(5), 682–690.

Vasilyeva, M., Duffy, S., & Huttenlocher, J. (2007). Developmental changes in the use of absolute and relative information: the case of spatial extent. *Journal of Cognition and Development, 8*(4), 455–471.

Wagner, S., Winner, E., Cicchetti, D., & Gardner, H. (1981). Metaphorical mapping in human infants. *Child Development, 52*(2), 728–731.

Wass, S. V. (2014). Comparing methods for measuring peak look duration: are individual differences observed on screen-based tasks also found in more ecologically valid contexts? *Infant Development and Behavior, 37*, 315–325.

Watson, J. S. (1966). Perception of object orientation in infants. *Merrill-Palmer Quarterly*, *12*(1), 73–94.

Wilcox, T., & Baillargeon, R. (1998). Object individuation in infancy: the use of featural information in reasoning about occlusion events. *Cognitive Psychology*, *37*, 97–155.

Wilcox, T., & Chapa, C. (2002). Infants' reasoning about opaque and transparent occluders in an individuation task. *Cognition*, *85*, B1–B10.

Wilcox, T., & Chapa, C. (2004). Priming infants to attend to color and pattern information in an individuation task. *Cognition*, *90*, 265–302.

Wynn, K. (1992). Addition and subtraction by human infants. *Nature*, *358*, 749–750.

Xu, F., & Carey, S. (1996). Infant metaphysics: the case of numerical identity. *Cognitive Psychology*, *30*, 111–153.

Xu, F., & Spelke, E. S. (2000). Large number discrimination in 6-month-old infants. *Cognition*, *74*(1), B1–B11.

Xu, F., Spelke, E. S., & Goddard, S. (2005). Number sense in human infants. *Developmental Science*, *8*(1), 88–101.

CHAPTER

2

Link Between Numbers and Spatial Extent From Birth to Adulthood

Maria Dolores de Hevia

University Paris-Descartes, Sorbonne Paris Cité, Paris; Laboratory of Psychology of Perception, Paris, France

OUTLINE

2.1 INTRODUCTION

The ability to discriminate and represent information of magnitude is foundational to human reasoning. We, and many species of nonhuman creatures, use this ability in everyday experiences, such as estimating the number of people attending a dinner; it allows us and other animals to decide whether the best bet when facing a group is to attack or to escape; and it also intervenes when one needs to calculate the time available to complete an action.

Continuous Issues in Numerical Cognition. http://dx.doi.org/10.1016/B978-0-12-801637-4.00002-0

The origin of these concepts is a long-standing debate. Are representations of quantity, like number, space, and time, available early in life so that they can intervene in shaping our understanding of the world, or are these representations rapidly extracted by learning mechanisms following exposure to an already structured physical world? Research investigating the developmental origins of these concepts can shed light on these questions.

A related fundamental cognitive ability is the propensity to use spatial information in order to represent other nonspatial concepts. For instance, we use a vertical spatial axis to represent levels of social status, of power, even to convey levels of self-esteem, and of morality. Most cultures also tend to picture the future as if it were ahead and the past behind, in reference to one's own body. These links are also evident in common measurement tools, where we use a spatial format to represent levels of temperature, pressure, and time. However, among many others, the number–space mapping is essential for human cognition, as it is well reflected in two branches of science: mathematics and geometry. In mathematics, all real numbers are referred to as the "real line," where each number corresponds to a point in a one-dimensional line; in geometry, numbers are used in order to study questions of shape, size, relative position of figures, and the properties of space, with the invention of the Cartesian coordinates as a clear example.

Noted early by psychologists like Francis Galton (Galton, 1880), who documented the propensity of consciously visualizing numbers in spatial configurations, it is now firmly established by cognitive science that adults' processing of numerical information is accomplished by spontaneously deploying visuospatial mental resources. While most researchers studying numerical cognition have focused on the computational properties of number, in particular tracing the ontogenetic and phylogenetic basis of one of its signatures, that is, the analog format (Cantlon & Brannon, 2007; Xu & Spelke, 2000), recent research has concerned the spatial signature of number (Opfer, Furlong, & Thompson, 2010). In fact, under the influential "number line" model (Restle, 1970), proposed in its most recent form by S. Dehaene (Dehaene, 1992), numbers are represented analogically along a continuum of magnitude which is spatially oriented. Until very recently, however, the origins and nature of this phenomenon were poorly understood. To address these questions, it has been put forward and developed a framework to study this number–space interaction (de Hevia, Girelli, & Vallar, 2006). On the one hand, numbers can be mapped onto space in the form of corresponding spatial extents depending on their magnitude (nondirectional mapping). On the other hand, numbers can be associated to different spatial positions, depending on their relative magnitude and/or order (directional mapping). In this chapter I will review recent research seeking to understand the origins, nature and properties of these two types of interactions between the numerical and spatial domains.

2.2 NUMBERS AND SPACE

The link between numbers and space is present in our mental imagery, brain, and behavior. The idea of visualizing numbers, initially conceived more than a century ago (Galton, 1880), received its first empirical evidence with the description of the SNARC effect [the Spatial–Numerical Association of Response Codes (Dehaene, Bossini, & Giraux, 1993)], which refers to the phenomenon by which subjects are faster responding to small numbers with their left hand and to large numbers with their right hand. The link between numbers and space has been well documented also at the neuroanatomical level. One of the first sources of evidence came from the Gerstmann syndrome, which is characterized by the tetrad of acalculia, left–right disorientation, finger agnosia, and agraphia, and is found in patients with inferior parietal damage (Gerstmann, 1940). Even more telling is recent research showing that overlapping areas in the parietal cortex are engaged in numerical and spatial processing (Fias, Lauwereyns, & Lammertyn, 2001; Nieder, 2005; Pinel, Piazza, Le Bihan, & Dehaene, 2004), neuropsychological disorders of visuospatial nature can extend also to certain numerical computations (Vuilleumier, Ortigue, & Brugger, 2004; Zorzi, Priftis, & Umiltà, 2002), and neural circuits dedicated to eye movements are recruited during arithmetic (Knops, Thirion, Hubbard, Michel, & Dehaene, 2009). Moreover, the posterior parietal cortex in the primates, which encloses quantity-selective neurons, contains accurate information about both continuous (length) and discrete (number of items) quantity (Tudusciuc & Nieder, 2007). Overall, these findings suggest that the representation of number does not involve a dedicated area, but only a fraction of IPS (intraparietal sulcus) neurons, highly distributed in the IPS, and intermingled with other representations of size, location, and other continuous spatial dimensions.

The description of an association between number and space in childhood was first described by Piaget (Piaget, 1952). In the number conservation task, where the experimenter shows two horizontal rows of tokens in one-to-one correspondence, 3- to 6-year-old children judge longer rows as being greater in number even when they have fewer tokens, and think that the number of objects in a row increases as the experimenter spreads the objects apart. In fact, although it was traditionally assumed that children develop an understanding of number slowly, and only after many years of experience and education, there is now a wide consensus based on research from the last few decades, that we humans possess a number sense (Dehaene, 1997), a system that extracts the cardinality of sets, that is evolutionary ancient (Vallortigara, Regolin, Chiandetti, & Rugani, 2010), preverbal (Xu & Spelke, 2000), and functional at birth (Coubart, Izard, Spelke, Marie, & Streri, 2014; Izard, Sann, Spelke, & Streri, 2009). A symbolic numerical system and formal mathematics are thought to be built upon this system.

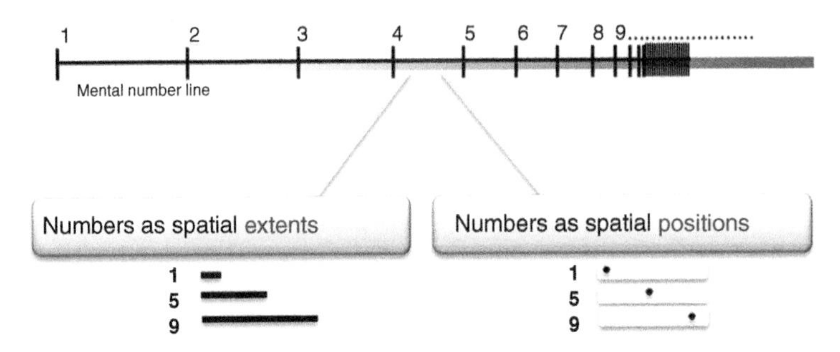

FIGURE 2.1 **The two forms the interaction between the domains of number and space can take.** The mental number line model encompasses two types of number-space mappings: on the one hand, numbers correspond to different spatial extents (nondirectional number-space mapping), and on the other hand numbers correspond to different spatial positions (directional number-space mapping).

Adults' representations of number are thought to correspond to analogical representations of magnitude in a spatially oriented continuum (Dehaene, Piazza, Pinel, & Cohen, 2003). Therefore, mature representations of number are built on a core system of numerical representation that connects to spatial representations in the form of a *mental number line*. Although it is not established yet whether number is spatial per se or whether a spatial numerical representation is only a short-term memory frame constructed on-task (van Dijck & Fias, 2011), in this chapter I will argue that the "mental number line" model encompasses the phenomenon of an intimate link between the domains of number and space (Fig. 2.1). This link can take place in two ways: numbers can be mapped onto corresponding spatial extents (nondirectional number–space mapping), and be associated to different spatial positions (directional number–space mapping) (de Hevia et al., 2006).

2.3 NONDIRECTIONAL NUMBER–SPACE MAPPING

In order to investigate the origins and developmental course of the number–space mapping, we suggested the existence of a nondirectional mapping between these dimensions (de Hevia, Girelli, Bricolo, & Vallar, 2008; de Hevia et al., 2006; de Hevia & Spelke, 2009). This mapping does not include information of orientation, but the two sources of information, numerical and spatial extent, interact with each other by virtue of both being continuous dimensions of magnitude. In fact, one of the main cognitive attributes characterizing representations of quantitative dimensions is their analog format (Gallistel & Gelman, 2000). This signature reflects the fact that discrimination conforms to Weber's law, where discrimination depends on the ratio rather than on the absolute values (Stevens, 1970). Thus, the cognitive and/or neural constraints appear to be similar in the representations of any attribute that can be formalized in "more than" or "less than" terms (Bonn & Cantlon, 2012), which might explain why the dimensions of number and spatial extent interact with each other.

Some experimental work addressed the existence of a nondirectional number–space mapping in adults. A series of experiments, adapting the line bisection task with numerical irrelevant cues (Fischer, 2001), investigated the impact of number on visuospatial processing. In these studies, participants are presented with one line flanked at each side by a number, and are instructed to set the midpoint of the line in a fast but accurate way. No instructions regarding the numbers are provided. A systematic bias toward the larger number irrespective of its position in the line was found and replicated in various experimental conditions (de Hevia et al., 2006). The biases were interpreted as resulting from a "cognitive illusion" of length, whereby the side ipsilateral to the larger number is perceived "larger" and/or the side ipsilateral to the smaller number is perceived "smaller" [in analogy to the optical illusions in the figures of Müller-Lyer and in the combined forms of the Opel-Kundt (Müller-Lyer, 1889)].

Further experiments tested the "cognitive illusion" hypothesis through a computerized task (de Hevia et al., 2008). Participants are presented for 1 s with two numbers on the screen, separated by a variable distance; afterwards, they are requested to reproduce the length of the space between the two numbers, with no time limits. Spatial extents delimited by two small numbers are systematically underestimated; those delimited by large numbers are overestimated. Crucially, when the numbers were different (one small and one large), there was no effect on performance, confirming previous observations with the line bisection paradigm, and providing further support for the hypothesis of a cognitive illusion brought about numerical size (Fig. 2.2). This paradigm has been recently adapted to test for the specificity of the numbers' effect onto spatial representation (Viarouge & de Hevia, 2013). We tested whether other types of ordinal information that do not convey information of magnitude, as it is the case with the letters of the alphabet, might exert an influence in visuospatial representations. Moreover, we tested whether numbers might influence reproduction of the level of brightness of a figure, a further dimension of magnitude. We found that numbers affect reproduction of spatial extent, replicating previous observations, but that they do not modulate reproduction of the level of brightness. Moreover, we found that it is the information of magnitude, and not the ordinal information, which is responsible for the illusory effects, as letters did not exert any influence in performance, either in the size or the brightness reproduction tasks.

Collectively, this work supports the idea that numbers correspond to different spatial lengths, varying parametrically with number magnitude (Restle, 1970). Moreover, irrelevant-to-the-task numbers can modulate visuospatial computations, and this effect critically depends on their

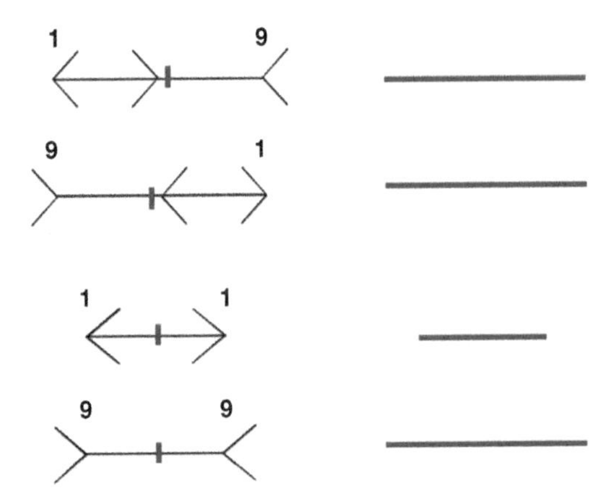

FIGURE 2.2 A "cognitive illusion of length" produced by numbers is presented in analogy to the visual illusions in the figures of Müller-Lyer and in the combined forms of the Opel-Kundt. Red segments correspond to the performance in bisection tasks (left) and reproduction tasks (right). In the line bisection tasks there is a bias toward the larger number when the flankers differ (left, up), but no lateral bias when then flankers are identical (left, bottom). In the reproduction task, spaces delimited by identical small numbers are underestimated and those delimited by identical large numbers overestimated (right, bottom), while there are no misestimations of length when the flankers differ (right, up).

information of magnitude. Therefore, the mapping of numbers onto space is not only in the form of an oriented mental number line, but numerical representations also link to spatial representations in the form of a nondirectional mapping. Where does this nondirectional number–space mapping come from? One possibility is that exposure to mathematics education, which makes explicit the link between numbers and space through graphs, rulers, and other measurement devices, is key to the development of this mapping. Another possibility is that the mapping between numbers and space precedes education and even extensive experience with these dimensions. In what follows, I will review evidence supporting the second possibility.

A series of experiments adapted the line bisection task to the evaluation of the effects of number on children's spatial processing. Previous research focusing on the number–space mapping during childhood had failed to disclose an early mapping of these dimensions when testing for the presence of the SNARC effect using symbolic number (Arabic digits). This effect refers to the advantage for responding to small numbers with the left hand and to large numbers with the right hand (Dehaene et al., 1993). However, since children do not access automatically information of magnitude from the Arabic symbol until the age of 6–7 years (Girelli, Lucangeli, & Butterworth, 2000; Rubinsten, Henik, Berger, & Shahar-Shalev, 2002), we reasoned that a nondirectional number–space mapping using nonsymbolic number (arrays of dots) might be observable in the childhood years.

After confirming that in adults the signature bias toward the larger Arabic number in the bisection of lines extends to the bisection of lines flanked by nonsymbolic numbers, we tested 7-year-old children with both types of stimuli: We found that a signature bias toward the larger number was found for lines flanked by nonsymbolic number, but no significant bias for lines flanked by symbolic number (de Hevia & Spelke, 2009).

Further experiments presented the same task to 5-year-old children, using only lines flanked by nonsymbolic number (de Hevia & Spelke, 2009). As for adults and 7-year-old children, 5-year-old children presented a systematic bias toward the side ipsilateral to the larger nonsymbolic number. Crucially, the analyses comparing this effect to the one present in adults and 7-year-old children were found to be similar (no effects of age range were found significant). A final set of experiments in 5-year-old children were conducted in order to verify that this effect was not related to any nonnumerical continuous dimension. Although the stimuli employed for the nonsymbolic condition with all our participants (adults, 7- and 5-year-old children) controlled for the overall area occupied by the single elements of the array, two further experiments controlled for the total contour length, the overall area occupied by the array configuration, and the space between the flanking array and the end of the line. In all these conditions, children systematically bisected the line toward the larger nonsymbolic number, and the effect was similar across conditions (no effects of experimental condition were found to be significant).

This work formed the basis for new research undertaken with an even younger population: preverbal human infants. We hypothesized that our minds might be predisposed to treat number and space as intrinsically related, also based on the evidence showing that even uneducated adults spontaneously map number onto space (Dehaene, Izard, Spelke, & Pica, 2008). We therefore tested infants' abilities to relate numerical and spatial information (de Hevia & Spelke, 2010). In a first experiment, we habituated 8-month-old infants to a series of either increasing or decreasing number, in the form of arrays of dots presented in a screen (following a 1:2 ratio). Once the looking time decreased significantly, signaling that habituation had been reached, we presented a series of lines lengths that differed in magnitude following the same 1:2 ratio, alternating increasing and decreasing order trials. If infants are able to transfer discrimination of the ordinal information embedded in the numerical series to the displays showing different line lengths, then infants habituated to increasing number should look longer to the lines following a decreasing order, and infants habituated to the decreasing numerical order should look longer to the series of lines following an increasing order. That was the pattern of results obtained: Infants generalized the habituation looking pattern (low looking times) to the series of lines that followed the familiar order, and looked significantly longer to the series of lines that followed the novel order at test. In a second series of experiments infants were presented with the information of number and space at the same time, but following no predictable order. One group of infants were presented with a rule that established a positive (congruent) relationship between number and space: the larger the number, the longer the line. They were then presented with two test trials containing new numbers and line lengths: In one trial, they were paired to follow the same (positive) rule, and in a different trial, a different (inverse) rule, where the larger the number, the shorter the line. Infants looked significantly longer to the test trial presenting the same (positive) rule, showing that they were able to learn the rule and apply it to new exemplars. A second group of infants tested with similar methods and materials was familiarized with a negative (inverse) relationship between number and space: the larger the number, the shorter the line. Note that this rule is equally predictable than the positive one, except that the two dimensions are paired incongruently. At test, infants received the same trials than the group familiarized to the positive rule. In contrast to those infants, however, infants did not show any preference for any of the two mappings at test, suggesting they were not able to learn the inverse rule and apply it to new exemplars during test. This series of experiments show that infants at 8 months of age do spontaneously map number onto space, and that this mapping has a privileged direction: Both sources of information must be positively aligned to each other. Similar findings were reported for the dimensions of size (Lourenco & Longo, 2010) and temporal duration (Lourenco & Longo, 2010; Srinivasan & Carey, 2010) with infants of similar age.

Further experiments addressed the following question: do infants have the ability to map number and space (and time) from birth? In particular, is it learned through exposure to the natural correlations between these variables in the environment (eg, more objects occupy more space), or is it given by an early predisposition to map these representations? Building on the evidence that from birth humans have the ability to discriminate numerical information (Izard et al., 2009), we tested whether 0- to 3-day-old newborns are able to link the information of number and time to the information of spatial extent. The paradigm had two phases: a familiarization phase (60 s) immediately followed by a test phase (Fig. 2.3). Across the two phases, all infants were presented with two values of numerosities and/or durations (one small and one large), and two values of length (one small and one large). First, during familiarization, one of the auditory stimuli was paired with one visual length. Second, during test, the auditory numerosity and/or duration was changed (from small to large or from large to small), and was paired with the familiar and the novel visual length in two successive trials. Therefore, compared to familiarization, test trials thus contained either one change (auditory change only) or two changes (auditory and visual changes). Crucially, two familiarization conditions were created such that in the two-change trials the

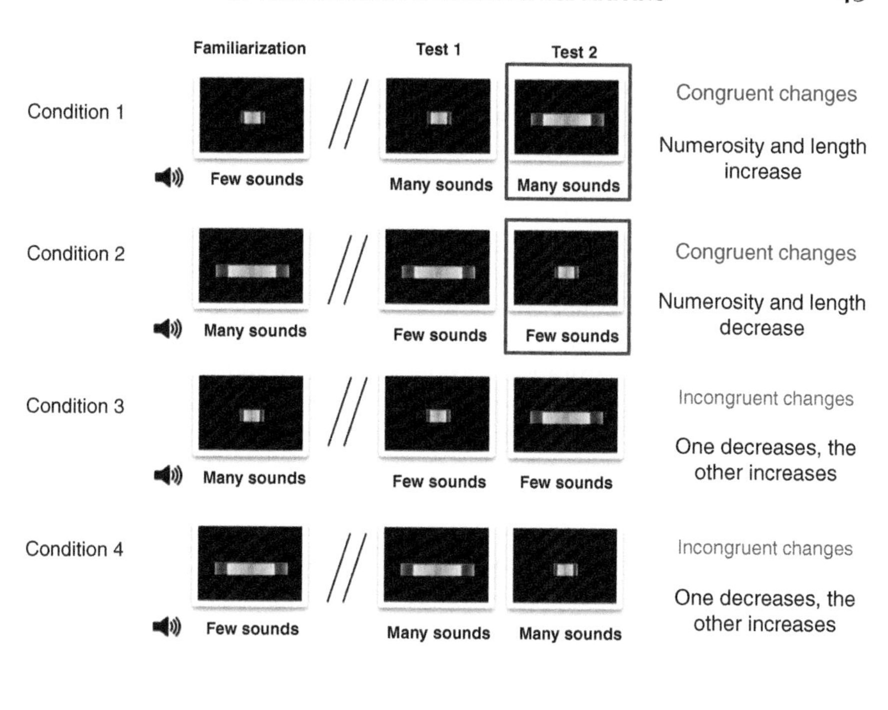

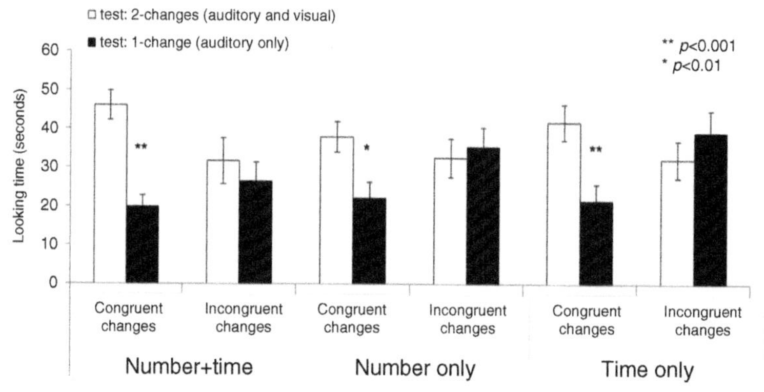

FIGURE 2.3 Different experimental conditions in which human newborns were tested in their ability to form mappings between auditory magnitude (number + time; number only; time only) and visual line lengths (top). Looking times during the two test trials where either one or two changes took place. Newborns systematically looked longer to two-change test trials provided the magnitude changes across dimensions were in the same direction (congruent changes: one increases, the other increases; or one decreases, the other decreases). *Adapted from de Hevia, M. D., Izard, V., Coubart, A., Spelke, E. S., & Streri, A. (2014). Representations of space, time and number in neonates. Proceedings of the National Academy of Sciences, 111 (13), 4809–4813.*

auditory numerosity/duration and visual length either changed in the same direction (either both increasing or both decreasing), or in opposite directions (one increasing, the other decreasing). We showed that neonates react to a simultaneous increase (or decrease) in spatial extent, and in duration or numerical quantity, but they do not react when the magnitudes vary in opposite directions; see Fig. 2.3 (de Hevia, Izard, Coubart, Spelke, & Streri, 2014).

To succeed in our tests, newborns have to build an expectation of congruency between magnitude-related changes in number, time, and space, from familiarization to test. Notably, neonates revealed this sensitivity both when numerical and temporal cues were simultaneously available, and when number and duration was presented in isolation (Fig. 2.3). Taken together, these findings show that the human mind thus may be predisposed to relate these three fundamental dimensions prior to extensive postnatal experience with the natural correlations between numbers of objects, spatial extents, and temporal durations. Therefore, the mapping is not a mere association learned through a statistical learning that can establish nonmeaningful associations, but the information of magnitude presented in the form of number and/or time and spatial length has to be congruent at an abstract level.

2.3.1 Specificity of the Nondirectional Number–Space Mapping

Following the finding that infants and newborns spontaneously map number, time, and space (de Hevia et al., 2014b; de Hevia & Spelke, 2010; Lourenco & Longo, 2010; Srinivasan & Carey, 2010), a further question we addressed was whether these mappings reflect a general ability to map any continuous dimension onto any other, or whether the number–time–space mappings reflect a privileged connection between these knowledge systems.

Some of these questions have been addressed through studies of cross-dimensional mappings with 4- to 5-year-old children (de Hevia, Vanderslice, & Spelke, 2012). In this study children were first asked to map the dimensions of number, line length, and level of brightness in a series of intradimensional mappings: they were presented with cards depicting different numbers, different line lengths, and different levels of brightness, and asked to map one series of cards to a second series of cards. The experimenter presented a certain mapping (within or between dimensions) and then asked the child to decide which of two alternatives (eg, two stimuli with different line lengths) matched a specific stimulus (eg, a stimulus of a given brightness). Children performed significantly above chance for the three dimensions for mappings within dimensions. Next, children were tested in the crucial, interdimensional mappings, where each dimension is mapped onto each other: number–line length, line length–brightness, and brightness–number. We tested one group of children with a positive mapping between each of these interdimensional mappings (eg, the longer the line, the brighter the object), a second group of children with positive interdimensional mappings with a different direction of the mapping (eg, from number to length vs. from length to number), and a final group of children with an inverse mapping between the dimensions (eg, the larger the number, the shorter the line). For the positive interdimensional mappings, only the number–length mappings (in both directions), and the length–brightness mappings (in one direction) were performed significantly above chance, while for the inverse interdimensional mappings, none of the mapping conditions were performed significantly above chance. These results suggest, first, that the number–length mappings might have a privileged status at the start of the school years; second, the mappings of one magnitude dimension, such as number and brightness, onto the dimension of length are intuitive for these children, suggesting that space might function as a privileged dimension onto which other sources of information are mapped; third, at the age tested in this study, not all the continuous dimensions map onto any other in the same way.

Further studies tested 8-month-old infants in their ability to create the mappings of number and level of brightness (de Hevia & Spelke, 2013) adopting the same methods that revealed successful number–space mappings (de Hevia & Spelke, 2010). For number–brightness mappings, we obtained a pattern of performance both quantitatively and qualitatively different from the number–space mappings. First, when infants are presented with a series of numerical order (either increasing or decreasing), and are later tested with a series of different levels of brightness (both in increasing and in decreasing order), infants do not show any significant difference in their looking times during the test trials, and therefore are not able to transfer the discrimination from a series of numerical displays to a series of brightness levels. Infants at the same age successfully transfer discrimination from a series of numerical displays to a series of line lengths. Second, when infants are familiarized with a positive number–brightness mapping, they look longer at the new rule at test, showing evidence that they have extracted the rule during familiarization, and have applied it to new exemplars during test. Third, when infants are familiarized with inverse number–brightness mapping, they are not able to extract the rule during familiarization. Although in some conditions infants are able to extract the information of a rule establishing a positive relationship between number and brightness, and between number and length (or space), and to use it productively, there are differences between the two mappings in infancy. While in the number–space mappings, infants look longer at the new information that follows a positive mapping, in the number–brightness mappings they do not show this preference for the positive mappings, suggesting that positive number–space mappings recruit attentional resources, whereas positive number–brightness mappings, even if learnable, do not recruit attention. These findings, together with previous evidence for successful mappings between brightness and loudness (Lewkowicz & Turkewitz, 1980) and failure to map loudness and spatial length (Srinivasan & Carey, 2010), suggest that dimensions referring to the intensity of stimulation (eg, brightness or loudness) are treated more similarly than the dimensions of number, spatial extent, and time. The mappings between quantitative dimensions might derive from biologically predisposed links between the dimensions of number, spatial extent, and time, which are functional from birth.

In summary, the fundamental, privileged mappings between number and spatial extent are functional at birth, before the acquisition of language, counting, symbolic knowledge, or before extensive experience with correlations between these dimensions, and remain constant throughout childhood and adulthood after the onset of education.

2.4 DIRECTIONAL NUMBER–SPACE MAPPING

The work described previously has considered as the object of study the nondirectional mapping between number and space. Numbers, however, map also onto a directional space, as represented by the SNARC effect, which refers to the advantage for responding to small numbers with the left hand and to large numbers with the right hand (Dehaene et al., 1993). This effect, and more generally spatial–numerical associations (SNA) (Patro, Nuerk, Cress, & Haman, 2014) have been thought to emerge over the course of the elementary school years (Berch, Foley, Hill, & Ryan, 1999; van Galen & Reitsma, 2008) or even during the preschool years (Ebersbach, Luwel, & Verschaffel, 2013; Hoffmann, Hornung, Martin, & Schiltz, 2013; Patro & Haman, 2012) as a by-product of the acquisition of counting (Opfer et al., 2010), and to be dependent on the specific reading direction (Shaki, Fischer, & Göbel, 2012). Recent work has investigated both the origins of this directional mapping as well as its specificity.

One series of experiments aimed at testing infants' ordinal abilities. In fact, we know from the adult literature that the information of order is highly relevant for the oriented spatial mapping: different numerical and nonnumerical ordinal series are spatially oriented (Gevers, Reynvoet, & Fias, 2003, 2004; Previtali, de Hevia, & Girelli, 2010; Rusconi, Kwan, Giordano, Umiltà, & Butterworth, 2006). Therefore, an investigation of the origins of the spatial orientation of numbers might benefit from the study of infants' ability to represent and discriminate ordinal information. We first tested 7-month-old infants on their abilities to extract ordinal numerical information with the use of featural cues that would help extract the ordinal rule (Kirkham, Slemmer, & Johnson, 2002): a group of infants was habituated to increasing numerical sequences, the other group to decreasing numerical sequences, and afterwards both groups were tested with both increasing and decreasing numerical sequences. Infants in both groups showed a significantly higher looking time to the series at test that followed the novel order, suggesting that infants at this age are able to extract the information of order embedded in a numerical series (Picozzi, de Hevia, Girelli, & Macchi Cassia, 2010).

A second series of experiments tested the hypothesis that at 7 months of age the information of number is spatially oriented. We employed the same methods and materials than the study previously described, except that we presented the numerical arrays oriented from left to right (for both the increasing and decreasing series). Under these conditions, both groups of infants, those habituated to increasing order (left-to-right oriented) as well as those habituated to decreasing order (left-to-right oriented), looked significantly longer to the new increasing (left-to-right oriented) numerical sequences at test. This pattern of performance was not replicated when the same numerical sequences were presented from right to left. In these conditions, in fact, infants did not discriminate the ordinal information embedded in the sequences. Therefore, these findings show that infants take into account the specific association between numerical order and space (Fig. 2.4). Moreover, infants display a preference for numerical arrays increasing in number and oriented from left to right (de Hevia, Girelli, Addabbo, & Macchi-Cassia, 2014). Collectively, these studies offer preliminary hints for an early spatial organization of numerical information along a left-to-right horizontal axis.

While these studies suggest that infants have a preference for increasing number left-to-right oriented, they do not directly establish whether there is an association between small numbers with the left side and large numbers with the right side, as the SNARC and related effects have shown in adults (Dehaene et al., 1993) and children (Patro & Haman, 2012). In a recent study, we used a Posner-like attentional task (Posner, 1980) similar to the one used in adults where targets on the left are detected faster when primed by a central small number (either in symbolic or nonsymbolic notations) and targets on the right are detected faster when primed by a large number (Bulf, Macchi-Cassia, & de Hevia, 2014; Fischer, Castel, Dodd, & Pratt, 2003). We investigated whether a nonsymbolic numerical cue presented in the center of the screen could drive the allocation of visual attention in 8- to 9-month-old infants in the same way as they do in adults. Infants' eye movements were registered while presenting lateralized targets that were preceded by central nonsymbolic numbers. Infants were faster at detecting targets appearing on the right when cued by large numbers (an array of nine dots), and targets appearing on the left when cued by small numbers (an array of two dots) (Bulf, de Hevia, & Macchi-Cassia, 2015). This finding indicates that, as in adults, nonsymbolic numbers

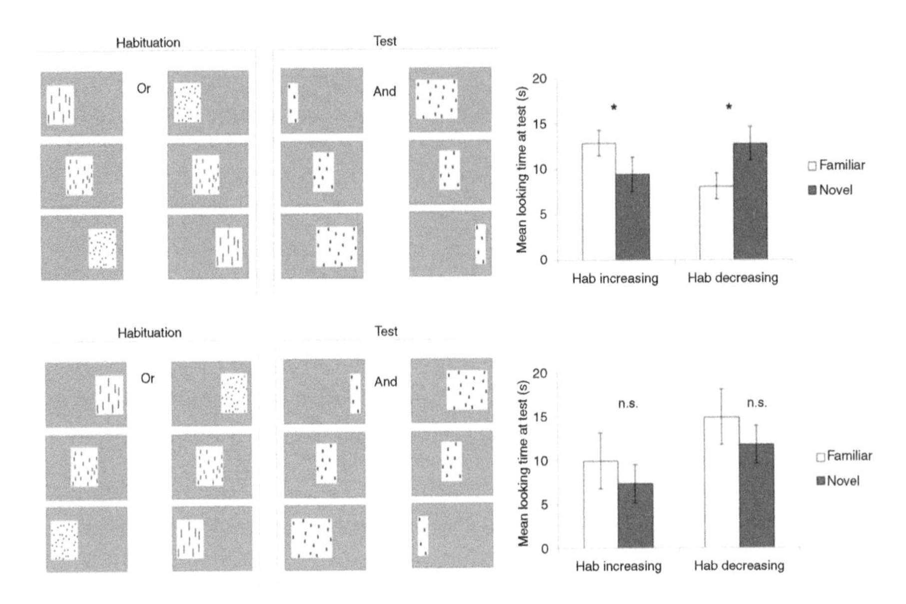

FIGURE 2.4 **Experimental stimuli used to test infants' ability to discriminate ordinal numerical information along a spatially oriented axis.** Infants are habituated to either increasing or decreasing order and are tested with new numbers in alternating increasing and decreasing order. One group of infants see all numerical series left-to-right oriented (top left). Infants' looking times varied during test trials for each habituation condition (increasing vs. decreasing) for familiar and novel order. All infants look significantly longer to increasing, left-to-right-oriented numerical sequences (top right). A second group of infants see all numerical series right-to-left oriented (bottom left). Infants do not discriminate at test between novel and familiar test trials in either of the two habituation conditions (bottom right). *Adapted from de Hevia, M. D., Girelli, L., Addabbo, M., & Macchi-Cassia, V. (2014). Human infants' preference for left-to-right oriented increasing numerical sequences. PLoS ONE, 9, e96412.*

induce attentional shifts toward a peripheral region of space that is congruent with the numbers' relative positions along a left-to-right-oriented mental number line, already during the first year of life. The effect was not present when using visual geometrical forms of different sizes as central cues (small vs. large, respecting the same ratio used with numerical cues). This finding leaves open the question, ripe for future research, whether at this age preverbal infants use an oriented spatial continuum only to represent numerical information, which might be later generalized to other types of magnitude-related and ordinal stimuli, as it is the case in adulthood. However, it is possible that the effect of nonsymbolic number onto spatial attention found in infants (Bulf et al., 2015a) and adults (Bulf et al., 2014) is different from the effects described when using symbolic number. In fact, some authors have casted doubt on the automaticity of the effect of symbolic number onto spatial attention, since a reliable effect has been observed only when adults intentionally attend to numerical magnitude (Galfano, Rusconi, & Umilta, 2006; Ristic, Wright, & Kingstone, 2006).

Although SNARC-like effects have been described for information of magnitude other than number, such as physical size, brightness, and conceptual size for adults (Ren, Nicholls, Ma, & Chen, 2011), not all dimensions have the same effects on visuospatial performance. Based on past research showing that perceiving numbers causes spatial shifts of attention to the left or right side, depending upon the numbers' magnitude (Fischer et al., 2003), we investigated whether this phenomenon extends to nonsymbolic numbers, as well as to the continuous dimensions of size and brightness. After a numerical (ie, symbolic Arabic digits or nonsymbolic arrays of dots) or a nonnumerical cue (ie, shapes of different size or different brightness level) was presented, participants' saccadic response to a target that could appear in the left or the right side of the screen was registered using an automated eye-tracker system. We found that, both in the case of Arabic digits and dot arrays, right targets were detected faster when preceded by large numbers, and left targets were detected faster when preceded by small numbers. Moreover, while participants were faster at detecting right targets when cued by large-sized shapes and left cues when cued by small-sized shapes, brightness cues did not modulate the detection of peripheral targets (Bulf et al., 2014).

2.4.1 Origins of a Directional Number–Space Mapping

While it was first thought that a nondirectional number–space mapping was the most fundamental link between number and space, with its orientation taking place only later with the onset of cultural influence (de Hevia & Spelke, 2009, 2010), the studies reviewed previously show that the orientation of this link takes place even before reading and writing habits are well acquired. It is still an empirical question, however, to determine the precise time frame when culture impacts the spatialization of information already in infancy, as parents' interactions with their infants might very well already instill a certain spatial directionality (McCrink, Birdsdall, & Caldera, 2011). Future work using similar methods as those reviewed earlier with infants in cultures where reading/writing direction is, for instance, right-to-left oriented, can shed much light on the developmental course of the onset of cultural influence.

Without denying the modulating effect of culture, recent research with nonhuman animals has opened the possibility that a universal left-to-right orientation might originate from biologically determined spatial asymmetries: A directional number–space mapping might be present very early in development as part of our evolutionary history. In fact, it has been recently shown that nonhuman animals do map number onto a left-to-right orientation. This work originated in the studies of newborn chicks and has now been extended to nonhuman primates, which are evolutionary closer to humans. In these studies, animals are first trained to touch on a screen (in the case of monkeys) or to physically approach (in the case of chicks) a particular element in a series of elements vertically arranged. During the test the elements' array is rotated into a horizontal configuration and the animal is tested in the detection of the element that corresponds to the trained ordinal position, so that the animal could succeed by detecting the position starting from the left or from the right side of the array. In these conditions, however, both newborn chicks (Rugani, Kelly, Szelest, Regolin, & Vallortigara, 2010) and trained monkeys (Drucker & Brannon, 2014) systematically use the left side of the array to detect the ordinal position. Moreover, it has been recently shown that in nonhuman animals (newborn chicks) there is a spontaneous association between relatively small with the left side of space and relatively large with the right side of space, as it has been shown with symbolic number in adults. This mapping is independent of other nonnumerical variables, and is flexible and context-dependent ("8" is small when compared to "18" but large when compared to "5") (Rugani, Vallortigara, Priftis, & Regolin, 2015).

Cerebral asymmetries have been taken as responsible for these findings in nonhuman animals (Rugani et al., 2010), arguing that the right hemisphere takes the control in spatial–numerical tasks and therefore a leftward bias appears. Along these lines, studies on adults' oriented number–space mapping suggest that the main determinants for the orientation are the attentional asymmetries, which are mainly driven by cultural habits such as reading and writing abilities (Dehaene et al., 1993; Shaki et al., 2012; Zebian, 2005). However, asymmetries in the deployment of attention do not only depend on our cultural habits. The well-known pseudoneglect phenomenon refers to the fact that adults tend to overestimate the left side of the visual hemispace, and is interpreted on the basis of biologically determined cerebral asymmetries (Jewell & McCourt, 2000). Moreover, it has been suggested that a timing asymmetry exists in the maturation of the cerebral hemispheres with a temporal advantage for the right hemisphere over the left one (Rosen, Galaburda, & Sherman, 1987). These architectural and spatiotemporal constraints on brain development may determine an early advantage of the left over the right visual hemispace, which may constrain both the exploration of external space and the organization of information along a representational space (de Hevia, Girelli, & Macchi-Cassia, 2012).

Recent research on ordinal abilities in infancy offer a second clue as to why it is increasing, as opposed to decreasing order, which is left-to-right oriented early in development. A series of studies with 4-month-old infants have investigated the properties of ordinal discrimination using nonnumerical magnitude information: size-based sequences. We have found that this ability is functional at 4 months of age, but critically, there is an asymmetry with successful discrimination of increasing sequences versus a failure for the discrimination of decreasing sequences (Macchi Cassia, Picozzi, Girelli, & de Hevia, 2012). This signature of asymmetrical order has been recently replicated at the same age with numerical sequences: Infants are able to discriminate increasing but not decreasing numerical order (de Hevia, Addabbo, Croci, Girelli, & Macchi-Cassia, in preparation). This asymmetry offers another piece of information to the understanding of the origins of the left-to-right-oriented number line: the advantage for the increasing order, together with the existence of biologically determined attentional biases (ie, more attention deployed to the left hemispace), could give rise to the spatially oriented numerical representation (de Hevia et al., 2012a).

In summary, our most sophisticated numerical abilities derive from a core foundational, and evolutionary ancient, sense of number. This sense of number seems to have evolved in such a way that it exploits visuospatial resources and appears to be left-to-right oriented early in human development, and is present across species. The question to address now is why this is so, why is this advantageous for us humans and other animals.

2.4.2 Use of Space to Represent Other Dimensions

Adults represent nonnumerical ordinal information as spatially oriented (Gevers et al., 2003; Gevers et al., 2004; Rusconi et al., 2006), with SNARC-like effects present in the classification of letters of the alphabet, months of the year, and pitch height, so that the first elements of the series are responded faster with the left hand, and the last elements with the right hand. We tested whether for an element series that is not overlearned, as opposed to the series employed in past studies, subjects would spontaneously organize information in a left-to-right-oriented continuum. We asked adult participants to memorize a list of unrelated words, and they were later tested on order-relevant and order-irrelevant tasks. We observed the presence of SNARC-like effects when subjects were asked to classify each word as coming before/after a given one (order-relevant task). Moreover, subjects showed SNARC-like effects when they had to decide whether in the word of an image there was a given phoneme (the phoneme monitoring task, which is an order-irrelevant task). Therefore, when adults learn a list of unrelated words (never seen before in that particular order) they spontaneously map them into an oriented spatial continuum. The activation of this spatial representation, moreover, was automatic, since participants would access it even when the order of the list was irrelevant to the task at hand (Previtali et al., 2010).

Further work with adult participants addressed the question whether any type of information, for which the order is relevant and for which we have evidence of a "spatialization," has a modulating effect on visuospatial and motor processing. We have evidence that the oriented spatial codes associated to number can modulate our visuospatial and motor computations (Andres, Davare, Pesenti, Olivier, & Seron, 2004;

Fischer, 2003). Would also, for instance, processing of the letters of the alphabet modulate our motor behavior? We tested a group of adult participants in a task where they were required to write (under copy or dictation, both in an ordered and unordered way) a series of one-digit numbers and a series of letters of the alphabet. We observed that under all conditions numbers caused shifts of the motor output: Small magnitude numbers were written more leftwards relative to large magnitude numbers, which were written more rightwards. This phenomenon was present even when numbers were presented in unordered conditions. However, the letters of the alphabet did not produce these effects on the motor behavior: The spatial position of the letters was not correlated with their position in the alphabet series (Perrone, de Hevia, Bricolo, & Girelli, 2010). These results suggest that even though other types of nonnumerical information have a spatial code associated to them, their processing does not have an impact on our visuospatial and motor representations, in contrast to numerical information.

What is the purpose of the spatialization of nonspatial concepts? Does it play a critical role in acquiring serial order and in memorizing contents? The idea that the specific spatial orientation might either facilitate or hinder the extraction of an ordinal rule is based on the findings that discrimination of numerical order is preserved when the numerical arrays are presented left-to-right oriented, but disrupted when presented right-to-left oriented (de Hevia et al., 2014a). We have recently investigated whether spatial orientation can help or hinder learning abstract ordinal rules depending on the specific orientation, left-right versus right-left. For this purpose we designed a situation where the information used has no magnitude content. We tested 7-month-old infants' ability to detect and generalize higher order, abstract relations (ABB or ABA) from a sequence of geometric shapes (ie, rule learning). Crucially, we manipulated the spatial orientation of the information. Generally, when presented from left-to-right, infants were able to learn and generalize both ABB and ABA rule patterns, whereas they failed when the same abstract rules were presented from right to left. There are two main findings to highlight. First, infants at this age have the ability to extract ABB rules when presented in the center of the screen (Johnson et al., 2009), they preserve this ability when the rules are presented from left to right, but it is lost when presented from right to left. Second, even if infants at this age are not able to learn ABA rules when presented in the center of the screen (Johnson et al., 2009), they still fail when presented from right to left, but critically succeed when presented from left to right (Bulf, Gariboldi, de Hevia, & Macchi-Cassia, 2015). These findings support the idea that the specific spatial orientation has a dramatic impact on ordinal learning and memory, and points to the idea that the use of a spatial directionality supports the learning of any serial information from very early in development.

2.5 CONCLUSIONS

The association between numbers and space, in the form of a "mental number line," is not a product of human invention. Spatial signatures of mature numerical representations are present very early in development: on the one hand, fundamental, nondirectional number–space mappings are present from birth and persist into childhood and adulthood; on the other hand, directional number–space mappings are present in human animals and other nonhuman species with a, possibly malleable, left-to-right orientation. This orientation can in turn impact learning and discrimination of abstract serial order. Although space is used for representing many nonquantitative dimensions, the mapping between numbers and space seems to be privileged from very early in development.

References

Andres, M., Davare, M., Pesenti, M., Olivier, E., & Seron, X. (2004). Number magnitude and grip aperture interaction. *Neuroreport, 15*(18), 2773–2777.

Berch, D. B., Foley, E. J., Hill, R. J., & Ryan, P. M. (1999). Extracting parity and magnitude from Arabic numerals: developmental changes in number processing and mental representation. *Journal of Experimental Child Psychology, 74*(4), 286–308.

Bonn, C., & Cantlon, J. F. (2012). The origins and structure of quantitative concepts. *Cognitive Neuropsychology, 29*, 149–173.

Bulf, H., Macchi-Cassia, V. M., & de Hevia, M. D. (2014). Are numbers, size and brightness equally efficient in orienting visual attention? Evidence from an eye-tracking study. *PLoS ONE, 9*, e99499.

Bulf, H., de Hevia, M. D., & Macchi-Cassia, V. (2015a). Small on the left, large on the right: numbers orient preverbal infants' visual attention onto space. *Developmental Science*, 1–8.

Bulf, H., Gariboldi, V., de Hevia, M. D., & Macchi-Cassia, V. (2015b). *Left-to-right orientation facilitates learning of abstract rules in 7-month-old infants.* Paper presented at the Budapest CEU Conference on Cognitive Development, Hungary.

Cantlon, J. F., & Brannon, E. M. (2007). Basic math in monkeys and college students. *PLoS Biology, 5*, e328.

Coubart, A., Izard, V., Spelke, E. S., Marie, J., & Streri, A. (2014). Dissociation between small and large numerosities in newborn infants. *Developmental Science, 17*, 11–22.

de Hevia, M. D., & Spelke, E. S. (2009). Spontaneous mapping of number and space in adults and young children. *Cognition, 110*, 198–207.

de Hevia, M. D., & Spelke, E. S. (2010). Number-space mapping in human infants. *Psychological Science, 21*, 653–660.

de Hevia, M. D., & Spelke, E. S. (2013). Not all continuous dimensions map equally: number-brightness mapping in infants. *PLoS ONE, 8*(11), e81241.

de Hevia, M. D., Girelli, L., & Vallar, G. (2006). Numbers and space: a cognitive illusion? *Experimental Brain Research, 168*, 254–264.

de Hevia, M. D., Girelli, L., Bricolo, E., & Vallar, G. (2008). The representational space of numerical magnitude: illusions of length. *Quarterly Journal of Experimental Psychology, 61*, 1496–1514.

de Hevia, M. D., Girelli, L., & Macchi-Cassia, V. (2012a). Minds without language represent number through space: origins of the mental number line. *Frontiers in Psychology, 3*, 466.

de Hevia, M. D., Vanderslice, M., & Spelke, E. S. (2012b). Cross-dimensional mapping of number, length and brightness by preschool children. *PLoS ONE, 7*, e35530.

de Hevia, M. D., Girelli, L., Addabbo, M., & Macchi-Cassia, V. (2014a). Human infants' preference for left-to-right oriented increasing numerical sequences. *PLoS ONE, 9*, e96412.

de Hevia, M. D., Izard, V., Coubart, A., Spelke, E. S., & Streri, A. (2014b). Representations of space, time and number in neonates. *Proceedings of the National Academy of Sciences, 111*(13), 4809–4813.

de Hevia, M. D., Addabbo, M., Croci, E., Girelli, L., & Macchi-Cassia, V. M. (in preparation). Orders are not all the same: infants' detection of increasing number comes first.

Dehaene, S. (1992). Varieties of numerical abilities. *Cognition, 44*(1–2), 1–42.

Dehaene, S. (Ed.). (1997). *The number sense: how the mind creates mathematics*. New York: Oxford University Press.

Dehaene, S., Bossini, S., & Giraux, P. (1993). The mental representation of parity and number magnitude. *Journal of Experimental Psychology: General, 122*(3), 371–396.

Dehaene, S., Piazza, M., Pinel, P., & Cohen, L. (2003). Three parietal circuits of number processing. *Cognitive Neuropsychology, 20*, 487–506.

Dehaene, S., Izard, V., Spelke, E. S., & Pica, P. (2008). Log or linear? Distinct intuitions of the number scale in Western and Amazonian indigene cultures. *Science, 320*(5880), 1217–1220.

Drucker, C. B., & Brannon, E. M. (2014). Rhesus monkeys (*Macaca mulatta*) map number onto space. *Cognition, 132*, 57–67.

Ebersbach, M., Luwel, K., & Verschaffel, L. (2013). Further evidence for a spatial-numerical association in children before formal schooling. *Experimental Psychology, 18*, 1–7.

Fias, W., Lauwereyns, J., & Lammertyn, J. (2001). Irrelevant digits affect feature-based attention depending on the overlap of neural circuits. *Cognitive Brain Research, 12*, 415–423.

Fischer, M. H. (2001). Number processing induces spatial performance biases. *Neurology, 57*(5), 822–826.

Fischer, M. H. (2003). Spatial representations in number processing: evidence from a pointing task. *Visual cognition, 10*(4), 493–508.

Fischer, M. H., Castel, A. D., Dodd, M. D., & Pratt, J. (2003). Perceiving numbers causes spatial shifts of attention. *Nature Neuroscience, 6*(6), 555–556.

Galfano, G., Rusconi, E., & Umilta, C. (2006). Number magnitude orients attention, but not against one's will. *Psychonomic Bulletin & Review, 13*, 869–874.

Gallistel, C. R., & Gelman, R. (2000). Non-verbal numerical cognition: from reals to integers. *Trends in Cognitive Sciences, 4*(2), 59–65.

Galton, F. (1880). Visualised numerals. *Nature, 21*, 252–256.

Gerstmann, J. (1940). Syndrome of finger agnosia, disorientation for right and left, agraphia, and acalculia. *Archives of Neurology and Psychiatry, 44*, 398–408.

Gevers, W., Reynvoet, B., & Fias, W. (2003). The mental representation of ordinal sequences is spatially organized. *Cognition, 87*, B87–B95.

Gevers, W., Reynvoet, B., & Fias, W. (2004). The mental representation of ordinal sequences is spatially organized: evidence from days of the week. *Cortex, 40*(1), 171–172.

Girelli, L., Lucangeli, D., & Butterworth, B. (2000). The development of automaticity in accessing number magnitude. *Journal of Experimental Child Psychology, 76*, 104–122.

Hoffmann, D., Hornung, C., Martin, R., & Schiltz, C. (2013). Developing number-space associations: SNARC effects using a color discrimination task in 5-year-olds. *Journal of Experimental Child Psychology, 116*, 775–791.

Izard, V., Sann, C., Spelke, E. S., & Streri, A. (2009). Newborn infants perceive abstract numbers. *Proceedings of the National Academy of Sciences, 106*, 10382–10385.

Jewell, G., & McCourt, M. E. (2000). Pseudoneglect: a review and meta-analysis of performance factors in line bisection tasks. *Neuropsychologia, 38*(1), 93–110.

Johnson, S. P., Fernandes, K. J., Frank, M. C., Kirkham, N. Z., Marcus, G. F., Rabagliati, H., & Slemmer, J. A. (2009). Abstract rule learning for visual sequences in 8- and 11-month-olds. *Infancy, 14*, 2–18.

Kirkham, N. Z., Slemmer, J. A., & Johnson, S. P. (2002). Visual statistical learning in infancy: evidence for a domain general learning mechanism. *Cognition, 83*, B35–B42.

Knops, A., Thirion, B., Hubbard, E. M., Michel, V., & Dehaene, S. (2009). Recruitment of an area involved in eye movements during mental arithmetic. *Science, 324*, 1583–1585.

Lewkowicz, D. J., & Turkewitz, G. (1980). Cross-modal equivalence in early infancy: auditory-visual intensity matching. *Developmental Psychology, 16*, 597–607.

Lourenco, S. F., & Longo, M. R. (2010). General magnitude representation in human infants. *Psychological Science, 21*, 873–881.

Macchi Cassia, V. M., Picozzi, M., Girelli, L., & de Hevia, M. D. (2012). Increasing magnitude counts more: asymmetrical processing of ordinality in 4-month-old infants. *Cognition, 124*, 183–193.

McCrink, K., & Opfer, J. E. (2015). Development of spatial-numerical associations. *Current Directions in Psychological Science, 23*, 439–445.

McCrink, K., & Wynn, K. (2009). Operational momentum in large-number addition and subtraction by 9-month-olds. *Journal of Experimental Child Psychology, 103*, 400–408.

McCrink, K., Birdsdall, W., & Caldera, C. (2011). *Parental transmission of left-to-right structure in early childhood.* Paper presented at the Cognitive Development Society, Philadelphia.

Müller-Lyer, F. C. (1889). Optische Urtheilstäuschungen. *Archiv für Anatomie und Physiologie Physiologische Abteilung (Supplement-Band), 2*, 263–270.

Nieder, A. (2005). Counting on neurons: the neurobiology of numerical competence. *Nature Reviews Neuroscience, 6*, 177–190.

Opfer, J. E., Furlong, E. E., & Thompson, C. A. (2010). Early development of spatial-numeric associations: evidence from spatial and quantitative performance of preschoolers. *Developmental Science, 13*, 761–771.

Patro, K., & Haman, M. (2012). The spatial-numerical congruity effect in preschoolers. *Journal of Experimental Child Psychology, 111*, 534–542.

Patro, K., Nuerk, H. C., Cress, U., & Haman, M. (2014). How number-space relationships are assessed before formal schooling: a taxonomy proposal. *Frontiers in Psychology, 5*, 419.

Perrone, G., de Hevia, M. D., Bricolo, E., & Girelli, L. (2010). Numbers can move our hands: a spatial representation effect in digits handwriting. *Experimental Brain Research, 205*, 479–487.

Piaget, J. (1952). *The child's conception of number.* New York: Basic Books.

Picozzi, M., de Hevia, M. D., Girelli, L., & Macchi Cassia, V. (2010). Seven-month-old infants detect ordinal numerical relationships within temporal sequences. *Journal of Experimental Child Psychology, 107*, 359–367.

Pinel, P., Piazza, M., Le Bihan, D., & Dehaene, S. (2004). Distributed and overlapping cerebral representations of number, size, and luminance during comparative judgments. *Neuron, 41*, 983–993.

Posner, M. I. (1980). Orienting of attention. *The Quarterly Journal of Experimental Psychology, 32*, 3–25.

Previtali, P., de Hevia, M. D., & Girelli, L. (2010). Placing order in space: the SNARC effect in serial learning. *Experimental Brain Research, 201*, 599–605.

Ren, P., Nicholls, M. E. R., Ma, Y., & Chen, L. (2011). Size matters: non-numerical magnitude affects the spatial coding of response. *PLoS ONE, 6*, e23553.

Restle, F. (1970). Speed of adding and comparing numbers. *Journal of Experimental Psychology, 83*, 274–278.

Ristic, J., Wright, A., & Kingstone, A. (2006). The number line effect reflects top-down control. *Psychonomic Bulletin & Review, 13*, 862–868.

Rosen, G. D., Galaburda, A. M., & Sherman, G. F. (1987). Mechanisms of brain asymmetry: new evidence and hypotheses. In D. Ottoson (Ed.), *Duality and unity of the brain* (pp. 29–36). New York: Plenum Press.

Rubinsten, O., Henik, A., Berger, A., & Shahar-Shalev, S. (2002). The development of internal representations of magnitude and their association with Arabic numerals. *Journal of Experimental Child Psychology, 81*, 74–92.

Rugani, R., Kelly, D. M., Szelest, I., Regolin, L., & Vallortigara, G. (2010). Is it only humans that count from left to right? *Biology Letters, 6*, 290–292.

Rugani, R., Vallortigara, G., Priftis, K., & Regolin, L. (2015). Number-space mapping in the newborn chick resembles humans' mental number line. *Science, 347*, 534–536.

Rusconi, E., Kwan, B., Giordano, B. L., Umiltà, C., & Butterworth, B. (2006). Spatial representation of pitch height: the SNARC effect. *Cognition, 99*(2), 113–129.

Shaki, S., Fischer, M. H., & Göbel, S. M. (2012). Direction counts: a comparative study of spatially directional counting biases in cultures with different reading directions. *Journal of Experimental Child Psychology, 112*, 275–281.

Srinivasan, M., & Carey, S. (2010). The long and the short of it: on the nature and origin of structural similarities between representations of space and time. *Cognition, 116*, 217–241.

Stevens, S. S. (1970). Neural events and the psychophysical law. *Science, 170*, 1043–1050.

Tudusciuc, O., & Nieder, A. (2007). Neuronal population coding of continuous and discrete quantity in the primate posterior parietal cortex. *Proceedings of the National Academy of Sciences, 104*, 14513–14518.

Vallortigara, G., Regolin, L., Chiandetti, C., & Rugani, R. (2010). Rudiments of mind: insights through the chick model on number and space cognition in animals. *Comparative Cognition and Behavior Reviews, 5*, 78–99.

van Dijck, J. P., & Fias, W. (2011). A working memory account for spatial-numerical associations. *Cognition, 119*, 114–119.

van Galen, M. S., & Reitsma, P. (2008). Developing access to number magnitude: a study of the SNARC effect in 7- to 9-year-olds. *Journal of Experimental Child Psychology, 101*, 99–113.

Viarouge, A., & de Hevia, M. D. (2013). The role of numerical magnitude and order in the illusory perception of size and brightness. *Frontiers in Cognition, 4*, 484.

Vuilleumier, P., Ortigue, S., & Brugger, P. (2004). The number space and neglect. *Cortex, 40*, 399–410.

Xu, F., & Spelke, E. S. (2000). Large number discrimination in 6-month-old infants. *Cognition, 74*(1), B1–B11.

Zebian, S. (2005). Linkages between number concepts, spatial thinking, and directionality of writing: the SNARC effect and the reverse SNARC effect in English and Arabic monoliterates, biliterates, and illiterate Arabic speakers. *Journal of Cognition and Culture, 5*, 165–190.

Zorzi, M., Priftis, K., & Umiltà, C. (2002). Brain damage: neglect disrupts the mental number line. *Nature, 417*, 138–139.

CHAPTER

3

Catching Math Problems Early: Findings From the Number Sense Intervention Project

Nancy C. Jordan, Nancy Dyson

University of Delaware, Newark, DE, United States of America

OUTLINE

Continuous Issues in Numerical Cognition. http://dx.doi.org/10.1016/B978-0-12-801637-4.00003-2

Number competencies are foundational to mathematics achievement. Although mathematics content in preschool and kindergarten (ie, 3–6 years of age) covers multiple topics, most researchers agree that number sense is of primary importance (National Research Council, 2009). Children who leave kindergarten with poorly developed number sense are at serious risk for developing mathematics difficulties or disabilities and may never catch up to their grade-level peers without intensive help (Jordan, Kaplan, Ramineni, & Locuniak, 2009). Kindergartners who perform below the 10th percentile on number-related tasks have a 70% chance of scoring below the 10th percentile 5 years later (Morgan, Farkas, & Wu, 2009).

At kindergarten entry, a wide range of individual differences in number sense already exists (Jordan et al., 2009). Although most preschoolers and kindergartners have the capacity to learn foundational mathematics competencies, many do not realize their potential because of lack of opportunities in school or at home (National Research Council, 2009). Catching number deficiencies early is vital for promoting school readiness in mathematics. This chapter provides an overview of a kindergarten (ie, children 5–6 years of age) number sense intervention our research team developed and tested over a 4-year period (Dyson, Jordan, Beliakoff, & Hassinger-Das, 2015; Dyson, Jordan, & Glutting, 2013; Dyson, Jordan, & Hassinger-Das, 2015; Hassinger-Das, Jordan, & Dyson, 2015; Jordan, Glutting, Dyson, Hassinger-Das, & Irwin, 2012). We designed the intervention for children who struggle with number competencies in kindergarten. Our approach was iterative in that we revised and retested the intervention over the study period to increase its effectiveness with the most resistant learners.

3.1 CONCEPTUAL FRAMEWORK

3.1.1 Number Sense in the 3- to 6-Year-Old Age Period Refers to Core Knowledge of Number, Number Relations, and Number Operations (Jordan, Fuchs, & Dyson, 2015; National Research Council, 2009)

Nonverbal knowledge of numerical magnitudes typically develops early without specific input or instruction (Jordan & Levine, 2009). This knowledge does not involve number words or symbols. Infants can discriminate between relatively large sets of objects with 2:1 (at 6 months) and 3:2 (at 9 months) ratios through an approximate number system (Wood & Spelke, 2005) and can precisely discriminate between small sets of one to four objects (Feigenson, Dehaene, & Spelke, 2004). Arguably, these nonverbal number abilities provide the basis for developing symbolic number sense, which is unique to humans (Siegler & Lortie-Forgues, 2014).

Learning number words and symbols and using them to label small sets greatly extend a preschoolers' numerical thinking (National Research Council, 2009). This new kind of learning depends on the input the child receives, typically at home or in day care or preschool settings.

3.1.1.1 *Number*

Number knowledge in early childhood involves several interrelated skills (Clements & Sarama, 2009; Sarnecka & Wright, 2013). These skills include subitizing and counting. Subitizing is the ability to recognize and verbally label small quantities to four or five, without counting. For example, if the child is shown three fingers, he or she can state "three" right away without counting his or her fingers from one. The child can also readily provide three objects when asked. A child who only knows two might give a caregiver exactly two pennies when asked for two, but give a pile of pennies when asked for three.

Counting enables children to extend their number sense beyond small numbers in the subitization range (Baroody, 1987). In order to enumerate a set of objects, the child needs to know the count words in sequential order and to count each object only once (ie, one-to-one correspondence) (Gelman & Gallistel, 1978). The cardinality principle states that the final number in the count always indicates how many items are in the set. Children also begin to learn to recognize one-digit numerals or written number symbols. Although most children learn basic number skills involving symbols in preschool, some with limited learning opportunities or with learning disabilities enter kindergarten without these foundations (Dyson et al., 2013, 2015b; Jordan et al., 2012b).

3.1.1.2 *Number Relations*

The ability to compare the exact magnitudes of two number words or numerals represents a significant developmental accomplishment. Kindergartners recognize that 5 is bigger than 4 or that 3 is smaller than 5, even when quantities are not present (Griffin, 2004). Children begin to see that as they move up a count list, the numbers represent larger quantities and as they move down, the numbers represent smaller quantities. Eventually, they construct a linear representation of number, that is, that each number is exactly one more than the previous one ($n + 1$). This principle underpins success with addition and subtraction operations. Deficits in discerning relations between numbers are a characteristic of mathematics learning disabilities (Rousselle & Noël, 2007).

3.1.1.3 *Number Operations*

Addition and subtraction operations are a focus of primary school mathematics instruction (Jordan et al., 2009), although these abilities start to develop much earlier. Many preschoolers successfully solve nonverbal calculations with object representations (eg, the child is shown 3 dots that are then hidden with a cover; two more dots are slid under the cover, and the child indicates that 5 objects are now under the cover) (Levine, Jordan, & Huttenlocher, 1992; Huttenlocher, Jordan, & Levine, 1994). Early facility with nonverbal calculations predicts later conventional addition and subtraction skills (Levine et al., 1992). In primary school, children learn to solve arithmetic combinations with totals to 10 and then to 20. They also solve story problems that vary in semantic complexity (Riley, Greeno, & Heller, 1983; Hanich, Jordan, Kaplan, & Dick, 2001).

Children make use of a variety of calculation strategies (Carpenter, Hiebert, & Moser, 1983). Counting is the predominant method for solving simple addition and subtraction problems. Counting allows children to calculate exactly how much less or how much more one number is from another (National Research Council, 2009). For combinations with totals of 10 or less, young children can represent each part of the problem on their fingers and then count all of the fingers to get the sum. By the end of kindergarten, some children see they can count on from the first or the larger addend to get a sum (eg, for 5 + 2, the child counts 6, 7 on two fingers to get 7), a more efficient method than counting out both addends (Baroody, Lai, & Mix, 2006). Children who count on their fingers in kindergarten develop fluency with number combinations earlier than children who do not count on their fingers (Jordan, Kaplan, Ramineni, & Locuniak, 2008). Children also learn that numbers can be broken into smaller sets of numbers, sometimes referred to as conceptual subitizing (Clements, 1999). For example, the number 4 can be seen as 1 and 3, or 2 and 2. Thinking about different "partners" for sums and relating them to the inverse operation of subtraction (1 + 3 = 4 and 4 − 1 = 3) encourages meaningful mental problem solving (Fuson, Grandau, & Sugiyama, 2001). As children become proficient with some arithmetic combinations, they can use this knowledge to derive solutions to other combinations (eg, 3 + 3 = 6, so 3 + 4 = 7). Children with mathematics difficulties do not make good use of adaptive counting calculation strategies and thus are more error prone (Geary, 1994; Jordan, Kaplan, Oláh, & Locuniak, 2006).

3.1.2 Number Sense Follows a Developmental Progression

Number sense from 3 to 6 years of age follows a developmental progression (Clements & Sarama, 2007; Frye et al., 2013; National Research Council, 2009). Children go from developing core knowledge of number (eg, subitizing and counting), to understanding relations between numbers, to operating on numbers. The cycle typically repeats as children learn to recognize and count larger numbers (National Research Council, 2009). As number sense develops there are distinct as well as overlapping skills that emerge at various age levels. For example, in preschool, children's knowledge of number magnitude is based primarily on the representation of physical objects. Children can state which group of objects has "more." Simultaneously, children learn to recite the count sequence as they would the alphabet and learn to count and state cardinal number words for small sets of objects. By age 6, children coordinate these two skills and develop the ability to compare magnitudes using numerals only (no objects present) (Case, 1996). The ability to enumerate increasingly larger collections and to form a fuller understanding of the count sequence continues to develop in kindergarten. In kindergarten, children typically use verbal and written forms of number representation with or without concrete sets being present. In addition to counting higher, kindergartners can take apart and put together numbers and use numerals to represent multidigit numbers. In first grade, children see the meaning of the ones, tens, and hundred places. They can count-on from different places in the sequence up to at least 100, without objects present.

3.1.3 Number Sense in Kindergarten Predicts Success in Mathematics

Core number competencies in kindergarten (especially knowledge of number operations) predict computational and problem-solving proficiency in third grade, even when controlling for reading, age, and general cognitive factors (Jordan et al., 2009; Jordan, Glutting, Ramineni, & Watkins, 2010; Locuniak & Jordan, 2008). Similar findings have been reported in shorter-term studies in kindergarten and first grade (Clarke & Shinn, 2004; Lembke & Foegen, 2009; Methe, Hintze, & Floyd, 2008). Number sense performance and growth during kindergarten account for a substantial portion of the variance in mathematics achievement at the end of first grade (Jordan, Kaplan, Locuniak, & Ramineni, 2007). Importantly, the mathematics achievement gap between low- and middle-income learners can be explained by their number competencies.

Students with weak number sense almost certainly will struggle with later mathematical concepts. Poor number sense leads to rote memorization, which in turn leads to inadequate problem-solving skills (Robinson, Menchetti, & Torgesen, 2002). Moreover, poor number sense impedes development of advanced topics, such as fractions and algebra (Wu, 1999; Jordan et al., 2013; Vukovic et al., 2014).

3.1.4 Deficiencies in Number Sense can be Identified Early

The high prevalence of mathematics difficulties in the United States has stimulated investment in early screening in schools. In a review of the predictive validity of math screeners, Gersten et al. (2012) found that the ability to name the larger of two numerals (Clarke, Baker, Smolkowski, &

Shard 2008; Seethaler & Fuchs, 2010), to name the missing numeral from a string of numerals (Clarke et al., 2008; Lembke & Foegen, 2009), and to add and subtract (Jordan et al., 2010b; Seethaler & Fuchs, 2010) were most predictive. Number screening measures also can reliably classify children who will go on to need additional help in mathematics (Geary, Bailey, & Hoard, 2009; Jordan et al., 2010a; Seethaler & Fuchs, 2010).

3.1.5 Number Sense is Malleable and Targeted Help in Number Sense Leads to Improved Math Achievement in School

There is growing evidence to suggest that number sense can be developed in most children when taught explicitly along a developmental progression (Frye et al., 2013). Frye et al. (2013) describe the teaching of number and operations using a developmental progression in the following example:

1. Provide opportunities for children to practice recognizing the total number of objects in small sets (1 to 4 items) and labeling them with a number word without counting the set. The teacher can also ask, "Give me two," "Give me four," and so forth.
2. Encourage one-to-one counting as a means of identifying the total number of items in larger sets of five or more (eg, count six objects and then indicate how many are in the set).
3. Once children can recognize and count sets, provide opportunities for children to compare quantities through counting (eg, 5 is one more than 4, one less than 6).
4. Once children develop these fundamental number skills, encourage them to solve basic operations problems (eg, 5 + 1 is the next number in the count sequence, 6).

At the preschool level, randomized controlled trials have shown positive effects of teaching number sense based on this kind of developmental progression (Baroody, Eiland, & Thompson, 2009; Clements & Sarama, 2007, 2008; Klein, Starkey, Sarama, Clements, & Iyer, 2008). Other researchers have effectively developed number competencies in kindergartners (Clarke et al., 2014) and first graders (Bryant et al., 2011).

3.2 NUMBER SENSE INTERVENTIONS

Our research with at-risk kindergartners (ie, children from low-income communities and/or who performed poorly on a screener) focuses on developing number competencies in small-group settings. We designed the interventions to address the three aforementioned components of number sense: number, number relations, and number operations. We tested the intervention in four randomized controlled trials (Dyson et al., 2013, 2015a, 2015b; Hassinger-Das et al., 2015; Jordan et al., 2012b). After each iteration of the intervention, the lessons were revised based on the data collected in the previous year.

3.2.1 Participants

For the first two trials, participants were recruited from every kindergarten class in five schools that primarily serve low-income children (ie, upward of 90%). For the third and fourth trials, we tested the effectiveness of the intervention only with the lowest-achieving children, based on a cutoff score on a number sense screening test administered in the first part of kindergarten. They were drawn from schools serving children with a range of social classes. Thus, in cohorts 3 and 4, we tested the intervention on children who displayed low number sense, regardless of their income status.

Table 3.1 shows participant demographics for each of the four cohorts. Income status was not collected for individual students in the first 3 years, although the majority of children in the schools were reported to be of low-income status. In year four, we also received permission to obtain income status per student.

3.2.2 Experimental Design

Table 3.2 shows the experimental design over the 4 years. In years one and two, 128 students were chosen at random from the pool of consenting children and randomly assigned to the experimental or control groups. In years three and four, all consenting children were screened on a number

TABLE 3.1 Demographics for Total Sample by Year

		Gender		Ethnicity (percentage of total group)				EL	Income status[b]	Age as of Sep. 1 (in months)
Year	*N*	Male	Female	AA	H	C	Other	(%)[a]	Eligible (%)	Mean (SD)
1	121	69	52	55	37	6	2	25	++	66 (4.0)
2	128	65	63	48	41	10	1	25	++	67 (4.4)
3	124	65	59	19	63	18	2	55	++	65 (3.6)
4	126	62	64	25	51	21	3	41	81.7	66 (3.5)

Note. *N*, number of participants; AA, African American; H, Hispanic; C, Caucasian; other, biracial or Asian; EL, English learners; and SD, standard deviation.

[a]*EL students were identified by enrolment in bilingual classrooms.*

[b]*Income status determined by Free or Reduced Lunch Eligibility; ++, data per student was not collected.*

TABLE 3.2 Experimental Design—Number of Children Randomly Assigned to Each Group by Year

			Control groups		Experimental groups	
Year	*N*	**Business as usual**	**Storybook vocabulary intervention**	**Storybook number intervention**	**Number sense with number-list practice**	**Number sense with number-fact practice**
1	121	65			56	
2	128	44	42		42	
3	124	40		42	42	
4	126	42			40	44

sense screener (the Number Sense Brief will be described more fully with the other measures) (Jordan et al., 2010b, 2012a; Jordan & Glutting, 2012). Only children who performed below the group mean (21 out of 44 items) on this screener were included in experimental and control groups.

The purpose of the control groups varied over the years. In year one, there was one control group that received business as usual (BAU). In years two and three, we added a second control group, in addition to BAU. In year two, the second control group, a small-group story book vocabulary intervention, was added to rule out that any observed group differences were due primarily to the use of special instruction in small groups. In year three, the storybook intervention control group was retained, but the intervention increased its focus on vocabulary targeted specifically to number concepts. In year four, we compared two experimental number sense intervention groups with differing practice conditions at the end of each lesson to a BAU control. Each of the two experimental groups received the identical number sense intervention for 25 min; the groups differed for the final 5 min of the 30-min intervention. One group ended with a number-list practice game (this condition replicated the number sense intervention of the previous years, which always ended in number-list game), while the other group ended with a number facts (addition and subtraction) practice game.

For each trial, we administered mathematics-related measures prior to the intervention (pretest), right after the intervention (posttest), and about 8 weeks after the intervention (delayed posttest). Number sense was assessed with a brief but valid untimed measure, the Number Sense Brief (NSB) (Jordan et al., 2010b, 2012a; Jordan & Glutting, 2012). Items assessed skills such as counting to 30; enumerating small sets; recognizing one- and two-digit numbers; comparing numerical magnitude; and adding and subtracting in different contexts (ie, nonverbal, story problems, and number combinations). Each item on the NSB was scored 0 (incorrect) or

1 (correct) with a total raw score of 44. As mentioned earlier, this measure was also used as a screener to select participants in years three and four. General mathematics achievement was measured with widely used *Woodcock-Johnson III Tests of Achievement* (WJ): Calculation subtest (Woodcock, McGrew, & Mather, 2007). Arithmetic fluency (in year four only) was measured using a 16-item timed task containing eight addition and eight subtraction items. Adapted from Jordan and Montani (1997), children were presented with a visually displayed combination (ie, 2 + 1, 3 − 2) and asked, "How much is two plus one (or three take away two)?" Children were required to produce an answer within 3 s.

Across the years, we always delivered the number sense interventions in a small-group setting (4 children per instructor) in school. Students received 24 lessons, 3 times a week for 8 weeks. Each lesson was 30 min long and highly scripted. Instructors were graduate students or undergraduate preservice teachers who had completed their student-teaching assignment. High fidelity of implementation was consistently documented.

Instructors attended teacher-training sessions prior to the intervention and weekly instructor meetings throughout the intervention. During these weekly meetings, upcoming lesson content, new lesson activities, important gestures, and so forth were reviewed and practiced. Instructors also discussed the lessons that had already been taught with respect to ease of implementation, student engagement, and student's ability to grasp the concept being introduced. This feedback led to future lesson revisions resulting in an intervention that was both engaging and effective.

3.2.3 Overview of the Number Sense Curriculum

As noted earlier, the interventions sought to strengthen core knowledge of number, number relations, and number operations. Four instructional approaches guided the lessons. First, activities made explicit the connections between various representations of number (visual models, oral naming, and written number symbols). Second, children were taught contrasting ideas simultaneously to emphasize inverse relationships (before/after, +1/−1, bigger/smaller). Third, the lessons incorporated a series of fast-paced game-like activities designed to hold the attention of kindergartners who may have short attention spans. Fourth, throughout the activities, children were encouraged to give quick, automatic, accurate answers when appropriate; if children could not produce a quick answer, we helped them use strategies to find an answer efficiently.

3.2.3.1 *Number*

The most basic number activities addressed students' ability to subitize or name small quantities of objects quickly without counting. Children were shown cards with one to four circles randomly placed (Fig. 3.1). They

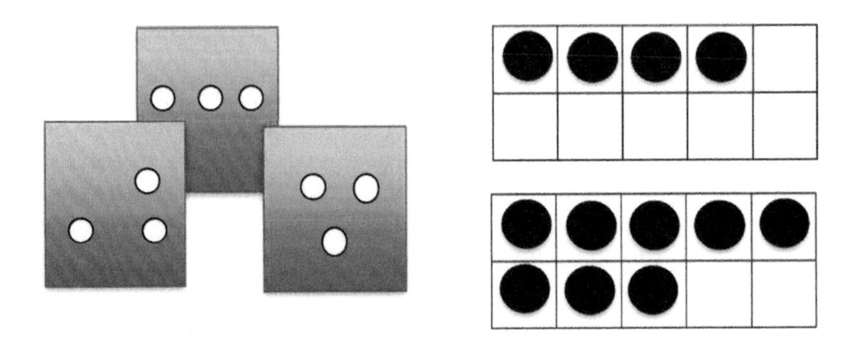

FIGURE 3.1 **The blue cards show representations for three used for the subitizing activity.** The child is shown the card for approximately 1 s and asked, "How many circles?" The black and white cards are ten frames showing the numbers four (top) and eight (bottom). This representation allows children to see eight as five and three more.

viewed the card for only 1 s and asked, "How many circles?" Children also were taught to conceptually subitize or recognize smaller sets within larger ones (Clements, 1999). During one conceptual subitizing activity, they saw a number from 1 to 10 in a ten-frame configuration (Fig. 3.1) and again were asked, "How many circles?" The ten-frame configuration shows two rows of five blocks. The blocks are filled in with dots beginning at the top left and moving across until the top row is filled. For numbers 1–5, the children can attend to the number of dots or the number of spaces in the top row, whichever is easier. For example, it may be difficult to subitize four circles in a row; however, it is easy to subitize the one blank box, alerting the child that the number is one less than 5. When the top row was filled (numbers 5–10), children only needed to attend to the bottom row. Through this activity the children developed fluency in recognizing quantities as 5 and so much more.

A coordinating activity had students quickly producing a number from 1 to 10 on their fingers. The instructor would say, "Show me eight fingers." Children were expected to show five fingers on one hand and three on the other. This activity was also done in reverse where the instructor would hold up a certain number of fingers and say, "How many fingers?" This directly coordinates with the ten-frame representation for eight (Fig. 3.1), which shows the top row of five filled and three dots in the bottom row. The activity lays a foundation for the corresponding addition combination, $5 + 3 = 8$.

Each lesson also focused on a number of the day or "magic number." Although the fourth version of the interventions began with the smallest numbers (0, 1, and 2), the first three versions of the number sense interventions began with the number 11. The numbers 11–19 can be particularly difficult for children in the United States because our English naming

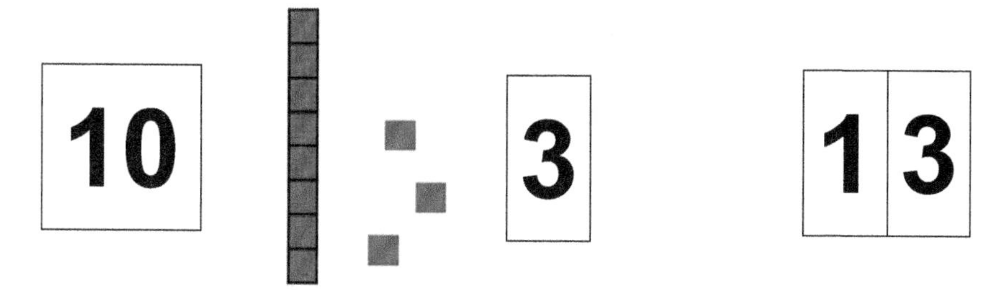

FIGURE 3.2 **Interlocking cubes are counted orally while children count along on their fingers.** When ten is reached (all their fingers are used), a stick of ten is made. Children continue to count, "11, 12, 13." Decade (10) and unit (3) cards are put next to the cubes while the instructor says, "Ten and three more is 13. Watch what I do!" The instructor then lays the unit card (3) on the decade card (10) and says again, "Ten and three more is 13." While lifting up the unit card, the instructor says, "See the ten hiding under the three? Ten and three more is 13." The instructor puts the unit card on 10 again and says, "Thirteen is our magic number today!"

system does not follow a normal pattern. We do not say ten-one, ten-two as is done in Asian languages, such as Chinese. Eleven and 12, two irregular number names, are followed by 13, not ten-three and so forth. It was thought that children had ample practice with the numbers 1–10 in their classrooms so we decided to make our first focus number 11 and include the numbers 1–10 in the activities. When we began working with the lowest achievers, we found that they needed more practice with the numbers 1–10 and modifications to the lessons (which will be discussed later in this chapter) were made.

Magic number activities were designed to connect numeral to quantity and to build base-ten understandings. For example, when the magic number was 13, students counted out thirteen blocks and made a stick of ten out of ten of the blocks, with three left over. Using decade and unit cards, the children "built" the number 13 as ten and three more by laying the unit card (3) on the ten-decade card (Fig. 3.2). Children sequenced number recognition cards (small playing cards with the numbers 1–13) and played a number recognition game where they were shown a number card (1–13), in turn, and asked to say the number. If the child misstated the number, the instructor corrected him or her right away and then showed the same card again. This gave students immediate feedback and allowed the turn to end in a correct answer.

For each subsequent lesson, the magic number was the next counting number. When the lesson focusing on the number 20 was reached, children were able to make two sticks of ten when counting with the interlocking blocks. The two blocks of ten were connected to the "2" in the 20 decade card. For lessons focusing on numbers 21–29, children built numbers as two sticks of ten and so many more and made the corresponding

1	2	3	4	5	6	7	8	9	10
11	12	13	14	15	16	17	18		20
21		23	24	25	26	27	28	29	30
31	32	33	34	35	36		38	39	40
41	42	43	44	45		47	48	49	50
51	52	53	54	55	56	57		59	60
61	62		64	65	66	67	68	69	70
	72	73	74	75	76	77	78	79	80
81	82	83	84		86	87	88	89	90
91	92	93		95	96	97	98	99	**100**

FIGURE 3.3 **A hundreds chart is a helpful way to represent counting numbers from 1 to 100.** For introductory hundreds chart activities, a chart such as this was shown to the group and children orally named the missing numbers. In later activities, children were given a paper-and-pencil activity and were asked to write in the missing numbers.

numeral using the 20 decade card and the appropriate unit card. Children sequenced the number cards, making a new row for each decade, replicating a hundreds chart. (Fig. 3.3.) In the final four lessons, children made numbers from 30 to 100 using decade and unit cards along with premade sticks of ten and unit blocks. A hundreds chart (Fig. 3.3) was used to reinforce base-ten understandings. During the final lessons, children were given a hundreds chart with numbers missing and asked to write in the missing number.

3.2.3.2 Number Relations

Children compared the magnitudes of numbers (1–10), first using a cardinality chart and then using a number list (Fig. 3.4). Eventually, they were asked to compare the magnitude of numbers without any visual representations. For example, instructors asked, "Which is bigger (or smaller), 3 or 8?" showing only the number symbols.

Lesson activities also included opportunities for children to find the number right before or the number right after a given number, ranging from 1 to 10. Children practiced the $n + 1/n - 1$ rules (+1 takes you to the number immediately after and −1 takes you to the number immediately

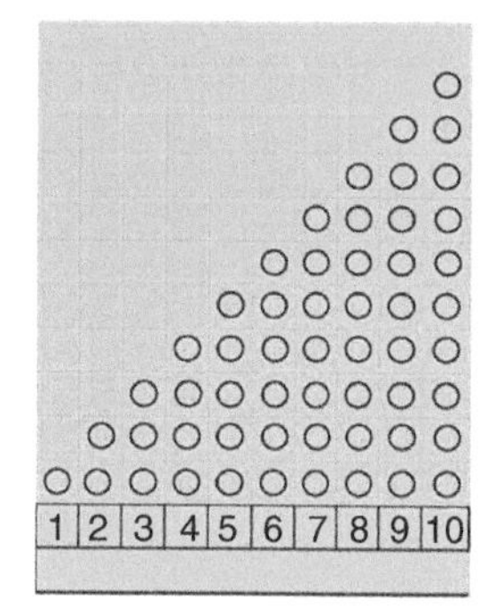

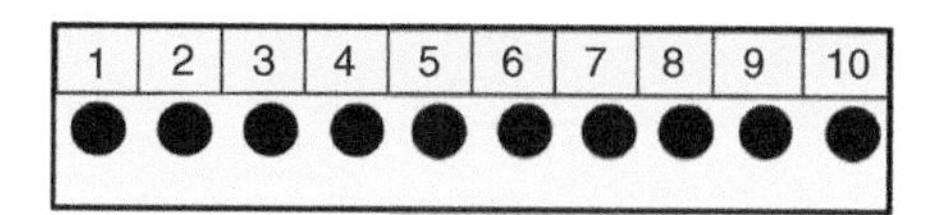

FIGURE 3.4 The cardinality chart (on the left) is used to show how quantities increase by 1 as you go up the number list. During the first ten lessons of the intervention, children build numbers 1–10 using interlocking blocks and place them on the cardinality chart above the appropriate number. From lesson 11 and forward, the children use the number list (shown on the right). Black tokens are added to the list for counting activities and children can see sets of numbers within a set. The number list also effectively models the + 1 property. For example, with 4 tokens on the number list, children can place "one more" and see that 4 and one more is 5.

before) using the number list. Children also played before/after games using the number recognition cards. They were each given two cards. The instructor then placed a number recognition card on the table and asked, for example, "Who has the number before six?" The student with the five card would put it down right before of six. The teacher continued to ask before/after questions until all the cards were down and placed in order.

3.2.3.3 *Number Operations*

Using black dots placed horizontally on a sheet of paper, children separated the dots into two groups or partners with a pencil (Fuson et al., 2001). By moving the pencil, new sets of partners are made for the same number of dots. For example, five dots can be separated into 1 and 4, 2 and 3, 3 and 2, or 4 and 1. Partner cards showed the groups of dots (Fig. 3.5) and children matched them to number sentences also presented on cards. Number sentences were presented with the solution visible except during fluency practice. Children learned specific strategies (eg, representing each addend on their fingers and then counting all of them, counting up from a number, and drawing models) to solve unknown combinations.

3.2.3.4 *Number-List Practice*

At the end of the sessions in the first 3 years of the intervention, children played a version of the Great Race Game (Ramani & Siegler, 2008). The game uses a game board with a number list to 10. The students spin a spinner with the numerals 1 and 2 and move their game piece up the

FIGURE 3.5 **Partner cards (upper left) show that numbers are made up of smaller sets.** Number sentence cards show the numerical representation associated with those sets. Children were presented with modeled story problems, "There was one apple in the tree and three on the ground. How many apples were there altogether?" and asked to match the appropriate partner card and number sentence card.

number list, saying the number as they land on it. For example, if a child is on 5 and spins a 2, they say "six, seven" rather than "one, two" as would be in most board games. This helps them to learn to count on from a number to solve addition problems. We adapted the game with minus one and minus two, as well as plus one and plus two, so children moved up and down the number list. Children stated the number sentence they were modeling as they moved their piece on the board, "Five and two more is six, seven."

3.2.4 Findings in the First 3 Years

ANCOVAs were run for each year of the intervention study, which produce adjusted means for the calculation of effect sizes. Details for the analyses can be found in Dyson et al. (2013) for year one, Jordan et al. (2012b) for year two, Hassinger-Das et al. (2015) for year three, and Dyson et al. (2015b) for year four. Following the guidelines of the *What Works Clearinghouse: Procedures and Standards Handbook* (Institute of Education Sciences, 2014), we used effect sizes (Hedge's *g*) to evaluate and compare effects of the intervention. Table 3.3 shows the effect sizes for each year by group for the outcome measures. An effect size greater than 0.25 (or one

TABLE 3.3 Effects Sizes (Hedge's g) by Group and Year for Mathematics Measures

	Number sense intervention with number-list practice						Storybook intervention						Number sense intervention with number-fact practice					
	Number sense		Calculation achievement		Fluency		Number sense		Calculation achievement		Fluency		Number sense		Calculation achievement		Fluency	
Year	**Post**	**Del**	**Post**	**Del**	**Post**	**Del**	**Post**	**Del**	**Post**	**Del**	**Post**	**Del**	**Post**	**Del**	**Post**	**Del**	**Post**	**Del**
1	0.65	0.66	0.24	0.23	NA	NA	—	—	—	—	—	—	—	—	—	—	NA	NA
2	1.08	0.76	0.99	0.45	NA	NA	0.12	0.06	0.03	−0.24	NA	NA	—	—	—	—	NA	NA
3	0.41	0.16	0.54	0.25	NA	NA	0.28	0.25	0.00	0.14	NA	NA	—	—	—	—	NA	NA
4	0.32	0.29	0.62	0.10	0.69	0.27	—	—	—	—	—	—	0.87	0.60	0.67	0.45	0.78	0.58

Note. NA, not administered; Post, posttest; and Del, delayed posttest.

quarter of the standard deviation of the sample) is a general marker of an effective intervention (Institute of Education Sciences, 2014). For each year, the comparison group was the BAU control.

We modified our number sense screener (NSB) slightly each year, so only the items ($n = 38$) common across the 4 years were used in determining the effect sizes, reported in Table 3.3. For years one and two, effect sizes for number sense were well above the 0.25 criterion and remained moderately strong at delayed posttest. In contrast, the vocabulary intervention group did not show gains significantly better than the BAU control group.

In year three, when only children scoring below a cutoff in the number sense pretest participated in the experiment, the effect sizes were smaller than those found in previous years. Interestingly, although the number sense intervention group made good gains, the BAU control group also made gains during the course of the study relative to those seen in previous years. This result may have been, at least partially, the result of school district mathematics curriculum changes. Participating schools changed from a more implicit, problem-based curriculum to one that provided more explicit help in number. Thus, the content of the kindergarten mathematics instruction the control BAU group received mirrored more closely the instruction the number sense intervention children received than it did in previous iterations of the trials. This may have contributed to the reduced effect sizes in year three.

3.2.5 Revisions in the Fourth Year Along With Findings

In the fourth trial, we wanted to see if we could boost number competencies further in very low-performing children (Dyson et al., 2015b). We observed in the previous iteration that some children, especially English learners (eg, children whose primary language at home was Spanish, rather than English), still struggled with the numbers 5 through 10 after the training. Some children even had trouble recognizing and naming quantities of two and three. Thus we revised our instruction to emphasize very small numbers in the beginning lessons to help children build core understandings. All aspects of the previous interventions were introduced in the beginning lessons, but using just the numbers 0, 1, and 2 initially. One block was shown to the children and put on the cardinality chart above the number 1 (Fig. 3.4). Then children saw two blocks. The instructor connected the two blocks and pointed out, "One and one more is two." The two blocks were put on the cardinality chart above the number 2. Children practiced quickly subitizing the quantities to two using the subitizing cards, fingers, and five frames (one row of a ten frame; Fig. 3.1). They also practiced number recognition, before/after, and bigger/smaller skills with the very small numbers. Reorganizing the activities this way allowed us to teach concepts to struggling learners using only numbers with which they were already comfortable. The later lessons were similar to the original ones in which children continued to practice their skills and applied their base-ten understanding to build and recognize numbers to 100. Number operations remained focused on combinations within 10.

In year four we also evaluated the effectiveness of two different practice conditions (Dyson et al., 2015b). In the previous years, the interventions ended with the number-list game as described earlier. The contrasting number-fact practice condition consisted of practicing number facts within five. The instructor showed the child a number fact that they had learned in the intervention and asked them to give the answer as quickly as they could or to "figure it out" using finger strategies. Students were timed individually to see how many facts they could answer correctly in 1 min. Facts completed were added together for a team score. Guessing would lead to error correction as described earlier. Since this slowed students down, they chose quick finger strategies over guessing.

In year four, there were substantial effects of the intervention relative to BAU controls (Table 3.3). However, there was a marked difference between the two intervention groups, both at posttest and delayed posttest. Not only did the number-facts intervention group score higher at posttest, but also they were also better able to retain their advantage over the control group 8 weeks after the intervention. Item analysis of the number sense measure showed that children in the number-facts intervention group performed better on items in many aspects of number sense, not just those requiring number operations. Further analysis also showed that the number sense intervention with number-fact practice was especially effective for English learners, although it is not clear why this was the case.

3.3 SUMMARY AND CONCLUSIONS

If children leave kindergarten with poor number sense, they are likely to encounter mathematics difficulties in subsequent grades. Number sense deficiencies lead to reliance on rote memorization and an inability to work meaningfully with numbers (Robinson et al., 2002). A weak understanding of whole numbers jeopardizes children's success with advanced topics, such as fractions and algebra, as they progress through school (Jordan et al., 2013; Wu, 1999) and potentially limits future educational and employment opportunities (National Mathematics Advisory Panel, 2008).

Our number sense intervention targets core understandings of number, number relations, and number operations in high-risk learners. Catching children's weaknesses in number sense early positions them for success. Over multiple trials, we obtained moderate to strong effect sizes, relative to controls, on a measure closely aligned to the intervention topics (ie, a number sense brief screener) as well as on a test of mathematics calculation achievement. Many of the gains were still present 2 months after the intervention ended, suggesting the results were stable. The effects of a general storybook intervention are not meaningful for improving mathematics outcomes (Jordan et al., 2012b), while a storybook intervention that emphasizes quantitative words produces modest gains but only on a number sense measure (Hassinger-Das et al., 2015).

Early proficiency with number operations, in particular, predicts later success with mathematics (Jordan et al., 2009), and this skill should be reinforced during instruction. Thus in the final trial, we tried to strengthen children's competencies through practice with addition and subtraction number facts (Dyson et al., 2015b). In a contrasting number-list condition, children solved combinations by counting up and down on a number-list game board. Although both number-list and number-fact counting strategies were taught during the core intervention, number-fact practice at the end of each lesson produced more gains than did the corresponding number-list practice. Fuchs et al. (2013) also find that number-fact practice is generally more effective than a game-board approach for improving first-graders' performance on calculation. Systematic instruction in number sense combined with speeded but meaningful practice of facts produces the best effects for kindergartners with serious delays.

Developing number sense early lays a foundation for mathematics learning, one that helps children access what is taught in the general classroom. Classroom teachers might use the types of activities described in our small-group intervention during whole-class instruction, such as making explicit the relationships between numbers, building fluency in quantity recognition, and building fluency in number facts through practice and use of strategies (Dyson et al., 2015b). Likewise, the activities might be adapted for one-on-one tutoring with older students with diagnosed learning and/or intellectual disabilities (Dyson et al., 2015a). Future research should define the specific number activities that are most effective for different types of learners.

Acknowledgment

This work was supported in part by Grant # HD059170 from the Eunice Kennedy Shriver National Institute of Child Health and Human Development. Correspondence concerning this article should be addressed to Nancy C. Jordan, School of Education, University of Delaware, Newark, Delaware 19716. Email: njordan@udel.edu.

References

Baroody, A. J. (1987). The development of counting strategies for single-digit addition. *Journal for Research in Mathematics Education*, *18*(2), 141–157.

Baroody, A. J., Lai, M. -L., & Mix, K. S. (2006). The development of young children's number and operation sense and its implications for early childhood education. In B. Spodek,

& O. Saracho (Eds.), *Handbook of research on the education of young children* (pp. 187–221). Mahwah, NJ: Lawrence Erlbaum Associates.

Baroody, A. J., Eiland, M., & Thompson, B. (2009). Fostering at-risk preschoolers' number sense. *Early Education and Development, 20*(1), 80–128.

Bryant, D. P., Bryant, B. R., Roberts, G., Vaughn, S., Hughes, K., Porterfield, J., & Gersten, R. (2011). Effects of an early numeracy intervention on the performance of first-grade students with mathematics difficulties. *Exceptional Children, 78*(1), 7–23.

Carpenter, T. P., Hiebert, J., & Moser, J. M. (1983). The effect of instruction on children's solutions of addition and subtraction word problems. *Educational Studies in Mathematics, 14*(1), 55–72.

Case, R. (1996). Introduction: reconceptualizing the nature of children's conceptual structures and their development in middle childhood. In R. Case & Y. Okamoto (Eds.), *The role of central conceptual structures in the development of children's thought. Monographs of the Society for Research in Child Development, Serial No. 246, Vol. 6* (pp. 1–26).

Clarke, B., Baker, S. K., Smolkowski, K., & Chard, D. (2008). An analysis of early numeracy curriculum-based measurement: examining the role of growth in student outcomes. *Remedial and Special Education, 29*(1), 46–57.

Clarke, B., & Shinn, M. (2004). A preliminary investigation into the identification and development of early mathematics curriculum-based measurement. *School Psychology Review, 33*, 234–248.

Clarke, B., Doabler, C. T., Smolkowski, K., Baker, S. K., Fien, H., & Strand Cary, M. (2014). Examining the efficacy of a tier 2 kindergarten intervention. *Journal of Learning Disabilities* doi: 10.1177/0022219414538514, Advanced online publication.

Clements, D. H. (1999). Subitizing: what is it? Why teach it? *Teaching Children Mathematics, 5*(7), 400–405.

Clements, D. H., & Sarama, J. (2007). Effects of a preschool mathematics curriculum: summative research on the Building Blocks project. *Journal for Research in Mathematics Education, 38*(2), 136–163.

Clements, D. H., & Sarama, J. (2008). Experimental evaluation of the effects of a research-based preschool mathematics curriculum. *American Educational Research Journal, 45*, 443–494.

Clements, D. H., & Sarama, J. (2009). *Learning and teaching early math: The learning trajectories approach*. New York, NY: Routledge.

Dyson, N. I., Jordan, N. C., & Glutting, J. (2013). A number sense intervention for low-income kindergartners at risk for mathematics difficulties. *Journal of Learning Disabilities, 46*(2), 166–181.

Dyson, N., Jordan, N. C., & Hassinger-Das, B. (2015a). The story of Kyle. *Teaching Children Mathematics, 21*(6), 354–361.

Dyson, N., Jordan, N. C., Beliakoff, A., & Hassinger-Das, B. (2015b). A kindergarten number-sense intervention with contrasting practice conditions for low-achieving children. *Journal of Research in Mathematics Education, 46*, 331–370.

Feigenson, L., Dehaene, S., & Spelke, E. S. (2004). Core systems of number. *Trends in Cognitive Sciences, 8*(7), 307–314.

Frye, D., Baroody, A. J., Burchinal, M., Carver, S. M., Jordan, N. C., & McDowell, J. (2013). *Teaching math to young children: A practice guide* (NCEE 2014-4005). Washington, DC: US Department of Education, Institute of Education Sciences, National Center for Education Evaluation and Regional Assistance. Available from http://ies.ed.gov/ncee/wwc/pdf/practice_guides/early_math_pg_111313.pdf

Fuchs, L. S., Geary, D. C., Compton, D. L., Fuchs, D., Schatschneider, C., Hamlett, C. L., & Changas, P. (2013). Effects of first-grade number knowledge tutoring with contrasting forms of practice. *Journal of Educational Psychology, 105*(1), 58–77.

Fuson, K. C., Grandau, L., & Sugiyama, P. A. (2001). Achievable numerical understandings for all young children. *Teaching Children Mathematics, 7*(9), 522–526.

Geary, D. C. (1994). *Children's mathematical development: Research and practical applications.* Washington, DC: American Psychological Association.

Geary, D. C., Bailey, D. H., & Hoard, M. K. (2009). Predicting mathematical achievement and mathematical learning disability with a simple screening tool: the Number Sets Test. *Journal of Psychoeducational Assessment, 27*, 265–279.

Gelman, R., & Gallistel, C. R. (1978). *The child's understanding of number.* Cambridge, MA: Harvard University Press.

Gersten, R., Clarke, B., Jordan, N. C., Newman-Gonchar, R., Haymond, K., & Wilkins, C. (2012). Universal screening in mathematics for students in the primary grades. *Exceptional Children, 78*, 423–445.

Griffin, S. (2004). Teaching number sense. *Educational Leadership, 61*(6), 39–42.

Hanich, L. B., Jordan, N. C., Kaplan, D., & Dick, J. (2001). Performance across different areas of mathematical cognition in children with learning difficulties. *Journal of Educational Psychology, 93*, 615–626.

Hassinger-Das, B., Jordan, N. C., & Dyson, N. (2015). Reading stories to learn math: mathematics vocabulary instruction for children with early numeracy difficulties. *Elementary School Journal, 116*, 242–264.

Huttenlocher, J., Jordan, N., & Levine, S. (1994). A mental model for early arithmetic. *Journal of Experimental Psychology: General, 123*(3), 284–296.

Institute of Education Sciences (2014). *What works clearinghouse: procedures and standards handbook*(Version3.0).Availablefromhttp://ies.ed.gov/ncee/wwc/pdf/reference_resources/wwc_procedures_v3_0_standards_handbook.pdf

Jordan, N. C., & Glutting, J. (2012). *Number sense screener* (Research ed.). Baltimore, MD: Paul H. Brookes.

Jordan, N. C., & Levine, S. C. (2009). Socio-economic variation, number competence, and mathematics learning difficulties in young children. *Developmental Disabilities Research Reviews, 15*, 60–68.

Jordan, N. C., & Montani, T. O. (1997). Cognitive arithmetic and problem solving: a comparison of children with specific and general mathematics difficulties. *Journal of Learning Disabilities, 30*(6), 624–634.

Jordan, N. C., Kaplan, D., Oláh, L. N., & Locuniak, M. N. (2006). Number sense growth in kindergarten: a longitudinal investigation of children at risk for mathematics difficulties. *Child Development, 77*(1), 153–175.

Jordan, N. C., Kaplan, D., Locuniak, M. N., & Ramineni, C. (2007). Predicting first-grade math achievement from developmental number sense trajectories. *Learning Disabilities Research & Practice, 22*(1), 36–46.

Jordan, N. C., Kaplan, D., Ramineni, C., & Locuniak, M. N. (2008). Development of number combination skill in the early school years: when do fingers help? *Developmental Science, 11*(5), 662–668.

Jordan, N. C., Kaplan, D., Ramineni, C., & Locuniak, M. N. (2009). Early math matters: kindergarten number competence and later mathematics outcomes. *Developmental Psychology, 45*(3), 850 867.

Jordan, N. C., Glutting, J., & Ramineni, C. (2010a). The importance of number sense to mathematics achievement in first and third grades. *Learning and Individual Differences, 20*(2), 82–88.

Jordan, N. C., Glutting, J., Ramineni, C., & Watkins, M. W. (2010b). Validating a number sense screening tool for use in kindergarten and first grade: prediction of mathematics proficiency in third grade. *School Psychology Review, 39*(2), 181–195.

Jordan, N. C., Glutting, J., & Dyson, N. (2012a). *Number sense screener user's guide.* Baltimore, MD: Brookes. Available from http://archive.brookespublishing.com/documents/NSS-technicalreport.pdf

Jordan, N. C., Glutting, J., Dyson, N., Hassinger-Das, B., & Irwin, C. (2012b). Building kindergartners' number sense: a randomized controlled study. *Journal of Educational Psychology, 104*(3), 647–660.

Jordan, N. C., Hansen, N., Fuchs, L. S., Siegler, R. S., Gersten, R., & Micklos, D. (2013). Developmental predictors of fraction concepts and procedures. *Journal of Experimental Psychology, 116*, 45–58.

Jordan, N. C., Fuchs, L. S., & Dyson, N. (2015). Early number competencies and mathematical learning: individual variation, screening, and intervention. In R. C. Kadosh & A. Dowker (Eds.), *Oxford handbook of numerical cognition*. Oxford, England: Oxford University Press.

Klein, A., Starkey, P., Sarama, J., Clements, D. H., & Iyer, R. (2008). Effects of a pre-kindergarten mathematics intervention: a randomized experiment. *Journal of Research on Educational Effectiveness, 1*, 155–178.

Lembke, E., & Foegen, A. (2009). Identifying early numeracy indicators for kindergarten and first-grade students. *Learning Disabilities Research & Practice, 24*(1), 12–20.

Levine, S. C., Jordan, N. C., & Huttenlocher, J. (1992). Development of calculation abilities in young children. *Journal of Experimental Child Psychology, 53*, 72–103.

Locuniak, M. N., & Jordan, N. C. (2008). Using kindergarten number sense to predict calculation fluency in second grade. *Journal of Learning Disabilities, 41*(5), 451–459.

Methe, S. A., Hintze, J. M., & Floyd, R. G. (2008). Validation and decision accuracy of early numeracy skill indicators. *School Psychology Review, 37*, 359–373.

Morgan, P. L., Farkas, G., & Wu, Q. (2009). Five-year growth trajectories of kindergarten children with learning difficulties in mathematics. *Journal of Learning Disabilities, 42*(4), 306–321.

National Mathematics Advisory Panel (2008). *Foundations for success: The final report of the National Mathematics Advisory Panel* (ED 00424P). Washington, DC: US Department of Education.

National Research Council. (2009). *Mathematics learning in early childhood: Paths toward excellence and equity*. Washington, DC: National Academies Press.

Ramani, G. B., & Siegler, R. S. (2008). Promoting broad and stable improvements in low-income children's numerical knowledge through playing number board games. *Child Development, 79*(2), 375–394.

Riley, M., Greeno, J., & Heller, J. (1983). Development of children's problem solving ability in arithmetic. In H. Ginsburg (Ed.), *The development of mathematical thinking* (pp. 153–196). New York: Academic Press.

Robinson, C. S., Menchetti, B. M., & Torgesen, J. K. (2002). Toward a two-factor theory of one type of mathematics disabilities. *Learning Disabilities Research & Practice, 17*(2), 81–89.

Rousselle, L., & Noël, M. -P. (2007). Basic numerical skills in children with mathematics learning disabilities: a comparison of symbolic vs non-symbolic number magnitude processing. *Cognition, 102*(3), 361–395.

Sarnecka, B. W., & Wright, C. E. (2013). The idea of an exact number: children's understanding of cardinality and equinumerosity. *Cognitive Science, 37*(8), 1493–1506.

Seethaler, P. M., & Fuchs, L. S. (2010). The predictive utility of kindergarten screening for math difficulty. *Exceptional Children, 77*, 37–59.

Siegler, R. S., & Lortie-Forgues, H. (2014). An integrative theory of numerical development. *Child Development Perspectives, 8*, 144–150.

Vukovic, R. K., Fuchs, L. S., Geary, D. C., Jordan, N. C., Gersten, R., & Siegler, R. S. (2014). Sources of individual differences in children's understanding of fractions. *Child Development, 85*, 1461–1476.

Wood, J. N., & Spelke, E. S. (2005). Infants' enumeration of actions: numerical discrimination and its signature limits. *Developmental Science, 8*(2), 173–181.

Woodcock, R. W., McGrew, K. S., & Mather, N. (2007). *Woodcock-Johnson tests of achievement*. Itasca, IL: Riverside Publishing. Available from http://www.riverpub.com/products/wjIIIComplete/index.html

Wu, H. (1999). *Some remarks on the teaching of fractions in elementary school*. Berkeley, CA: Department of Mathematics #3840, University of California. Available from http://www.math.berkeley.edu/~wu/

CHAPTER

4

Contextual Sensitivity and the Large Number Word Bias: When Is Bigger Really More?

Michele Mazzocco, Jenny Yun-Chen Chan, Maria Sera

Institute of Child Development, University of Minnesota, Minneapolis, MN United States of America

OUTLINE

Continuous Issues in Numerical Cognition. http://dx.doi.org/10.1016/B978-0-12-801637-4.00004-4

I recently observed a toddler traveling by train with an adult caregiver, presumably a grandparent. Seated in a stroller, the child used a one-handed pincer grasp to individually select and eat bite-size crackers from a cup. The caregiver coordinated finger movements and utterances with the toddler's behavior, first holding up a thumb while saying, "one," then sequentially increasing the number of raised digits while saying, "two, three, four, five." Using the nonfeeding hand, the child repeated these finger movements and number word pairings.

After counting "five," the caregiver and child each raised only their respective smallest digit (pinkie), and jointly uttered, "six," followed by raising sequentially increasing numbers of digits ending with the thumb and the number word, "ten"! A clever strategy, perhaps, to accommodate the one-handed feeding (by the child) or one-handed counting while the other hand grips a stroller in a moving railcar (by the adult). Yet it was clear that this was a well-established routine. What, I wondered, does the word "six" mean to this particular child, who associates "six" with a single, small, raised digit?

4.1 WHY ARE NUMBER WORDS CONFUSING?

Number words can be confusing, especially across the varied situations in which they occur. Unlike other instances of word learning, word-learning strategies such as the shape bias, the whole object constraint, and mutual exclusivity do not support learning the various meanings of number words. An exception may hold for mapping number words and numeral symbols, the latter of which maintain their shape across many contexts. Indeed, English-speaking children map number words and Arabic numerals at around 18 months of age, long before they acquire the cardinal and ordinal meanings of number words (Mix, 2009), but after they are able to rapidly learn names for many solid objects (Woodward, Markman, & Fitzsimmons, 1994).

The reason for this protracted development of number word vocabulary may be due to the fact that number words may refer to either discrete quantities (four apples) or to continuous quantities that are themselves noncountable but that are often paired with countable units of measurement (four inches of ribbon, four cups of water). Children often hear adults use number words when counting tangible sets of discrete objects (four crayons) or labeling cardinal values, and this type of input is associated with children's cardinal knowledge in preschool years (Gunderson & Levine, 2011; Mix, Sandhofer, Moore, & Russell, 2012). But this may not hold as strongly for continuous quantities, like water. English-speaking children may initially omit the units of measurement when quantifying continuous quantities (Mix, 2009), such that two bottles of milk become "two milks." They may also ignore units when interpreting measurement phrases such as interpreting "two-pound apples" as two apples (Syrett, 2013).

Likewise, number words may refer to units of time (four years old, four o'clock) or events (jumped two times, sneezed four times, took three train rides). The latter may refer to durative (walking, crying) or punctuated (clapping, kicking) intervals (as detailed by Barner, Wagner, & Snedeker, 2008). Toddlers may label and enumerate actions or events, such as, for instance, when a 2-year-old says, "Ow. Ow. Two ows," after falling twice in a row (Mix, 2009). Whether and how adults attend to number when quantifying events differs for durative but not for "punctual" (more iterative) events (Barner et al., 2008), such that six brief durative events are considered to be "more" than two comparable long events, even if their total duration is less than that of the two longer events (eg, six *short* walks are more *walks* than two long walks but involve less walking than two long walks). Barner and colleagues found that punctuated events are not as likely to be subject to this contrast (eg, "six short claps" refers to more claps and to more clapping than "two long claps"). These examples indicate that number word interpretation is complex, context dependent, and subject to syntactic influences.

Number words may also indicate ordinal position rather than set size (fourth in line, bin number four, Four Maple Drive). Compared to the cardinal interpretation of number words, identifying ordinal positions is more difficult for 3- to 5-year-olds, perhaps due to less experience with ordinal versus enumerating contexts (Colomé & Noël, 2012). Indeed, even before three years of age, most of parents' number talk refers to the cardinal value of object sets (although number talk itself is relatively infrequent; Gunderson & Levine, 2011). Still, the acquisition of cardinality precedes and correlates with children's understanding of ordinality, suggesting that cardinal knowledge may support the development of ordinality. Regardless of their relation with one another, there is evidence that cardinality and ordinality are independent in developmental trajectories and function (Wasner, Moeller, Fischer, & Nuerk, 2015).

The examples noted thus far implicate that situational and linguistic contexts influence number word meanings and their interpretation. Number words also appear in syntactic forms that support different interpretations (five apples, five of the apples, five girls eat two apples). These interpretations are subject to developmental differences. Whereas adults give collective interpretations (five girls together eat two apples) to ambiguous numerical statements, preschoolers tend toward distributive interpretations (five girls each eat two apples; Syrett & Musolino, 2013). Number words may also be paired with combinatorial nouns (four bags of apples, four pairs of socks, four dozen brownies) to imply a completely different number than the one uttered. The number words may be used to be descriptive (four dogs) or included in directives that refer to objects (give me four cookies) or actions that can be easily enumerated (tap four times) or not (wait four minutes). Because the meanings of number words vary as a function of the other words with which they appear, the intended meaning of number words may be ambiguous.

In this chapter, we do not attend to the processes by which children acquire number word meanings from such contrasting instances (addressed elsewhere; eg, Bloom & Wynn, 1997; Condry & Spelke, 2008; Davidson, Eng, & Barner, 2012; Gunderson, Spaepen, & Levine, 2015; Le Corre & Carey, 2007; Sarnecka & Carey, 2008; Wagner & Johnson, 2011). Instead, we propose that children's responses to ambiguities in the meaning of number words—*their response to number words in different contexts*—may be meaningful indicators of their number word and concept development. First, we briefly review some research on learning number word meanings relevant to this notion. We consider the prerequisites to number word understanding, and then specifically attend to studies that focus on the effects of context in children's number word knowledge and the prerequisite skills that underlie this knowledge. Next, we consider whether number words are a somewhat specialized form of lexical ambiguity. We then present some evidence that children's awareness of the multiple meanings of number words and thus their response to this ambiguity varies during and after the preschool years. Finally, we propose that this variation is one element of a child's "number sense" in that it reflects developing concepts about quantities, and that this aspect of number sense may play a role in the developmental course of mathematical learning difficulties.

4.2 NUMBER WORDS AND NUMBER SENSE

Despite a rich literature on children's acquisition and use of English number *words*, issues surrounding *number sense* and its development remain unresolved. Even less is known about how number words and concepts are learned in other languages, many of which have different number words, different syntactic forms involving number words, or both, compared to English. In most languages, the number words themselves do not differ as a function of the referents with which they are paired, but meanings differ across contexts. Of particular relevance to this chapter is children's sense of number and the pressing need to further delineate subcomponents of number sense, identify the developmental and individual differences in those skills, and understand which elements of number sense are most indicative of risk for and the nature of mathematical learning difficulties, at least in countries where learning disabilities are studied. Note that we do not equate number and mathematics, but rather recognize number skills as a foundation of concurrent and later mathematical thinking and learning, with number concepts an integral component of number sense.

Researchers' efforts to delineate number sense are underway, but far from complete. These efforts stem from emerging evidence that select aspects of number sense may play a more significant role in the development of children's mathematical thinking than do others (see Chapter 15), and that some of these skills may be more appropriate targets of early mathematics interventions than others (Chapter 3). Still, current definitions of number sense remain highly variable across research and applied settings, in terms of the specificity versus breadth of the cognitive processes or behaviors to which they refer (see Berch, 2005, for a review). Examples of measurable number sense subcomponents include preverbal, nonsymbolic representations of small exact numbers (ie, subitizing) and encoding of large approximate numerosities. These abilities are arguably evident in human infants and nonhuman primates, although for another view see Chapter 1. They are often manifested via magnitude comparison tasks that require differentiating numerosities across two sets (reviewed by Feigenson, Dehaene, & Spelke, 2004), and are also revealed through nonverbal arithmetic judgments (Wynn, 1992a; McCrink & Wynn, 2004). In contrast to these preverbal abilities, other aspects of number sense build on symbolic representations of number (eg, Arabic numerals, number words) introduced through informal and formal instruction, and these support the development of numbering skills (eg, counting, estimation), number relations (eg, ordinality, number comparison), and operations (eg, addition/subtraction, composition/decomposition, equivalence; Purpura & Lonigan, 2013), each of which is delineated further. Furthermore, the intuitive understanding of numbers, operations, and their interconnected relationship, and the ability to estimate and apply mathematical knowledge in flexible ways, are also considered aspects of number sense (Mcintosh, Reys, & Reys, 1992; Baroody, Lai, & Mix, 2006) and of mathematical proficiency. These constructs of symbolic number overlap with intentional versus automatic number sense (Bugden & Ansari, 2011), constructs that emphasize specific attentional contributions and the contribution of language and executive function skills to understanding, using, or operating with numbers. Importantly, associations between number sense and mathematics achievement vary as a function of *which* aspect of number sense is measured, and *when* during development it is measured (eg, Girelli, Lucangeli, & Butterworth, 2000). Fundamental to most if not all of these symbolic number skills is an understanding of the referents of these number words and symbols—in other words, number word meanings.

4.3 NUMBER WORD MEANINGS

What are the precursors to understanding number words? To accurately extract meaning from number words, children must first recognize that numerosities exist. If nonverbal infants make numerosity judgments, this aspect of number sense may be intuitive (Izard, Sann, Spelke, &

Streri, 2009; Xu, Spelke, & Goddard, 2005). Beyond infancy, young children attend to information about quantity even before they receive any formal numerical instruction. Their attention or awareness of number—whether spontaneous or guided—appears influenced by context, such as collection size and competing features (Baroody, Li, & Lai, 2008), and number awareness also shows developmental and individual differences (Hannula & Lehtinen, 2005). This awareness of numerosities is required in order for children to acquire number word meanings.

Children must also recognize that numerosities can be labeled by number words, and that number words refer to specific numerosities, that is, that number words are symbolic representations of exact quantities. This recognition is no small feat, and the underlying processes that support this understanding remain a topic of debate (eg, Davidson et al., 2012; Nuñez, forthcoming). Children initially learn number word meanings with marked constraints. Wynn (1992b) argued that even 2-year-olds know that number words refer to numerosities, albeit without knowledge of which specific numerosities map on to which words, and limited to knowledge of a subset of number words; for a period of time children may understand only the exact meaning of the number word "one." Accordingly, young children may, for instance, briefly overgeneralize number words for which they do not have an exact referent, such that one label (eg, "two") may refer to multiple distinct quantities (eg, any number > 1; Wynn, 1992b). When children begin to understand that each number word refers to a specific numerosity (at about three or four years of age), they report the same number word (even if it is not the *correct* number word) when enumerating a set that undergoes a transformation, as long as the transformation does not involve an explicit change in quantity (such as rearranging four peaches on a plate). However, only children who demonstrate knowledge of the cardinal principle also demonstrate knowledge that equal sets (four peaches, four hats) have the same number of items and are described with the same number word (Sarnecka & Gelman, 2004; Slusser & Sarnecka, 2011).

Conceptual understanding of number words has been attributed to knowers of this cardinal principle. For example, Sarnecka and Wright (2013) found that when preschoolers are asked to judge which of two numbers is "more," only cardinal principle knowers are accurate at comparing small exact numbers that differ slightly (eg, comparing sets of 5 vs. 6 items). There are two important aspects of this and related findings. First, as described extensively in the field (eg, Sarnecka & Carey, 2008), acquiring the cardinal principle is not an all-or-nothing process; it occurs in a predictable developmental sequence, although not without some variation. For instance, children may initially know the actual, precise meaning of only a subset of the number words (eg, "one" to "three") even if they reliably utter many more numbers (eg, up to ten or beyond) in the correct verbal counting sequence (LeCorre, Van de Walle, Brannon, & Carey, 2006), on their way to understanding the cardinal principle. Moreover, these subset knowers fail to understand the "later-greater principle," that is, that numbers that come later in the count list are greater (more) than numbers that precede them, whereas cardinal principle knowers do (Sarnecka & Carey, 2008). Second, we should still be cautious in the extent to which we attribute number word understanding to cardinal principle knowers, given the variation observed among preschool cardinal principle knowers' later-greater principle, with some performing at chance (Sarnecka & Carey, 2008), particularly with numbers greater than 5 (Davidson et al., 2012). Considered together, findings such as these reveal the complexities in the development of number word understanding throughout the preschool years. On the one hand, they illustrate why counting principles are a logical source of support for learning the number word meanings (Gelman & Gallistel, 1978; Mix, Sandhofer, & Baroody, 2005; Slusser & Sarnecka, 2011), since a tendency to enumerate sets may promote mapping numerosities onto number words. On the other hand, since accurate counting of item sets on demand does not guarantee cardinal meanings of number words (Baroody & Price, 1983), counting performance does not guarantee mastery of (nor depend upon) deeper number knowledge. Some researchers pose the possibility that "learning to generate and label sets may be a precursor to discovering the relationship between numbers, rather than the opposite" (Davidson et al., 2012, p. 164).

In addition to associating number words with numerosities, children also recognize number words as names of visual symbols (eg, Arabic numerals), such as that "3" is labeled "three." They might not, however, initially understand it refers (at least in some contexts) to a set of three (Mix, 2009). Establishing connections between number words, Arabic numerals, and numerosities is initially effortful. Although these associations gradually become more automated (at around 6–8 years of age; Girelli et al., 2000), it appears these associations are never fully automatized because even adults' magnitude comparison performance is diminished (in terms of response time) when the representations of number being compared include both symbolic and nonsymbolic formats. For instance, when asked to select which of two stimuli is the bigger number, adults' response times are significantly longer for cross-format comparisons, such as 6 and ■■, versus comparisons of representations that are only nonsymbolic (eg, ■■■■■■ and ■■) or only symbolic (eg, 6 and two; Lyons, Ansari, & Beilock, 2012). This cross-format decrement in response time demonstrates the remarkable complexity inherent in associations between number symbols and meanings that persists throughout development.

Thus far we have briefly summarized a sampling of the evidence that children seem intuitively drawn to numerosities and to comparing them, and that they appreciate the general nature of number words well before schooling. It is also the case that, from an early age, children discriminate between quantities and demonstrate an understanding of "more." Even before they can count or assign symbols to quantities, infants recognize when numerosities increase (or decrease; Brannon, 2002), and once their counting skills begin to emerge, preschoolers successfully compare the larger of two small quantities (items up to 5 or 6; Brannon & Van de Walle, 2001), although not consistently (Davidson et al., 2012). Children as young as 3 years of age use "more" in similar ways when comparing numerical quantities and when comparing areas, suggesting that initial knowledge of the word "more" taps into the underlying approximate number system (Odic, Pietroski, Hunter, Lidz, & Halberda, 2013), as does knowledge of other early learned dimensional adjectives such as *big*, which is correctly used by children as young as 2 years of age (Sera & Smith, 1987). When numerosity is pitted against size, 4-year-olds and adults judge discrete objects based on numerosity (3 little forks is more than 1 big fork), and a continuous substance based on size (1 big pile of sand is more than 3 small piles of sand; Barner & Snedeker, 2005). However, 2-year-olds, unlike older children and adults, also systematically use "more" to refer to the dark end of a continuum of darkness—a dimension that is not quantified by English-speaking adults (Smith & Sera, 1992). Thus, interpreting "more" does not only depend on exact number skills—since exact number knowledge promotes accuracy of more/less judgments when comparing numerosities—but also may depend on distinguishing numerical from nonnumerical dimensions.

The abilities described to this point may be viewed as precursors to conceptually based magnitude judgments, but they do not reflect mastery of the number words children hear or use. Although children engage in behaviors that both rely on and involve number word usage, they may lack an understanding of some or all of the number words they hear or utter. The aforementioned ambiguity associated with numbers across context may be one obstacle to children forming logical number–number word associations, and sources of this ambiguity are considered next.

4.4 SOURCES OF AMBIGUITY IN NUMBER WORD MEANINGS

That number words pose challenges for word learning is well known (eg, Wynn, 1990; 1992b; Sullivan & Barner, 2013), and this challenge is reflected by children's inconsistent number use. Context affects the interpretation of word meanings, as Fuson and Hall (1983) described in their summary of number use development, and number recitations may initially be meaningless phonetic sequences. For instance, children may recite the count list as early as 2 or 3 years of age, but some children produce unconventional or unstable ordered sequences and others intermix numbers, letters, and color names in their sequence production, suggesting an element of rote activity (Baroody & Price, 1983; Fuson & Hall, 1983; Gelman & Mecks, 1983). (In principle, this rote recitation may later extend to reciting "math facts" and other components of early or complex mathematics.) Children's use of number words becomes more relational when children count, but the words may still be uttered out of sequence or simply not tied to cardinal principles. This sequence maps onto the recognition that number words refer to numerosities in general, before understanding that the numerosities to which specific number words refer are themselves specific. It is not for some time that meaningful enumerating emerges, but even then it is subject to wide individual variation in the timing of these and other number learning sequences (Wynn, 1992b) and in the development of number concepts.

Several sources of the ambiguity of numbers are well described, even if their contribution to number word acquisition is unclear. Here we review a sample of such sources, to illustrate that the ambiguity of number words derives from the nature of the words themselves, from the syntactic and morphological structure of words they accompany, and lexical context.

4.4.1 Sets Versus Individuation

One source of ambiguity in the meaning of number words concerns the simple fact that number words refer to a property of item *sets* rather than to the individual items themselves or to the properties of those items (Wynn, 1992b). This fact conflicts with children's biases to interpret new words as referring to individual items (Markman, 1989) and their general tendency to count whole objects versus sets or parts (Shipley & Shepperson, 1990). Although children understand that number words refer to the number of discrete objects (eg, three apples in a bowl) instead of the amount of a continuous substance (eg, water in a bowl; Slusser, Ditta, & Sarnecka, 2013), this tendency to focus on individual items may sometimes lead to somewhat meaningless comparisons of "more" or "how many."

A similar distinction emerges when comparisons occur across units. For instance, if asked to judge whether two halves of one apple is more than, less than, or the same as one intact apple, adults are more likely to interpret quantity relative to partial versus whole units (in this case, equating two halves with one whole apple), whereas preschool-aged children are more likely to enumerate (or compare quantities of) whatever units appear in the set, regardless of whether the units are the same or, as first reported by Shipley and Shepperson (1990), different segments of similar objects (Brooks, Pogue, & Barner, 2011). Thus, children are more likely to judge that two halves (or even two-thirds) of an apple is "more than" one whole apple, and are also likely to count from 1 to 4 when asked to enumerate a set comprised of three pieces of an apple and one whole apple. Children will count a set of four objects as six if one item in the set appears in three separate pieces. Such enumeration behavior suggests that, at least for some children, counting is a highly routinized behavior that overrides children's whole object bias. It is unclear whether this illogical enumeration reflects foundational understanding of proportions, fractions, and part-whole distinctions per se or the context-bound ambiguity of relational language.

Contexts may constrain when children are sensitive to the units of individuation, particularly the use of familiar versus unfamiliar labels of part and whole objects. Brooks and colleagues showed that preschoolers are likely to correctly count two cojoined items as two items (not as one item, despite their being physically connected and despite the fact that the items move in unison), so long as the items remain distinct (eg, two dolls glued together). Children will also refer to explicitly labeled components of a whole (eg, two wheels of a bicycle) as the associated whole (eg, in this example, as one bicycle; Brooks et al., 2011). In these cases, children behave like adults, that is, they enumerate items based on the familiar whole (a bicycle) rather than on the individuated parts (two wheels). However, if children are asked to focus on the parts, they do so (such as noting that an animal has two feet) so long as the label used to denote the parts is familiar (as is the case for "feet"; Giralt & Bloom, 2000). When labels are less familiar (such as fork handles), children tend to treat each discrete piece as one unit (eg, the handle and tines of a fork are considered two units if they are presented separately) rather than the whole object (eg, the handle and tines, even if separated, are one fork) as the units of quantification (Brooks et al., 2011; first reported by Shipley & Shepperson, 1990).

4.4.2 Syntax and Morphology

Additional sources of number word ambiguity are linked to the syntactic or morphological structure of sentences that accompany number words in oral language. For instance, the syntactic structure of statements including number words often follows a pattern similar to that observed for modifiers of individual items (eg, "see the *white* dogs, see the *furry* bears") despite the fact that, in the case of number words, what is specified pertains to the set and not the individuals (eg, "see the two pigs"). In these examples, all the dogs are white and all the bears are furry, but we do not say that both pigs are "two." This inconsistent syntactic structure may interfere with young children's early attempts to derive word meanings, if number words are viewed as modifiers of the former type.

Bloom and Wynn (1997) described four specific syntactic constraints that appear to facilitate acquiring number word meanings. One constraint concerns the fact that number words are not used with modifiers (eg, the structure "very two" is extremely unusual and thus unlikely to be uttered), which helps to differentiate them from adjectives that *are* used with modifiers (eg, very white, very furry, very soft, very loud). This pattern may help children overcome the similarity in structure previously mentioned. Moreover, when modifiers and number words appear in the same phrase, number words may appear before or after an adjective ("three red flowers" in English vs. "three flowers red" in Italian), with varying frequency across languages. This structure further differentiates number words from modifiers, but may nevertheless introduce constraints on number word learning (see Ramscar, Dye, Popick, & O'Donnell-McCarthy, 2011, discussed in the next paragraph). Another constraint concerns the tendency observed when both young children and their caregivers use number words with countable objects more so than with mass substance per se such as pudding or water (Slusser et al., 2013). However, mass substances are quantifiable by countable units, which we label using the same number words (eg, two cups of water, four bowls of pudding). Finally, although infrequent, number words and quantifiers may occur with a partitive frame (eg, "five of the apples"), which is not typically observed with adjectives. Children are more likely to interpret a novel word as denoting quantity when it appears in a partitive frame (dom of cars) than a nonpartitive frame (dom cars), suggesting that syntactic cues may help children identify that number words refer to quantities (Syrett, Musolino, & Gelman, 2012).

Ambiguity of the intended meaning of number words is also affected by the relative positioning of number words and their associated nominals. Ramscar et al. (2011) tested whether postnominal construction of sentences (eg, "see the bears, there are three") would facilitate number word learning more so than prenominal construction (eg, "see, there are the three bears"). Both their computer simulation studies and experimental treatment studies with 2- and 3-year-olds supported this hypothesis. It appears that emphasizing the target set to be enumerated (eg, the bears) prior to eliciting enumeration directs children's attention to the set, which offers some constraints on their interpretation of the corresponding number word. (This effect need not be limited to number words, but may be especially facilitative to learning number words given their unique features.)

Ramscar's and others' findings support the notion that cultural variation in the timing of number word acquisition may stem, at least in part, from cross-linguistic differences in syntax (eg, Barner, Libenson, Cheung, & Takasaki, 2009) or morphology (Almoammer et al., 2013). Barner and colleagues found that, compared to Japanese-speaking children, English-speaking children acquire small number word meanings earlier, perhaps due to the greater similarity in syntactic structures of numerals and quantifiers in English, which may have facilitative effects on learning the quantity denoting meaning of numerals. Almoammer and coworkers compared the rate of learning number words "one," "two," and "three" among members of cultures with singular–plural markers (eg, in English, one *bear*, two *bears*, three *bears*, etc.). Here the morphological form for plural (ie, *s*) does not change as a function of "how many," and its use signifies only "more than one." That is, knowing that there are "bears in the woods" does not reveal exactly how many bears are present. In addition to these singular–plural markers, some languages also have a dual marker used specifically with sets of two. In this case, the plural indicates "more than two" and the dual indicates "exactly two." These researchers found that speakers of a language with this dual marker acquired the meaning of "two" faster (at a younger age) than did speakers of languages without a dual marker. Findings also indicate that the names of some numbers such as *eleven* and *twelve* may pose a bottleneck for English speakers in comparison to languages in which the names are more transparent, such as *ten and one*, *ten and two*, etc. (Miller & Stigler, 1987). These examples illustrate how syntactic and morphological features may contribute to, or resolve, ambiguity in number word meanings.

4.4.3 Lexical Ambiguity

Number words are subject to specific types of lexical ambiguity. At the level of phrases, number words can be paired with one of two types of units of measurement, attributive (eg, two-pound apples) or pseudopartitive (eg, two pounds of apples). The attributive phrases refer to a property of individual items (each apple weighs two pounds), whereas the pseudopartitive phrases refer to a property of a set (combined, the apples weigh two pounds). (Note that in either case, there is no indication of the actual number of apples; we will return to this point in our conclusion.) Four-year-olds are likely to interpret both pseudopartitive and attributive phrases as referring to a property of a set. Furthermore, when asked "do I still have two-pound apples?" after seeing the experimenter taking away some apples from the set, some children respond "No, just one ... one apple," suggesting that 4-year-olds apply cardinal interpretation to measurement phrases and that this may be a result of frequent experience with counting items in a set (Syrett, 2013). Number words may be ambiguous at the sentence level also. For instance, the sentence "two people saw a bear" can be interpreted as a collective statement (together, two people saw a bear; in this case, there is one bear) or as a distributive statement (two people each saw one bear; in this case, there are two bears). When the words "together" or "each" are absent from these sentences, adults are more likely to make collective interpretations, whereas 4-year-olds are more likely to make distributive interpretations (Syrett & Musolino, 2013). When the words "together" or "each" are present, children and adults are both more likely to appropriately infer distributive meanings of statements referring to "each," and collective meanings of statements referring to "together." However, the effect size for this difference is much smaller in children than it is for adults (Syrett & Musolino, 2013), which suggests that young children show only limited contextual sensitivity to number word interpretation from sentences.

There are at least five possible explanations for what may unite these examples of number word ambiguity. The first is that numerosities themselves are not easily perceived. In view of evidence of numerosity differentiation in human infants and other species, this notion is unlikely to account for much variation in number word ambiguity, except in extreme cases such as dyscalculia (Mazzocco, Feigenson, & Halberda, 2011; Piazza et al., 2010). We refer to this as *numerosity deficits driving number word ambiguity*. The second explanation is that number *words* themselves are lexically ambiguous, as a unique form of lexical ambiguity tied specifically to the challenges associated with number concepts. In this case, number concepts affect, and are reflected by, number word interpretation. We refer to this notion as *number words driving ambiguity*. The third explanation concerns the fact that *enumeration* is conceptually ambiguous, which underlies the confusion in number words. The precision of numbers, especially when combined with other terms such as "dozens" might make numbers more difficult to interpret. We refer to this notion as the *enumeration driving number word ambiguity*. Fourth, perhaps number word interpretation difficulty is no different than the challenge children face with other (nonnumerical) polysemous words, and that the ambiguity inherent in number word interpretation is not tied to numerical properties per se. We refer to this notion as the *number words as homonyms*. Finally, perhaps numerosity and number words do present unique challenges (as in the first three previous explanations) that are nonetheless mediated by metacognitive or metalinguistic skills that support successful interpretation of other forms of ambiguity (those implicated in the fourth explanation). We refer to this notion as *awareness of numerical ambiguity*. Note that these explanations are not completely mutually exclusive, although any one may be a primary mechanism underlying ambiguity for some individuals, and combinations of these notions may interact with each other.

These explanations lead to different hypotheses about children's probable responses to the ambiguity of number words. The first, *numerosity deficits* notion, implicates perceptual mechanisms that may underlie incorrect and perhaps even nonnumerical interpretations. The next two notions—that number words, or that enumeration, drive numerical ambiguity suggest that children's awareness and interpretation of numerically ambiguous information—*reflects the nature of their number concepts*. In this case, children's interpretations of number words in ambiguous contexts would be correlated with their number knowledge performance but not with nonnumerical homonym interpretation performance. The fourth notion (*numbers as homonyms*) suggests that children's number word learning *reflects their overall ability to deal with lexical ambiguity*, in which case children's interpretations of number words in ambiguous contexts *would* be correlated with homonym interpretation, and potentially with executive function skills (cognitive flexibility, metacognition) believed to be associated with homonym interpretation (Mazzocco, Myers, Thompson, & Desai, 2003). The last explanation based on *awareness of numerical ambiguity* suggests that emerging number concepts are likely to be influenced by the unique challenges of number words of which children are aware, but that the effects of these challenges are mediated by metalinguistic and metacognitive abilities, including learning strategies.

Support for any of these explanations depends on a measure of response to numbers in ambiguous contexts. In view of the evidence that children do make incorrect interpretations of number words across the wide range of circumstances described to this point, we propose that systematically exploiting this tendency may reveal aspects of developing number concepts and/or response to lexical ambiguity.

Why would we expect children to respond to numerically ambiguous information in a manner consistent with their response to nonnumerical lexical ambiguity? In a sense, number words are polysemous because they can refer to a range of values [discrete set size, ordinal position, single items (3 bananas) or mass (3 bunches of bananas) objects or even intangible referents (3 songs) or movements (3 hops)], the meaning of which is highly contextualized. These examples represent a complementary polysemy unlike the contrastive polysemy of homonyms (eg, Nerlich & Clarke, 2003), but the extent to which number words are perceived as polysemous likely varies with development because the commonality across situations in which numbers are used only gradually becomes apparent and eventually becomes automatized. This notion is important because, if ambiguity is a feature of number words early in their acquisition, it may be a source for misconceptions about numbers and number words that affect the trajectory of children's later mathematical learning. Although this type of polysemy is similar to that of other dimensional terms such as *big* (eg, a big insect is much smaller than a small elephant), the case of number words might be more complex because they are combined with other nouns that also refer to numbers and precise quantities (*two dozen cans* equals 24 cans; but *a dozen big cans* does not really tell you precisely how much fluid there is).

If children's response to numerical ambiguity follows that of other types of polysemy, like homonyms (Doherty, 2004; Mazzocco, 1999), we might expect a similar developmental trajectory: literal interpretations early in the preschool years, diminishing significantly but not completely by age 7 years. In many of the examples of numerical ambiguity described earlier, the number words per se are not as ambiguous as is the referent associated with the number word. For example, in the phrase, "a two pound apple," two refers to how many pounds, not how many apples, but 4-year-olds show a bias toward interpreting "two" apples (Syrett, 2013). Likewise, when interpreting the ambiguous sentence, "two people saw a bear," 4-year-olds (but not adults) are more likely to conclude that there are two bears (Syrett & Musolino, 2013). These response tendencies may reflect the routines of counting, as the authors suggest, or may be a form of "literal" interpretations of numerically ambiguous statements.

4.5 MEASURING CHILDREN'S RESPONSES TO NUMERICAL AMBIGUITY: THE LARGE NUMBER WORD BIAS

What might a "literal" interpretation of a number word indicate, and what would it look like behaviorally? One possibility is that children interpret the numbers they hear based on their value alone, without regard to the meaning they take on in combination with other words. There is evidence that this contextually stripped interpretation occurs, based on children's errors in identifying or interpreting the referents in numerically ambiguous phrases or sentences, as discussed previously (Syrett, 2013; Syrett & Musolino, 2013, respectively). During a magnitude comparison task, this contextually independent focus on number may lead to a "large number word bias"—children might base magnitude judgments on number words uttered (eg, *four*, *two*) instead of on the situational or linguistic context and referents associated with those number words (eg, "*four* books" vs. "*two* boxes of books"). In the former case, four is always more than two; in the latter case, either *four* or *two* may be associated with "more books"—depending on the evidence presented. If the boxes are closed, especially if they are small enough to question whether they may even contain more than two books each, the correct answer may be unclear. In that ambiguous case, it would be correct to respond, "I don't know," when asked, "who has more?" But a large number word bias (LNWB) may prompt children to attribute "more" to the person associated with the large number word; in this case, the child with four books.

We tested for the LNWB within fifty-three 4- to 7-year-olds (mean age 5 years 11 months) who had no known risk for poor mathematics achievement (Mazzocco, Chan, & Praus, 2015). These children were tested individually with a storybook task designed to elicit number comparisons in different contexts. Warm-up trials that preceded the story task were designed to ensure adequate vocabulary for number and container words presented in the task, assess counting and number knowledge for quantities up to six, and convey information about task demands and response options. During these warm-up trials, children were explicitly taught that an "I don't know" response was possible, acceptable, and expected. They were given four opportunities to accurately report that they did not have sufficient information to respond (by expressing "I don't know" in any form, including a shoulder shrug). All fifty-three participants included in our final analyses had passed the numerical warm-up tasks, and had also passed at least one of the "I don't know" prompts by correctly reporting, "I don't know" when such a response was clearly appropriate (eg, such as when asked how many items were in an illustrated opaque box). Children who failed the I don't know warm-up were omitted from the study.

During the experimental trials, the participating children listened to an examiner read the story, and were asked to make "who has more" judgments in numerically congruent versus incongruent conditions. These conditions pertained to whether the larger of two uttered number words was linked to the larger (congruent) or smaller (incongruent) set of items being compared. We also varied whether visual cues provided evidence of set size (visible and invisible conditions, balanced within congruent and incongruent contexts), and we varied the composition of sets being compared (comparisons between different types of referents, described in the following discussion). We included nonnumerical "filler" items in the story task (eg, What is Ann's mom's name? What animals did Ann and Joe see?) to monitor attention and promote task engagement.

While the story was read aloud, the examiner asked the children a series of magnitude judgment questions in the following form: "Ann has four books. Joe has two boxes of books. Who has more?" Illustrations accompanied the story, but only some of these illustrations provided explicit evidence of the correct answer to these questions.

To test for the LNWB, we examined performance on four subsets of trials for which the larger number word was associated with an *incorrect* response. One subset of trials was comprised of the items from the incongruent condition that were paired with an illustration revealing why the *smaller* number word was the correct answer (eg, pictures depicting that Ann had *four books* and that Joe's *two boxes of books* each contained three or four books, such that Joe had, in fact, seven books). In this case, there was sufficient evidence to make an accurate determination of who had more, so the "who has more" query was appropriate. Second, on other trials an "I don't know" response was warranted for one of several reasons: either the illustrations did not reveal the actual quantity associated with combinatorial language (in contrast with the sets of books described earlier), or the magnitude comparisons were *implicitly* illogical (asking "who has more?" after comparing conflicting units such as two books and sneezing twice) or *explicitly* illogical (eg, asking "who has more stickers?" when comparing four vs. two books). Across 24 items, scores ranged from 0 (thus no responses conformed to the large number bias) to 24 (responses consistently conformed to a large number bias across conditions).

Many children showed an LNWB, but the frequency of LNWB responses did not vary with age, at least in our cross-sectional study. The frequency did vary across children, with some making LNWB errors on only a few (20% of all possible) responses and others making many such errors (91%). This frequency range, coupled with no age differences in frequency, suggests that the LNWB is subject to individual differences more than to developmental differences, at least among 4- to 7-year-olds.

What might elicit a large number word bias? The overall number of LNWB errors was not correlated with children's age (as aforementioned) nor with overall vocabulary or counting performance. In fact, most participants were cardinal principle knowers (based on warm-up trials). Other child characteristics that may contribute to the LNWB, but which were not measured in this study, include executive function (cognitive flexibility), listening comprehension, or attention. Based on their responses to engagement-oriented questions, however, participants were clearly attentive to and engaged with the activity, so inattentiveness to story contexts is an unlikely explanation for the LNWB responses.

As aforementioned, we manipulated trials in order to test the conditions under which an LNWB response is more likely. Item analyses across these trials were worth pursuing because all children made at least some LNWB responses and yet also avoided an LNWB response on at least some trials.

First, we found (not surprisingly) that the presence of illustrations that counter the LNWB diminishes the likelihood of an LNWB response. Whereas 66% of the children *never* gave an LNWB response when illustrations countered the error, all children gave at least one LNWB response when the illustrations were either absent or inconclusive. Despite this effect, it is curious that some children showed an LNWB even when there was clear evidence to counter it.

Second, we included trials involving an implicitly illogical comparison of tangible and nontangible items. If nontangible items were (mis)represented in some tangible way (such as illustrating "3 years of age" with a picture of a child holding up three fingers, or illustrating "5 jumps" with five distinct images of the same child jumping appeared on one page), this increased the likelihood of an LNWB response, compared to story segments without any illustrations. Surprisingly, LNWB responses were quite common whether illustrations did or did not accompany the illogical "tangible/intangible items" trials. Across trials, when children were asked "who has more," on average, 75% gave LNWB responses if there were *no* illustrations of the tangible and intangible items being compared, only slightly less than the 86% who gave LNWB responses when there were illustrations depicting the intangible items as sets. In fact, when the illustrations were present, most children (64%) consistently gave LNWB responses for all relevant trials; and without illustrations, fewer but still many (42%) children consistently gave LNWB responses. Thus, LNWB responses were common under both conditions.

Finally, the LNWB persisted even when the comparisons we asked children to make were explicitly illogical, but less so compared to logical comparisons (as described previously). The explicitly illogical condition involved conflicting labels between the referents being compared (eg, "Joe has two books and Anne has four books") and referents in the query (eg, "who has more butterflies?"). This illogical condition led many children to correctly indicate that an answer was not possible (eg, "I don't know" or "what do you mean?" or even a shrug of the shoulders). Across all instances in which children failed to correctly give an "I don't know" type of response, 74% of children's responses were consistent with the LNWB. In another variation of this illogical item, no numbers were actually uttered (eg, "Joe has books and Anne has books. Who has more butterflies?"). The average percentage of children responding correctly (eg, saying "I don't know") was only slightly higher when number words were not uttered (78%), than when number words were uttered (71%).

4.6 IMPLICATIONS OF THE LARGE NUMBER WORD BIAS

These results (together with some past work) offer evidence of an LNWB. At issue, then, is what this LNWB means. First, if the large number word bias proves to be a reliable phenomenon, it may be that researchers and others can exploit contextual effects to infer a deeper understanding of a child's number knowledge, or at least their number word knowledge. But this presumes a meaningful interpretation of the effect, and that it reflects numerical ability, skills, or knowledge; or the intersection of numerical and linguistic or metalinguistic skills and contextual sensitivity. Second, if the phenomenon is reliable and meaningful in early childhood, the LNWB may be an indicator of risk for poor mathematics outcomes. Perhaps the LNWB is an indicator of early misconceptions that may foreshadow an atypical number concept trajectory. Third, another alternative is that the LNWB does not reveal what children *know* about numbers, but only that they *respond* in a manner that is aligned with a decontextualized or an absolute understanding of number words, which may reflect responses to lexical or relational ambiguity in general. This is relevant to the notion that the LNWB reflects weak number concepts to begin with, and that a routinized response to the question, "who has more" reflects rote answers rather than reflective consideration. Finally, if the LNWB is a manifestation of contextual insensitivity to numerical interpretation in early childhood, of interest is whether differences in contextual sensitivity persist beyond early childhood and, if so, how they are manifested in older children or adults.

4.7 CONCLUSIONS AND FUTURE DIRECTIONS

In this chapter, we described why number words might be confusing to young children, even beyond the early stages of number word acquisition, and proposed that researchers can exploit this confusion to gain knowledge of children's emerging number concepts. Building upon the extensive developmental literature on the process, timing, and influences on the rate of children's number word acquisition, we proposed that children's awareness of the inherent ambiguity of some numerical statements, and children's ability to respond accordingly, may be a unique, overlooked aspect of their broader "number sense." That is, individual differences in children's ability to extract meaning from number words in contexts may reflect the nature of their emerging number concepts. How and whether this particular aspect of number sense influences mathematical learning is a topic of future studies.

There is nearly universal acceptance that number sense (broadly defined here, inclusive of basic and higher-order number skills) contributes to later mathematics achievement, but this acceptance is coupled with concern that the nature, number, and specification of number sense components are neither firmly established nor consistently measured or described. This complicates studies of the mechanisms underlying the association between early number sense and later mathematics achievement. In this chapter we have provided initial evidence that individual differences in responses to numerical ambiguity may be systematically measured by testing children's sensitivity to number words in various contexts, and that these measures may contribute one component to the quest for a comprehensive understanding of number sense. Further research is needed to develop and test such measures, and to develop models of potential causal pathways linking measures of numerical ambiguity to formal early mathematics. One possibility is that some of the misconceptions about mathematics that abound during the formal school years are preceded by early misconceptions about the meanings of number words across contexts, or by routinized procedures that lead to a literal interpretation of numbers devoid of contextual influence. Either of these may affect learning trajectories during the earliest years of schooling.

Our proposed "literal" interpretations of numbers in context are similar in nature to several examples shared earlier in this chapter—Syrett's (2013) examples of children interpreting attributive or pseudopartitive phrases with numbers as enumerating items that were not enumerated at all (eg, neither "two-pound apples," nor "two pounds of apples" enumerates apples, but children tend to interpret the phrases as referring to "two apples"), or inferring two bears were seen when presented with the sentence, "two people saw a bear." These responses to number may parallel what Davidson et al. (2012, p. 164) described as "another blind procedure for applying numerals to sets" on tests of cardinal principle knowledge. In these examples, one issue is what the numbers actually enumerate. If the target of enumeration is unclear, the meaning of the number word may be ambiguous. If the target is misunderstood, the number word meaning might be misinterpreted. The constrained or biased performance patterns observed in our assessment of the large number word bias may also reflect routinized behavior, may be associated with numerical skills per se, may reflect broader problem-solving tendencies also evident in responses to nonnumerical lexical ambiguity (Ramscar, Dye, & Klein, 2013), or may result from the intersection of numerical concepts and linguistic or executive function skills. Whether these lead to individual differences in later mathematical thinking and learning warrants further study.

References

Almoammer, A., Sullivan, J., Donlan, C., Marušič, F., Žaucer, R., O'Donnell, T., & Barner, D. (2013). Grammatical morphology as a source of early number word meanings. *Proceedings of the National Academy of Sciences, 110*(46), 18448–18453.

Barner, D., & Snedeker, J. (2005). Quantity judgments and individuation: evidence that mass nouns count. *Cognition, 97*(1), 41–66.

Barner, D., Wagner, L., & Snedeker, J. (2008). Events and the ontology of individuals: verbs as a source of individuating mass and count nouns. *Cognition, 106*(2), 805–832.

Barner, D., Libenson, A., Cheung, P., & Takasaki, M. (2009). Cross-linguistic relations between quantifiers and numerals in language acquisition: evidence from Japanese. *Journal of Experimental Child Psychology, 103*(4), 421–440.

Baroody, A. J., & Price, J. (1983). The development of the number-word sequence in the counting of three-year-olds. *Journal for Research in Mathematics Education, 14*(5), 361–368.

Baroody, A. J., Lai, M., & Mix, K. S. (2006). The development of young children's early number and operation sense and its implications for early childhood education. In B. Spodek, & O. N. Saracho (Eds.), *Handbook of research on the education of young children* (2nd ed., pp. 187–221). Mahwah, NJ: Lawrence Erlbaum Associates Publishers.

Baroody, A. J., Li, X., & Lai, M. (2008). Toddlers' spontaneous attention to number. *Mathematical Thinking and Learning, 10*(3), 240–270.

Berch, D. B. (2005). Making sense of number sense: implications for children with mathematical disabilities. *Journal of Learning Disabilities, 38*(4), 333–339.

Bloom, P., & Wynn, K. (1997). Linguistic cues in the acquisition of number words. *Journal of Child Language, 24*(3), 511–533.

Brannon, E. M. (2002). The development of ordinal numerical knowledge in infancy. *Cognition, 83*(3), 223–240.

Brannon, E. M., & Van de Walle, G. A. (2001). The development of ordinal numerical competence in young children. *Cognitive Psychology, 43*(1), 53–81.

Brooks, N., Pogue, A., & Barner, D. (2011). Piecing together numerical language: children's use of default units in early counting and quantification. *Developmental Science, 14*(1), 44–57.

Bugden, S., & Ansari, D. (2011). Individual differences in children's mathematical competence are related to the intentional but not automatic processing of Arabic numerals. *Cognition, 118*(1), 32–44.

Colomé, À., & Noël, M. (2012). One first? Acquisition of the cardinal and ordinal uses of numbers in preschoolers. *Journal of Experimental Child Psychology, 113*, 233–247.

Condry, K. F., & Spelke, E. S. (2008). The development of language and abstract concepts: the case of natural number. *Journal of Experimental Psychology: General, 137*(1), 22–38.

Davidson, K., Eng, K., & Barner, D. (2012). Does learning to count involve a semantic induction? *Cognition, 123*(1), 162–173.

Doherty, M. J. (2004). Children's difficulty in learning homonyms. *Journal of Child Language, 31*(01), 203–214.

Feigenson, L., Dehaene, S., & Spelke, E. (2004). Core systems of number. *Trends in Cognitive Sciences, 8*(7), 307–314.

Fuson, K., & Hall, J. (1983). The acquisition of early number word meanings: a conceptual analysis and review. In H. P. Ginsburg (Ed.), *The development of mathematical thinking* (pp. 47–107). New York, NY: Academic Press.

Gelman, R., & Gallistel, C. R. (1978). *The child's understanding of number*. Cambridge, MA: Harvard University Press.

Gelman, R., & Meck, E. (1983). Preschoolers' counting: Principles before skill. *Cognition, 13*(3), 343–359.

Giralt, N., & Bloom, P. (2000). How special are objects? Children's reasoning about objects, parts, and holes. *Psychological Science, 11*(6), 497–501.

Girelli, L., Lucangeli, D., & Butterworth, B. (2000). The development of automaticity in accessing number magnitude. *Journal of Experimental Child Psychology, 76*(2), 104–122.

Gunderson, E. A., & Levine, S. C. (2011). Some types of parent number talk count more than others: relations between parents' input and children's cardinal-number knowledge. *Developmental Science, 14*(5), 1021–1032.

Gunderson, E. A., Spaepen, E., & Levine, S. C. (2015). Approximate number word knowledge before the cardinal principle. *Journal of Experimental Child Psychology, 130*, 35–55.

Hannula, M. M., & Lehtinen, E. (2005). Spontaneous focusing on numerosity and mathematical skills of young children. *Learning and Instruction, 15*(3), 237–256.

Izard, V., Sann, C., Spelke, E. S., & Streri, A. (2009). Newborn infants perceive abstract numbers. *Proceedings of the National Academy of Sciences of the United States of America, 106*(25), 10382–10385.

Le Corre, M., & Carey, S. (2007). One, two, three, four, nothing more: An investigation of the conceptual sources of the verbal counting principles. *Cognition, 105*(2), 395–438.

Le Corre, M., Van de Walle, G. A., Brannon, E. M., & Carey, S. (2006). Re-visiting the performance/competence debate in the acquisition of counting principles. *Cognitive Psychology, 52*(2), 130–169.

Lyons, I. M., Ansari, D., & Beilock, S. L. (2012). Symbolic estrangement: evidence against a strong association between numerical symbols and the quantities they represent. *Journal of Experimental Psychology: General, 141*(4), 635–641.

Markman, E. M. (1989). *Categorization and naming in children: Problems of induction*. Cambridge, MA: MIT Press.

Mazzocco, M. M. (1999). Developmental changes in indicators that literal interpretations of homonyms are associated with conflict. *Journal of Child Language, 26*(02), 393–417.

Mazzocco, M. M., Myers, G. F., Thompson, L. A., & Desai, S. S. (2003). Possible explanations for children's literal interpretations of homonyms. *Journal of Child Language, 30*(4), 879–904.

Mazzocco, M. M., Feigenson, L., & Halberda, J. (2011). Impaired acuity of the approximate number system underlies mathematical learning disability (dyscalculia). *Child Development, 82*(4), 1224–1237.

Mazzocco, M. M., Chan, J. Y., & Praus, T. L. (2015). Children's judgments of numbers in context reveal emerging number concepts. Poster presented at the Biennial Meeting of the Society for Research in Child Development. Philadelphia, PA.

McCrink, K., & Wynn, K. (2004). Large-number addition and subtraction by 9-month-old infants. *Psychological Science, 15*(11), 776–781.

Mcintosh, A., Reys, B. J., & Reys, R. E. (1992). A proposed framework for examining basic number sense. *For the Learning of Mathematics, 12*(3), 2–8.

Miller, K. F., & Stigler, J. W. (1987). Counting in Chinese: cultural variation in a basic cognitive skill. *Cognitive Development, 2*(3), 279–305.

Mix, K. S. (2009). How Spencer made number: first uses of the number words. *Journal of Experimental Child Psychology, 102*(4), 427–444.

Mix, K. S., Sandhofer, C. M., & Baroody, A. J. (2005). Number words and number concepts: the interplay of verbal and nonverbal quantification in early childhood. In R. V. Kail (Ed.), *Advances in child development and behavior*. (Vol. 33, pp. 305–346). San Diego, CA: Elsevier Inc.

Mix, K. S., Sandhofer, C. M., Moore, J. A., & Russell, C. (2012). Acquisition of the cardinal word principle: The role of input. *Early Childhood Research Quarterly, 27*(2), 274–283.

Nerlich, B., & Clarke, D. D. (2003). Polysemy and flexibility: introduction and overview. In B. Nerlich, Z. Todd, V. Herman, & D. D. Clarke (Eds.), *Polysemy: Flexible patterns of meaning in mind and language* (pp. 3–30). Berlin, Germany: Walter de Gruyter GmbH & Co.

Nuñez, R. (forthcoming). How much mathematics is 'hard-wired,' if any at all: biological evolution, development, and the essential role of culture. In M. Sera, M. Maratsos, & S. Carlson (Eds.), *Culture and Developmental Systems, Minnesota Symposia on Child Psychology* (Vol. 38). Hoboken, NJ: Wiley.

Odic, D., Pietroski, P., Hunter, T., Lidz, J., & Halberda, J. (2013). Young children's understanding of "more" and discrimination of number and surface area. *Journal of Experimental Psychology: Learning, Memory, and Cognition, 39*(2), 451–461.

Piazza, M., Facoetti, A., Trussardi, A. N., Berteletti, I., Conte, S., Lucangeli, D., Dehaene, S., & Zorzi, M. (2010). Developmental trajectory of number acuity reveals a severe impairment in developmental dyscalculia. *Cognition, 116*, 33–41.

Purpura, D. J., & Lonigan, C. J. (2013). Informal numeracy skills: the structure and relations among numbering, relations, and arithmetic operations in preschool. *American Educational Research Journal, 50*(1), 178–209.

Ramscar, M., Dye, M., Popick, H. M., & O'Donnell-McCarthy, F. (2011). The enigma of number: why children find the meanings of even small number words hard to learn and how we can help them do better. *PLoS One, 6*(7), e22501.

Ramscar, M., Dye, M., & Klein, J. (2013). Children value informativity over logic in word learning. *Psychological Science, 24*(6), 1017–1023.

Sarnecka, B. W., & Carey, S. (2008). How counting represents number: what children must learn and when they learn it. *Cognition, 108*(3), 662–674.

Sarnecka, B. W., & Gelman, S. A. (2004). Six does not just mean a lot: preschoolers see number words as specific. *Cognition, 92*(3), 329–352.

Sarnecka, B. W., & Wright, C. E. (2013). The idea of an exact number: children's understanding of cardinality and equinumerosity. *Cognitive Science, 37*(8), 1493–1506.

Sera, M., & Smith, L. B. (1987). Big and little: "Nominal" and relative uses. *Cognitive Development, 2*(2), 89–111.

Shipley, E. F., & Shepperson, B. (1990). Countable entities: developmental changes. *Cognition, 34*(2), 109–136.

Slusser, E. B., & Sarnecka, B. W. (2011). Find the picture of eight turtles: a link between children's counting and their knowledge of number word semantics. *Journal of Experimental Child Psychology, 110*(1), 38–51.

Slusser, E., Ditta, A., & Sarnecka, B. (2013). Connecting numbers to discrete quantification: a step in the child's construction of integer concepts. *Cognition, 129*(1), 31–41.

Smith, L. B., & Sera, M. D. (1992). A developmental analysis of the polar structure of dimensions. *Cognitive Psychology, 24*(1), 99–142.

Sullivan, J., & Barner, D. (2013). How are number words mapped to approximate magnitudes? *The Quarterly Journal of Experimental Psychology, 66*(2), 389–402.

Syrett, K. (2013). The role of cardinality in the interpretation of measurement expressions. *Language Acquisition, 20*(3), 228–240.

Syrett, K., & Musolino, J. (2013). Collectivity, distributivity, and the interpretation of plural numerical expressions in child and adult language. *Language Acquisition, 20*(4), 259–291.

Syrett, K., Musolino, J., & Gelman, R. (2012). How can syntax support number word acquisition? *Language Learning and Development, 8*(2), 146–176.

Wagner, J. B., & Johnson, S. C. (2011). An association between understanding cardinality and analog magnitude representations in preschoolers. *Cognition, 119*(1), 10–22.

Wasner, M., Moeller, K., Fischer, M. H., & Nuerk, H. C. (2015). Related but not the same: ordinality, cardinality and 1-to-1 correspondence in finger-based numerical representations. *Journal of Cognitive Psychology, 27*(4), 426–441.

Woodward, A. L., Markman, E. M., & Fitzsimmons, C. M. (1994). Rapid word learning in 13- and 18-month-olds. *Developmental Psychology, 30*(4), 553–566.

Wynn, K. (1990). Children's understanding of counting. *Cognition, 36*(2), 155–193.

Wynn, K. (1992a). Addition and subtraction by human infants. *Nature, 358*(6389), 749–750.

Wynn, K. (1992b). Children's acquisition of the number words and the counting system. *Cognitive Psychology, 24*(2), 220–251.

Xu, F., Spelke, E. S., & Goddard, S. (2005). Number sense in human infants. *Developmental Science, 8*(1), 88–101.

CHAPTER

5

Learning, Aging, and the Number Brain

Marinella Cappelletti

Department of Psychology, Goldsmiths College University of London, London, United Kingdom; University College London, Institute of Cognitive Neuroscience, London, United Kingdom

OUTLINE

Learning is the acquisition of a new skill or the improvement of an existing one, a process which is typically accelerated during childhood and that continues throughout life (Park & Reuter-Lorenz, 2009). A set of skills of vital importance for the quality of life of individuals and of society is maths, the subject of substantial learning at school and in everyday life. This is because mastering mathematical concepts is linked to higher standard of living, of employability and health. Indeed people with low numeracy skills tend to earn less and spend less; they are also more likely to be sick, to be in trouble with the law, and need more help at school (Butterworth, Varma, & Laurillard, 2011; Parsons & Bynner, 1997, 2005).

Continuous Issues in Numerical Cognition. http://dx.doi.org/10.1016/B978-0-12-801637-4.00005-6

Learning maths is not just important per se, but the learning process itself also offers a window into the understanding of how the brain changes in the course of acquiring or improving a cognitive skill in general. This information is particularly critical for characterizing the fast changes that occur in the aging brain. This chapter will first describe the main features of the number system across the lifetime, and it will focus on learning in the context of numerical and arithmetical abilities.

5.1 NUMBER SYSTEM ACROSS THE LIFESPAN

5.1.1 Number System in Development and Young Adulthood

Whether learning or improving mathematical concepts is possible in healthy aging and what exactly can be learned are yet open questions. This is because, until relatively recently, the question of how numerical abilities and mathematical concepts change with age was largely unexplored. Certainly more attention has been given to studying these abilities and concepts in developing populations or in aging participants but with neurological disorders following brain lesions (Cappelletti & Cipolotti, 2011; Libertus & Brannon, 2009). Collectively, these previous studies contributed to show that humans are equipped with a system defined as the approximate number system (ANS) (Feigenson, Dehaene, & Spelke, 2004; Halberda, Mazzocco, & Feigenson, 2008). This system is thought to be based on encoding numerosities as analog magnitudes (eg, Izard, Dehaene-Lambertz, & Dehaene, 2008; Stoianov & Zorzi, 2012), and it is often measured in terms of "number acuity," or the ability to discriminate numerosities (eg, which set has more elements; eg, Halberda et al., 2008). Number acuity is typically expressed as Weber fraction (*wf*), reflecting the amount of noise in the underlying approximate number representation (Halberda et al., 2008; Halberda, Lya, Wilmerb, Naimana, & Germine, 2012; Piazza, Izard, Pinel, Le Bihan, & Dehaene, 2004). The precision of the ANS, namely the extent to which the numerosity of a set of items can be discriminated from the numerosity of another set, is highly variable across individuals and refines gradually from infancy to adulthood (Halberda et al., 2008, 2012; Halberda & Feigenson, 2008; Libertus & Brannon, 2009; Lipton & Spelke, 2003; Piazza et al., 2004, 2010; Wynn, 1998; Xu, Spelke, & Goddard, 2005). Some recent research aiming to characterize the features of the ANS suggested the importance of this system as a cognitive foundation for higher-level mathematical abilities (De Wind & Brannon, 2012; Halberda et al., 2008, 2012; Lyons & Beilock, 2011; but see Butterworth, 2010 and Carey, 2001 for a different opinion), and indicated that congenital impairments in the understanding of mathematical concepts tend to correlate with poor ANS (Mazzocco, Feigenson, & Halberda, 2011; Piazza et al., 2010).

Besides the ANS, humans also possess an exact, symbolic system for calculation that is typically learned at school and is used in everyday life for essential activities such as banking and shopping. Processing non-symbolic and symbolic magnitudes relies on the prefrontal cortex (PFC) and especially the parietal cortices and the intraparietal sulcus (IPS), which some researchers suggest hosting a number module or number sense (Dehaene, Piazza, Pinel, & Cohen, 2003; Feigenson et al., 2004; Fias, Lammertyn, Reynvoet, Dupont, & Orban, 2003; Venkatraman, Ansari, & Chee, 2005). Other brain regions besides the parietal and the frontal ones are important for numerical processing, at least in the developing brain. For instance, a recent study monitored the development of arithmetic strategies and their neuronal underpinning in 7 to 9-year-old children (Qin et al., 2014). It was found that the transition across strategies, specifically from procedure-based to memory-based problem solving, corresponded to increased hippocampal and decreased prefrontal–parietal engagement. Moreover, increased hippocampal–neocortical functional connectivity predicted longitudinal improvements in retrieval-strategy use (Qin et al., 2014).

5.1.2 Number System in Aging

Does our ability to use numbers and mathematical concepts change with aging? Are these changes specific to math or do they rather reflect decline of more general cognitive skills such as attention or executive functions? Little is known about the impact of healthy aging on numerical skills, and the few initial studies that investigated this issue focused mostly on arithmetical abilities, that is, the cognitive processes required when solving problems such as 8×9 or $243 + 39$ (eg, Duverne & Lemaire, 2005; Geary & Lin, 1998; Salthouse & Kersten, 1993). These studies concurred to show that older people are as good as, or even better, than younger people in solving simple arithmetical operations such as "$8 \times 4 = ?$" (Allen, Smith, Jerge, & Vires-Collins, 1997; Duverne, Lemaire, & Michel, 2003; Geary, Frensch, & Wiley, 1993; Geary & Whiley, 1991). They also indicated that older participants may have a similar repertoire of strategies to solve simple arithmetical problems and they may even be able to learn new ones (Allen et al., 1997; Geary & Lin, 1998; Geary & Wiley, 1991; Geary et al., 1993). However, older participants show differences in strategy distribution and in the efficiency used to select among them compared to younger participants (eg, Duverne & Lemaire, 2005). For example, both younger and older adults have a multistrategy approach when solving multidigit arithmetical operations, although older participants tend to use fewer strategies, and are also generally slower in their mathematical performance (Geary & Wiley, 1991; Geary et al., 1993; Lemaire & Lecacheur, 2001; Lemaire, Arnaud, & Lecacheur, 2004; Siegler & Lemaire, 1997).

Besides characterizing age-related effects in the use of arithmetical strategies at the behavioral level, some studies also explored the corresponding brain electrical activity (El Yagoubi, Lemaire, & Besson, 2005). In one of these studies, older and younger participants were asked to verify the result of two-digit addition problems. Event-related potential (ERP) data showed that arithmetical performance corresponded to a hemispheric asymmetry in younger but not in older adults; namely, the left hemisphere advantage in younger adults was reduced in older adults (El Yagoubi et al., 2005).

Some studies have highlighted the more general issue in the literature of cognitive aging about the sources of age-related decline in cognition. Some general mechanisms have been proposed, for instance, reduced processing speed, working memory capacities, or inhibitory abilities (eg, Craik & Salthouse, 2000; Salthouse, 1991, 2004; Hasher, Lustig, & Zacks, 2007). This would mean that age-related differences in performing a cognitive task occur because these abilities are generally impoverished in older adults. This view is supported by the observation that changes in the use of strategies are not specific for arithmetic, but that instead age-related differences in strategy repertoire, distribution, and selection have also been observed in other cognitive domains such as memory (Dunlosky & Hertzog, 1998, 2000, 2001; Rogers, Hertzog, & Fisk, 2000), sentence verification (Reder, Wible, & Martin, 1986), as well as decision-making and problem-solving tasks (Hartley, 1986; Johnson, 1990). An alternative view suggests that age-related differences in cognition may be related to the decline of a specific set of concepts or abilities, for instance, mathematical ones. This decline may be reflected in age differences in the use of strategies to accomplish a specific cognitive task, for example, math, but without necessarily generalizing to other cognitive tasks.

Trying to disentangle the cause of cognitive decline as reflecting general mechanisms or instead domain-specific processes may be difficult when employing mathematical tasks. This is because these tasks are typically multicomponential, requiring several skills such as the retrieval of simple arithmetic facts, the correct use of procedures, and the monitoring of the steps of an arithmetic problem (Cappelletti & Cipolotti, 2011). It may therefore be difficult to isolate which specific component may be affected by aging, and whether an impairment is specific to that component or instead reflects more general age-related decline.

A different approach to test the impact of aging on numeracy is to assess other simpler skills (sometimes referred to as "biologically primary skills," Geary & Lin, 1998), which are thought to be foundational to more complex education- and language-based numerical and arithmetic abilities. As previously mentioned, one such foundational skill is our capacity to represent an approximate number, which relies on the approximate number system (ANS, Feigenson et al., 2004). The ANS refines progressively from infancy to adulthood (Lipton & Spelke, 2003; Halberda et al., 2008, 2012; Piazza et al., 2010), but whether it continues to improve with age is an open question. Longer exposure to numbers may refine the ANS further, similarly to what happens to other cognitive abilities like vocabulary and semantic memory (eg, Hedden & Gabrieli, 2004). Alternatively, the ANS may decline similar to other cognitive domains like memory or inhibitory skills, and this decline may also impact on mathematic abilities.

A few previous studies have focused on how healthy aging participants are able to represent approximate large numerosities (ie, more than 10 elements), which has usually been reported to be well maintained (eg, Gandini, Lemaire, & Dufau, 2008; Gandini, Lemaire, & Michel, 2009; Lemaire & Lecacheur, 2007; Trick, Enns, & Brodeur, 1996; Watson, Maylor, & Manson, 2002). Some of these studies have specifically focused on the strategies used to accomplish approximate quantification tasks (eg, Crites, 1992; Gandini et al., 2008; Luwel, Verschaffel, Onghena, & De Corte, 2003; Luwel, Lemaire, & Verschaffel, 2005; Siegel, Goldsmiths, & Madson, 1982). For instance, Gandini et al. (2008) found that younger and older adults used six different strategies to estimate arrays of items (specifically: benchmark, anchoring, decomposition/recomposition, approximate counting, exact counting). Although these strategies were associated with different levels of frequency, performance, and eye movement patterns, younger and older adults similarly preferred the benchmark and anchoring strategies. The benchmark strategy mainly relies on the retrieval of a numerical representation, whereas the anchoring strategy includes multistep arithmetic processes, such as counting (one by one or in groups) and addition.

Other studies have tried to characterize the neuronal correlates of these different strategies used for quantification. For instance, Gandini et al. (2008) showed that relative to younger people, older participants activated the dorsolateral prefrontal cortices during both the anchoring and the benchmarking strategies. They also activated the parietal and temporal regions more strongly during benchmark execution as compared to anchoring execution. Instead, relative to benchmark execution, anchoring-specific increases in activation were only observed in older adults in the right putamen and the right lingual gyrus (Gandini et al., 2008).

Some of these previous results do not always offer a comprehensive account of approximate abilities in elderly. This is because in some cases the focus was mainly on the strategies used to perform the numerosity task, rather than in terms of finer psychophysical measures of older participants' performance like the *wf* (eg, Gandini et al., 2008, 2009; Lemaire & Lecacheur, 2007; Watson, Maylor, & Bruce, 2005a, 2005b, 2007). Likewise, the experimental designs in other studies did not control for continuous variables that inevitably vary when varying the numerosity of the display, like the total area covered by the dot stimuli, that is, the cumulative area (eg, Piazza et al., 2004). If these continuous variables are not taken into account, for example, if in all trials an increase in numerosity always corresponds to an increase in cumulative area (eg, Gandini et al., 2008), it is unclear whether participants judged changes in numerosity or in these continuous variables (Gebuis & Reynvoet, 2012).

In contrast to studies suggesting well-maintained numeracy skills in aging, a recent Internet-based study showed that the ability to discriminate numerosities (as indexed by the *wf*) may actually be sensitive to aging (Halberda et al., 2012). This age-related deterioration may reflect either the decline of the ANS itself or of more peripheral cognitive processes, such as executive functions, which are involved in discriminating numerosities. The role of working memory in numerosity estimation has been explicitly hypothesized by Piazza et al. (2004), who suggested that in numerosity discrimination, an initial parietal lobe–based abstraction of the numerosity of a set is followed by an estimation of the numerosity itself, whereby items would be continuously accumulated in working memory and then compared, a process of active maintenance of information known to rely on prefrontal cortices (Danker & Anderson, 2007; Carpenter, Just, & Reichle, 2000).

Likewise, inhibitory abilities have been suggested to play a critical role in numerosity discrimination since they are needed to ignore task-irrelevant information, for instance represented by fewer but large-size dots (Szűcs, Nobes, Devine, Gabriel, & Gebuis, 2013). This is the case of "incongruent" numerosity trials where some continuous variables change orthogonally to numerosity, like when sets of fewer large-size dot stimuli have a bigger cumulative area than sets of smaller-size but more numerous dot stimuli. In this case, the task-irrelevant but salient information about cumulative area (Hurewitz, Gelman, & Schnitzer, 2006) has to be ignored in order to correctly discriminate numerosity. Since working memory and inhibitory processes—among others—tend to decline with age (Hedden & Gabrieli, 2004; Grady, 2012; Nyberg, Lövdén, Riklund, Lindenberger, & Bäckman, 2012; Salthouse, Atkinson, & Berish, 2003; "Inhibitory Deficit" Theory, Hasher & Zacks, 1988; Hasher, Zacks, & May, 1999, 2007; May & Hasher, 1998; Kane & Engle, 2003), they may in turn affect performance in numerosity discrimination tasks.

Following this logic, a recent study in older and younger participants explored the precision of their number acuity using a simple parametrically modulated numerosity discrimination task (Cappelletti, Didino, Stoianov, & Zorzi, 2014). Besides systematically measuring number acuity, this study also aimed to determine what underlies patterns of spared or impaired numerosity processing, namely whether any changes may be attributed to general decline, or instead to processes more specific to numerosity. It was reasoned that if number acuity did not differ between older and young participants, this may be suggestive of maintained ANS. In contrast, age-related differences in number acuity may reflect impairments specific to the number system, or, alternatively, decline of more general cognitive processes. To distinguish between these two possibilities, all participants were first examined with a set of established neuropsychological measures investigating arithmetical abilities. Since some theories suggest that these abilities are linked to the ANS (Halberda et al., 2008), they may therefore be impaired if the ANS is impaired. The integrity of older participants' inhibitory abilities was also assessed using dedicated and well-established tasks; this is because these abilities might contribute to any age-related difference in numeracy.

It was found that older participants had a significantly larger *wf* relative to young participants, in line with the results of Halberda et al. (2012)' Internet study, and which may indicate impaired numerosity processing. However, a more detailed analysis of older participants' performance showed that this larger *wf* reflected a specific inability to process trials in which numerosity was incongruent with other measures of continuous quantity (eg, where the least numerous set contained dot stimuli of larger size). Instead, congruent trials (eg, those where the more numerous set had a larger cumulative area) were performed equally accurately in younger and older participants. This impairment in incongruent trials did not correspond to difficulties in other numerical or arithmetic problems that were exceptionally well maintained in older participants. Instead, it was suggested that aging participants may have used a numerosity or quantity-based strategy to perform the numerosity task, and their larger *wf* was explained by impoverished inhibitory processes. This hypothesis was supported by the observation that elderlies' inhibitory abilities were significantly worse than younger participants, and that they correlated with performance in the incongruent numerosity trials. Therefore, the more difficult it was to inhibit task-irrelevant information the worse was the performance in the incongruent numerosity trials. The hypothesis that performance in incongruent numerosity trials may depend on the integrity of inhibitory processes was further tested using an established computational model (Stoianov & Zorzi, 2012). This showed that a specific degradation of the inhibitory components of the numerosity model resulted in a significantly larger impairment in processing incongruent numerosity trials, which resembled the older participants' performance (Cappelletti et al., 2014).

The results of aging participants' maintained performance in congruent numerosity trials but impaired in incongruent ones reinforces the idea that there are two fundamental processes intrinsic to numerosity discrimination. One is the abstraction of numerosity in a set and the other is the inhibition of task-irrelevant information (Gebuis & Reynvoet, 2012; Piazza et al., 2004; Szűcs et al., 2013). The results of Cappelletti et al. (2014), and more recently of Norris, McGeown, Guerrini, & Castronovo (2015), suggest that relative to younger participants the abstraction of numerosity can be maintained in aging. In contrast, impoverished inhibitory processes in aging participants seemed to affect performance in the incongruent numerosity trials. Such distinction between abstraction and inhibitory processes in numerosity discrimination is likely to be undetected in younger populations since their inhibitory abilities are likely to be well preserved. Likewise, this distinction may not be obvious even in elderly people if the experimental design does not allow distinguishing between congruent and incongruent trials, or if performance measures do not allow finer analyses.

Another recent study extended the investigation of nonsymbolic quantity processing in aging. This was done by examining whether the ability to process the continuous dimensions of time and space may be maintained in healthy aging (Lambrechts, Karolis, Garcia, Obende, & Cappelletti, 2013). Time and space processing were assessed in a two-choice task whereby participants had to indicate which of two horizontal lines was longer in either duration or spatial extension in different blocks. Numerosity discrimination and mathematical abilities were also tested, as well as the integrity of older participants' memory, attention, and executive processes, which might reflect or contribute to any age-related difference in quantity processing. It was found that although elderly participants showed typical age-related decline in memory, attention, and executive functions, time and space discrimination was processed equally accurately in young and older people. It was argued that the resilience of quantity processing in aging may reflect the stability of primitive resources dedicated to this type of processing. Continuous quantity processing seems therefore resilient to aging similar to other types of quantity such as numerosity and simple math as well as semantic memory, and vocabulary (Hedden & Gabrieli, 2004; Lambrechts et al., 2013).

Collectively, the observation that quantification skills (both nonsymbolic skills like numerosity, time, and space, and symbolic skills like numeric and arithmetic skills) are relatively resilient to normal aging may be explained by the suggestion that they rely on a primitive and ANS (Cappelletti et al., 2014; Hasher & Zacks, 1979). This is consistent with the view that primitive systems tend to be more robust to aging (Trick et al., 1996; Lemaire & Lecacheur, 2007) because thy are acquired earlier in life. This may put them in a stronger position relative to more complex and recently acquired skills (Trick et al., 1996). It is also possible that the quantity system relies on brain areas that are naturally less or later affected by aging (Nyberg et al., 2012).

Whether being more resilient to aging may also mean that numerical abilities are also malleable to learning and improving is an open question. This is important because it provides information on more dynamic features of the number system. It can also clarify the extent to which learning or improving numeracy skills may rely on cognitive processes that are peripheral to numeracy such as inhibitory abilities. The next section illustrates some recent behavioral, neuroimaging, and brain stimulation studies that explored learning in the number brain across the lifespan.

5.2 LEARNING AND THE QUANTITY SYSTEM

5.2.1 Training the Arithmetic Abilities in the Developing and Adult Brain

Several training studies have been successful at improving children's arithmetic abilities (Beirne-Smith, 1991; Fuchs et al., 2008, Fuchs, Fuchs, & Compton, 2012; Johnson & Bailey, 1974 Rittle-Johnson & Koedinger, 2009) or other abilities supporting arithmetic (Bergman-Nutley & Klingberg, 2014). A recent study added further information by trying to characterize which behavioral and neural factors predict individual differences in arithmetic skill acquisition (Supekar et al., 2013). In this study, twenty-four 8- to 9-year-old children underwent structural and resting-state functional MRI scans pretutoring, followed by 8 weeks of one-to-one math tutoring. A significant shift in arithmetic problem-solving strategies from counting to fact retrieval was reflected in an increase of speed and accuracy in arithmetic problem solving following tutoring. Not surprisingly, some children improved significantly more than others, although no behavioral measures, including intelligence quotient, working memory, or mathematic abilities, predicted these individual differences. Instead, performance improvements were predicted by pretutoring hippocampal volume as well as intrinsic functional connectivity of the hippocampus with dorsolateral and ventrolateral prefrontal cortices and the basal ganglia (Supekar et al., 2013).

Similar posttraining behavioral improvements have been observed in young adults. A number of recent studies showed that following extensive training of simple or complex arithmetic problems, accuracy and speed improve in trained problems relative to untrained ones (Delazer et al., 2003; Grabner, Ansari, Koschutnig, Reishofer, Ebner, & Neuper, 2009; Ischebeck et al., 2006, Ischebeck, Zamarian, Egger, Schocke, & Delazer, 2007, Ischebeck, Zamarian, Schocke, & Delazer, 2009; Núñez-Peña, 2008; Pauli et al., 1994; Zamarian, Ischebeck, & Delazer, 2009). These studies also successfully identified the neuronal correlates of learning, and indicated the parietal and frontal regions as those most strongly reflecting training-related changes. For instance, Delazer et al. (2003) observed greater activation in the left intraparietal sulcus and in the left inferior frontal regions when participants performed untrained relative to trained complex multiplication problems. In contrast, trained relative to untrained problems resulted in stronger activation in the left angular gyrus. This was interpreted as suggesting a training-related shift in activation from left intraparietal regions—more strongly involved in magnitude processing—to the left angular gyrus, considered important for retrieving information coded in verbal format (Delazer et al., 2003). The angular gyrus has been interpreted as having a domain-general, rather than a domain-specific role in arithmetical processing. This is because activation increases in this brain area are not specific to training-related arithmetic improvement, and they have been interpreted as indicating more general learning-based processes (Grabner et al., 2009; Wu et al., 2009).

Further studies examined the extent to which training-related effects are specific for the type of arithmetical problem tested, namely whether comparable effects can be observed when looking at subtraction or multiplication problems (Ischebeck et al., 2006). It was found that although the left angular gyrus showed an effect of training when multiplication problems were used, no such effect was present when the training was based on subtraction problems (Ischebeck et al., 2006).

Another study explored whether training-related changes may depend on the type of method used to train participants. Specifically "training by drill," whereby participants learned the result of a two-operant problem, was compared with "training by strategy," in which participants learned to apply an instructed algorithm (Delazer et al., 2005). Solving arithmetic problems using the training by drill strategy corresponded to stronger activation of the angular gyrus relative to solving the problems using the strategy algorithm. Thus, the type of training method and the specific arithmetical operation trained can dynamically modulate the relative activation of the temporoparietal regions during mathematic processing (Delazer et al., 2003; Zamarian et al., 2009).

5.2.2 Training the Number System in the Young and Aging Brain Also Coupled With Brain Stimulation

A few recent studies examined whether the ANS can be further refined with training (De Wind & Brannon, 2012; Park & Brannon, 2013). In one such study, young adults received an intense four-session training of a numerosity discrimination task (ie, identifying how many items are in a set or which set has more items). They improved significantly following the first training session, and remained stable during the rest of the training (De Wind & Brannon, 2012). Another study, based on training nonsymbolic arithmetic in young adults specifically investigated the extent to which this improvement may generalize to symbolic arithmetic (Park & Brannon, 2013). It was found that participants improved in performing nonsymbolic arithmetic, and this transferred to symbolic ones. This is an important result for future intervention programs because children with low numerical competence may be trained with nonsymbolic number knowledge in order to improve their understanding of symbolic math (Park & Brannon, 2013).

A few recent investigations have explored whether numerical abilities can be boosted with training associated with brain stimulation. This is a safe and noninvasive method known to facilitate brain plasticity (Fertonani, Pirulli, & Miniussi, 2011; Terney, Chaieb, Moliadze, Antal, & Paulus, 2008). One such study reported a remarkable learning effect in eight young adults. Cohen Kadosh, Soskic, Iuculano, Kanai, and Walsh (2010) trained participants with artificial symbols indicating numerical information for 6 days while they received either sham or anodal transcranial direct current (tDC) stimulation to the right parietal lobe and cathodal tDC stimulation to the left parietal lobe. Participants receiving stimulation showed a pattern of performance consistent with numerate adults, while those who received sham stimulation or stimulation in the opposite configuration (right parietal cathodal and left parietal anodal) showed a performance that resembled that of people with more elementary numerical skills. The effect of numerical learning combined with stimulation lasted about 6 months, similar to what has been shown in other domains like motor skills (Reis et al., 2009) or planning (Dockery, Hueckel-Weng, Birbaumer, & Plewnia, 2009). These results of numerical learning have been interpreted as indicating that tDCS could lead to faster mastery and maintenance of learned material (Cohen Kadosh et al., 2010).

Another recent study investigated the extent to which number acuity can be boosted with training combined with brain stimulation in the form of transcranial random noise stimulation, tRNS (Terney et al., 2008). Forty young participants were extensively trained for 5 consecutive days with an established numerosity discrimination task (see Halberda et al., 2008) while receiving real or sham tRN stimulation to the parietal lobes bilaterally. All participants significantly improved relative to pretraining; the improvement was larger in people who received parietal stimulation associated with training relative to training only (sham), stimulation only, or training associated with stimulation to brain areas not involved in numerosity discrimination,that is, the bilateral motor regions (Cappelletti et al., 2013). Moreover, a significant transfer was observed to cognitive abilities that are thought to be cognitively and anatomically linked to numerosity, namely space and time discrimination (Bueti & Walsh, 2009; Cantlon, Platt, & Brannon, 2009; Walsh, 2003). Indeed, participants receiving parietal tRNS and numerosity training improved also in their ability to discriminate space and time, but not other more general abilities like attention and memory (Cappelletti et al., 2013). This effect was not observed in the other stimulation conditions, which was interpreted as suggesting that parietal stimulation coupled with training is likely to have boosted a magnitude system common to number, time, and space, such that improvement in one of these magnitude dimensions extended to the others (Cappelletti et al., 2013).

A further study based on the same experimental and stimulation design was used in healthy aging participants (Cappelletti, Pikkat, Upstill, Speekenbrink, & Walsh, 2015). It aimed at examining whether learning may reflect distinct strategies, which may in turn differ across the age spectrum. In particular, we asked whether learning in young (ages 19–35) and older people (ages 60–73) may reflect enhanced ability to integrate information required to perform a cognitive task and more specifically numerosity discrimination, or whether it may instead reflect the ability to inhibit task-irrelevant information for successful task performance. This could be tested because the numerosity discrimination task allows distinguishing between cue-integration and inhibitory abilities. Cue integration is typically required to process congruent numerosity trials in which performance benefits from integrating converging information from numerosity and the cumulative area covered by the items. Instead, in this task inhibitory abilities are required to process incongruent numerosity trials because information about the stimuli size needs to be ignored.[a] It was found that both age groups who received training associated with parietal stimulation became more skilled at the numerosity task following 5 days of training, but subsequent analyses showed that they improved for different reasons. Younger people improved by becoming better at integrating different types of numerical information, but older individuals improved by suppressing less important information. These results indicate that successful learning can depend on different, age-dependent cognitive processes, which reflect fundamentally distinct ways in which the brain works at different ages (Cappelletti et al., 2015).

This chapter presented the main behavioral and neuronal features characterizing the developing and the aging number brain. It also discussed recent studies investigating whether numerical abilities can be boosted with training, also associated with brain stimulation in young and aging participants. These studies successfully demonstrated that nonsymbolic and arithmetical abilities can improve following training, especially if combined with brain stimulation. Together the evidence presented suggests that the number system is strong, generally resilient to aging, and sufficiently malleable to allow further improvement.

[a]Note that the distinction between cue integration and inhibition of irrelevant information for congruent and incongruent trials, respectively, may depend on the task used. For instance, in the Stroop task whereby participants ignore word meaning and attend only to the ink color, there are indications that conflictual information, which typically relies on inhibitory abilities, affects performance even in the congruent condition (eg, Kalanthroff, Goldfarb, Henik, 2013).

Acknowledgments

This work was supported by Royal Society Dorothy Hodgkin Fellowship and British Academy research grants.

References

Allen, P. A., Smith, A. F., Jerge, K. A., & Vires-Collins, H. (1997). Age differences in mental multiplication: evidence for peripheral but not central decrements. *Journal of Gerontology: Psychological Sciences, 52b*, 81–90.

Beirne-Smith, M. (1991). Peer tutoring in arithmetic for children with learning disabilities. *Except Child, 57*, 330–337.

Bergman-Nutley, S., & Klingberg, T. (2014). Effect of working memory training on working memory, arithmetic and following instructions, *Psychological Research, 78*(6), 869–877.

Bueti, D., & Walsh, V. (2009). The parietal cortex and the representation of time, space, number and other magnitudes. *Philosophical Transactions of the Royal Society B: Biological Sciences, 364*, 1831–1840.

Butterworth, B. (2010). Foundational numerical capacities and the origins of dyscalculia. *Trends in Cognitive Sciences, 14*, 534–541.

Butterworth, B., Varma, S., & Laurillard, D. (2011). Dyscalculia: from brain to education. *Science, 332*, 1049–1053.

Cantlon, J. F., Platt, M. L., & Brannon, E. M. (2009). Beyond the number domain. *Trends in Cognitive Science, 13*, 83–91.

Cappelletti, M., & Cipolotti, L. (2011). The neuropsychological assessment and treatment of calculation disorders. In K Halligan, U Kischka, & J Marshall (Eds.), *Handbook of Clinical Neuropsychology*. Oxford: Oxford University Press.

Cappelletti, M., Gessaroli, E., Hithersay, R., Mitolo, M., Didino, D., Kanai, R., Cohen Kadosh, R., & Walsh, V (2013). Neuroenhancement: greatest, long-term and transferable quantity judgment induced by brain stimulation combined with cognitive training. *The Journal of Neuroscience, 33*, 14899–14907.

Cappelletti, M., Didino, D., Stoianov, I., & Zorzi, M. (2014). Number skills are maintained in healthy aging. *Cognitive Psychology, 69C*, 25–45.

Cappelletti, M., Pikkat, H., Upstill, E., Speekenbrink, M., & Walsh, V. (2015). Learning to integrate vs. inhibiting information is modulated by age. *The Journal of Neuroscience, 35*, 2213–2225.

Carey, S. (2001). Cognitive foundations of arithmetic: evolution and ontogenesis. *Mind & Language, 16*, 37–55.

Carpenter, P., Just, M. A., & Reichle, E. D. (2000). Working memory and executive function: evidence from neuroimaging. *Current Opinion in Neurobiology, 10*, 195–199.

Cohen Kadosh, R., Soskic, S., Iuculano, T., Kanai, R., & Walsh, V. (2010). Modulating neuronal activity produces specific and long-lasting changes in numerical competence. *Current Biology, 20*, 2016–2020.

Craik, F. I. M., & Salthouse, T. A. (2000). *The handbook of aging and cognition* (2nd ed.). Mahwah, NJ: Erlbaum.

Crites, T. (1992). Skilled and less skilled estimators' strategies for estimating discrete quantities. *The Elementary School Journal, 92*, 601–619.

Danker, J., & Anderson, F., Jr. (2007). The roles of prefrontal and posterior parietal cortex in algebra problem solving: a case of using cognitive modeling to inform neuroimaging data. *NeuroImage, 35*, 1365–1377.

De Wind, N. K., & Brannon, E. M. (2012). Malleability of the approximate number system: effects of feedback and training. *Frontiers in Human Neuroscience, 6*, 68.

Dehaene, S., Piazza, M., Pinel, P., & Cohen, L. (2003). Three parietal circuits for number processing. *Cognitive Neuropsychology, 20*, 487–506.

Delazer, M., Domahs, F., Bartha, L., Brenneis, C., Lochy, A., Trieb, T., & Benke, T. (2003). Learning complex arithmetic—an fMRI study. *Brain Research Cognitive Brain Research, 18*, 76–88.

Delazer, M., Ischebeck, A., Domahs, F., Zamarian, L., Koppelstaetter, F., Siedentopf, C. M., Kaufmann, L., Benke, T., & Felber, S. (2005). Learning by strategies and learning by drill—evidence from an fMRI study. *NeuroImage, 25*, 838–849.

Dockery, C. A., Hueckel-Weng, R., Birbaumer, N., & Plewnia, C. (2009). Enhancement of planning ability by transcranial direct current stimulation. *Journal of Neuroscience, 29*, 7271–7277.

Dunlosky, J., & Hertzog, C. (1998). Aging and deficits in associative memory: what is the role of strategy production? *Psychology and Aging, 13*, 597–607.

Dunlosky, J., & Hertzog, C. (2000). Updating knowledge about encoding strategies: a componential analysis of learning about strategy effectiveness from task experience. *Psychology and Aging, 15*, 462–474.

Dunlosky, J., & Hertzog, C. (2001). Measuring strategy production during associative learning: the relative utility of concurrent versus retrospective reports. *Memory and Cognition, 29*, 247–253.

Duverne, S., & Lemaire, P. (2005). Arithmetic split effects reflect strategy selection: an adult age comparative study in addition verification and comparison tasks. *Canadian Journal of Experimental Psychology, 59*, 262–278.

Duverne, S., Lemaire, P., & Michel, B. F. (2003). Alzheimer's disease disrupts arithmetic facts retrieval processes but not arithmetic strategy selection. *Brain and Cognition, 52*, 302–318.

El Yagoubi, R., Lemaire, P., & Besson, M. (2005). Effects of aging on arithmetic problem solving: an event-related brain potential study. *Journal of Cognitive Neuroscience, 17*, 37–50.

Feigenson, L., Dehaene, S., & Spelke, E. (2004). Core systems of number. *Trends in Cognitive Science, 8*, 307–314.

Fertonani, A., Pirulli, C., & Miniussi, C. (2011). Random noise stimulation improves neuroplasticity in perceptual learning. *The Journal of Neuroscience, 31*, 15416–15423.

Fias, W., Lammertyn, J., Reynvoet, B., Dupont, P., & Orban, G. A. (2003). Parietal representation of symbolic and nonsymbolic magnitude. *Journal of Cognitive Neuroscience, 15*, 47–56.

Fuchs, L. S., et al. (2008). Remediating computational deficits at third grade: a randomized field trial. *Journal of Research on Educational Effectiveness, 1*, 2–32.

Fuchs, L. S., Fuchs, D., & Compton, D. L. (2012). The early prevention of mathematics difficulty: its power and limitations. *Journal of Learning Disabilities, 45*, 257–269.

Gandini, D., Lemaire, P., & Dufau, S. (2008). Older and young adults' strategies in approximative quantification. *Acta Psychologica, 129*, 175–189.

Gandini, D., Lemaire, P., & Michel, B. F. (2009). Approximate quantification in young, healthy older adults, and Alzheimer patients. *Brain and Cognition, 70*, 53–61.

Geary, D. C., & Lin, J. (1998). Numerical cognition: age-related differences in the speed of executing biologically primary and biologically secondary processes. *Experimental Aging Research, 24*, 101–137.

Geary, D. C., & Whiley, J. G. (1991). Cognitive addition: strategy choices and speed-of-processing differences in young and elderly adults. *Psychology and Aging, 6*, 474–483.

Geary, D. C., Frensch, P. A., & Wiley, J. G. (1993). Simple and complex mental subtraction: strategy choice and speed-of-processing differences in younger and older adults. *Psychology and Aging, 8*, 242–256.

Gebuis, T., & Reynvoet, B. (2012). The role of visual information in numerosity estimation. *PLoS One, 7*, e37426.

Goldfarb, L., & Henik, A. (2007). Evidence for task conflict in the Stroop effect. *Journal of Experimental: Human Perceptual Performance, 33*, 1170–1176.

Grabner, R. H., Ansari, D., Koschutnig, K., Reishofer, G., Ebner, F., & Neuper, C. (2009). To retrieve or to calculate? Left angular gyrus mediates the retrieval of arithmetic facts during problem solving. *Neuropsychologia, 47*, 604–608.

Grady, C. (2012). The cognitive neuroscience of ageing. *Nature Review Neuroscience, 20*, 491–505.

Halberda, J., & Feigenson, L. (2008). Developmental change in the acuity of the "number sense": the approximate number system in 3-, 4-, 5-, and 6-year-olds and adults. *Developmental Psychology, 44*, 1457–1465.

Halberda, J., Mazzocco, M., & Feigenson, L. (2008). Individual differences in non-verbal number acuity correlate with maths achievement. *Nature, 455*, 665–668.

Halberda, J., Lya, R., Wilmerb, J. B., Naimana, D. Q., & Germine, L. (2012). Number sense across the lifespan as revealed by a massive Internet-based sample. *Proc Natl Acad Sci USA, 109*(28), 11116–11120.

Hartley, A. A. (1986). Instruction, induction, generation, and evaluation of strategies for solving search problems. *Journal of Gerontology, 41*, 650–658.

Hasher, L., & Zacks, R. T. (1979). Automatic and effortful processes in memory. *Journal of Experimental Psychology: General, 108*, 356–388.

Hasher, L., & Zacks, R. T. (1988). Working memory, comprehension, and aging: a review and a new view. In G. H. Bower (Ed.), *The psychology of learning and motivation* (pp. 193–225). (Vol. 22). New York, NY: Academic Press.

Hasher, L., Zacks, R. T., & May, C. P. (1999). Inhibitory control, circadian arousal, and age. In D. Gopher, & A. Koriat (Eds.), *Attention & performance, XVII, cognitive regulation of performance: interaction of theory and application* (pp. 653–675). Cambridge, MA: MIT Press.

Hasher, L., Lustig, C., & Zacks, R. T. (2007). Inhibitory mechanisms and the control of attention. In A. Conway, C. Jarrold, M. Kane, A. Miyake, & J. Towse (Eds.), *Variation in working memory* (pp. 227–249). New York: Oxford University Press.

Hedden, T., & Gabrieli, J. (2004). Insight into the aging mind: a view from cognitive neuroscience. *Nature Reviews Neuroscience, 5*, 87–97.

Hurewitz, F., Gelman, R., & Schnitzer, B. (2006). Sometimes area counts more than number. *Proceedings of the National Academy of Sciences of the United States of America, 103*, 19599–19604.

Ischebeck, A., Zamarian, L., Siedentopf, C., Koppelstatter, F., Benke, T., Felber, S., & Delazer, M. (2006). How specifically do we learn? Imaging the learning of multiplication and subtraction. *NeuroImage, 30*, 1365–1375.

Ischebeck, A., Zamarian, L., Egger, K., Schocke, M., & Delazer, M. (2007). Imaging early practice effects in arithmetic. *NeuroImage, 36*, 993–1003.

Ischebeck, A., Zamarian, L., Schocke, M., & Delazer, M. (2009). Flexible transfer of knowledge in mental arithmetic—an fMRI study. *NeuroImage, 44*, 1103–1112.

Izard, V., Dehaene-Lambertz, G., & Dehaene, S. (2008). Distinct cerebral pathways for object identity and number in human infants. *PLoS Biology, 6*, e11.

Johnson, M. M. S. (1990). Age differences in decision making: a process methodology for examining strategic information processing. *Journal of Gerontology: Psychological Sciences, 45*, 75–78.

Johnson, M., & Bailey, J. S. (1974). Cross-age tutoring: fifth graders as arithmetic tutors for kindergarten children. *Journal of Applied Behavioural Analysis, 7*, 223–232.

Kalanthroff, E., Goldfarb, L., & Henik, A. (2013). Evidence for interaction between stop signal and Stroop task conflict. *Journal of Experimental Psychology: Human Perception and Performance, 39*, 579.

Kane, M. J., & Engle, R. W. (2003). Working memory capacity and the control of attention: the contributions of goal neglect, response competition, and task set to Stroop interference. *Journal of Experimental Psychology: General, 132*, 47–70.

Lambrechts, A., Karolis, S., Garcia, S., Obende, J., & Cappelletti, M. (2013). Age does not count: resilience of quantity processing in healthy ageing. *Front Psychology, 10*, 865.

Lemaire, P., & Lecacheur, M. (2001). Older and younger adults' strategy use and execution in currency conversion tasks: insights from French franc to euro and euro to French franc conversions. *Journal of Experimental Psychology: Applied, 7*, 195–206.

Lemaire, P., & Lecacheur, M. (2007). Aging and numerosity estimation. *Journal of Gerontology: Psychological Sciences, 62*, 305–312.

Lemaire, P., Arnaud, L., & Lecacheur, M. (2004). Adults' age-related differences in adaptivity of strategy choices: evidence from computational estimation. *Psychology and Aging, 19*, 467–481.

Libertus, M., & Brannon, E. (2009). Behavioural and neural basis of number sense in infancy. *Current Directions in Psychological Science, 12*, 73.

Lipton, J. S., & Spelke, E. S. (2003). Origins of number sense: large number discrimination in human infants. *Psychological Science, 15*, 396–401.

Luwel, K., Verschaffel, L., Onghena, P., & De Corte, E. (2003). Strategic aspects of numerosity judgment: the effect of task characteristics. *Experimental Psychology, 50*, 63–75.

Luwel, K., Lemaire, P., & Verschaffel, L. (2005). Children's strategies in numerosity judgment. *Cognitive Development, 20*, 448–471.

Lyons, I., & Beilock, S. (2011). Numerical ordering ability mediates the relation between number-sense and arithmetic competence. *Cognition, 121*, 256–261.

May, C. P., & Hasher, L. (1998). Synchrony effects in inhibitory control over thought and action. *Journal of Experimental Psychology: Human Perception and Performance, 24*, 363–379.

Mazzocco, M. M. M., Feigenson, L., & Halberda, J. (2011). Impaired acuity of the approximate number system underlies mathematical learning disability (dyscalculia). *Child Development, 82*, 1224–1237.

Norris, J. E., McGeown, W., Guerrini, C., & Castronovo, J. (2015). Aging and the number sense: preserved basic non-symbolic numerical processing and enhanced basic symbolic processing. *Frontiers In Psychology, 6*, 999.

Núñez-Peña, M. I. (2008). Effects of training on the arithmetic problem-size effect: an event-related potential study. *Experimental Brain Research, 190*, 105–110.

Nyberg, L., Lövdén, M., Riklund, K., Lindenberger, U., & Bäckman, L. (2012). Memory aging and brain maintenance. *Trends in Cognitive Sciences, 16*, 292–305.

Park, J., & Brannon, E. M. (2013). Training the approximate number system improves math proficiency. *Psychological Science, 24*, 2013–2019.

Park, D. C., & Reuter-Lorenz, P. (2009). The adaptive brain: aging and neurocognitive scaffolding. *Annual Review Psychology, 60*, 173–196.

Parsons, S., & Bynner, J. (1997). Numeracy and employment. *Education & Training, 39*, 43–51.

Parsons, S., & Bynner, J. (2005). *Does numeracy matter more?* London: NRDC.

Pauli, P., Lutzenberger, W., Rau, H., Birbaumer, N., Rickard, T. C., Yaroush, R. A., & Bourne, L. E., Jr. (1994). Brain potentials during mental arithmetic: effects of extensive practice and problem difficulty. *Cognitive Brain Research, 2*, 21–29.

Piazza, M., Izard, V., Pinel, P., Le Bihan, D., & Dehaene, S. (2004). Tuning curves for approximate numerosity in the human intraparietal sulcus. *Neuron, 44*, 547–555.

Piazza, M., Facoetti, A., Trussardi, A. N., Berteletti, I., Conte, S., Lucangeli, D., Dehaene, S., & Zorzi, M. (2010). Developmental trajectory of number acuity reveals a severe impairment in developmental dyscalculia. *Cognition, 116*, 33–41.

Qin, S., Cho, S., Chen, T., Rosenberg-Lee, M., Geary, D., & Menon, V. (2014). Hippocampal-neocortical functional reorganization underlies children's cognitive development. *Nature Neuroscience, 17*, 1263–1269.

Reder, L. M., Wible, C., & Martin, J. (1986). Differential memory changes with age: exact retrieval versus plausible inference. *Journal of Experimental Psychology: Learning, Memory, and Cognition, 12*, 72–81.

Reis, J., Schambra, H. M., Cohen, L. G., Buch, E. R., Fritsch, B., & Zarahn, E. (2009). Noninvasive cortical stimulation enhances motor skill acquisition over multiple days through an effect on consolidation. *Proceedings of the National Academy of Sciences of the United States of America, 106*, 1590–1595.

Rittle-Johnson, B., & Koedinger, K. (2009). Iterating between lessons on concepts and procedures can improve mathematics knowledge. *British Journal of Educational Psychology, 79*, 483–500.

Rogers, W. A., Hertzog, C., & Fisk, A. (2000). An individual differences analysis of ability and strategy influences age-related differences in associative learning. *Journal of Experimental Psychology: Learning, Memory, and Cognition, 26*, 359–394.

Salthouse, T. A. (1991). Mediation of adult age differences in cognition by reductions in working memory and speed of processing. *Psychological Science, 2*, 179–183.

Salthouse, T. A. (2004). What and when of cognitive aging. *Current Directions in Psychological Science, 13*, 140–144.

Salthouse, T. A., & Kersten, A. W. (1993). Decomposing adult age differences in symbol arithmetic. *Memory & Cognition, 2*, 699–710.

Salthouse, T. A., Atkinson, T. M., & Berish, D. E. (2003). Executive functioning as a potential mediator of age-related cognitive decline in normal adults. *Journal of Experimental Psychology: General, 132*, 566–594.

Siegel, A. W., Goldsmiths, L. T., & Madson, C. R. (1982). Skills in estimation problems of extent and numerosity. *Journal for Reseacrh in Mathematical Education, 13*, 211–232.

Siegler, R. S., & Lemaire, P. (1997). Older and younger adults' strategy choices in multiplication: testing predictions of ASCM using the choice/no-choice method. *Journal of Experimental Psychology: General, 126*, 71–92.

Stoianov, I., & Zorzi, M. (2012). Emergence of a "visual number sense" in hierarchical generative models. *Nature Neuroscience, 15*, 194–196.

Supekar, S., Swigart, A. G., Tenison, C., Jolles, D. D., Rosenberg-Lee, M., Fuchs, L., & Menon, V. (2013). Neural predictors of individual differences in response to math tutoring in primary-grade school children. *Proceedings of the National Academy of Sciences of the United States of America, 110*, 8230–8235.

Szűcs, D., Nobes, A., Devine, A., Gabriel, F., & Gebuis, T. (2013). Visual stimulus parameters seriously compromise the measurement of approximate number system acuity and comparative effects between adults and children. *Frontiers in Psychology, 4*, 444.

Terney, D., Chaieb, L., Moliadze, V., Antal, A., & Paulus, W. (2008). Increasing human brain excitability by transcranial high-frequency random noise stimulation. *The Journal of Neuroscience, 28*, 14147–14155.

Trick, L. M., Enns, J. T., & Brodeur, D. A. (1996). Life span changes in visual enumeration: the number discrimination task. *Developmental Psychology, 32*, 925–932.

Venkatraman, V., Ansari, D., & Chee, M. W. (2005). Neural correlates of symbolic and non-symbolic arithmetic. *Neuropsychologia, 43*, 744–753.

Walsh, V. (2003). A theory of magnitude: common cortical metrics of time, space and quantity. *Trends in Cognitive Science, 7*, 483–488.

Watson, D. G., Maylor, E. A., & Manson, N. J. (2002). Aging and enumeration: a selective deficit for the subitization of targets among distractors. *Psychology and Aging, 17*, 496–504.

Watson, D. G., Maylor, E. A., & Bruce, L. A. M. (2005a). Effects of age on searching for and enumerating targets that cannot be detected efficiently. *Quarterly Journal of Experimental Psychology, 58*, 1119–1142.

Watson, D. G., Maylor, E. A., & Bruce, M. A. (2005b). Search, enumeration, and aging: eye movement requirements cause age-equivalent performance in enumeration but not in search tasks. *Psychology and Aging, 20*, 226–240.

Watson, D. G., Maylor, E. A., & Bruce, L. A. M. (2007). The role of eye movements in subitizing and counting. *Journal of Experimental Psychology: Human Perception and Performance, 33*, 1389–1399.

Wu, S. S., Chang, T. T., Majid, A., Caspers, S., Eickhoff, S. B., & Menon, V. (2009). Functional heterogeneity of inferior parietal cortex during mathematical cognition assessed with cytoarchitectonic probability maps. *Cerebral Cortex, 19*, 2930–2945.

Wynn, K. (1998). Psychological foundations of number: numerical competence by human infants. *Trends in Cognitive Sciences, 2*, 296–303.

Xu, F., Spelke, E. S., & Goddard, S. (2005). Number sense in human infants. *Developmental Science, 8*, 88–101.

Zamarian, L., Ischebeck, A., & Delazer, M. (2009). Neuroscience of learning arithmetic—evidence from brain imaging studies. *Neuroscience and Biobehavioral Reviews, 33*, 909–925.

CHAPTER

6

Development of Counting Ability: An Evolutionary Computation Point of View

Gali Barabash Katz, Amit Benbassat**, Moshe Sipper***

*Department of Cognitive Sciences, Ben-Gurion University of the Negev, Beer-Sheva, Israel; **Department of Computer Science, Ben-Gurion University of the Negev, Beer-Sheva, Israel

OUTLINE

Continuous Issues in Numerical Cognition. http://dx.doi.org/10.1016/B978-0-12-801637-4.00006-8

6.1 INTRODUCTION

Size perception (estimating an area of a blob without counting) is shared by humans (Brannon, Lutz, & Cordes, 2006) and animals (Cantlon & Brannon, 2007). Counting (to report the exact number of items in an array), on the other hand, is a more complex ability, which is currently known to be specific to humans. Although these two systems are different, current literature finds interesting overlaps, which might shed light on the development of counting ability through evolution.

First, there is early evidence of shared functional information between different systems through evolution in the work of Piaget (1955) and Flavell (1963), where they suggested that cognitive structures are content independent and domain general. Kashtan and Alon (2005) demonstrated a similar idea by simulating evolution of artificial neural networks (ANNs) in an environment with modularly varying goals, and showed that the final networks had more reusable building blocks than the control networks that evolved in an environment with a single fixed goal.

When it comes to numerical abilities, it is commonly suggested that the basic numerical intuitions that are shared between humans and animals are supported by an evolutionarily ancient approximate number system (ANS) where the number of discrete objects are represented as a continuous mental magnitude (Cantlon, Platt, & Brannon, 2009; Dehaene, 2001). This core system enables the representation of the approximate number of items in visual or auditory arrays without verbally counting (Halberda, Mazzocco, & Feigenson, 2008) and might be the root for high-level human numerical abilities such as arithmetic (Dehaene, 2001).

Second, early imaging studies that were conducted in order to test the hypothesis of a common substrate for processing symbolic and nonsymbolic stimuli revealed a site in the left intraparietal sulcus (IPS) that showed greater activation during numerical compared to nonnumerical judgments (Fias, Lammertyn, Reynvoet, Dupont, & Orban, 2003).

Subsequent studies compared stimuli for size, luminance, or number and found that the right hIPS (horizontal segment of the IPS) is not devoted exclusively to number processing but is engaged whenever subjects attend to the dimension of size—whether numerical or physical (Pinel, Piazza, Le Bihan, & Dehaene, 2004). Moreover, in one of the studies (Cohen Kadosh et al., 2005), a largely overlapping network of frontal, parietal, and occipitotemporal areas of both hemispheres was found, thus confirming the view that many of the neural resources used for number comparison are shared by other comparison tasks as well.

Today, it seems that the left IPS mainly responds to numerical stimuli in a format-specific manner—specifically, to symbolic stimuli. The right IPS, on the other hand, is activated regardless of stimuli format and is also activated by nonnumerical magnitudes. These findings led researchers to suggest that the right IPS might support a general magnitude system, used to process both numerical and nonnumerical magnitudes, rather than an abstract approximate number system (Chapter 15).

Third, there is also evidence that nonsymbolic number and cumulative area representation contribute shared and unique variance to mathematical competence (Lourenco, Bonny, Ferandez, & Rao, 2012). College students were asked to estimate which array was greater in number or cumulative area, after which they completed a battery of standardized math tests. The authors found that individual differences in both number and cumulative area precision were correlated with interindividual differences in arithmetic and geometry. Moreover, whereas number precision contributed unique variance to advanced arithmetic, cumulative area precision contributed unique variance to geometry. Based on their results, Lourenco et al. (2012) suggested that uniquely human branches of mathematics interface with an evolutionarily primitive general magnitude system, which includes partially overlapping representations of numerical and nonnumerical magnitude.

When thinking about the evolutionary approach, we can identify similar patterns between primate approximate counting and human exact counting. Recently, Cantlon, Piantadosi, Ferrigno, Hughes, and Barnard (2015) reported that nonhuman primates exhibit a cognitive ability that is algorithmically and logically similar to human counting. In their experiment, monkeys were given the task of choosing between two food caches. First, they saw one cache baited with peanuts, one at a time. Then, the second cache was baited, one peanut at a time. At the point when the second cache was approximately equal to the first one, the monkeys spontaneously moved to choose the second cache even before that cache was completely baited. By using a novel Bayesian analysis, the authors showed that monkeys use an approximate counting algorithm for comparing quantities in sequence that is incremental, iterative, and condition controlled, similar to formal counting in humans.

Developmental dyscalculia [DD; a severe difficulty in learning and making simple mathematical calculations (Kosc, 1974), with an estimated prevalence of about 5–7% (Shalev, 2007)] might also serve as a link between the ANS and symbolic arithmetic. Today, there is growing evidence that the relationship between numerical distance and IPS activity is disrupted in children with developmental dyscalculia (Fias et al., 2003), and many known effects are compromised in DD (eg, size congruity effect: Rubinsten & Henik, 2005, 2006; distance effect: Price, Holloway, Räsänen, Vesterinen, & Ansari, 2007). The cause of the deficit might be a network failure between continuous magnitude processing (ie, the size perception system) and discrete magnitude processing (ie, the counting system). Children as well as adults who suffer from these specific learning disabilities are at a disadvantage in both academic and everyday life situations, especially when it comes to handling money (Henik, Rubinsten, & Ashkenazi, 2014). Thus, a better evolutionary understanding of the development of counting in relation to size perception can help researchers think of novel ways to improve day-to-day life of people with arithmetic disabilities.

Currently, the suggested computerized models explaining how the counting system functions use artificial neural networks along with different unsupervised learning techniques (ie, teaching a predefined network to successfully perform a required task such as counting). Verguts and Fias (2004) used an unsupervised learning technique for teaching ANNs to process nonsymbolic and symbolic inputs. First, they presented the ANNs a nonsymbolic input (eg, a collection of dots) in a comparison task and showed the distance and size effects. Then, they presented symbolic and nonsymbolic inputs simultaneously and found that the ANNs used the number-selective neurons that were already available, when processing symbolic stimuli. Their findings present a possible linkage between higher-order numerical cognition and more primitive numerical abilities.

Stoianov and Zorzi (2012) presented binary images (sets of black shapes on a white background) to ANNs in a comparison task (ie, "Which is larger?") against a given reference number, and after an unsupervised learning process, the ANNs excelled in the task. The given stimuli differed by the number of objects presented and their cumulative area (the authors controlled both properties). They suggested a "deep network model" (which developed through unsupervised learning without using evolutionary techniques), containing one visual input layer and two hidden layers, in order to explain the numerosity estimation process. According to their model, there are two types of neurons in hidden layer 1: neurons that deal with visual processing (center surround neurons) and neurons that deal with cumulative area, and hidden layer 2 neurons deal with numerosity processing. In addition, hidden layer 2 neurons' activity can be approximated by a linear combination of the activity of both types of hidden layer 1 neurons. The hierarchical model generated the same results in a numerosity comparison task as did human adults, thus can serve as the key to understanding the neural mechanism underlying numerosity perception.

Based on the researches described earlier, we consider the following two potential hypotheses regarding the development of counting. One is that the counting system evolved on the basis of a primitive system designed to perceive size and evaluate amount of substance (Cantlon et al., 2009; Henik, Leibovich, Naparstek, Diesendruck, & Rubinsten, 2012; Lourenco et al., 2012). The other is that both systems evolved separately, in different periods of time.

In addition, as previously mentioned, exact counting is a complex task. Kaufman, Lord, Reese, and Volkmann (1949) suggested that the entire enumeration process is carried out first by subitizing, (ie, the phenomenon of giving a rapid, confident and accurate report of the amount of up to four presented items) and then by counting (from five items and above). Thus, subitizing might be a stepping-stone in the development of the exact counting system—a theory we examine in the current research.

Herein we present a novel approach of computerized simulation with ANNs for examining the development of counting ability, in a process known as evolutionary computation.

6.2 EVOLUTIONARY COMPUTATION

Evolutionary computation (EC) is a subfield in computer science that is inspired by biology. An evolutionary algorithm (EA) enables solving complex problems by using an evolutionary process metaphor.

An evolutionary process usually includes:

- An environment with limited resources, which is translated to the problem one wants to solve.
- Individuals, which are translated into candidate solutions.
- A concept of fitness (ie, the compatibility of the individual to survive and reproduce in the environment), which is translated to a probability of generating new solutions.

In nature, the competition for limited resources causes a selection of those who fit better to the environment (ie, natural selection), and as a result, the fitness of the population improves over time. The same happens in EAs, and as in nature, the fit individuals create a new generation (ie, new candidate solutions) by using parameters of recombination and mutation (see Box 6.1 for an example of a typical EA).

Genetic algorithms (GAs, Goldberg, 1989) are a simple variant of evolutionary algorithms that use a simple alphabet (eg, a binary alphabet "0," "1") and can solve complex problems with a simple model representation.

BOX 6.1

EXAMPLE OF A TYPICAL EA:

Begin

1. *Initialize* the population with random candidate solutions;
2. *Evaluate* each candidate;
3. Repeat until (*termination condition* is satisfied)
   - **3.1** *Select* parents
   - **3.2** *Recombine* pairs of parents
   - **3.3** *Mutate* the resulting offspring
   - **3.4** *Evaluate* new candidates
   - **3.5** *Select* individuals for the next generation

End

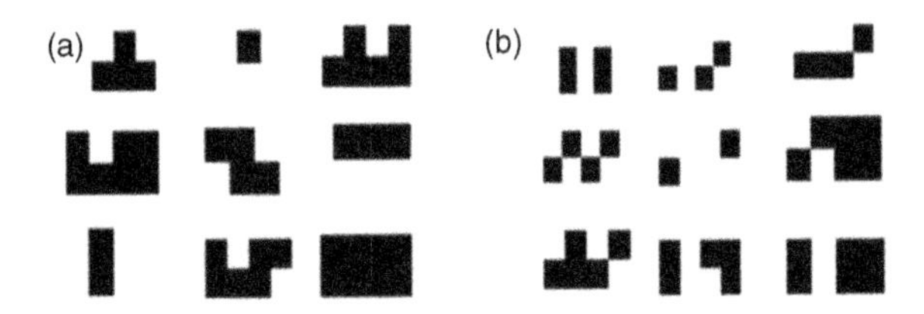

FIGURE 6.1 **Examples of continuous and discrete 2 × 4 binary arrays.** (a) Continuous stimuli and (b) discrete stimuli. We define a stimulus as continuous if there is a path from each visible cell in the array to every other visible cell that passes through visible cells (in single up/down/left/right steps; Katz et al., 2013).

It should be noted that even though GAs use binary discrete inputs, they can represent size, which is continuous, as can be seen in Fig. 6.1a.

Evolutionary algorithms are stochastic, meaning that even for identical parameter setups, results may vary, but they usually share the same trend. Genetic operators (eg, mutation and recombination) can lead to new, unexpected, and creative solutions to complex problems.

6.3 CURRENT STUDY (OR HOW CAN EVOLUTIONARY ALGORITHMS HELP IN UNDERSTANDING THE DEVELOPMENT OF THE COUNTING SYSTEM)?

In this study we computationally examined whether individuals that evolved a proficiency in size perception (ie, evolved to solve a classification problem of BIG vs. SMALL; Table 6.1) have an advantage in learning

TABLE 6.1 Size Perception, Counting, and Control Groups' Genetic Algorithms With Their Fitness Functions and Outputs

#	Evolution	Fitness and output
1	SP	A given input is classified as BIG if its number of ones ≥ array.length/2, and otherwise as SMALL. The same logic is applied to the output. BIG is an output with a number of ones that is ≥ [array.length/2]. SMALL is an output of a number of ones that is < [array.length/2].
2	C	For a given input, the exact number of ones given is expected to be in the output (without order considerations).
3	SP-C	The size perception fitness function is switched to the counting fitness function mid-run after 25 generations (the entire size perception run took an average of 48.5 generations).
4	C1	This is a random classification task. The set of inputs was divided randomly into two groups: one group expected a BIG output and the other a SMALL one, similar to SP (see table item 1).
5	C1-C	The control 1 fitness function switched to the counting fitness function after 25 generations. We expected an output of the exact number of ones, similar to C (see table item 3).

Note: SP, evolved to perceive size; C, evolved to count; SP-C, evolved first to perceive size and then to count; C1, evolved to excel in control 1 task; and C1-C, evolved first to excel in control 1 task and then to count (Katz et al., 2013).

to count. In order to test our hypothesis, we chose genetic algorithms to develop artificial neural networks through an evolutionary process. We used this technique to evolve ANNs that could excel in a size perception task and then further evolve the same networks to count. Our main goal was to examine whether these ANNs have an advantage in evolving the ability to count over new learners of counting (ie, ANNs that did not first evolve to perceive size).

6.4 NEUROEVOLUTION OF AUGMENTING TOPOLOGIES (NEAT)

NEAT is a method for evolving artificial neural networks using genetic algorithms, developed by Stanley and Miikkulainen (2002). It simulates evolution by starting with small, simple networks that become increasingly complex through evolution. Just as organisms in nature increased in their complexity through evolution, so do neural networks in NEAT. This process of continual elaboration allows finding highly sophisticated and complex neural networks. NEAT evolves both the connection weights and architecture (by adding and removing connections and nodes) of the neural networks. The networks start with minimal topologies (this is analogous to natural evolution that begins from simple forms) and gradually become more complex (Stanley & Miikkulainen, 2002).

We chose to use NEAT[a] in order to evolve ANNs to perceive size and to count. We analyzed the complexities of the final networks created by evolution by examining the number of inner nodes added to the networks as the task became more complex.

6.5 METHODS

6.5.1 Stimuli

In order to represent the visual input, we used a 2 by 4 two-dimensional binary array. A total of 256 (2^8) possible stimuli could be produced by this array; some of these are discrete and some are continuous. We defined a stimulus as continuous if there was a path from each visible cell in the array to every other visible cell that passed through visible cells (in single up/down/left/right steps). According to this definition, of the 256 possible stimuli, 147 are discrete and 109 are continuous (Katz, Benbassat, Diesendruck, Sipper, & Henik, 2013; Fig. 6.1).

We divided the stimuli into three sets of stimuli:

1. The "continuous" set.
2. The "discrete" set.
3. Both continuous and discrete, which was called the "all" set.

Sets (1) and (2) were composed of 108 stimuli randomly divided into training and test sets of 54 stimuli[b] each (no repetitions). Set (3) was composed of a training set of 128 (256/2) stimuli and a test set, which included the remaining 128 stimuli.

Prior to being inserted into the NEAT system, the stimulus was flattened into a one-dimensional array of 0s and 1s (Fig. 6.2).

All the evolutionary simulations that we performed included a comparison to a reference number. In each trial the given blob (ie, binary string) was compared to the half size of the string array (in all three simulations it was $8/2 = 4$ for a Boolean array of 2×4), meaning: "Is the number of 'ones' in the current blob larger than 4?"

[a]Specifically, the NEAT4J Java implementation.

[b]This value was chosen because there were 147 discrete and 109 continuous stimuli out of 256, and we wanted to keep an equal number of stimuli in the training and test groups; thus, we chose the minimal group (i.e., 109) and divided it into two equal groups of 54 stimuli (one for training and one for testing).

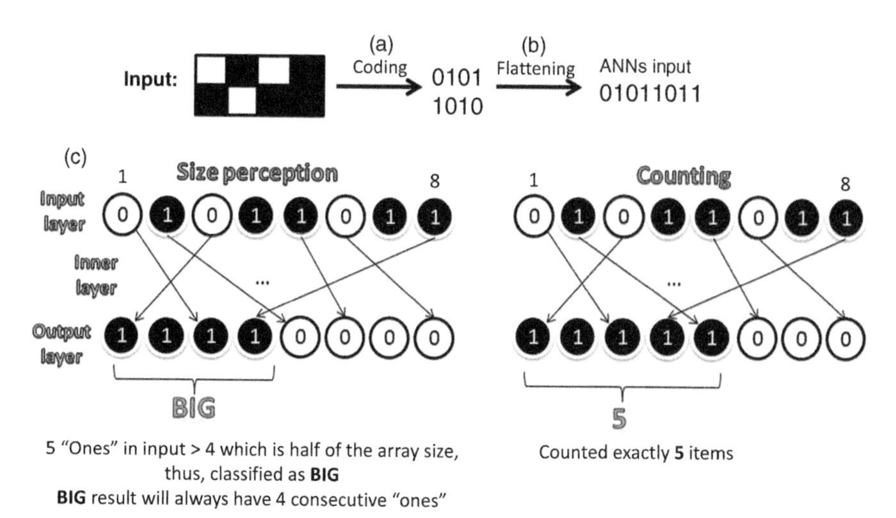

FIGURE 6.2 **Example of encoding a stimulus into a binary string.** (a) Encoding a single discrete stimulus of 2 × 4 to a binary representation. (b) Flattening the two-dimensional array into a one-dimensional array. (c) A demonstration of how an ANN can perform tasks of size perception and counting on the same input (Katz et al., 2013).

6.5.2 Procedure

Several different types of evolutionary procedures were tested during the study. Every *evolutionary run* was performed 30 times using each of the 3 stimuli sets (*continuous*, *discrete*, and *all*), resulting in a total of 90 runs per type. The procedure was similar for all run types. Every run began with a training stage, where the population was trained to succeed in a certain task (ie, classify BIG/SMALL blobs or counting the exact number of squares of the given blob) until its average fitness score (ie, the score of the function that should be optimized in order to continue to the next generation in evolution; see Fig. 6.3 for fitness calculation formula) exceeded

$$\text{Fitness} = \sum_{i=1}^{\text{Inputs}} \text{Score}, \ \text{Max} = \text{Output} \cdot \text{Length}$$

$$\text{Score} = \frac{\text{Max} - \sum_{i=0}^{\text{Max}-1} |\text{Expected}_{\text{Sorted}}[i] - \text{Observed}_{\text{Sorted}}[i]|}{\text{Max}}$$

FIGURE 6.3 **Fitness calculation.** The *fitness* is the summation of all scores of the inputs (ie, the stimuli presented to the current individual). The *score* is calculated according to the distance of each digit from the expected array to the observed array (after sorting both arrays).

10%, and more specifically, the value 0.999.[c] If this condition was not met, the algorithm halted after 3000 generations. Next, an evaluation of the population took place. During this test stage, each individual was evaluated in relation to each of the tasks relevant to the fitness functions used; an additional test on counting accuracy was performed in all runs. For example, the "size perception" population trained to perceive size (ie, classify BIG/SMALL blobs) was tested on size perception (ie, again asked to classify BIG vs. SMALL blobs) but also tested on counting (ie, asked to count the number of squares of the given blob, see Box 6.2 for the procedure schema).

BOX 6.2

FROM SIZE PERCEPTION TO COUNTING SIMULATION SCHEMA

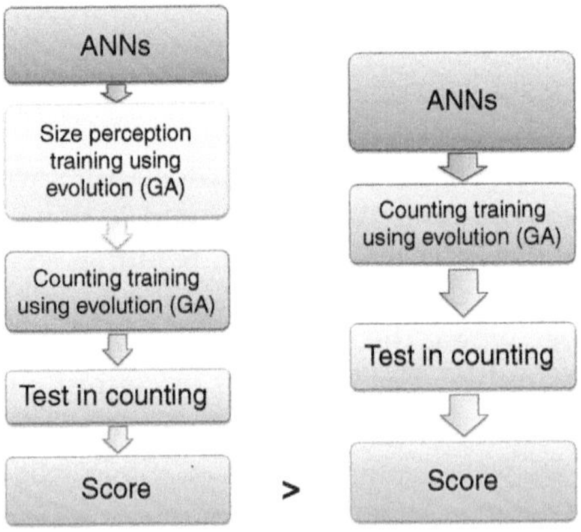

ANNs
Size perception training using evolution (GA)
Counting training using evolution (GA)
Test in counting
Score
>
ANNs
Counting training using evolution (GA)
Test in counting
Score

By using evolutionary computation techniques, we generated artificial neural networks that excelled in size perception and presented a significant advantage in evolving the ability to count over those that evolved this ability from scratch.

[c]The value 0.999 in size perception and in counting, and 0.998 in subitizing, as this was found to be sufficient in order to excel in this task in the testing stage.

6.5.3 Genetic Algorithm Parameters

We used the same evolutionary parameters in all the simulations that will be described here:

- Population size of 100 individuals
- Termination condition: fitness of 0.999 or 0.998[c] or 3000 generations if previous condition was not met
- Mutation probability $P_m = 0.25$ (ie, when creating a new generation of ANNs, this is the probability of adding/removing connections and nodes in the new ANN offspring)
- Crossover probability $P_c = 1$ (ie, the probability that the new ANN offspring will end up having half the genes from one parent and half from the other)

6.5.4 Calculation of Fitness Function

The *fitness* was the summation of all scores of the inputs (ie, the stimuli presented to the current individual). The *score* was calculated by the summation of the differences between the expected array index and the observed array index, and it was normalized by *max,* which is the output length—in our case, 8 (Katz et al., 2013).

Each evolutionary run defined the expected output it got from NEAT. When a result from NEAT was received, the fitness function checked if the expected output was equal to the observed one. If so, it assigned a 100% score, otherwise it calculated a score according to the distance of each digit from the expected array to the observed one (after sorting both arrays) as follows:

6.6 SIMULATIONS

The following two Simulations #1 and #2 were first presented briefly in Katz et al. (2013) and are being described here with great detail in order to set the context for the *new* Simulation #3.

6.6.1 Simulation 1: From Size Perception to Counting (Katz et al., 2013)

We created five evolutionary runs in order to test our hypothesis. The goal was to train a set of ANNs in a certain task and then switch to a different task by changing the fitness function mid-run. The algorithm performed the switch from one fitness function to the other after a predefined number of generations, which was chosen after several trial-and-error runs. We opted for this approach instead of waiting for the networks to excel in the tasks, in order to avoid the "overfitting" phenomenon in the consecutive runs (ie, bloated networks with too many inner nodes that excelled specifically on the training inputs rather than solving the more general problem). Thus, we switched from the size classification tasks to the counting task after 25 generations (Table 6.1).

6.6.2 Results

We conducted 3 different two-way ANOVA (analysis of variance) tests, for 3 different dependent variables: counting score, number of generations, and number of inner nodes in an ANN.

The 3 (stimuli type) × 5 (evolutionary runs) design was between subjects, since each evolutionary run produced a different population of ANNs. The following 5 evolutionary runs were performed:

1. Size perception (SP)—the ANNs evolved to perceive size.
2. Counting (C)—the ANNs evolved to count.
3. Size perception and then counting (SP-C)—the ANNs evolved first to perceive size and then to count.
4. Control group (C1)—the ANNs evolved to excel in a control 1 task (a classification task; Table 6.1).
5. Control and then counting (C1-C)—the ANNs evolved first to excel in the control 1 task and then to count.

Each of the 5 evolutionary runs (see previous list) ran 3 times—each time with a new stimuli type:

1. Continuous stimuli only
2. Discrete stimuli only
3. Both continuous and discrete stimuli

In the following analyses, we present the results that are relevant to our theoretical considerations, with significance level of $p < 0.05$. As previously mentioned, these results were first reported in Katz et al. (2013) and are being explained here with great detail to set the context for the new results of Simulation 3).

6.6.2.1 *Counting Score (Score of the Final Counting Test)*

The counting score (based on the *fitness calculation*, see Fig. 6.3) was higher in networks that first evolved to perceive size or to perform other classification tasks than networks that evolved to count independently (eg, *SP-C vs. C*: $F\,(1, 435) = 58.49$, $MSE = 0.0062$, $p < 0.01$, $\eta_p^2 = 0.118$, and also *C1-C vs. C*: $F\,(1, 435) = 385.49$, $MSE = 0.0062$, $p < 0.01$, $\eta_p^2 = 0.47$).

In addition, when the tasks were easy (ie, the networks evolved to excel in a single task, eg, SP, C1, and not two tasks as in SP-C or C1-C), the score for discrete stimuli was higher than for continuous stimuli [eg, *continuous vs. discrete in SP*: F (1, 435) = 84.47, MSE = 0.0062, $p < 0.01$, $\eta_p^2 = 0.162$, and also *continuous vs. discrete in C1*: F (1, 435) = 215.89, MSE = 0.0062, $p < 0.01$, $\eta_p^2 = 0.33$]. However, when the tasks became more complex, the opposite pattern was observed and the ANNs with continuous stimuli had better scores at counting [*continuous vs. discrete in C1-C*: F (1, 435) = 6.146, MSE = 0.0062, $p < 0.05$, $\eta_p^2 = 0.0139$; in *C* and *SP-C* the differences were not significant, see Fig. 6.4].

6.6.2.2 Generations

SP and *C1* runs were faster (ie, had a smaller number of generations) than all other tasks (*SP*: M = 48.5, SD = 9.21; *C1*: M = 44.9, SD = 5.15; *C*: M = 308.22, SD = 4.32; *SP-C*: M = 330.7, SD = 16.33; *C1-C*: M = 334.7, SD = 24.33). More generations were required to evolve the networks to count after evolving to perform any other task [*C vs. C1-C*: F (1, 435) = 183.49, MSE = 172.51, $p < 0.01$, $\eta_p^2 = 0.296$].

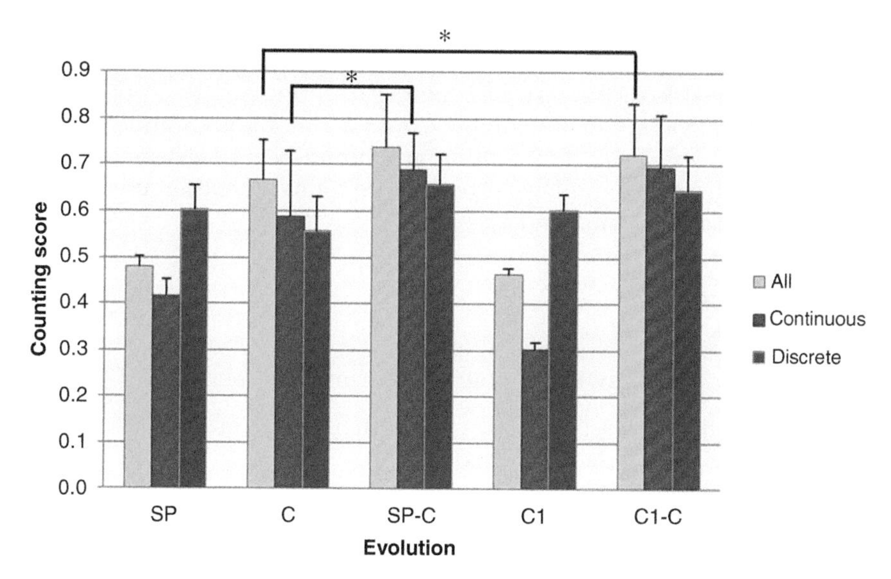

FIGURE 6.4 **Counting scores results of Simulation 1—from size perception to counting.** The counting score of the ANNs that first evolved to perceive size and then to count (SP-C) is higher than the counting score of ANNs that evolved to count (C) from scratch (Table 6.1; Katz et al., 2013). *C1*, control group with a random classification task; *yellow*, ANNs that evolved and were tested with both types of stimuli (continuous and discrete); *purple*, ANNs that evolved and were tested with continuous stimuli; and *red*, ANNs that evolved and were tested with discrete stimuli. In all evolutionary runs, the actual stimuli were different between the training (by evolution) and testing sets. The "*" represents significant differences: the counting score when evolving first to perceive size and then to count is significantly higher than the counting score received when evolving to count from scratch, with continuous stimuli; and the counting score when evolving to count after first learning a random classification task is significantly higher than the score when evolving to count from scratch, with both stimuli types.

6.6.2.3 *Inner Nodes (A Measurement for Network Complexity)*

Complex runs (ie, runs that had a complex task to learn such as counting) generated ANNs with more inner nodes than the simpler ones (complex runs: *C*: $M = 11.81$, $SD = 4.71$; *SP-C*: $M = 12.39$, $SD = 4.6$; and *C1-C*: $M = 12.05$, $SD = 5.04$ and simpler runs: *SP*: $M = 3.64$, $SD = 1.64$; *C1*: $M = 3.96$, $SD = 1.37$).

In addition, during the complex runs the ANNs that *evolved and were tested* on discrete stimuli contained *more inner nodes* than the ones evolved and tested with continuous stimuli [*discrete vs. continuous in C1-C*: $F(1, 435) = 25.25$, $MSE = 13.87$, $p < 0.01$, $\eta_p^2 = 0.05$, and *discrete vs. continuous in SP-C*: $F(1, 435) = 5.55$, $MSE = 13.87$, $p < 0.05$, $\eta_p^2 = 0.002$, but *discrete vs. continuous in* C was not significant].

6.6.3 Simulation 2: Continuous Versus Discrete (Katz et al., 2013)

In the evolutionary runs discussed earlier, we trained and tested ANNs on a certain stimuli type (eg, evolved with continuous and tested with continuous), the results of which seemed to imply that continuous stimuli were more suitable to the task of evolving counting ability. In order to examine if this was so, or if the high score for continuous stimuli in the counting test was just due to continuous stimuli being less complex to process than discrete stimuli, we proceeded to train on continuous stimuli and test on discrete stimuli and vice versa.

6.6.4 Results

We conducted an ANOVA with an array of 2 (evolutionary runs) × 4 (stimuli type). The following 2 evolutionary runs were performed:

1. Size perception (SP)—the ANNs evolved to perceive size.
2. Counting (C)—the ANNs evolved to count.

Both evolutionary runs had a final test in Counting. Each evolutionary run was performed 4 times, each time with a new stimuli type:

1. Evolved with continuous stimuli and tested with discrete stimuli
2. Evolved with discrete stimuli and tested with continuous stimuli
3. Evolved with continuous stimuli and tested with continuous stimuli
4. Evolved with discrete stimuli and tested with discrete stimuli

Similar to previous analyses, we present the results relevant to our theory, with a significance level of $p < 0.05$. As previously mentioned, these results were first reported in Katz et al. (2013) and are being explained here with great detail to set the context for the new results of Simulation 3.

6.6.4.1 Counting Score

The interaction between stimuli and evolutionary run was significant, F (3, 232) = 15.16, MSE = 0.0084, $p < 0.01$, $\eta_p^2 = 0.16$. In both SP and C runs, the counting score when *evolved with continuous but tested with discrete* was *higher* than when evolved and tested with continuous stimuli, F (1, 232) = 26.6, MSE = 0.0084, $p < 0.01$, $\eta_p^2 = 0.09$).

Different results were observed with discrete stimuli[d]: When *evolved with discrete and tested with continuous,* the counting score was *lower* than the counting score of the control groups (ie, evolving and testing with discrete stimuli, F (1, 232) = 108.87, MSE = 0.0084, $p < 0.01$, $\eta_p^2 = 0.32$). In addition, the ANNs that *evolved to count with discrete* stimuli were *worse* in counting than those that *evolved to perceive size with discrete stimuli* (but were tested in counting): F (1, 232) = 4.14, MSE = 0.0084, $p < 0.05$, $\eta_p^2 = 0.017$. Another interesting result was that in SP runs, when *evolving ANNs with continuous stimuli and testing with discrete stimuli,* the counting score was *higher* than the other way around (ie, evolving ANNs with discrete stimuli and testing with continuous stimuli): *SP*: F (1, 232) = 41.09, MSE = 0.0084, $p < 0.01$, $\eta_p^2 = 0.15$, and *counting*: F (1, 232) = 70.97, MSE = 0.0084, $p < 0.01$, $\eta_p^2 = 0.23$.

Finally, the results when evolving with discrete stimuli and testing with continuous stimuli were the worst among all other combinations in both size perception and counting runs, F (1, 232) = 37.89, MSE = 0.0084, $p < 0.01$, $\eta_p^2 = 0.14$, and the counting score in this case was the same for size perception and counting, F (1, 232) = 3.65, MSE = 0.0084, $p = ns$, $\eta_p^2 = 0.02$ (Fig. 6.5).

6.6.4.2 Generations

Training ANNs on continuous stimuli and testing on discrete ones required the same number of generations as training ANNs on discrete stimuli and testing on continuous ones. In addition, this number of generations was equal to the number of generations of the control groups (for all four SP runs it took about 46 generations: M = 46.13, SD = 7.41, and for all four counting runs it took about 308 generations: M = 308.45, SD = 4.09).

6.6.4.3 Inner Nodes

The interaction between stimuli and evolutionary run was significant, F (3, 232) = 5.321, MSE = 13.32, $p < 0.01$, $\eta_p^2 = 0.064$. In the *counting* runs, more inner nodes were produced when the ANNs evolved with discrete stimuli (M = 15.66, SD = 5.99) than when evolved with continuous stimuli (M = 11.46, SD = 3.702, F (1, 232) = 10.25, MSE = 13.32, $p < 0.05$, $\eta_p^2 = 0.04$). Moreover, the number of inner nodes in *counting* runs in ANNs that evolved with discrete stimuli and tested with continuous stimuli was

[d]There might be a different explanation for the result differences between discrete and continuous stimuli because there were no discrete 0s, 1s, 7s, or 8s, and a minority of the 6s were discrete, which might have caused a bias toward the subitizing range numbers.

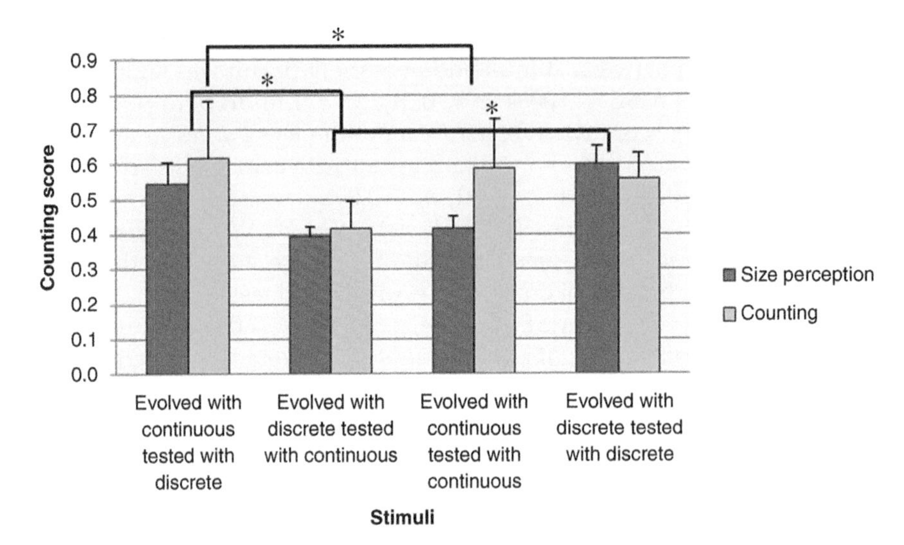

FIGURE 6.5 **Counting scores results of Simulation 2—continuous versus discrete.** The ANNs that were evolved to count with continuous stimuli and were tested with discrete stimuli presented better counting skills than the other group, which evolved to count with discrete stimuli but was tested with continuous stimuli. The "*" represents significant differences: the counting score in both SP and C tasks when evolved with continuous and tested with discrete is significantly higher than the counting score received when evolved on discrete and tested on continuous, and is also significantly higher than the score of the control group (ie, evolved and tested on continuous stimuli); but when evolving on discrete stimuli and tested on continuous, the counting score is significantly lower than counting score of the control group (ie, evolved on discrete and tested on discrete stimuli). (Katz et al., 2013).

higher than all other stimuli combinations, $F(1, 232) = 4.21$, $MSE = 13.32$, $p < 0.05$, $\eta_p^2 = 0.02$. In *SP* runs, the number of inner nodes stayed the same for all stimuli combinations ($M = 3.43$, $SD = 1.72$).

6.6.5 Simulation 3a: Adding a Subitizing Task

In order to improve the counting score (that stood at approximately 60% accuracy in the counting tests in the previous simulations), six additional evolutionary run types containing subitizing tasks were created (Table 6.2).

6.6.6 Results

An ANOVA with an array of 3 (stimuli types) × 6 (evolutionary runs) was conducted. The following 6 evolutionary runs were performed:

1. Size perception (SP)—the ANNs evolved to perceive size.
2. Counting (C)—the ANNs evolved to count.
3. Subitizing (Sub)—the ANNs evolved to subitize.
4. SP-Sub—the ANNs evolved first to perceive size and then to subitize.

TABLE 6.2 Subitizing-Related Genetic Algorithms and Their Fitness Function

#	Evolution	Fitness
1	Sub	For a given input with 1–4 ones, the expected output contains the exact same number of ones. For inputs with $N > 4$ ones, an output of $N - 1$ or $N + 1$ ones is also acceptable.
2	SP-Sub	The size perception fitness function switched to the subitizing fitness function mid-run after 25 generations.
3	Sub-C	The subitizing fitness function switched to the counting fitness function mid-run after 200 generations (number of generations before switch chosen experimentally).
4	C	For a given input, the exact number of ones given is expected to be in the output (without order considerations).
5	SP-Sub-C	The size perception fitness function switched to the subitizing fitness function after 25 generations, and after 200 additional generations there was another switch to the counting fitness function.
6	C1-Sub-C	The control 1 (random classification task—see Table 6.1 item 4) fitness function switched to the subitizing fitness function after 25 generations, and after 200 additional generations there was another switch to the counting fitness function.

Note: SP, evolved to perceive size; C, evolved to count; Sub, evolved to excel in the subitizing task; SP-Sub, evolved first to perceive size and then to excel in subitizing; SP-Sub-C, evolved first to perceive size, then evolved to excel in subitizing and finally evolved to count; and C1-Sub-C, evolved first to excel in control 1 task, then evolved to excel in subitizing and finally evolved to count.

5. SP-Sub-C—the ANNs evolved first to perceive size, then evolved to subitize, and finally evolved to count.
6. C1-Sub-C—the ANNs evolved first to excel in the control 1 task, then evolved to subitize and finally evolved to count.

Each of the 6 evolutionary runs ran 3 times—each time with a new stimuli type:

1. Continuous stimuli only
2. Discrete stimuli only
3. Both continuous and discrete stimuli

6.6.6.1 Counting Score

The interaction between stimuli and evolution was significant, F (10, 522) = 2.68, $MSE = 0.01$, $p < 0.05$, $\eta_p^2 = 0.05$. All the combinations of *subitizing with counting* led to *better counting scores* than counting independently, $F(1, 522) = 162.54$, $MSE = 0.01$, $p < 0.01$, $\eta_p^2 = 0.24$. Most interesting were the following significant comparisons: *SP-Sub-C vs. C*: F (1, 522) = 106.04, $MSE = 0.01$, $p < 0.01$, $\eta_p^2 = 0.16$, and *C1-Sub-C vs. C*: F (1, 522) = 108.62, $MSE = 0.01$, $p < 0.01$, $\eta_p^2 = 0.17$, which is the classification control group.

In the majority of the runs, ANNs that evolved with continuous stimuli attained better counting scores than ANNs that evolved with discrete stimuli, F (1, 522) = 39.75, $MSE = 0.01$, $p < 0.01$, $\eta_p^2 = 0.07$ (Fig. 6.6).

6.6.6.2 Generations

A significantly *higher number of generations* were required to achieve peak performance in the counting task with a subitizing stage than without it [ie, *Sub-C* ($M = 518.4$, $SD = 36.44$) *vs.* C ($M = 308.22$, $SD = 4.32$)].

6.6.6.3 Inner Nodes

Discrete stimuli ANNs were composed of more inner nodes than continuous stimuli ANNs, meaning that the networks specialized in discrete stimuli were larger and more complex than the ones specialized in continuous stimuli, F (1, 522) = 52.88, $MSE = 30.098$, $p < 0.01$, $\eta_p^2 = 0.09$.

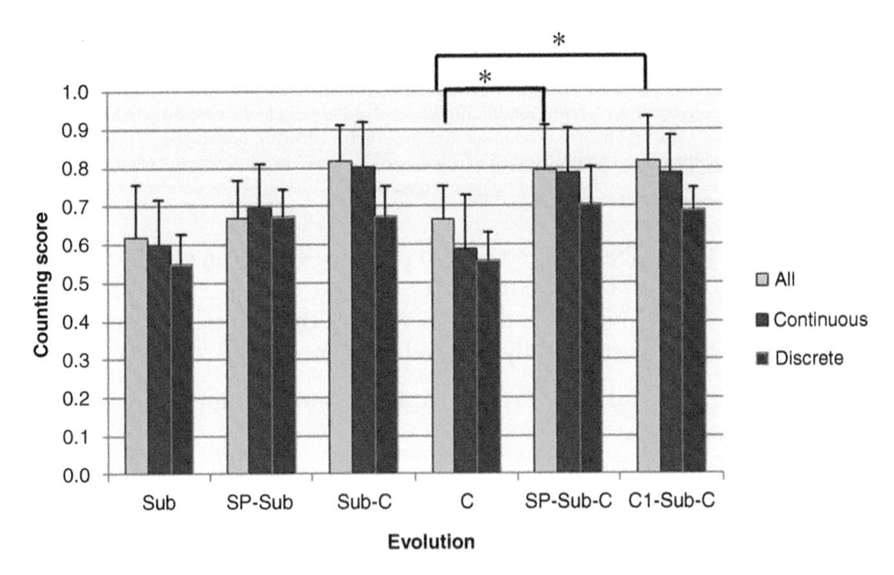

FIGURE 6.6 **Counting scores result of Simulation 3a—adding a subitizing task.** All the combinations of *subitizing with counting* led to better counting scores than counting independently, and in the majority of the runs, ANNs that evolved with continuous stimuli attained better counting scores than when evolved with discrete stimuli. *Sub*, subitizing; *SP*, size perception; *C*, counting; *C1*, control group with random classification task; *yellow*, ANNs that evolved and were tested with both types of stimuli (continuous and discrete); *purple*, ANNs that evolved and were tested with continuous stimuli; and *red*, ANNs that evolved and were tested with discrete stimuli. In all evolutionary runs, the actual stimuli were different between the training (by evolution) and testing sets. The "*" represents significant differences: the counting score in both SP-Sub-C and C1-Sub-C runs were significantly higher than the counting score recived in the C runs.

6.6.7 Simulation 3b: Continuous Versus Discrete With Subitizing

As in Simulation 2, in order to make sure that the high score for continuous stimuli in the counting test was not just due to continuous stimuli being less complex to process than discrete stimuli were, we performed another set of evolutionary runs, now with a subitizing task, and evolved the ANNs on continuous stimuli and tested them on discrete stimuli and vice versa.

6.6.8 Results

6.6.8.1 Counting Score

An ANOVA with an array of 4 (stimuli types) × 4 (evolutionary runs) was conducted. The following 4 evolutionary runs were performed:

1. Size perception (SP)—the ANNs evolved to perceive size.
2. Subitizing (Sub)—the ANNs evolved to subitize.
3. Counting (C)—the ANNs evolved to count.
4. SP-Sub-C—the ANNs evolved first to perceive size, then evolved to subitize, and finally evolved to count.

Each evolutionary run was performed 4 times, each time with a new stimuli type, as follows:

1. Evolved with *continuous* stimuli and tested with *discrete* stimuli
2. Evolved with *discrete* stimuli and tested with *continuous* stimuli
3. Evolved with *continuous* stimuli and tested with *continuous* stimuli
4. Evolved with *discrete* stimuli and tested with *discrete* stimuli

As in Simulation 2, we noted *significantly higher* counting scores when evolving ANNs with continuous stimuli and testing with discrete stimuli than the other way around, $F\ (1, 464) = 341.68$, $MSE = 0.009$, $p < 0.01$, $\eta_p^2 = 0.42$. In addition, the ANNs that evolved with continuous stimuli but tested with discrete stimuli were *better* than their control group (ie, that were evolved and tested on continuous stimuli), $F\ (1, 464) = 11.29$, $MSE = 0.009$, $p < 0.01$, $\eta_p^2 = 0.03$.

When evolved with discrete stimuli, the ANNs in the control group (that evolved and were tested with discrete stimuli) were *better in counting* than those who evolved with discrete stimuli and were tested with continuous stimuli, $F\ (1, 464) = 244.31$, $MSE = 0.009$, $p < 0.01$, $\eta_p^2 = 0.34$.

Finally, when comparing the ANNs that evolved with continuous stimuli (regardless of the type of stimuli in the counting tests) to the ones that evolved with discrete stimuli, we can see that SP-Sub-C was *significantly better in counting* than all other evolutionary runs, $F\ (1, 464) = 72.25$, $MSE = 0.009$, $p < 0.01$, $\eta_p^2 = 0.13$ (Fig. 6.7).

Note that the difference in counting score between the groups that evolved with continuous stimuli in the SP-Sub-C run was not significant,

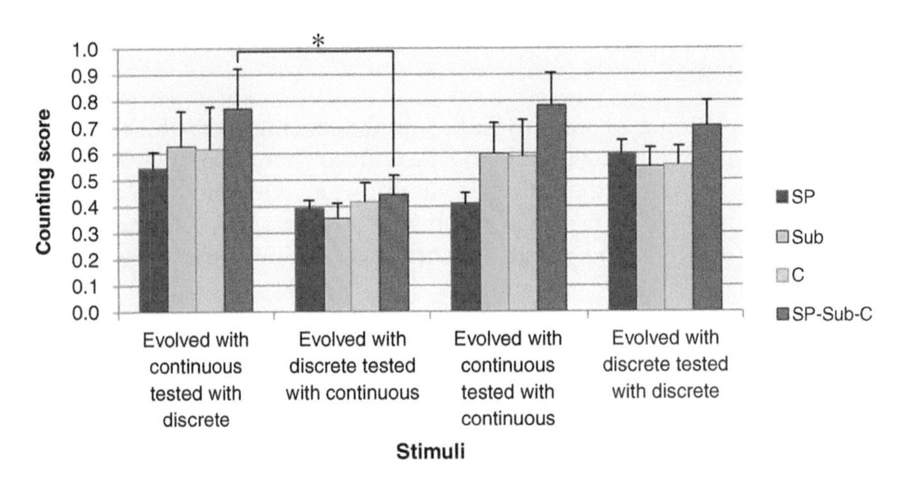

FIGURE 6.7 **Counting scores of Simulation 3b—continuous versus discrete with subitizing.** *SP*, ANNs that evolved to excel in size perception; *Sub*, ANNs that evolved to excel in subitizing; *C*, ANNs that evolved to excel in counting; and *SP-Sub-C*, ANNs that evolved to excel in all three tasks (size perception, subitizing, and counting). The "*" represents significant differences: the counting score in SP-Sub-C run was significantly higher when evolved with continuous and tested with discrete stimuli than the counting score received when evolved with discrete and tested with continuous stimuli.

F (1, 464) = 0.23, MSE = 0.009, p = 0.63, η_p^2 = 0.0005. The average counting score in the counting run when evolved with continuous stimuli and tested with discrete stimuli was M = 0.61, SD = 0.16. The improved counting score after adding the subitizing task between the size perception and the counting tasks was M = 0.77, SD = 0.15.

6.7 SUMMARY OF MAIN RESULTS

The results of Simulation 1—*From Size Perception to Counting*—indicated that the counting skills can be improved [ie, higher counting scores, less complex structure of the net (less inner nodes and less links), but it took more generations] if ANNs were first evolved to perform another, simpler, classification task (eg, size perception or some other classification task) and then evolved further to count (Katz et al., 2013).

In Simulation 2—*Continuous Versus Discrete*—we found that training with continuous stimuli resulted in significantly better counting skills than training with discrete stimuli, despite the reasonable assumption that discrete stimuli would lend themselves better to the counting task.[d]

In addition, evolving with discrete stimuli resulted in larger and more complex networks (ie, more inner nodes and links) than when evolving with continuous stimuli. Moreover, it seems that a certain division between continuous and discrete stimuli appears to be useful when training ANNs to improve their counting skills (Katz et al., 2013).

Finally, our results in Simulations 3a and 3b—*Adding a Subitizing Task and Continuous Versus Discrete With Subitizing*—indicated that subitizing was indeed a key stage in the evolution of counting systems.

It is important to mention that the division we did between discrete and continuous stimuli might not reflect nature. As mentioned in the Stimuli section (under Methods), we defined a stimulus as continuous if there was a path from each visible cell in the array to every other visible cell that passed through visible cells (in single up/down/left/right steps). Nevertheless, the division might have implications for the development of rehabilitation methods. For example, it is known that dyscalculic people have poor counting skills, but based on their compromised size effects (eg, size congruity effect: Rubinsten & Henik, 2005, 2006), they might also have deficits in the ANS or in the networks interfacing ANS with the exact counting system. Thus, training in size perception tasks with continuous stimuli might result in improving their counting skills.

6.8 DISCUSSION

In the current research, we examined the following two hypotheses: (1) the counting system developed through evolution from a more primitive size perception system and (2) both systems evolved independently in different epochs of time. According to our results, the counting system that evolves from scratch fails to excel at counting; *thus it is possible that the counting system we are familiar with might have evolved on the back of some less precise system, instead of evolving independently*.

However, it appears that better counting systems can evolve from different kinds of primitive systems (classification systems in our case) with no specific relation to size perception. Yet, this finding may be due to the size perception task that we chose (the BIG/SMALL decision) or because of the small dataset.

In addition, it took more generations to evolve a proficiency in counting in all the evolutionary runs that contained at least two tasks than when evolving only to count from scratch. This might be because the networks had to change their structure in order to adjust to a new task in the middle of the run (eg, to switch from size perception to counting, or from subitizing to counting).

The work of Kashtan, Noor, and Alon (2007) might shed some light on the reason for high counting scores that were as probable when evolving from populations previously trained for control tasks as they were when evolving from populations trained for size perception. Kashtan and coworkers made the ANNs switch between goals during an evolutionary learning process and found that as each new goal shared some subproblems with the previous goal, the networks evolved to be more modular and developed modules that were useful for both tasks. They found that

(a)

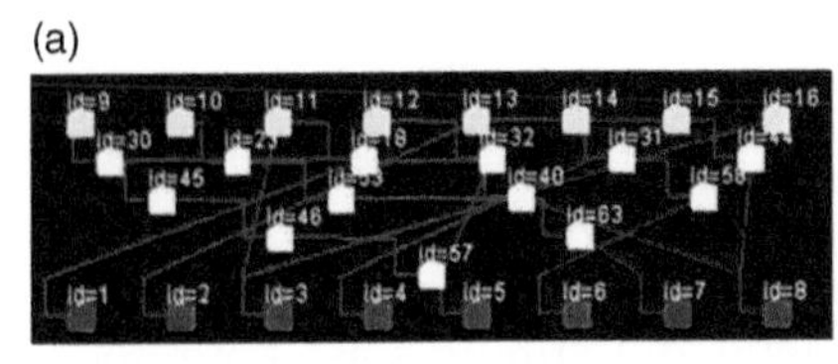

(b)

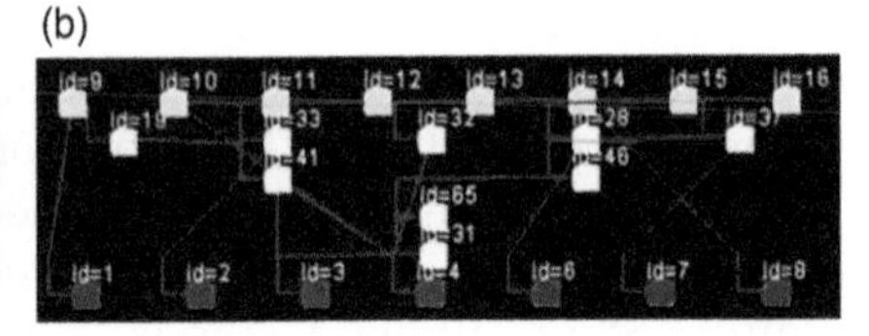

FIGURE 6.8 **ANN's structure after evolutionary process is done.** (a) The structure of the best ANN individual who evolved to count from scratch with discrete stimuli and tested with continuous stimuli, achieving a counting score of 57.4. (b) The structure of the best ANN individual who evolved to count from scratch with continuous stimuli and tested with discrete stimuli, achieving a counting score of 98.1. Interestingly, this individual survived with a mutation that removed input #5 node through evolution *Yellow*, output neurons; *Purple*, input neurons; *White*, inner neurons. We can see in (a) that there are more inner nodes when evolved to count with discrete and tested with continuous stimuli than the other way around (b), meaning the problem was more complex for the ANN in (a) thus additional inner nodes were required in order to perform the counting task in this case. (Katz et al., 2013).

modularly varying goals could push populations away from local fitness maxima, thus guiding them toward evolvable and modular solutions; in addition, the harder the problem, the faster the speedup. Although, in the current study, the goal was switched between size perception and counting only once, in future research we may switch the goals alternately until the ANNs excel at both tasks.

6.8.1 Complexity of the Net

Overall, when we examined the networks structure, we saw that ANNs who evolved to count with discrete stimuli had more complex networks than when evolving with continuous stimuli (ie, more inner nodes, more links between nodes). In addition, evolving with continuous but testing with discrete led to better counting scores than the other way around. When we examined the structure of the ones who evolved on discrete, we saw an economical (ie, less inner nodes, less links), well-organized structure. Fig. 6.8 shows the structures of two ANN individuals from the counting (C) runs. The bottom layer is the input layer, the top layer is the output layer, and the inner nodes are found in between them. Note that each individual ANN at the beginning of the evolutionary process has only 8 inputs and 8 outputs. The hidden nodes and the network's connections change during the evolutionary process resulting in the final networks.

References

Brannon, E. M., Lutz, D., & Cordes, S. (2006). The development of area discrimination and its implications for number representation in infancy. *Developmental Science, 9*, F59–F64.

Cantlon, J. F., & Brannon, E. M. (2007). How much does number matter to a monkey? *Journal of Experimental Psychology: Animal Behavior Processes, 33*, 32–41.

Cantlon, J. F., Platt, M. L., & Brannon, E. M. (2009). Beyond the number domain. *Trends in Cognitive Sciences, 13*, 83–91.

Cantlon, J. F., Piantadosi, S., Ferrigno, S., Hughes, K., & Barnard, A. (2015). The origins of counting algorithms. *Psychological Science, 26*(6), 853–865.

Cohen Kadosh, R., Henik, A., Rubinsten, O., Mohr, H., Dori, H., Van de Ven, V., Zorzi, M., Hendler, T., Goebel, R., & Linden, D. (2005). Are numbers special? The comparison systems of the human brain investigated by fMRI. *Neuropsychologia, 43*, 1238–1248.

Dehaene, S. (2001). Is the number sense a patchwork? *Memory and Language, 16*, 89–100.

Fias, W., Lammertyn, J., Reynvoet, B., Dupont, P., & Orban, G. A. (2003). Parietal representation of symbolic and nonsymbolic magnitude. *Journal of Cognitive Neuroscience, 15*, 47–56.

Flavell, J. H. (1963). *The developmental psychology of Jean Piaget*. New York: Van Nostrand Reinhold.

Goldberg, D. E. (1989). *Genetic algorithms in search, optimization and machine learning*. Reading, MA: Addison-Wesley.

Halberda, J., Mazzocco, M. M. M., & Feigenson, L. (2008). Individual differences in nonverbal number acuity correlate with maths achievement. *Nature, 455*, 665–668.

Henik, A., Leibovich, T., Naparstek, S., Diesendruck, L., & Rubinsten, O. (2012). Quantities, amounts, and the numerical core system. *Frontiers in Human Neuroscience, 5*, 186.

Henik, A., Rubinsten, O., & Ashkenazi, S. (2014). Developmental dyscalculia as a heterogeneous disability. In R. Cohen Kadosh, & A. Dowker (Eds.), *The Oxford handbook of numerical cognition* (pp. 662–677). Oxford: Oxford University Press.

Kashtan, N., & Alon, U. (2005). Spontaneous evolution of modularity and network motifs. *Proceedings of the National Academy of Sciences, 102*, 13773–13778.

Kashtan, N., Noor, E., & Alon, U. (2007). Varying environments can speed up evolution. *Proceedings of the National Academy of Sciences, 104*, 13711–13716.

Katz, G., Benbassat, A., Diesendruck, L., Sipper, M., & Henik, A. (2013). From size perception to counting: an evolutionary computation point of view. In: *Proceedings of the 15th annual conference companion on genetic and evolutionary computation (GECCO 2013) companion* (pp. 1675–1678). New York: ACM.

Kaufman, E. L., Lord, M. W., Reese, T. W., & Volkmann, J. (1949). The discrimination of visual number. *American Journal of Psychology, 62*, 498–525.

Kosc, L. (1974). Developmental dyscalculia. *Journal of Learning Disabilities, 7*, 159–162.

Lourenco, S. F., Bonny, J. W., Fernandez, E. P., & Rao, S. (2012). Nonsymbolic number and cumulative area representations contribute shared and unique variance to symbolic math competence. *Proceedings of the National Academy of Sciences, 109*, 18737–18742.

Piaget, J. (1955). *The language and thought of the child*. Cleveland, OH: World Publishing Company.

Pinel, P., Piazza, M., Le Bihan, D., & Dehaene, S. (2004). Distributed and overlapping cerebral representations of number, size, and luminance during comparative judgments. *Neuron, 41*, 983–993.

Price, G. R., Holloway, I., Räsänen, P., Vesterinen, M., & Ansari, D. (2007). Impaired parietal magnitude processing in developmental dyscalculia. *Current Biology, 17*, R1042–R1043.

Rubinsten, O., & Henik, A. (2005). Automatic activation of internal magnitudes: a study of developmental dyscalculia. *Neuropsychology, 19*, 641–648.

Rubinsten, O., & Henik, A. (2006). Double dissociation of functions in developmental dyslexia and dyscalculia. *Journal of Educational Psychology, 98*, 854–867.

Shalev, R. S. (2007). Prevalence of developmental dyscalculia. In D. B. Berch, & M. M. M. Mazzocco (Eds.), *Why is math so hard for some children? The nature and origins of mathematical learning difficulties and disabilities* (pp. 49–60). Baltimore, MD: Paul H. Brookes Publishing Co.

Stanley, K. O., & Miikkulainen, R. (2002). Evolving neural networks through augmenting topologies. *Evolutionary Computation, 10*(2), 99–127.

Stoianov, I., & Zorzi, M. (2012). Emergence of a "visual number sense" in hierarchical generative models. *Nature Neuroscience, 15*(2), 194–196.

Verguts, T., & Fias, W. (2004). Representation of number in animals and humans: a neural model. *Journal of Cognitive Neuroscience, 16*(9), 1493–1504.

SECTION II

ANIMAL STUDIES

CHAPTER

7

Number Versus Continuous Quantities in Lower Vertebrates

*Christian Agrillo****, Maria Elena Miletto Petrazzini**, Angelo Bisazza*****

*Department of General Psychology, University of Padova, Padova, Italy; **Cognitive Neuroscience Center, Padova, Italy

OUTLINE

7.1 INTRODUCTION

The capacity of estimating, storing in memory, and comparing quantities is of prime importance for all aspects of the relationship of an animal with its environment. For example, this ability is essential for movement and orientation in space and to make optimal foraging decisions. A scout honeybee (*Apis mellifera*) needs to evaluate the distance to a newly discovered patch of flowers and transfer this information to hive mates because successful exploitation of an abundant resource depends on rapid recruitment of many other foragers (von Frisch, 1967). Striped field mice (*Apodemus agrarius*) prefer to catch red wood ants (*Formica polyctena*) in small groups because ants may bite them when they prey on large aggregates

Continuous Issues in Numerical Cognition. http://dx.doi.org/10.1016/B978-0-12-801637-4.00007-X

(Panteeleva, Reznikova, & Vygonyailova, 2013). The decision to initiate a conflict often depends on the precise assessment of the relative size of the opponent. Male oscar fish (*Astronotus ocellatus*) can visually estimate the relative size of an opponent and promptly attack only rivals that are smaller than themselves (Beeching, 1992). Among chimpanzees (*Pan troglodytes*) that cooperatively defend territories, the decision to retreat or enter an intergroup contest depends on the estimation of the relative numerosity of the opponent group (Wilson, Britton, & Franks, 2002).

Quantity estimation is also very important for optimal resource allocation during reproduction. Females of the freshwater goby (*Padogobius bonelli*) choose males whose nests have the greatest surface available for egg deposition (Bisazza, Marconato, & Marin, 1989), and female swordtails (*Xiphophorus helleri*) prefer males with the longest caudal fin (Basolo, 1990). To ensure hatch synchrony with the host, females of the brood parasite wood ducks (*Aix sponsa*) examine several hosts' nests and lay eggs in nests with smaller clutches because this indicates that a host female has probably not initiated incubation (Odell & Eadie, 2010). Finally, a precise estimation of quantities is often crucial to evade predation. Individual ostriches (*Struthio camelus*), for example, increase time devoted to vigilance as the number of individuals foraging in the same patch decreases (Bertram, 1980) and banded killifish (*Fundulus diaphanous*) form larger shoals when exposed to the odor of their predator because larger groups better protect the individual fish from being captured (Hoare, Couzin, Godin, & Krause, 2004).

From the previous examples it is clear that in many cases the quantity to be assessed is continuous (eg, the area of the nest, the size of the rival, or the length of male caudal fin), while in other cases, animals need to estimate the number of discrete items contained in a set (the number of eggs in the nest, ants in a patch or social companions in a group). It is interesting to note that even in the latter cases, the numerosity of items can often be estimated using continuous quantities as proxy. Clutch size could be inferred, for instance, by the overall volume occupied by eggs and the quantity of ant prey from the cumulative area occupied or from the total amount of movement observed. Because the capacity to make optimal decisions is dependent on the type of cognitive mechanism involved, it is important to understand the exact process through which an animal estimates the number of objects, that is, if it enumerates objects, uses nonnumerical cues that covary with number or bases its assessment on multiple mechanisms. To discriminate among these alternatives is often complex and is achieved through complex laboratory experiments in which the different types of variables are controlled for (eg, Beran & Beran, 2004; Gómez-Laplaza & Gerlai, 2013).

There is a long tradition of experimental investigation in mammals and birds for discrimination of both continuous quantities and numerical abilities (Koehler, 1951; Kuroda, 1931; Lashley, 1912), whereas the study of such skills in other vertebrates began only in recent years. Many examples of reptiles, amphibians, and fish capable of estimating continuous quantities (length, area, volume, weight, duration, etc.) can be found in behavioral biology (eg, Andersson, 1994; Pyke, Pulliam, & Charnov, 1977). For example, female river bullheads (*Cottus gobio*) select mates primarily on the basis of their body size (Bisazza & Marconato, 1988); among sand lizards (*Lacerta agilis*), males were less likely to attack opponents in which the area of the nuptial green coloration was experimentally enlarged, while females were more likely to mate with these males, suggesting that both sexes were able to assess the size of the badge (Olsson, 1994). With very few exceptions, none of these studies investigated the limits of discrimination capacities or the mechanisms underlying these faculties (Mark & Maxwell, 1969). Conversely, in the last 10 years there has been considerable interest in studying numerical abilities of amphibians and particularly of fish, whose abilities to discriminate different numerical quantities appear in many respects comparable to those observed in many mammals and birds.

In this chapter we will review the literature about numerical cognition in fish and amphibians. However, because in most cases a numerical task can be solved by using information on continuous measures, this aspect will always be taken into account. Initially, we will summarize the literature based on free choice tests. These studies will be divided based on the type of stimulus presented (food/social companions) and the type of control for continuous quantities (no control/continuous quantities sequentially equated/item-by-item procedure to prevent the use of multiple continuous quantities). Then, we will present the studies using operant conditioning in which subjects are trained to discriminate between two groups of inanimate objects (eg, dots) controlled for continuous quantities. The last part of the chapter will be devoted to delineating some hypotheses on why mixed data are reported in the literature on the use of number over continuous quantities by lower vertebrates.

7.2 METHODOLOGIES FOR THE STUDY OF QUANTITY DISCRIMINATION

Two main approaches have been adopted in the literature to study numerical abilities: free choice tests that require little or no training and tests with more extensive training procedures. Several authors believe that these two methodologies might recruit partially different neural circuits and hence lead to different results (reviewed in Agrillo & Bisazza, 2014). Due to the primary relevance of these methodological questions in the issue of "number versus continuous quantity," we will first summarize the existing literature by taking into account the different methodological approaches.

7.2.1 Free Choice Tests

Free choice tests typically take advantage of a natural tendency of a species to prefer more or less of something. An animal is generally presented with two sets containing different numbers of biologically relevant stimuli. Choice of different quantities of food items is the most common situation used in studies with mammals and birds. For instance, one study tested the spontaneous ability of elephants to choose the larger quantity of carrots presented on two separate plates (Perdue, Talbot, Stone, & Beran, 2012). In order to get one of the two options, the subjects had to reach for the preferred plate. The assumption underlying this type of study is that, if subjects are able to make relative quantity judgments, they are expected to spontaneously select the larger quantity.

For lower vertebrates, two types of stimuli are used, social companions (shoal quantity discrimination in fish) and food items (food quantity discrimination in fish and amphibians).

7.2.1.1 Shoal Quantity Discrimination

Sociality is the main antipredatory strategy for many fish and the advantages of living in a group generally increase as the number of individuals in the group increases. For this reason, in an unknown environment individual fish tend to join other conspecifics, and if two shoals are present, individuals exhibit a preference for the larger shoal (Buckingham, Wong, & Rosenthal, 2007). Several studies have taken advantage of this tendency to assess the limits of quantity discrimination. Typically, a single fish is moved into an unfamiliar empty tank where it can see two groups of social companions. The proportion of time spent near the larger of the two groups is recorded as a measure of discrimination (Fig. 7.1a).

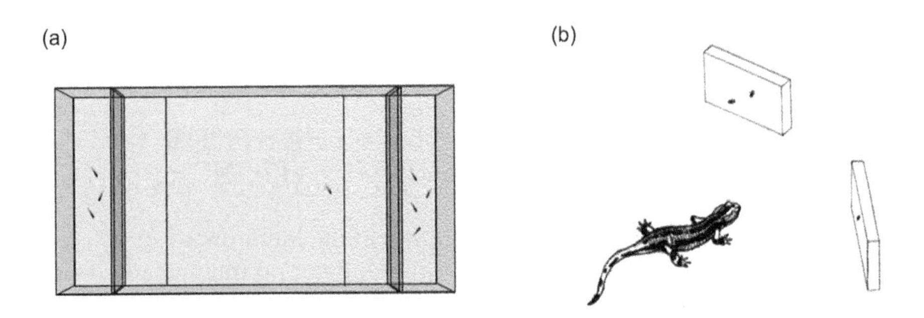

FIGURE 7.1 **Stimuli used in free choice tests are biologically relevant items, such as food items or prey.** For instance, guppies are observed in their preference for the largest group of social companions (a) and salamanders in their spontaneous preference for reaching the largest group of prey (b). Due to the nature of the stimuli (living organisms), the control of continuous quantities is hardly possible in free choice tests.

Experiments using this procedure demonstrated that mosquitofish (*Gambusia holbrooki*) discriminate between groups differing by one unit up to four items (1 vs. 2, 2 vs. 3, and 3 vs. 4, but not 4 vs. 5; Agrillo, Dadda, Serena, & Bisazza, 2008). A closely related species, the guppy (*Poecilia reticulata*), exhibited the same limit when tested in comparable conditions (Agrillo, Piffer, Bisazza, & Butterworth, 2012), while angelfish (*Pterophillum scalare*) were found to discriminate 1 versus 2 and 2 versus 3 but not 3 versus 4 fish (Gómez-Laplaza & Gerlai, 2011). Larger quantities can also be discriminated but only by increasing the numerical ratio between the smaller and the larger quantity: mosquitofish could discriminate 8 versus 16 (0.50 ratio) but not 8 versus 12 (0.67). Guppies (Agrillo et al., 2012b) and swordtails (*X. helleri*, Buckingham et al., 2007) were able to discriminate a 0.50 ratio and angelfish a 0.56 ratio (Gómez-Laplaza & Gerlai, 2011). In guppies, mosquitofish, and swordtails, accuracy of discrimination of numbers greater than four appears to decrease with increasing numerical ratio, in agreement with Weber's law.

Numerical abilities were also studied in male mosquitofish in a situation of mate choice. Males were found to select the larger group of females in 1 versus 3 and 2 versus 4 comparisons (Agrillo, Dadda, & Serena, 2008). When the same number of females was visible, male mosquitofish preferred a group not containing males; however, no preference for the more favorable sex ratio was observed when the number of males per group was varied (2 males + 3 females vs. 4 males + 3 females). This suggests that male mosquitofish probably cannot simultaneously assess the quantity of males and females within a group.

As readers could have noted, in these types of tests all stimulus fish were visible at the time of subjects' choice, and fish could have used continuous quantities, such as the cumulative surface area of the stimuli, their density or the total activity within the shoals. In short, it is not possible to establish the exact mechanism adopted by fish to select the larger quantity in this task.

7.2.1.2 Control for Continuous Quantities in Shoal Quantity Discrimination

Some studies have tried to investigate the mechanisms involved in shoal numerosity discrimination by controlling one continuous quantity at a time. Agrillo et al. (2008b) studied spontaneous preferences of mosquitofish in a 2 versus 3 and 4 versus 8 discrimination in which larger individuals were placed in the smaller shoals so that the overall volume occupied by stimulus fish was the same for each numerical contrast. Interestingly, when the overall volume of the stimuli was equated, subjects showed no significant choice preference in either numerical contrast, highlighting the importance of the overall volume in this type of discrimination. More recently, Gómez-Laplaza and Gerlai (2012) investigated whether the capacity of angelfish to select the larger shoal was affected by the density in the two stimulus shoals (5 vs. 10 discrimination). In this experiment density was kept constant, by reducing the volume of the stimulus compartment by 50% for the smaller of the two stimulus shoals. In this condition angelfish were unable to select the larger shoal, indicating the spontaneous use of density information at least in this species.

The degree of movement of stimulus fish is another continuous quantity that might reveal the numerosity of a shoal. In one study the use of this cue was investigated by varying the temperature of water of the stimulus aquaria in a shoal choice task done in zebrafish (Pritchard, Lawrence, Butlin, & Krause, 2001). Like many other fish, zebrafish live in a range of temperatures and their activity increases as water temperature increases. This study found that, when the two stimulus shoals were at the same temperature, subjects generally preferred the larger shoal (the discrimination was 2 vs. 4). However, this preference was abolished when the temperature of the larger shoal was reduced reducing fish activity, suggesting that perhaps the subject fish preferred more active shoals as well as more numerous ones. More recently, Agrillo et al. (2008b) showed that in mosquitofish the difference in temperature affects shoal choice in a discrimination of 4 versus 8 but not in a discrimination of 2 versus 3.

An alternative procedure to equate the degree of movement consists in confining each stimulus fish in a single sector that restricts its movement. Using this procedure, Gómez-Laplaza and Gerlai (2012) found that angelfish continue to discriminate 5 versus 10 social companions after the total movement of the two schools was equated. This study also replicated with angelfish the procedure used with zebrafish and mosquitofish and found that controlling total movement with temperature affects the capacity to discriminate the larger group, as in the other two species. The two experiments on angelfish thus found different outcomes depending on which method was used to control for total movement of the shoals. This result highlights the importance of using multiple experimental strategies to control for continuous quantities before drawing conclusions about the mechanisms involved in relative quantity judgments of nonhuman animals.

The aforementioned procedures are useful to assess the contribution of different continuous quantities, but they can potentially introduce other confounding factors. For instance, when the experimental procedure controls for body size of stimuli, one must consider that fish often prefer to shoal with similar-sized individuals in order to minimize their conspicuousness in case of predatory attack (Landeau & Terborgh, 1986). When total quantity of movement of the stimulus shoals is controlled for, the behavior of stimulus fish in the larger shoal might appear less attractive for the subjects. As several fish species often remain immobile to minimize the risk of being detected by predators (an antibehavior strategy called "freezing," (see Chivers & Smith, 1994, 1995), reduced movement of stimuli in one group might function as an alarm cue on that side of the tank. Hence, the lack of preference for the larger shoal reported in these studies does not necessarily imply that volume or total activity are the only factors that determined fish choice.

Another possible strategy to reduce the use of continuous variables involves simultaneously presenting two groups of conspecifics differing in numerosity, but making the two groups visually unavailable at the time of choice. Using this procedure, Stancher, Sovrano, Potrich, and Vallortigara (2013) found that males of redtail splitfin (*Xenotoca eiseni*) are able to spontaneously select the larger number of females of the same species (sexual stimuli) in groups of 1 versus 2, 2 versus 3, and 3 versus 4. No ability for discrimination was reported with larger numerosities. However, this procedure does not entirely exclude the possibility that animals use continuous quantities, as subjects can potentially compare the different areas occupied by stimuli when they are visible and then remember the position occupied by the larger amount prior to its disappearance.

7.2.1.3 Preventing Access to Continuous Quantities

Some of the limitations to controlling for continuous quantities in free choice experiments can be overcome using an experimental strategy called "item-by-item presentation." This procedure has been frequently adopted with mammals (ie, chimpanzees: Beran, McIntyre, Garland, & Evans, 2013; rhesus monkeys: Hauser, Carey, & Hauser, 2000; elephants: Perdue et al., 2012) and birds (New Zealand robins: Hunt, Low, & Burns, 2008; chicks: Rugani, Fontanari, Simoni, Regolin, & Vallortigara, 2009) and is based on the sequential presentation of items within each set. For example, in one study (Hauser et al., 2000) macaques could observe experimenters who placed pieces of apple, one at a time, into each of two opaque containers before they were actually allowed to choose one of the two containers. To solve the task, animals had to build a representation of the contents of the two containers on the basis of the items that came sequentially into view. Then they had to compare the two representations and choose the more profitable one. In this way, a subject is prevented from having a global view of the entire contents of the groups and using continuous variables such as the total surface area occupied by stimuli.

A variation of the item-by-item procedure was recently applied using social companions as stimuli to assess whether mosquitofish use a spontaneous number representation when choosing between shoals differing in numerosity. Researchers used a modified version of the shoal choice test in which stimulus fish were singly confined in separate compartments of the stimulus tanks. Inside the subject tank, several opaque screens prevented the possibility that subjects saw more than one stimulus fish at a time. In this way, subjects were required to enumerate stimuli on one side, make the same operation on the other side, and finally compare the two numerosities. In this experiment mosquitofish spent more time near the larger shoal in 2 versus 3 and 4 versus 8 discriminations. Density of individuals and the overall space occupied by the stimulus shoals were controlled in two further tests. As mosquitofish were able to select the larger shoal also in those control tests, it suggests that fish had a spontaneous number representation of the two shoals (Dadda, Piffer, Agrillo, & Bisazza, 2009). In a subsequent study using the same apparatus and procedure, newborn guppies proved able to discriminate 2 versus 3 individuals and juvenile fish (40-day-old guppies) discriminated 4 versus 8 fish (Bisazza, Piffer, Serena, & Agrillo, 2010).

7.2.1.4 Food Quantity Discrimination

Food is certainly the most common stimulus used in numerical cognition studies with mammals (Hauser et al., 2000), and birds (Bogale, Aoyama, & Sugita, 2014). As stated in Section 7.2.1, one of the simplest procedures adopted in comparative psychology to assess whether animals display the capacity to discriminate between quantities consists of simultaneously presenting two groups of food items that remain visible until the time of choice. As the animals try to optimize food intake, the assumption is that they should select the larger quantity of food.

This type of methodology was used for the first time in lower vertebrates by Uller, Jaeger, Guidry, and Martin (2003) with salamanders (*Plethodon cinereus*). When given a choice between tubes containing fruit flies differing in numerosity (Fig. 7.1b), salamanders selected the larger group in tests of 1 versus 2 and 2 versus 3, but not 3 versus 4. Similar results have been reported in another amphibian. Stancher, Rugani, Regolin, and Vallortigara (2015) found that bombinas (*Bombina orientalis*) were able to discriminate 1 versus 2 and 2 versus 3, but not 3 versus 4 mealworms. They could also discriminate larger quantities provided that the ratio between the smaller and the larger quantity was at least equal to 0.50 (3 vs. 6, 4 vs. 8, but not 4 vs. 6).

Concerning fish species, spontaneous choice of different quantities of food items has been studied only recently. Lucon-Xiccato, Miletto Petrazzini, Agrillo, and Bisazza (2015) presented adult guppies with the choice of two plastic cards to which small pieces of food flakes had been previously glued. Guppies were shown the following numerical contrasts: 1 versus 4, 2 versus 4, 2 versus 3, and 3 versus 4 food items of the same size. They were able to select the larger quantity in 1 versus 4 and 2 versus 4, but no discrimination was found in the other two contrasts.

However, as stated in Section 7.2.1.1, the nature of quantitative abilities is unclear in these studies. It is possible that the subjects were selecting the larger group on the basis of cumulative surface area or density. In addition, in the salamanders' study, the group containing more flies had a higher probability of having at least one insect active, thus increasing the probability of it being detected by the amphibians' eyes. Hence, experimental strategies are necessary to assess the relevance of continuous quantities in food choice tasks.

7.2.1.5 Control for Continuous Quantities in Food Quantity Discrimination

As with shoal choice experiments, the easiest experimental strategy to assess possible use of continuous quantities is to control one continuous quantity at a time. For example, to shed light on the mechanisms adopted by salamanders, Krusche, Uller, and Dicke (2010) observed spontaneous behavior of salamanders when 8 and 16 potential prey were presented. Stimuli in this experiment were live crickets, videos of live crickets, or images animated by a computer program. When no control was exerted, salamanders discriminated the larger group, but they chose randomly when the experimenter controlled for the total movement of the stimuli. This suggests that a continuous quantity, such as the total activity of prey, is a dominant feature in salamanders' foraging behavior (Krusche et al., 2010).

With respect to amphibians, Stancher et al. (2015) made a control test to assess the influence of movement and volume in bombinas' quantity discrimination. Motion cues were controlled in two different ways: in the first test (1 vs. 2 mealworms), the single-item stimulus was composed of a living (hence moving) mealworm, while the two-item stimulus was composed of a living mealworm and a dead one. In the second test, subjects were presented with living mealworms: the single-item stimulus was immobilized in a single central position and both the anterior and the posterior parts of the body could oscillate; on the contrary, mealworms included in the two-item group were each fixed in two parts (center and posterior parts). In this way, they could move only the anterior part of their body, thus approximately matching the total number of movements made by the single-item stimulus. Volume was controlled in a third test in which opaque partitions selectively hide some portions of the stimulus body from the subject's visual perspective, while the single mealworm was entirely visible to the subject. Bombinas selected the larger group in test 2 and test 3, thus showing that volume and movement did not play a crucial role in their relative quantity judgments. The lack of a choice preference reported in test 1 was ascribed by the authors to the fact that dead prey was probably neglected by the subjects, because bombinas usually prefer to catch moving prey (Stancher et al., 2015). Even though bombinas do not seem to use volume when the alternative prey differ in numerosity, they are still capable of discriminating between volume of prey when numerical information is made irrelevant, as subjects appeared to select the most advantageous option between a smaller and a larger mealworm with a 0.5 ratio.

Regarding further research on guppies, Lucon-Xiccato et al. (2015) conducted a second experiment in which they controlled for the overall quantity of food by equating the cumulative surface area in a 2 versus 4 comparison (a comparison that guppies were able to discriminate when number and volume were congruent). When stimuli were controlled for volume, guppies no longer selected the group containing the larger number of food items. This result might appear surprising at first, as it would seem that, when continuous variables are controlled for, guppies fail to choose the largest quantity of food items even in numerical discriminations, such as 2 versus 4, that they easily discriminate in other contexts using numerical information only. However, this outcome is perfectly explainable if one takes into account the evolutionary pressures that likely have shaped decisional mechanisms of foraging behavior of most species. Natural selection in fact is expected to favor mechanisms that maximize the amount of food (ie, calories) gained and thus the total volume rather than the number of items of food. Some of these controls ensure identical calorie amounts in both choice tests, so there is no reason for the animal to discriminate between them. If two sets differ in number, but the amount of food and the handling and consumption times are identical for both choices, we should not expect a preference from the animal. In brief, controlling for continuous quantities in numerical experiments with choice of food items appears very difficult when both arrays are simultaneously available as the predicted outcome of these experiments is a lack of a preference based on number of items irrespective of whether animals are able to discriminate numerosity or not.

Actually in the experiment we just described (Lucon-Xiccato et al., 2015) guppies significantly selected the smaller set composed by larger items even if the two options contained the same total amount of food, a result that was not predicted even on the basis of the hypothesis that natural selection is optimizing the choice of the largest total quantity of food. One possible explanation for this result is that guppies were attracted by the largest available food item, which, as a consequence of controlling for total surface area, was more often present in the set with a smaller number of food items. However, food items often showed small differences in area, and one should hypothesize that guppies are much better at discriminating the size of food items than their number. To examine this hypothesis, fish were given the choice between two food items differing in area by the same ratios adopted in the numerical experiment (0.25, 0.50, 0.67, 0.75). In this task guppies were extremely accurate, being able to discriminate between two food items that differ by 0.75 in surface area though showing a ratio-dependent performance as if they were following Weber's law. Thus, guppies discriminated at least up to a 0.75 ratio between two areas, and up to a 0.5 ratio if these quantities were fragmented into more units as they were in the numerical experiment.

This result also has indirect implications for the debate about the existence of one or several independent magnitude systems. Several studies in humans (eg, Agrillo & Piffer, 2012; Bueti & Walsh, 2009; Vicario, 2011) suggest the existence of a common magnitude system for nonsymbolic estimation of time, space, and number, which has led to "a Theory of Magnitude" (ATOM, Walsh, 2003). Recent evidence supports this view in nonhuman animals as well (Mendez, Prado, Mendoza, & Merchant, 2011). However, if a single magnitude system underlies numerical and spatial abilities, one should expect the same accuracy when processing both types of magnitude. The observation that guppies were much more accurate in discriminating areas than quantities of objects speaks against the existence of a single magnitude system in this species, although we must acknowledge the possibility that, as numerical information does not seem to be the most salient cue in food choice, subjects might have largely ignored this type of information.

A strong preference for large food items was indeed demonstrated in another experiment in which guppies were given the choice between six equally small pieces of food and two pieces of food, one large and one small. Fish showed a strong preference for the latter set even if the total quantity of food was twice as much in the former option. Why do guppies exhibit such a strong preference for selecting the group containing the largest food item even when it leads them to choose a less profitable option? It is worth noting that guppies in the wild forage in groups and compete for prey items, hence natural selection might have favored cognitive mechanisms allowing a rapid and efficient choice of the most profitable piece of food. In short, while an individual is processing one food item in the patch, shoal mates will probably consume the residual items. In this condition, there is probably a strong advantage for the individual that detects and consumes the larger item of food immediately when a new patch is encountered. This hypothesis is supported by the observation that the same pattern of preference was observed in a distantly related species, the chimpanzee, that forages in groups and competes for food (Boysen, Berntson, & Mukobi, 2001). For instance, when presented with two quantities of food items, chimpanzees usually selected the largest amount of food, but they showed a preference for the smaller set if this contained the single largest food item (Beran, Evans, & Harris, 2008).

It is interesting to note that guppies can discriminate between three and four social companions (0.75 ratio, Agrillo et al., 2012b) but their discriminative threshold did not exceed a 0.50 ratio when comparing different sets of food items (2 vs. 4, Lucon-Xiccato et al., 2015). The comparison of different studies in guppies suggests that this species possesses a sophisticated mechanism to estimate the quantities of social companions, but that this system is specific to the social domain and cannot be used to easily represent numbers in other contexts. This points toward the existence of distinct quantification systems in guppies that are characterized by domain and task specificity, each system operating largely independently from the others, instead of the existence of a single supramodal and domain-independent quantitative system.

7.2.2 Training Procedures

The use of free choice tests for studying numerical abilities presents two main limits. First, motivation plays an important role, and null results do not necessarily imply a lack of discrimination. For instance, if a salamander does not show a significant preference in 12 versus 16 fruit flies, this might occur simply because subjects are not motivated to select the larger group because both patches provide more food than the animal can handle. Second, many experimental procedures have been devised to control for continuous quantities, but such controls remain complicated when dealing with living organisms (such as prey or social companions) as stimuli, and, as noted, there is also the difficulty of making animals attend to number when other dimensions are more salient in terms of maximizing food reward.

As an alternative, classical and operant conditioning have often been adopted to study numerical cognition, especially in mammals and birds (Beran, 2006; Davis, 1984). Typically subjects are trained to learn a numerical rule in order to receive a reward. Neutral stimuli are associated with a food reward on the basis of some rules such as choosing the larger item, or the larger number of items. For instance, a monkey can be trained to select the larger group of dots presented on the monitor in order to obtain a single food pellet, thus dissociating how large the stimulus arrays are from how much food is ingested for selecting those arrays. The subject is first trained to use the apparatus and undergoes a shaping procedure in order to learn the association of reward with the correct response. The subject is then presented with a sequence of numerical discriminations until it reaches a learning criterion—normally 75 or 80% correct choices in two consecutive sessions (Bogale, Kamata, Mioko, & Sugita, 2011; Emmerton & Renner, 2009; Jordan, MacLean, & Brannon, 2008; Vonk & Beran, 2012). Then, critical tests and training trials are presented until sufficient data are collected to assess performance.

Currently, studies using training procedures with lower vertebrates are limited and are entirely confined to fish. Both food and social rewards have been used. Agrillo, Dadda, Serena and Bisazza (2009) trained mosquitofish to discriminate between different numbers of two-dimensional figures using access to conspecifics as a reward. Subjects were singly inserted into an unfamiliar square environment and provided with two doors at the opposite corners, one associated with three figures and the other associated with two figures (randomly determined across trials).

Only the door associated with a given numerosity permitted access to another compartment containing social companions, and subjects needed to discriminate the two numerosities in order to obtain the social reward. In the first experiment of this study, mosquitofish were trained to discriminate between two and three figures with no control for continuous quantity until they learned the task. Subjects were then retested while controlling for a continuous quantity in various ways: Their performance was unaffected by controlling the sum of perimeters and controlling total brightness but dropped to chance level when stimuli were matched for the cumulative surface area or for the overall space occupied by the arrays, indicating that these latter cues had been spontaneously used by the fish during the learning process. In a second experiment, fish were trained in a condition in which only numerical information could be used, meaning that cumulative surface area, luminance of the sets, density, and overall space occupied by the most lateral stimuli were simultaneously controlled for. Mosquitofish proved able to learn the discrimination, showing an ability to enumerate the figures without the aid of nonnumerical cues. Apparently, the learning rate did not differ in the two experiments, although we must note that methodological differences between the two experiments did not permit a fine comparison of learning processes in the two conditions (number + continuous quantities vs. number only). This issue was specifically investigated in a subsequent study (see Agrillo, Piffer, & Bisazza, 2011; Section 7.3).

Another study (Agrillo, Piffer, & Bisazza, 2010) using the same methodology investigated whether a spontaneous use of continuous quantities also occurs for larger numbers, such as 4 versus 8. Results showed that mosquitofish primarily attended to cumulative surface area but could also use discrete numerical information when cumulative surface area and other continuous quantities were controlled.

In the aforementioned studies (Agrillo et al., 2009), cumulative surface area was controlled for by using proportionally smaller figures in the more numerous set. This introduces another problem, however, because as a by-product of this type of control, smaller-than-average figures were more frequent in the set containing fewer items (and vice versa) and fish could use this nonnumerical cue to discriminate between the two groups. This forces the experimenter to devise an additional control experiment to rule out this other possibility, such as by including, at the end of the experiment, control trials in extinction in which figures had the same size for all stimuli (Agrillo et al., 2009). In other studies, a different type of control for cumulative surface area was performed (Agrillo, Miletto Petrazzini, Piffer, Dadda, & Bisazza, 2012; Agrillo, Miletto Petrazzini, Tagliapietra, & Bisazza, 2012; Agrillo, Miletto Petrazzini, & Bisazza, 2014). For example, to control for cumulative surface area in one-third of the stimuli used for the training phase, the two numerosities were 100% equated for area, in

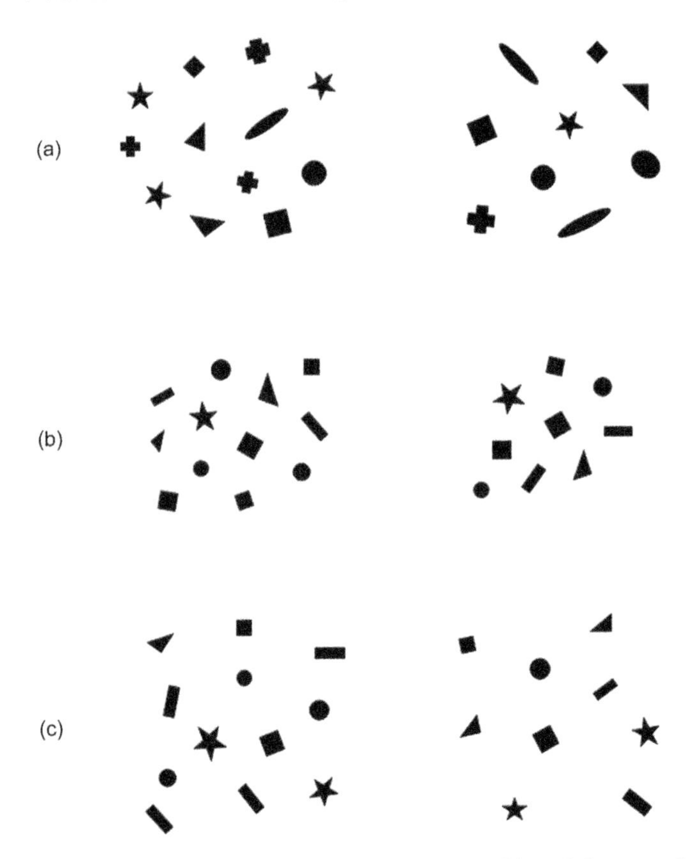

FIGURE 7.2 **Schematic representation of the stimuli used in training procedures.** The use of inanimate objects permits a fine control of different continuous quantities. Here we depict an example of a 9 versus 12 contrast used in a guppy study, with cumulative surface area controlled to 100% (a), to 85% (b), and to 70% (c). In (a) and (c) stimuli are controlled for the overall space encompassed by the most lateral figures, whereas in (b) they are controlled for density. *Adapted from Agrillo et al. (2014).*

one-third area was controlled to 85%; and, in the remaining third, it was controlled to 70% (Fig. 7.2). In this way the largest figure could be placed in the more and in the less numerous sets with equal frequency. In probe trials, cumulative surface area was perfectly matched. As a consequence, the subject could use neither individual figure size (an unreliable cue in the training phase) nor cumulative surface area (matched to 100% in probe trials) to solve the task.

Recently, a different training procedure was devised to use food as a reward in which mosquitofish were placed individually in rectangular tanks (Agrillo et al., 2012a). At intervals, two stimuli (sets of two-dimensional figures of different numerosity) were introduced at the opposite ends and food was provided near the stimulus to be reinforced

FIGURE 7.3 **Schematic presentation of the apparatus used to train fish.** Subjects are housed in an experimental tank for the duration of the experiment, and stimuli are repeatedly presented at the two ends of the tank using two PC monitors or two plastic cards submerged at the two bottoms of the tank. *Adapted from Agrillo et al. (2014).*

(Fig. 7.3). The proportion of time spent near the positive stimulus in probe trials was used as a measure of discrimination performance. The results showed that fish can make use of numerical information only, to discriminate both small quantities (eg, 2 vs. 4 items) and large quantities (eg, 100 vs. 200 items).

This procedure also was used to compare numerical abilities in the same tasks in five different teleost fish (Agrillo et al., 2012b): guppies, zebrafish, angelfish, redtail splitfin, and Siamese fighting fish (*Betta splendens*). Fish were initially trained to discriminate between 5 versus 10 and 6 versus 12 (0.50 ratio), and they were subsequently tested in their ability to generalize the numerical rule to more difficult ratios (0.67 and 0.75) or to a larger (25 vs. 50) or a smaller (2 vs. 4) total set size. Stimuli were controlled for continuous variables as described earlier. With only a few exceptions, the five species showed similar performance. No species was able to discriminate the 0.75 ratio (9 vs. 12); all species but angelfish were able to discriminate the 0.67 ratio (8 vs. 12). All species tended to easily generalize the learned discrimination to a smaller set size (2 vs. 4) and no species was able to generalize to a larger set size (25 vs. 50). Overall, there were more similarities than differences across these species, which suggests that different fish have similar numerical systems. The five species encompassed a broad spectrum of ecological specializations, with some species being highly social and others basically solitary, and some living in open areas and others in dense vegetation. The finding that all these species process numerical information more or less in the same way is more in agreement with the presence of ancient numerical systems inherited from a common ancestor than with separate evolution of numerical abilities in each species.

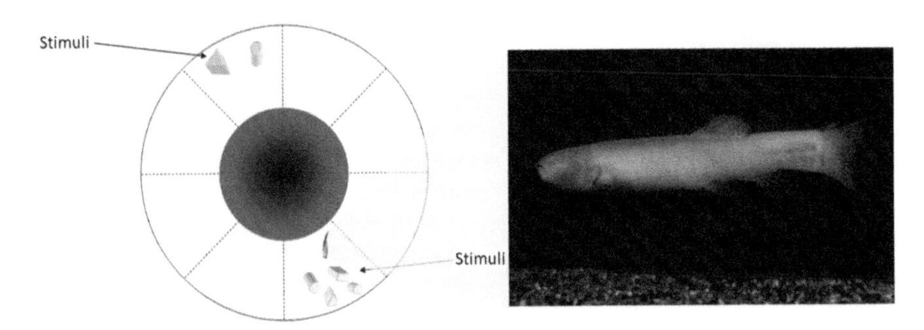

FIGURE 7.4 **The blind cavefish (*P. andruzzii*).** This species evolved for millions of years in a homogeneous environment in absence of natural predators and with a scarcity of food resources. The experimental tank used by Bisazza et al. (2014b) was divided into eight equal sectors: stimuli were placed in opposite sectors and food was provided only in correspondence to the reinforced quantity. Cavefish proved able to discriminate between two and four objects when continuous quantities were controlled for, suggesting the existence of a core number system shared by all vertebrates. *Adapted from Bisazza et al. (2014b); Photo credit: Luca Scapoli.*

A partially similar procedure has been adopted to study quantitative abilities in a cavefish as well (Bisazza, Tagliapietra, Bertolucci, Foà, & Agrillo, 2014). *Phreatichthys andruzzii* is a cave-dwelling species that evolved in the phreatic layer of the Somalia desert. This species (Fig. 7.4) shows eye degeneration and lives in an environment in the absence of predators, making it an ideal model for investigating the capacity to process number and continuous quantities in a species that evolved in extreme environmental conditions. Fish were trained to receive a food reward for discriminating between two groups of vertical sticks placed in opposite positions of their home tank (Fig. 7.4). In one experiment, the discrimination was 2 versus 4 with half of the stimuli controlled for continuous quantities and half not controlled for continuous quantities. Cavefish discriminated only when number and continuous quantities were simultaneously available, indicating that they make spontaneous use of continuous quantities when left free to use both number and continuous quantities. This finding aligns with the results previously reported in trained mosquitofish (Agrillo et al., 2009, 2010). However, when stimuli were controlled from the beginning for continuous quantities, cavefish proved able to solve the task, suggesting that the availability of multiple cues in the first part of the training is crucial in determining which information (numerical/continuous) will be adopted by fish. In short, if number and continuous quantities are available in the initial phase of the learning process, fish tend to use continuous quantities or a combination of numerical and continuous quantities; if forced to use numerical information only, they can solve the task as well, as they are likely to possess the cognitive mechanisms to elaborate both types of information. However, another recent study showed that guppies, that were initially trained to discriminate between two quantities of dots in which both numerical and continuous information were available, solved the task by using numerical information only. This showed that they were able to extract numerical information from the initial phase of the learning process (Dadda, Agrillo, Bisazza, & Brown, 2015). At present, therefore, it is difficult to establish whether the opposite results reported in these two studies are due to interspecific differences, the use of different procedures or the involvement of a different sensory modality.

Training procedures also permit researchers to address some issues that otherwise could not be investigated by the mere observation of spontaneous behavior. For instance, if animals are trained to discriminate which circle is darker than another one (lighter circle: negative stimulus, S−; darker circle: positive stimulus, S+) and then, after reaching the learning criterion, they are presented with S+ together with a stimulus darker than S+, which kind of choice do they exhibit? Do they select S+ again or, instead, do they select the novel stimulus, showing to have learned a "relation" (select "darker")? This phenomenon, defined as "transposition" by Köhler (1938), has been previously investigated in mammals and birds in different cognitive tasks (Hanggi, 2003; Lazareva, 2012). For what concern numerical cognition, when subjects learn to discriminate between two values of numerosity (eg, 5 versus 10 figures controlled for continuous quantities), they could be applying two kinds of numerical strategies. First, they might learn a rule about a specific value of numerosity: "select 10 objects," a sort of "absolute numerosity" rule. Alternatively, they might learn to select the larger (or smaller) group regardless of the absolute numerosity of the set, what we might call a "relative numerosity" rule. In both cases, the behavioral output is the same. This issue was recently investigated by Miletto Petrazzini, Agrillo, Izard, and Bisazza (2015) in guppies. Fish were initially trained to select one quantity such as 12 in a 6 versus 12 item discrimination and, after reaching the learning criterion, they were presented with a 12 versus 24 items in nonreinforced trials. Guppies spontaneously selected the larger numerosity (24 in this example) and not the specific numerosity (12 items) that was previously reinforced, showing that they previously had solved the training task using a relative numerosity rule, instead of an absolute one. However, in another experiment in which they were trained to always select a set of four items against smaller or larger numbers, guppies learned to choose this quantity above chance level, showing that they can flexibly use either relative or absolute decision criteria, depending on the context.

In conclusion, free choice tests and training procedures differ in many respects and might reveal different aspects of number and continuous quantity processing. In free choice tests, animals are thought to exhibit their natural behavioral repertoire in the presence of biologically relevant stimuli. Hence, if an animal appears to use continuous quantities in a free choice task, this probably means that it at least partially uses such a cue in its natural environment. In contrast, some have suggested that extensive training procedures may lead to recruitment of neurocognitive systems that normally have different functions and to use cues that are normally not involved in quantity processing under natural conditions (Barnard et al., 2013; Hauser & Spelke, 2004).

Other issues, however, speak in favor of using training procedures. As stated at the beginning of this section, motivation is a critical aspect in all free choice tests and a lack of preference for two large quantities of food, such as 12 versus 16 prey, does not necessarily mean a lack of discrimination. With biological objects as stimuli, it is usually difficult to separate the contributions of the different sensorial modalities. Social companions, for example, have a shape, produce smells and noises, move in space and change orientation. But the most important limit of a free choice approach is complicatedness of controlling for continuous quantities. Even when this is technically possible, the risk of artifacts remains high. For instance, in shoal choice tests, the attempt to control for total surface area using conspecifics of different size may collide with natural preference for the size of shoal mates in that species. Instead, training procedures usually employ two-dimensional, static figures drawn or presented on a monitor. Today, specific software can generate groups of two-dimensional figures accurately controlled for all continuous quantities at the same time (see Gebuis & Reynvoet, 2011).

We believe it is not possible to clearly depict the relation between number and continuous quantities in lower vertebrates without taking into account the methodology adopted. In some cases researchers have used artificial stimuli finely controlled for nonnumerical cues, whereas others have used ecological stimuli that cannot be entirely controlled for all nonnumerical cues. Drawing conclusions by using only one of these approaches might bring us to neglect half of the story.

7.3 RELATIVE SALIENCE OF CONTINUOUS VERSUS NUMERICAL INFORMATION

Some of the aforementioned studies suggested that amphibians (Krusche et al., 2010) and fish (Agrillo et al., 2009, 2010; Bisazza et al., 2014b) often use continuous quantities in their relative quantity judgments instead of numbers. This might indirectly suggest that numerical information is more cognitively demanding for lower vertebrates. This idea was initially advanced by Davis and his colleagues with respect to mammals and birds.

The authors argued that numerical discrimination is difficult for animals and they rely on number only as a "last resort" strategy when no alternative solution is available to accomplish the task (Davis & Pérusse, 1988; Davis & Memmott, 1982). The traditional explanation for this "last resort" hypothesis is that number requires more effortful processing compared with continuous quantity and thus overall perceptual cues are the quickest and easiest indicators of numerosity.

As a test of this hypothesis, Agrillo et al. (2011) trained mosquitofish to discriminate two from three geometric figures in three different conditions: in one condition, subjects could only use numerical information to distinguish between the quantities; in a second condition fish could use only continuous quantities (1 vs. 1 figures, ratio between the areas equal to two-thirds), while in the latter condition both numerical and continuous information were available. If numerical information is more difficult to process, one would expect the subject to need a larger number of trials to learn the task when only numerical information was available compared to the condition in which continuous quantities or both types of information were available. Greater accuracy was indeed found when subjects could use numerical and continuous quantities together, suggesting that the presence of multiple cues represents the easiest condition for fish. In nature, a combination of numerical and continuous information is the rule: larger social groups have a larger cumulative surface and more pieces of food normally occupy more space (Agrillo et al., 2008b; Beran et al., 2008; McComb, Packer, & Pusey, 1994). Not surprisingly, many species have evolved mechanisms that use multiple cues to estimate quantity and animals accessing multiple cues show higher performance than those using just a single cue. Interestingly, in this study no difference was found in the capacity to learn the discrimination based on numerical or continuous information alone, suggesting that, at least for mosquitofish, processing number is not more complex than processing nonnumerical information, even if the combined cues clearly best support learning.

If number and continuous quantities can both be processed with apparently the same cognitive effort, why do some comparative studies report the use of numerical information while others reported the use of continuous quantities? Are there contextual factors that determine the relative salience of one type of information over the other? At least four hypotheses could be advanced about this issue:

1. *Interspecies variation*. It is possible that the relative salience of numerical information over continuous quantities varies among species. For instance, for the visual system of an amphibian (which is highly sensitive in order to detect any movement in the visual field), the quantity of movement can be the main cue to determine the choice. Indeed, salamanders (Krusche et al., 2010) proved to be

sensitive to the quantity of movement when choosing between groups of crickets differing in numerosity. On the contrary, this continuous quantity may be less relevant in other species. An interspecific variation in the mechanism used to discriminate between quantities has also been suggested in birds (Emmerton & Renner, 2009). The cognitive systems of the species might be shaped by evolution on the basis of different pressure selections, with some species preferentially attending a range of continuous quantities and others not doing so.

2. *The use of numerical information is age dependent.* It is possible that the developmental trajectories of number and continuous quantity processing are not the same. Some abilities may develop before others. For example, newborn and juvenile animals might use a range of continuous quantities, while adult individuals might use numerical information in the same quantity task. In support of this hypothesis, it was found that newborn guppies can discriminate between two and four inanimate objects by using continuous quantities (Miletto Petrazzini, Agrillo, Piffer, & Bisazza, 2014), while adults of the same species can also solve the same numerical contrast by using numerical information only (Agrillo et al., 2014).
The comparison of these two studies also has an implication for the ATOM debate (Bueti & Walsh, 2009). Indeed, if a single magnitude system underlies numerical and spatial abilities, one should expect the same developmental trajectories in the capacity to process both magnitudes. For instance, a recent study in our species showed that 9-month-old infants can transfer associative learning across time, space, and number, showing that the ability to process these magnitudes is simultaneously available in 9-month-olds (Lourenco & Longo, 2010). The fact that the developmental trajectory of numerical and continuous quantities is different in guppies reinforces the idea, previously advanced in Section 7.2.2, of separate cognitive systems for processing number and space in fish.

3. *Salience of numerical information is stimulus dependent.* We often assume that if subjects successfully discriminate between quantities when number and continuous quantities are both available and do not subsequently select the larger/smaller group when continuous information is controlled for, they are using continuous quantities. This reasoning, though sound in theory, does not take into account the possibility that the relative salience of numerical information can vary as a function of the type of stimulus. In this chapter we have used the example of the food choice task. As stated, it is not unexpected that natural selection shaped the quantification systems in order to maximize the amount of food retrieved rather than the number of pieces of food retrieved. While total area and volume may be the most salient information in several food choice tasks, numerical

information may be the most salient information in other contexts. For instance, when searching for potential mates, a male may gain an advantage by joining the larger number of females (and not the group of females encompassing the larger space). Hence, the differences in the relative salience of numerical information over continuous quantities reported in the literature may be at least partially due to the nature of the stimuli. In line with this hypothesis, two recent studies showed that newborn guppies can use numerical information to discriminate between two and three social companions (Bisazza et al., 2010), while they tend to use continuous quantities when required to discriminate between two and four inanimate objects (Miletto Petrazzini et al., 2014).

4. *Salience of numerical information is task dependent.* Experiments on infants using a habituation paradigm suggest that they can attend to number and continuous quantities simultaneously. The relative salience of these two dimensions would depend on the type of task, with infants preferentially attending to number over continuous quantities when large sets of objects were presented (reviewed in Cordes & Brannon, 2008).
   It is possible that the relative salience of numerical information in animal species varies depending on the experimental conditions (numerical range, ratio, type of paradigm, etc.). As noted in the chapter, different procedures commonly determine different results in numerical cognition studies (Agrillo & Bisazza, 2014). For instance, it was shown that the performance in a quantity discrimination task in goldbelly topminnows is influenced by the type of procedure, with fish able to discriminate 3 versus 2 only in one of two different procedures (Agrillo & Dadda, 2007).

7.4 CONCLUSIONS AND FUTURE DIRECTIONS

After one decade of studies, considerable progress has been made in the understanding of quantitative abilities of amphibians and fish. Lower vertebrates have revealed surprising abilities in estimating and comparing sets of objects. The fact that lower vertebrates are capable of true numerical information processing is worth noting, as it may imply that the origin of number concepts is more ancient than we have previously thought. Also, we know that lower vertebrates are capable of using a wide range of continuous information. In this chapter, we have reviewed the existing studies surrounding the relation between numerical and continuous information in amphibians and fish.

Unfortunately, no firm conclusion can be reached at this stage about the exact mechanisms used by amphibians and fish. Different studies have employed different methods and stimuli, thus making interspecies comparisons difficult. In addition, some species have not been tested with procedures largely used in mammals. For instance, no study has used the sequential presentation of the stimuli in food choice tests and the use of training procedures has been entirely confined to fish. Until we develop proper protocols to train amphibians as well, we cannot effectively compare quantitative abilities among lower vertebrates.

It is interesting to note that little attention has been devoted to study area discrimination in amphibians and fish. While we have reported several findings about the discriminative thresholds in numerical tasks (eg, 0.80 ratio in guppies after extensive training, see Bisazza, Agrillo, & Lucon-Xiccato, 2014), we are still far from understanding whether the same limits would occur when subjects are required to discriminate which object is larger. A recent study of guppies showed a different discriminative threshold in numerical and area discrimination using food items as stimuli (Lucon-Xiccato et al., 2015), in contrast with ATOM's prediction according to which subjects should have the same discriminative threshold for the two magnitudes, but we need to collect more data on area discrimination with different sets of stimuli (eg, dots) and paradigms (eg, training procedures) before drawing a firm conclusion. Strictly related to this issue, it would be interesting to assess whether different discriminative thresholds for space and number exist in other fish species too.

Also, while quantitative abilities in mammals have been tested using visual stimuli (Biro & Matsuzawa, 2001), auditory stimuli (Beran, 2012), and olfactory stimuli (Horowitz, Hecht, & Dedrick, 2013), the study of amphibians and fish has been almost entirely confined to visual tasks. Evidence of a similar numerical acuity in mammals for visual and nonvisual stimuli has led some authors to hypothesize the existence of a single, evolutionarily ancient, modality independent, numerical system (Beran, 2008; Feigenson, 2007; Jordan et al., 2008). In contrast, the results reported in the cavefish study (a lower performance in the presence of nonvisual stimuli) may indirectly suggest the existence of multiple core number systems in fish in which visual and nonvisual numerosities are mentally represented with different signal variabilities, but of course further investigation is necessary to shed light on this issue.

As a final note, it is worth mentioning the current lack of studies in reptiles; we have no information as to whether snakes, turtles, or crocodiles display at least the capacity to discriminate different amounts of food. This is also the case of cartilaginous fish (sharks and rays), a group of vertebrates that is increasingly used in cognition research (Schluessel, 2015). Only after we collect information on these organisms will we presumably come to have a better picture of how quantitative mechanisms evolved in vertebrates.

Acknowledgments

We would like to thank Avishai Henik for inviting us to make this contribution and Michael J. Beran for his comments. Financial support was provided by FIRB grant "*Futuro in ricerca 2013*" (prot.: RBFR13KHFS) from "Ministero dell'Istruzione, Università e Ricerca" (MIUR, Italy) to Christian Agrillo.

References

Agrillo, C., & Bisazza, A. (2014). Spontaneous versus trained numerical abilities. A comparison between the two main tools to study numerical competence in non-human animals. *Journal of Neuroscience Methods*, *234*, 82–91.

Agrillo, C., & Dadda, M. (2007). Discrimination of the larger shoal in the poeciliid fish *Girardinus falcatus*. *Ethology Ecology & Evolution*, *19*(2), 145–157.

Agrillo, C., & Piffer, L. (2012). Musicians outperform non-musicians in magnitude estimation: evidence of a common processing mechanism for time, space and numbers. *Quarterly Journal of Experimental Psychology*, *65*(12), 2321–2332.

Agrillo, C., Dadda, M., & Serena, G. (2008a). Choice of female groups by male mosquitofish (*Gambusia holbrooki*). *Ethology*, *114*(5), 479–488.

Agrillo, C., Dadda, M., Serena, G., & Bisazza, A. (2008b). Do fish count? Spontaneous discrimination of quantity in female mosquitofish. *Animal Cognition*, *11*(3), 495–503.

Agrillo, C., Dadda, M., Serena, G., & Bisazza, A. (2009). Use of number by fish. *PLoS One*, *4*(3), e4786.

Agrillo, C., Piffer, L., & Bisazza, A. (2010). Large number discrimination by mosquitofish. *PLoS One*, *5*(12), e15232.

Agrillo, C., Piffer, L., & Bisazza, A. (2011). Number versus continuous quantity in numerosity judgments by fish. *Cognition*, *119*(2), 281–287.

Agrillo, C., Miletto Petrazzini, M. E., Piffer, L., Dadda, M., & Bisazza, A. (2012a). A new training procedure for studying discrimination learning in fish. *Behavioural Brain Research*, *230*(2), 343–348.

Agrillo, C., Miletto Petrazzini, M. E., Tagliapietra, C., & Bisazza, A. (2012b). Inter-specific differences in numerical abilities among teleost fish. *Frontiers in Psychology*, *3*, 483.

Agrillo, C., Piffer, L., Bisazza, A., & Butterworth, B. (2012c). Evidence for two numerical systems that are similar in humans and guppies. *PLoS One*, *7*(2), e31923.

Agrillo, C., Miletto Petrazzini, M. E., & Bisazza, A. (2014). Numerical acuity of fish is improved in the presence of moving targets, but only in the subitizing range. *Animal Cognition*, *17*(2), 307–316.

Andersson, M. (1994). *Sexual selection*. Princeton, NJ: Princeton University Press.

Barnard, A. M., Hughes, K. D., Gerhardt, R. R., DiVincenti, L., Bovee, J. M., & Cantlon, J. F. (2013). Inherently analog quantity representations in olive baboons (*Papio anubis*). *Frontiers in Psychology*, *4*, 253.

Basolo, A. L. (1990). Female preference for male sword length in the green swordtail, *Xiphophorus helleri* (Pisces: Poeciliidae). *Animal Behaviour*, *40*(2), 332–338.

Beeching, S. C. (1992). Visual assessment of relative body size in a cichlid fish, the Oscar, *Astronotus ocellatus*. *Ethology*, *90*, 177–186.

Beran, M. J. (2006). Quantity perception by adult humans (*Homo sapiens*), chimpanzees (*Pan troglodytes*), and rhesus macaques (*Macaca mulatta*) as a function of stimulus organization. *International Journal of Comparative Psychology*, *19*, 386–397.

Beran, M. J. (2008). The evolutionary and developmental foundations of mathematics. *PLoS Biology*, *6*(2), 221–223.

Beran, M. J. (2012). Quantity judgments of auditory and visual stimuli by chimpanzees (*Pan troglodytes*). *Journal of Experimental Psychology: Animal Behavior Processes*, *38*(1), 23–29.

Beran, M. J., & Beran, M. M. (2004). Chimpanzees remember the results of one-by-one addition of food items to sets over extended time periods. *Psychological Science, 15*(2), 94–99.

Beran, M. J., Evans, T. A., & Harris, E. H. (2008). Perception of food amounts by chimpanzees based on the number, size, contour length and visibility of items. *Animal Behaviour, 75*(5), 1793–1802.

Beran, M. J., McIntyre, J. M., Garland, A., & Evans, T. A. (2013). What counts for "counting"? Chimpanzees, *Pan troglodytes*, respond appropriately to relevant and irrelevant information in a quantity judgment task. *Animal Behaviour, 85*(5), 987–993.

Bertram, B. C. R. (1980). Vigilance and group size in ostriches. *Animal Behaviour, 28*(1), 278–286.

Biro, D., & Matsuzawa, T. (2001). Use of numerical symbols by the chimpanzee (*Pan troglodytes*): cardinals, ordinals and the introduction of zero. *Animal Cognition, 4*, 193–199.

Bisazza, A., & Marconato, A. (1988). Female mate choice, male-male competition and parental care in the river bullhead, *Cottus gobio L.* (Pisces, Cottidae). *Animal Behaviour, 36*(5), 1352–1360.

Bisazza, A., Marconato, A., & Marin, G. (1989). Male competition and female choice in *Padogobius martensi* (Pisces, Gobiidae). *Animal Behaviour, 38*(3), 406–413.

Bisazza, A., Piffer, L., Serena, G., & Agrillo, C. (2010). Ontogeny of numerical abilities in fish. *PLoS One, 5*(11), e15516.

Bisazza, A., Agrillo, C., & Lucon-Xiccato, T. (2014a). Extensive training extends numerical abilities of guppies. *Animal Cognition, 17*(6), 1413–1419.

Bisazza, A., Tagliapietra, C., Bertolucci, C., Foa, A., & Agrillo, C. (2014b). Non-visual numerical discrimination in a blind cavefish (*Phreatichthys andruzzii*). *Journal of Experimental Biology, 217*, 1902–1909.

Bogale, B. A., Kamata, N., Mioko, K., & Sugita, S. (2011). Quantity discrimination in jungle crows, *Corvus macrorhynchos*. *Animal Behaviour, 82*(4), 635–641.

Bogale, B. A., Aoyama, M., & Sugita, S. (2014). Spontaneous discrimination of food quantities in the jungle crow, *Corvus macrorhynchos*. *Animal Behaviour, 94*, 73–78.

Boysen, S. T., Berntson, G. G., & Mukobi, K. L. (2001). Size matters: impact of item size and quantity on array choice by chimpanzees (*Pan troglodytes*). *Journal of Comparative Psychology, 115*, 106–110.

Buckingham, J. N., Wong, B. B. M., & Rosenthal, G. G. (2007). Shoaling decisions in female swordtails: how do fish gauge group size? *Behaviour, 144*, 1333–1346.

Bueti, D., & Walsh, V. (2009). The parietal cortex and the representation of time, space, number and other magnitudes. *Philosophical Transaction of the Royal Society B, 364*(1525), 1831–1840.

Chivers, D. P., & Smith, R. J. F. (1994). The role of experience and chemical alarm signalling in predator recognition by fathead minnows, *Pimephales promelas*. *Journal of Fish Biology, 44*, 273–285.

Chivers, D. P., & Smith, R. J. F. (1995). Fathead minnows, *Pimephales promelas*, learn to recognize chemical stimuli from high risk habitats by the presence of alarm substance. *Behavioral Ecology, 6*, 155–158.

Cordes, S., & Brannon, E. M. (2008). Quantitative competencies in infancy. *Developmental Science, 11*(6), 803–808.

Dadda, M., Piffer, L., Agrillo, C., & Bisazza, A. (2009). Spontaneous number representation in mosquitofish. *Cognition, 112*(2), 343–348.

Dadda, M., Agrillo, C., Bisazza, A., Brown, C. (2015). Laterality enhances numerical skills in the guppy, *Poecilia reticulata*. *Frontiers in Behavioral Neuroscience, 9*, 285.

Davis, H. (1984). Discrimination of the number three by a raccoon (*Procyon lotor*). *Animal Learning and Behavior, 12*, 409–413.

Davis, H., & Memmott, J. (1982). Counting behavior in animals: a critical evaluation. *Psychological Bulletin, 92*(3), 547–571.

Davis, H., & Pérusse, R. (1988). Numerical competence in animals: definitional issues, current evidence, and a new research agenda. *Behavioural and Brain Science, 11*, 561–651.

Emmerton, J., & Renner, J. C. (2009). Local rather than global processing of visual arrays in numerosity discrimination by pigeons (*Columba livia*). *Animal Cognition, 12*(3), 511–526.

Feigenson, L. (2007). The equality of quantity. *Trends in Cognitive Sciences, 11*(5), 185–187.

Gebuis, T., & Reynvoet, B. (2011). Generating nonsymbolic number stimuli. *Behavior Research Methods, 43*(4), 981–986.

Gómez-Laplaza, L. M., & Gerlai, R. (2011a). Can angelfish (*Pterophyllum scalare*) count? Discrimination between different shoal sizes follows Weber's law. *Animal Cognition, 14*(1), 1–9.

Gómez-Laplaza, L. M., & Gerlai, R. (2011b). Spontaneous discrimination of small quantities: shoaling preferences in angelfish (*Pterophyllum scalare*). *Animal Cognition, 14*(4), 565–574.

Gómez-Laplaza, L. M., & Gerlai, R. (2012). Activity counts: the effect of swimming activity on quantity discrimination in fish. *Frontiers in Psychology, 3*, 484.

Gómez-Laplaza, L. M., & Gerlai, R. (2013). Quantification abilities in angelfish (*Pterophyllum scalare*): the influence of continuous variables. *Animal Cognition, 16*(3), 373–383.

Hanggi, E. B. (2003). Discrimination learning based on relative size concepts in horses *Equus caballus*. *Applied Animal Behaviour Science, 83*, 201–213.

Hauser, M.D., & Spelke, E.S. (2004). Evolutionary and developmental foundations of human knowledge. *The cognitive neurosciences III*. Cambridge: MIT press.

Hauser, M. D., Carey, S., & Hauser, L. B. (2000). Spontaneous number representation in semi-free-ranging rhesus monkeys. *Proceedings of the Royal Society of London Series B: Biological Sciences, 267*(1445), 829–833.

Hoare, D. J., Couzin, I. D., Godin, J. G. J., & Krause, J. (2004). Context-dependent group size choice in fish. *Animal Behaviour, 67*, 155–164.

Horowitz, A., Hecht, J., & Dedrick, A. (2013). Smelling more or less: investigating the olfactory experience of the domestic dog. *Learning and Motivation, 44*(4), 207–217.

Hunt, S., Low, J., & Burns, K. C. (2008). Adaptive numerical competency in a food-hoarding songbird. *Proceedings of the Royal Society B: Biological Sciences, 275*(1649), 2373–2379.

Jordan, K. E., MacLean, E. L., & Brannon, E. M. (2008). Monkeys match and tally quantities across senses. *Cognition, 108*(3), 617–625.

Koehler, O. (1951). The ability of birds to "count". *Animal Behaviour, 9*, 41–45.

Köhler, W. (1938). Simple structural functions in the chimpanzee and in the chicken. In W. D. Ellis (Ed.), *A source book of Gestalt psychology* (pp. 217–227). London: Routledge & Kegan Paul.

Krusche, P., Uller, C., & Dicke, U. (2010). Quantity discrimination in salamanders. *The Journal of Experimental Biology, 213*, 1822–1828.

Kuroda, R. (1931). On the counting ability of a monkey (*Macacus cynomolgus*). *Journal of Comparative Psychology, 12*, 171–180.

Landeau, L., & Terborgh, J. (1986). Oddity and the "confusion effect" in predation. *Animal Behaviour, 34*(5), 1372–1380.

Lashley, K. S. (1912). Visual discrimination of size and form in the albino rat. *Journal of Animal Behavior, 2*, 310–331.

Lazareva, O. F. (2012). Relational learning in a context of transposition: a review. *Journal of the Experimental Analysis of Behavior, 97*(2), 231–248.

Lourenco, S. F., & Longo, M. R. (2010). General magnitude representation in human infants. *Psychological Science, 21*(6), 873–881.

Lucon-Xiccato, T., Miletto Petrazzini, M.E., Agrillo, C., & Bisazza, A. (2015). Guppies (*Poecilia reticulata*) prioritise item size over total amount in a food quantity discrimination task. *Animal Behaviour*, 107, 183–191.

Mark, R. F., & Maxwell, A. (1969). Circle size discrimination and transposition behaviour in cichlid fish. *Animal Behaviour, 17*, 155–158.

McComb, K., Packer, C., & Pusey, A. (1994). Roaring and numerical assessment in contests between groups of female lions, *Panthera leo*. *Animal Behaviour*, *47*(2), 379–387.

Mendez, J. C., Prado, L., Mendoza, G., & Merchant, H. (2011). Temporal and spatial categorization in human and non-human primates. *Frontiers in Integrative Neuroscience*, *5*, 50.

Miletto Petrazzini, M. E., Agrillo, C., Piffer, L., & Bisazza, A. (2014). Ontogeny of the capacity to compare discrete quantities in fish. *Developmental Psychobiology*, *56*(3), 529–536.

Miletto Petrazzini, M.E., Agrillo, C., Izard, V., & Bisazza, A. (2015). Relative versus absolute numerical representation in fish: can guppies represent 'fourness'? *Animal Cognition*, *18*(5), 1007–1017.

Odell, N. S., & Eadie, J. M. (2010). Do wood ducks use the quantity of eggs in a nest as a cue to the nest's value? *Behavioral Ecology*, *21*(4), 794–801.

Olsson, M. (1994). Uptial coloration in the sand lizard, *Lacerta agilis*: an intra-sexually selected cue to lighting ability. *Animal Behaviour*, *48*, 607–613.

Panteleeva, S., Reznikova, Z., & Vygonyailova, O. (2013). Quantity judgments in the context of risk/reward decision making in striped field mice: first "count," then hunt. *Frontiers in Psychology*, *4*, 53.

Perdue, B. M., Talbot, C. F., Stone, A. M., & Beran, M. J. (2012). Putting the elephant back in the herd: elephant relative quantity judgments match those of other species. *Animal Cognition*, *15*(5), 955–961.

Pritchard, V. L., Lawrence, J., Butlin, R. K., & Krause, J. (2001). Shoal choice in zebrafish, *Danio rerio*: the influence of shoal size and activity. *Animal Behaviour*, *62*, 1085–1088.

Pyke, G. H., Pulliam, H. R., & Charnov, E. L. (1977). Optimal foraging: a selective review of theory and tests. *The Quarterly Review of Biology*, *52*, 137–154.

Rugani, R., Fontanari, L., Simoni, E., Regolin, L., & Vallortigara, G. (2009). Arithmetic in newborn chicks. *Proceedings of the Royal Society B: Biological Sciences*, *276*(1666), 2451–2460.

Schluessel, V. (2015). Who would have thought that "Jaws" also has brains? Cognitive functions in elasmobranchs. *Animal Cognition*, *18*(1), 19–37.

Stancher, G., Sovrano, V. A., Potrich, D., & Vallortigara, G. (2013). Discrimination of small quantities by fish (redtail splitfin, *Xenotoca eiseni*). *Animal Cognition*, *16*(2), 307–312.

Stancher, G., Rugani, R., Regolin, L., & Vallortigara, G. (2015). Numerical discrimination by frogs (*Bombina orientalis*). *Animal Cognition*, *18*(1), 219–229.

Uller, C., Jaeger, R., Guidry, G., & Martin, C. (2003). Salamanders (*Plethodon cinereus*) go for more: rudiments of number in an amphibian. *Animal Cognition*, *6*(2), 105–112.

Vicario, C. M. (2011). Perceiving numbers affects the subjective temporal midpoint. *Perception*, *40*(1), 23–29.

von Frisch, K. (1967). *The dance language and orientation of bees*. Cambridge, MA: Harvard University Press.

Vonk, J., & Beran, M. J. (2012). Bears "count" too: quantity estimation and comparison in black bears (*Ursus americanus*). *Animal Behaviour*, *84*, 231–238.

Walsh, V. (2003). A theory of magnitude: common cortical metrics of time, space and quantity. *Trends in Cognitive Sciences*, *7*(11), 483–488.

Wilson, M. L., Britton, N. F., & Franks, N. R. (2002). Chimpanzees and the mathematics of battle. *Proceedings of the Royal Society London B: Biological Sciences*, *269*, 1107–1112.

CHAPTER

8

Going for More: Discrete and Continuous Quantity Judgments by Nonhuman Animals

Michael J. Beran, Audrey E. Parrish

Department of Psychology and Language Research Center, Georgia State University, Atlanta, GA, United States of America

OUTLINE

Imagine a monkey walking along the forest floor, approaching two trees in the distance, each with ripe fruit on its branches, and with the monkey having access to one tree or the other, but not both. Or, imagine a guppy finding itself isolated from its conspecifics, and with a choice of one of two shoals to swim toward so as to regain the relative safety of "numbers." Or imagine a pair of chimpanzees walking the edge of their home range, on patrol for intruders, when they hear the calls of other, unfamiliar chimpanzees in the distance. Finally, imagine yourself in an airport about to spend money on a bagel in a coffee shop, and think about how you might choose which one to pick from the available assortment. These are normal, frequent, and important choices and they reflect how organisms use quantity and amount to guide decision making in a variety of

Continuous Issues in Numerical Cognition. http://dx.doi.org/10.1016/B978-0-12-801637-4.00008-1

species-relevant situations. And for more than 100 years, they have been a central research focus in comparative psychology, with the understanding that these capacities for choosing among amounts of "stuff" are likely as evolutionarily old and phylogenetically broad as almost any other perceptual or cognitive capacities.

Broadly defined, these kinds of quantity judgments, and the mechanisms that underlie them, are subsumed under the research domain known as numerical cognition. However, they should more accurately be considered as cases of quantitative cognition because they do not have to rely on the use of abstract numerical concepts or a "number sense" (Dehaene, 1997), even if the mechanisms at work in these cases relate back to numerical cognition in a critical way, as we discuss later. In each of the previous examples, the actor in question was making a relative judgment—choosing or discriminating between amounts or numbers of things. The first case would show clear adaptive value, as obtaining more food versus less food is critical to survival and reproductive success, and perhaps no other test has been as widely used in this research area as has food quantity judgment. The second case of the guppy also shows a clear instance with immediate survival value—being in larger shoals offers more protection, and so finding a way to put oneself into those shoals is important and relies on the ability to discriminate between the choices on the basis of the relative number of fish in them (eg, Gomez-Laplaza & Gerlai, 2011; Piffer, Agrillo, & Hyde, 2012). The third case also is a relational judgment, as those two chimpanzees must assess whether their group size exceeds that of the intruders they hear, and then they must adjust their behavior in terms of approaching or avoiding those intruders. Chimpanzees do that (Wilson, Hauser, & Wrangham, 2001), and other animals also appear capable of making similar comparisons of group size (eg, Benson-Amram, Heinen, Dryer, & Holekamp, 2010; McComb, Packer, & Pusey, 1994). Finally, we know that humans are the ultimate number users, with mathematical capabilities that exceed anything seen elsewhere in the animal kingdom, but the example was given to show that we still face the same kinds of daily choices that involve getting the most of a good thing as do other animals, and as did our hominid ancestors. We will show that humans, like other animals, also rely on similar mechanisms in many of those instances, especially when we are prevented from using more advanced mathematical abilities (such as counting and arithmetic) and instead must rely on our shared, analogical system for representing quantities. We also will show that even with that system in place, there are times when illusions occur in our quantity judgments, and we fail to choose optimally because we fail to represent discrete or continuous quantities accurately. These failures are another aspect of quantity representation that we share with nonhuman animals.

In a research area that is as broad and as long-standing in the history of comparative psychology as is nonhuman animal numerical cognition research, one is forced to be selective in what can be summarized. Given that the purpose of this chapter is to reflect on what kinds of quantities animals can represent and use, and how they do that, we will focus primarily on relational judgments about amounts or numbers of various kinds of stimuli. These stimuli can be discrete (eg, food items such as grapes, dots on a computer screen) or continuous (eg, liquids, sand), and in some cases, responding is under the control of the true numerical properties of the stimuli (eg, when all other factors such as density, size, area, and other aspects that covary with number are controlled). We will overview the kinds of tests given to nonhuman animals to assess their relational judgment abilities for amounts and numbers of things, and the manner in which continuous and discrete quantities are represented in similar (or dissimilar) ways by nonhuman animals and by humans. This includes a discussion of the potential mechanisms that support quantitative cognition. These quantitative skills found in nonhuman animals are relevant precursors to more number-specific and advanced mathematical abilities that emerge in our own species, and understanding the extent and limits of nonhuman animal quantity discrimination capacities improves our understanding of the evolutionary basis of human numerical cognition. As our focus is on quantity judgments, we begin with a critical distinction among the different kinds of judgments one can assess with animals because this distinction is important in ensuring that we know whether animals are making judgments of "how large," "how much," or "how many."

8.1 RELATIVE QUANTITY VERSUS RELATIVE NUMEROUSNESS JUDGMENTS

An important assessment of any test given to animals is whether they are making judgments of the number of items in sets (a numerousness judgment), or whether these are judgments only of the amount of sets (a quantity judgment). Or, stated differently, the question is whether animals are choosing on the basis of a "how many" rule or a "how much" rule. Unfortunately, this distinction is not always made clear in how researchers report what they assessed, in part because of a long-standing use of the term *relative numerousness judgment* (Davis & Perusse, 1988). This term often is applied in cases where numerousness (ie, the cardinal numerical value of a stimulus set) is not the only cue to which animals (or humans) can respond. Those cases, in which other stimulus dimensions such as area, size, or density may be confounded with number of items, are better labeled as *relative quantity judgments*. That would reserve the relative numerousness judgment label for a different class of comparison, one in which *only* number could be used a reliable stimulus dimension to guide responding. The term "relative" here means that the subject is discriminating one quantity in relation to another (as with two sets of discrete or continuous food items or two sets of dots), and it is contrasted with "absolute" judgments in which the exact size or quantity must be discriminated (as when one is asked whether a set of dots holds three dots or does not).

In this framework, it would be difficult to argue that any participant would perform a relative numerousness judgment when choosing between discrete sets of food items. If the items were all identical in size, the number of items would be confounded with amount (or area) of those items, thereby *not requiring* any encoding or representation of the numerosity of the set to generate accurate comparisons. There is not a reason to assume that an animal sees six grapes and eight grapes, for example, as a comparison of "six" and "eight" versus a more continuous representation of those amounts of food ("less" vs. "more"). And, if items within the choice sets differed in size, this would create a motivational conflict about using number of items when number does not covary with amount (eg, as with the choice between one whole banana and three slices of a banana, where the first option is more overall food). Thus, it is important when designing a comparison task to determine what one wants to assess. If the goal is to assess numerical discrimination, different stimuli must be used than if one instead is satisfied to assess quantity comparisons more broadly. We now focus on each of these classes of relational judgments, and the evidence for how nonhuman animals make those judgments.

8.2 RELATIVE FOOD QUANTITY JUDGMENTS BY ANIMALS AND CHOICE BIASES

In one of the most typically used tasks in the comparative "numerical cognition" literature, animals are presented with two (or more) food items or sets of food items, and are allowed to indicate which they prefer to receive. This kind of test using real food items is also known as the natural choice paradigm because of the clearly intuitive nature of the task, which allows one to see the limits of performance in animals given that they are attempting to maximize their food intake (Beran, Ratliff, & Evans, 2009; Silberberg, Widholm, Bresler, Fujita, & Anderson, 1998). The natural choice paradigm has several advantages including the requirement of little to no training due to the use of motivating edible stimuli. Additionally, the immediate distribution of the selected food for consumption establishes high spatial contiguity between stimulus and reward, and further allows for the establishment of rapid and reliable performances in quantity discrimination (Menzel, 1961).

The proficiency with which some species can discriminate food amounts allows us to ask questions about how different environmental contexts may disrupt such abilities, and even to relate these "decisional errors" to those made by humans. For example, we know that adult humans misperceive food portions as being larger than they really are when they are presented on a small plate in comparison to the same-sized portion presented on a large plate (Van Ittersum & Wansink, 2012). We recently documented similar perceptual-decisional biases in chimpanzees in their food magnitude judgments. When given two identically sized pieces of food, chimpanzees preferred the food presented on a small plate rather than the same amount of food presented on a large plate (Parrish & Beran, 2014a). Sometimes, they even preferred the *smaller* food item when it was on the smaller plate in comparison to the larger food item on the larger plate (Fig. 8.1). Similar errors occur in chimpanzee choice behavior

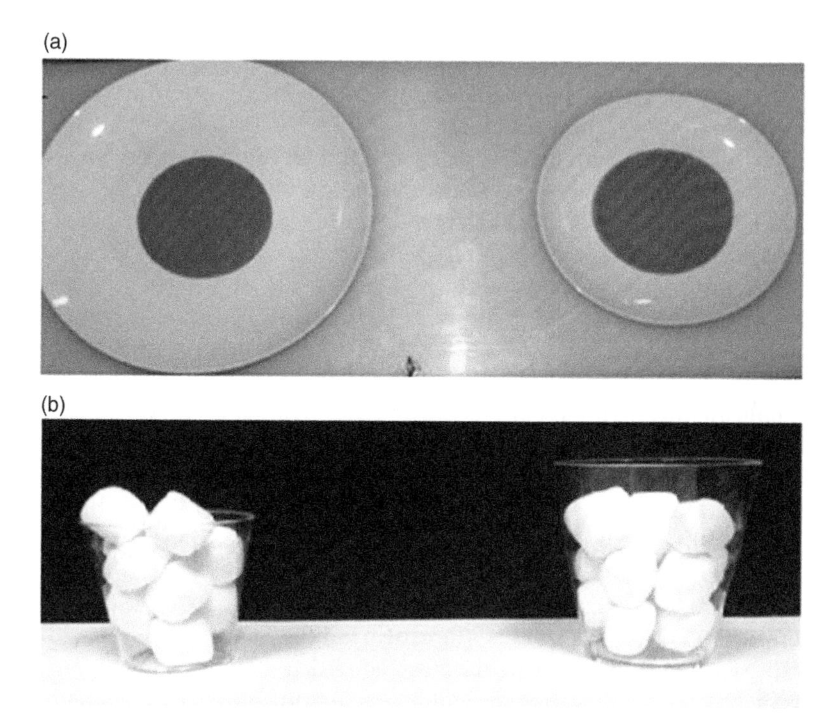

FIGURE 8.1 **Stimuli presented in contexts that make relational judgments more difficult.** (a) Despite being identically sized slices of food, many people see the slice at right as being larger because it is on a smaller plate. Chimpanzees also showed this choice bias. (b) Despite there being fewer marshmallows in the container on the left, chimpanzees sometimes chose this container, presumably because they saw that container as having more items in it due to its fuller appearance. When both containers were the same size, the chimpanzees were highly successful in choosing the larger quantity. Thus, container size impacted perception of quantity. *Part a: adapted from Parrish and Beran (2014a). Part b: adapted from Parrish and Beran (2014b).*

when containers were presented as more or less full of food. When less food was presented in a container that was nonetheless fuller looking (ie, more of its volume was filled), chimpanzee chose that option over a truly larger amount of food that was less full looking (Fig. 8.1; Parrish & Beran, 2014b). The manipulation of other set features also impacts quantity discrimination in chimpanzees, including the cohesiveness or brokenness of food sets. Chimpanzees preferred whole food items over broken food items even when the whole-item sets consisted of a smaller overall quantity of food than the sets containing broken items (Parrish, Evans, & Beran, 2015). And, the features of individual items within a set can impact quantity discrimination in chimpanzees. Other studies have demonstrated errors in maximization by chimpanzees when the smaller of two sets contained the overall largest single item of food within it (eg, Beran, Evans, & Harris, 2008a; Boysen, Berntson, & Mukobi, 2001). Thus, it is clear that in some cases quantity comparisons are not made solely on the basis of total amount of food, but also on the basis of individual food elements, and how those elements look. Animals can make errors in discriminating between food quantities based on features of food items and the context in which those items are presented.

Variations of these relative quantity judgments of food as well as nonfood items have been given to many species, including fish (eg, Agrillo, Dadda, Serena, & Bisazza, 2008; Agrillo, Piffer, & Bisazza, 2011; Dadda, Piffer, Agrillo, & Bisazza, 2009; Gomez-Laplaza & Gerlai, 2011, 2013; Piffer et al. 2012), amphibians (Krusche, Uller, & Dicke, 2010; Uller, Jaeger, Guidry, & Martin, 2003), birds (eg, Al Aïn, Giret, Grand, Kreutzer, & Bovet, 2009; Emmerton 1998; Garland, Low, & Burns, 2012; Pepperberg, 2006; Rugani, Regolin, & Vallortigara, 2007, 2008), and a large variety of nonprimate mammals including dogs (Ward & Smuts, 2007), cats (Pisa & Agrillo, 2009), voles (eg, Ferkin, Pierce, Sealand, & delBarco-Trillo, 2005), sea lions (eg, Abramson, Hernández-Lloreda, Call, & Colmenares, 2011), beluga whales (Abramson, Hernández-Lloreda, Call, & Colmenares, 2013), dolphins (eg, Jaakkola, Fellner, Erb, Rodriguez, & Guarino, 2005; Kilian, Yaman, von Fersen, & Gunturkun, 2003), hyenas (Benson-Amram et al., 2010), coyotes (Baker, Shivik, & Jordan, 2011), horses (Uller & Lewis, 2009), bears (Vonk & Beran, 2012), and elephants (Perdue, Talbot, Stone, & Beran, 2012), as well as many nonhuman primates (eg, Addessi, Crecimbene, & Visalberghi, 2008; Anderson et al., 2005; Anderson, Stoinski, Bloomsmith, & Maple, 2007; Barnard et al., 2013; Beran, 2001, 2004, 2012; Call, 2000; Evans, Beran, Harris, & Rice, 2009; Hanus & Call, 2007; Lewis, Jaffe, & Brannon, 2005). Animals also have shown good discrimination abilities for continuous quantities. For example, Al Aïn et al. (2009) showed that African gray parrots were good at choosing the larger of two discrete (seeds) or two continuous (parrot formula) quantities. Beran (2010) also reported that chimpanzees accurately selected the larger of two amounts of liquid poured into separate opaque containers, and vanMarle, Aw, McCrink, and Santos (2006) showed that capuchin monkeys could discriminate differences in poured amounts of banana puree. One can highlight the typical results of food quantity judgments by looking at research with chimpanzees. For example, chimpanzees were observed to discriminate between very slight differences in food sizes to a degree that rivaled and sometimes exceeded the abilities of human adults (eg, Menzel, 1960, 1961; Menzel & Davenport, 1962). More recent work extended those earlier investigations, demonstrating the proficiency of chimpanzees in choosing the larger quantity of food from two or more choices and across a range of presentation styles (eg, Beran, 2001, 2004; Beran & Beran, 2004; Boysen & Berntson, 1995; Hanus & Call, 2007; Rumbaugh, Savage-Rumbaugh, & Hegel, 1987) and after lengthy time delays (1–2 days) while incorporating several key features of the alternative sets (eg, quantity, quality, and handling time; Sayers & Menzel, 2012).

The performances of many animals in these relative quantity judgments shows what is called the *ratio effect*—performance in choosing the larger amount of food declined as the ratio of the small set divided by the large set increased. This effect is consistent with Weber's law, and is a recurring outcome of this kind of experimental test with animals (see Brannon & Roitman, 2003; Gallistel & Gelman, 2000). Weber's law states that a constant level of discrimination of two magnitudes requires a constant *relative* difference in those quantities even as their absolute magnitudes vary. Stated differently, Weber's law says that discrimination of sets becomes more difficult for a fixed difference as the magnitude of those sets increases. Thus, a subject's ability to discriminate 8 from 10 items but not 9 from 10 items means that subject likely also will discriminate 16 from 20 items but not 18 from 20 items. From this, one sees that it is the ratio of sets to each other that predicts discriminability. In this example, ratios of 0.8 can be discriminated but ratios of 0.9 cannot.

One can generate equivalent ratio effects in children (eg, Huntley-Fenner, 2001; Huntley-Fenner & Cannon, 2000) and even in adult humans, provided they cannot use formal counting routines (eg, Cordes, Gelman, Gallistel, & Whalen, 2001; Whalen, Gallistel, & Gelman, 1999). For example, Beran, Taglialatela, Flemming, James, and Washburn (2006) gave adult humans a computer task that was designed for monkeys (Beran, 2007; see next section). In the task, a digital hand on the screen "dropped" items into two onscreen containers, mimicking the one-by-one food delivery task used with primates and other animals (eg, Beran, 2001, 2004; Beran, Evans, Leighty, Harris, & Rice, 2008b; Garland et al., 2012; Hanus & Call, 2007), and humans had to select the larger set. When they had to repeat the alphabet aloud through the trials (to prevent subvocal counting), their performance on any given comparison was best predicted by the ratio of the sets presented in that comparison, matching the monkeys' performance, and that of many other animals as noted previously. All of these results converge on the idea of a phylogenetically widespread system for the approximate representation of quantities, without need of numerical symbols or knowledge of cardinal values. This system is sometimes called the approximate number system (ANS). It conforms to Weber's law by showing two behavioral effects—performance improves in judging between quantities as the difference between those quantities increases, and when the difference between quantities is held constant, performance is better for comparisons of smaller magnitudes rather than larger magnitudes (Brannon & Roitman, 2003; Dehaene, 1997; Gallistel & Gelman, 2000).

Although a consensus has emerged that many animals show the ratio effect that reflects the ANS in certain conditions, there is still some debate about what that effect reflects by way of the mechanism (or mechanisms) that generate quantity comparison judgments in animals (and humans). That debate will be discussed later. First, however, we must address the question of whether any of these relative quantity judgments are ever made on the basis of *only* numerical information, a necessary criterion for any claim that an animal is making relative numerousness judgments, or true judgments of "how many" rather than "how much."

8.3 RELATIVE NUMEROUSNESS JUDGMENTS BY ANIMALS

To assesses the role that number concepts play in stimulus discrimination and choice behavior requires a different approach than the use of identical food items (or any identical items, where the amount can covary with the number). The difficulty in doing this is that so many other factors can covary with number. Take, for example, the approach of training monkeys to choose the larger of two numbers of stimuli in two arrays on a computer screen. If all items are the same size, and the monkeys perform well, they could be using the number of items, but more likely are using the total area of the stimuli. One could control for this, by always making the larger numerical array have the smaller amount, but then item size (or average item size) operates as a nonnumerical cue, where the monkey could learn to pick the array with the items that are individually smaller, on average. This is not a numerical judgment. Or, the monkeys could learn to use the interitem distances, using density of stimuli as a cue. Or they could use total perimeter area, and so forth. One sees the difficulty, then, in establishing numerical judgments by animals and the difficulty in establishing exactly when relative numerousness judgments might be occurring. Without explicit evidence for a counting routine being applied to each array, and then the evaluation of the resulting cardinal values for those two counts, it is hard to establish that number is the controlling stimulus dimension that underlies discrimination and choice in these kinds of tests.

But it can be done, and when appropriate methods are presented, the results indicate that animals can judge "how many" just as well as they judge "how much." The approach that is needed is one in which, across trials, different confounds are controlled. This is necessary because it is not possible, at least in the visual modality, to present numerical sets that are not in some way also discriminable on the basis of some other feature. For example, if one controls the amount of two sets, the larger number of items must also have the smaller average item size within that set. If one controls the total area of a set, then item density covaries with number, and so on (for more discussion, see Gebuis, Gevers, & Cohen Kadosh, 2014; Leibovich & Henik, 2014; Mix, Huttenlocher, & Levine, 2002).

One of the most systematic and convincing efforts to show numerical judgments comes from the work of Brannon and her colleagues (eg, Brannon & Terrace, 1998, 2000). In this task, monkeys were presented with two sets, each consisting of items that varied in color, size, distance from each other, and a host of other features, and the monkeys were rewarded for choosing either the numerically smaller or larger set of items. Importantly, the task varied all of the individual item features across trials, so that none of the nonnumerical features consistently covaried with number in all trials. The only rule that guaranteed 100% accuracy across all trial types when applied correctly was the numerical rule—picking the larger or smaller number of items. Rhesus monkeys passed this test, showing that they used number as the relevant cue to which they responded. These numerical discriminations extended to other kinds of tests, allowing these researchers to examine things such as the influence of reference points on number judgments (eg, Brannon, Cantlon, & Terrace, 2006; Jones, Cantlon, Merritt, & Brannon, 2010), the role of semantic congruity and stimulus heterogeneity on these judgments (Cantlon & Brannon, 2005, 2006a), and other issues regarding what mechanisms underlie such judgments (eg, Cantlon & Brannon, 2006b, 2007). Other species also have been presented with this methodology, and in general have shown the same ability to make these true numerical comparisons and judgments, including baboons and squirrel monkeys (Smith, Piel, & Candland, 2003), capuchin monkeys (Judge, Evans, & Vyas, 2005), lemurs (Merritt, MacLean, Crawford, & Brannon, 2011), and pigeons (Scarf, Hayne, & Colombo, 2011).

In the computerized task discussed earlier in which monkeys chose between sequentially presented sets of items that dropped on the screen (Beran, 2007), there was also evidence that these judgments could be made as true numerical judgments. In one condition, items varied in size, so that the monkeys could not use the total amount of stimuli material they saw in each set as a cue, and in this method items were never visually available at the same time onscreen, so density, contour length, and other cues could not be used to discriminate sets. Even in that case, the monkeys succeeded in selecting the larger number of dots, although there was a clear ratio effect. There is also some evidence that numerical attributes of stimuli control responding even when two arrays of moving items must be compared by capuchin monkeys, rhesus monkeys, and adult humans (Beran, 2008), gorillas (Vonk et al., 2014), and black bears (Vonk & Beran, 2012).

These numerousness judgment tests can be given in noncomputerized formats as well, provided careful attention is given to constructing the stimuli. For example, a series of studies showed that squirrel monkeys would use the number of stimuli as the relevant dimension when asked to compare sets of as many as eight versus nine items. In these manual tests, the monkeys viewed circles and polygons printed on cards and were trained to choose the most numerous, least numerous, or even the intermediate-sized array, and they proved successful in doing this (Terrell & Thomas, 1990; Thomas & Chase, 1980; Thomas, Fowlkes, & Vickery, 1980).

These demonstrations are important with regard to the broader topic of numerical cognition, and particularly its evolutionary emergence. First, they show that number is relevant to nonhuman animals. But, in some ways, the idea that number is a "last resort" (Davis & Memmott, 1982) is also accurate. As discussed, using the number of food items is not necessarily the most effective approach when the goal is to maximize the amount of food obtained. Animals (including humans) face many situations in which number is but one feature of an array, and often it is not even a relevant one with regard to the motivation at choice time, a motivation that could differ across instances and individuals (eg, choosing between one large food patch or three smaller patches when foraging). But, this does not mean that number cannot be used when it is the necessary feature to which one must respond. That animals do use number shows that it is likely not a "first resort" but also likely not a "last resort." Context matters to a great extent, in terms of whether using number naturally produces the best response strategy. For example, using the amount of predators in an area (in terms of their total weight, size, or volume) likely is less adaptive than telling apart the number of predators since that factor makes survival more or less likely. But, telling apart the specific numbers of food items of varying size in two food patches is not more adaptive than telling apart the amount of food at each source, at least in many circumstances.

The second important implication of this research pertains to how number can be used by nonhuman animals. Number concepts for animals appear to be much "fuzzier" than they are for humans, especially those over the age of 4 years. Evidence for this again comes from ratio effects and similar measures of relative degrees of difference between the choices that are seen even in these tasks in which true numerical ("how many") judgments are being generated. Animals struggle with generating the discrete, cardinal numerical values that underlie advanced human mathematics. Humans know that the number 109 is as different from 111 as the number 2 is different from the number 4 (ie, both are a difference of two units). Nonhuman animals, however, could not discriminate between those two large quantities as easily as the two small ones because of the representational limits of the ANS. For animals, as true set size (ie, the number of things or amount of stuff being observed) increases, the perception and representation of that amount becomes more variable, and this leads to difficulties in ether reporting the exact amount (as when asked "how many is this") or, as we have discussed here, in comparing options ("which has more"). However, there is an ongoing debate as to whether nonhuman animals have access *only* to this approximate number system when they quantify things, or whether a second system also exists for more precisely representing small quantities.

8.4 MECHANISM(S) FOR REPRESENTING QUANTITY

Magnitude, quantity, and number clearly are relevant stimulus features to many species, but how are these represented? There are two prevalent models that have received attention in the broadly defined area of comparative numerical cognition. The first of these is the ANS, which was discussed earlier (also see Cantlon, Platt, & Brannon, 2009). By now, the reader is (we hope) convinced that evidence for this system is widespread, and it likely impacts all areas in which inexact, analog representation of quantities occurs. There are still questions to be resolved about this system, particularly with regard to how exactly quantity is represented or scaled (Dehaene, 2003; Roberts, 2005; Siegler & Opfer, 2003). One idea is that these representations are linear but with an increase in the variability of the discrimination proceeding in step with increases in the magnitude of the quantity itself (eg, scalar variability; Gallistel & Gelman, 1992; Meck & Church, 1983). Another possibility is that representations are logarithmic, in which case the scaling space between small numbers is larger than the space between large numbers, thereby making discrimination of small sets easier to discriminate even with the same absolute difference (eg, 2 vs. 4 and 10 vs. 12; see Dehaene, 2003; Roberts, 2005).

A key theoretical issue is whether there is *only* one system at work when animals represent quantity and number. A second candidate system for numerical representation has been proposed, and it is based on models of attention and preattentive object indexing. This model is often called the *object file* model, and the idea is that small numbers of items can be individually indexed and stored with high fidelity, thereby allowing for precise representation of small sets, with easy comparison between small sets. Although there is some inconsistency in determining the limits of this system, it is generally believed to be three or four items at maximum, so that when one enumerates or encodes individual elements within a set, those are stored in separate object files (Feigenson, Carey, & Hauser, 2002; Feigenson, Dehaene, & Spelke, 2004). Object files then operate as representations of items within the array (Simon, 1997), and there is a precision in these representations that is consistent across all of those set sizes (ie, "2" is as well represented and stored as "4" and comparing those two numbers is no more difficult than comparing "1" and "2" or "2" and "3").

Within the developmental literature, there have been mixed reports for the existence of this precise small number system, with some studies showing evidence for it (eg, Feigenson & Carey, 2005; Xu, 2003) but others showing mainly evidence of only the ANS (eg, Cantlon, Safford, & Brannon, 2010; Cordes & Brannon, 2009). Within the comparative literature, the object file model initially drew support from data reported by Hauser, Carey, and Hauser (2000) in which monkeys were presented with two sets of apple pieces, presented one piece at a time, into two opaque containers, and then were allowed to retrieve one container. Some comparisons involved all smaller numbers of apple pieces (1 vs. 2, 2 vs. 3, 3 vs. 4, and 3 vs. 5). Monkeys were reported to be good on all of these comparisons. However, for comparisons with larger numbers (4 vs. 5, 4 vs. 6, 4 vs. 8, and 3 vs. 8), the monkeys made more errors, suggesting to Hauser et al. a difference in performance between trials with small sets only and trials with large sets only or one large and one small set. The argument was that these latter types might have been too difficult for monkeys because their object files were exhausted for the larger set, and so the monkeys could not accurately store the number of items in that set and use that information in making the comparative judgment. Thus, it has been a focus in some studies to determine whether animals can discriminate larger sets from smaller ones at the same degree of sensitivity, or whether greater discriminatory precision is evident for comparing small sets relative to large sets even when both comparisons are of the same ratio (eg, comparing 2 and 3 vs. comparing 8 and 12).

Other testing paradigms, including manual and computerized tests, have failed to show any such set-size limits as would be predicted by the object file model in other primates. When the same chance-level comparisons from the Hauser et al. (2000) paper were given to other monkeys, these sets were discriminated (eg, Beran, 2007). Discrimination of these sets also was seen with four great ape species (gorillas, bonobos, chimpanzees, and orangutans; eg, Beran, 2004; Hanus & Call, 2007) and capuchin monkeys (Evans et al., 2009). And, ratio effects tend to be evident without any set-size limit effects in many other species ranging from sea lions (Abramson et al., 2011) to parrots (Al Aïn et al., 2009) to pigeons (Roberts, 2010). At present, there is some evidence for set-size limits in some species of fish (eg, Agrillo et al., 2012; Gomez-Laplaza & Gerlai, 2011; Piffer et al., 2012), birds (eg, Garland et al., 2012) and one report with

Asian elephants (Irie-Sugimoto, Kobayashi, Sato, & Hasegawa, 2009), but the latter outcome was not replicated in a subsequent study with African elephants (Perdue et al., 2012). Thus, nearly all studies with mammalian species (and particularly primates) show the signature effects of the ANS, but there is much more limited evidence of a second core system for number representation in animals that is precise but limited.

8.5 CONCLUSIONS

The comparative literature on numerical cognition indicates that many species perceive and process quantitative information, particularly with regard to comparison tasks. Evidence for numerical representations can be more difficult to establish, but with the right methodology, true numerical processing has been found in some species. These data point to the existence of an innate quantitative cognition in these species with the attendant abilities to perceive and manipulate discrete and continuous quantities. The research also indicates that these capacities in animals suffer from some of the same perceptual and decisional biases that humans demonstrate. And, the comparative literature continues to contribute to the ongoing question of whether one or two "core" systems of number exist. At present, the safe conclusion is that all species (including humans) share an approximate system for quantity representation, and that this system operates to encode and represent continuous and discrete quantities, and even "fuzzy" numerical information. Whether additional systems exist is less clear, and whether animals are capable of exhibiting the boundless, cardinality-based system for exact representation of number concepts is also unclear, and perhaps unlikely. Despite this limitation in their numerical cognition, our present understanding of animals' perception of quantities and their ability to compare things serve as important precursors to the formal numerical abilities that our species relies upon for so much of its cultural, scientific, and industrial development.

References

Abramson, J. Z., Hernández-Lloreda, V., Call, J., & Colmenares, F. (2011). Relative quantity judgments in South American sea lions (*Otaria flavescens*). *Animal Cognition, 14*, 695–706.

Abramson, J. Z., Hernández-Lloreda, V., Call, J., & Colmenares, F. (2013). Relative quantity judgments in the beluga whale (*Delphinapterus leucas*) and the bottlenose dolphin (*Tursiops truncatus*). *Behavioural Processes, 96*, 11–19.

Addessi, E., Crescimbene, L., & Visalberghi, E. (2008). Food and token quantity discrimination in capuchin monkeys (*Cebus apella*). *Animal Cognition, 11*, 275–282.

Agrillo, C., Dadda, M., Serena, G., & Bisazza, A. (2008). Do fish count? Spontaneous discrimination of quantity in female mosquitofish. *Animal Cognition, 11*, 495–503.

Agrillo, C., Piffer, L., & Bisazza, A. (2011). Number versus continuous quantity in numerosity judgments by fish. *Cognition, 119*, 281–287.

Agrillo, C., Piffer, L., Bisazza, A., & Butterworth, B. (2012). Evidence for two numerical systems that are similar in humans and guppies. *PLoS ONE, 7*, e31923.

Al Aïn, S. A., Giret, N., Grand, M., Kreutzer, M., & Bovet, D. (2009). The discrimination of discrete and continuous amounts in African grey parrots (*Psittacus erithacus*). *Animal Cognition, 12*, 145–154.

Anderson, U. S., Stoinski, T. S., Bloomsmith, M. A., Marr, M. J., Smith, A. D., & Maple, T. S. (2005). Relative numerousness judgment and summation in young and old Western Lowland gorillas. *Journal of Comparative Psychology, 119*, 285–295.

Anderson, U. S., Stoinski, T. S., Bloomsmith, M. A., & Maple, T. S. (2007). Relative numerousness judgment and summation in young, middle-aged, and old adult orangutans (*Pongo pygmaeus abelii* and *Pongo pygmaeus pygmaeus*). *Journal of Comparative Psychology, 121*, 1–11.

Baker, J. M., Shivik, J., & Jordan, K. E. (2011). Tracking of food quantity by coyotes (*Canis latrans*). *Behavioural Processes, 88*, 72–75.

Barnard, A. M., Hughes, K. D., Gerhardt, R. R., DiVincenti, L. J., Bovee, J. M., & Cantlon, J. F. (2013). Inherently analog quantity representations in olive baboons (*Papio anubis*). *Frontiers in Psychology, 4*, 253.

Benson-Amram, S., Heinen, V. K., Dryer, S. L., & Holekamp, K. E. (2010). Numerical assessment and individual call discrimination by wild spotted hyaenas *Crocuta crocuta*. *Animal Behaviour, 82*, 743–752.

Beran, M. J. (2001). Summation and numerousness judgments of sequentially presented sets of items by chimpanzees (*Pan troglodytes*). *Journal of Comparative Psychology, 115*, 181–191.

Beran, M. J. (2004). Chimpanzees (*Pan troglodytes*) respond to nonvisible sets after one-by-one addition and removal of items. *Journal of Comparative Psychology, 118*, 25–36.

Beran, M. J. (2007). Rhesus monkeys (*Macaca mulatta*) enumerate large and small sequentially presented sets of items using analog numerical representations. *Journal of Experimental Psychology: Animal Behavior Processes, 33*, 55–63.

Beran, M. J. (2008). Monkeys (*Macaca mulatta* and *Cebus apella*) track, enumerate, and compare multiple sets of moving items. *Journal of Experimental Psychology: Animal Behavior Processes, 34*, 63–74.

Beran, M. J. (2010). Chimpanzees (*Pan troglodytes*) accurately compare poured liquid quantities. *Animal Cognition, 13*, 641–649.

Beran, M. J. (2012). Quantity judgments of auditory and visual stimuli by chimpanzees (*Pan troglodytes*). *Journal of Experimental Psychology: Animal Behavior Processes, 38*, 23–29.

Beran, M. J., & Beran, M. M. (2004). Chimpanzees remember the results of one-by-one addition of food items to sets over extended time periods. *Psychological Science, 15*, 94–99.

Beran, M. J., Taglialatela, L. A., Flemming, T. J., James, F. M., & Washburn, D. A. (2006). Nonverbal estimation during numerosity judgements by adult humans. *Quarterly Journal of Experimental Psychology, 59*, 2065–2082.

Beran, M. J., Evans, T. A., & Harris, E. H. (2008a). Perception of food amounts by chimpanzees based on the number, size, contour length and visibility of items. *Animal Behaviour, 75*, 1793–1802.

Beran, M. J., Evans, T. A., Leighty, K., Harris, E. H., & Rice, D. (2008b). Summation and quantity judgments of sequentially presented sets by capuchin monkeys (*Cebus apella*). *American Journal of Primatology, 70*, 191–194.

Beran, M. J., Ratliff, C. L., & Evans, T. A. (2009). Natural choice in chimpanzees (*Pantroglodytes*): perceptual and temporal effects on selective value. *Learning & Motivation, 40*, 186–196.

Boysen, S. T., & Berntson, G. G. (1995). Responses to quantity: perceptual versus cognitive mechanisms in chimpanzees (*Pan troglodytes*). *Journal of Experimental Psychology: Animal Behavior Processes, 21*, 82–86.

Boysen, S. T., Berntson, G. G., & Mukobi, K. L. (2001). Size matters: impact of item size and quantity on array choice by chimpanzees (*Pan troglodytes*). *Journal of Comparative Psychology, 115*, 106–110.

Brannon, E. M., & Roitman, J. D. (2003). Nonverbal representations of time and number in animals and human infants. In W. H. Meck (Ed.), *Functional and neural mechanisms of interval timing* (pp. 143–182). Boca Raton, FL: CRC Press.
Brannon, E. M., & Terrace, H. S. (1998). Ordering of the numerosities 1 to 9 by monkeys. *Science, 282*, 746–749.
Brannon, E. M., & Terrace, H. S. (2000). Representation of the numerosities 1–9 by rhesus macaques (*Macaca mulatta*). *Journal of Experimental Psychology: Animal Behavior Processes, 26*, 31–49.
Brannon, E. M., Cantlon, J. F., & Terrace, H. S. (2006). The role of reference points in ordinal numerical comparisons by rhesus macaques (*Macaca mulatta*). *Journal of Experimental Psychology: Animal Behavior Processes, 32*, 120–134.
Call, J. (2000). Estimating and operating on discrete quantities in orangutans (*Pongo pygmaeus*). *Journal of Comparative Psychology, 114*, 136–147.
Cantlon, J. F., & Brannon, E. M. (2005). Semantic congruity affects numerical judgments similarly in monkeys and humans. *Proceedings of the National Academy of Sciences, 102*, 16507–16511.
Cantlon, J. F., & Brannon, E. M. (2006a). The effect of heterogeneity on numerical ordering in rhesus monkeys. *Infancy, 9*, 173–189.
Cantlon, J. F., & Brannon, E. M. (2006b). Shared system for ordering small and large numbers in monkeys and humans. *Psychological Science, 17*, 401–406.
Cantlon, J. F., & Brannon, E. M. (2007). How much does number matter to a monkey (*Macaca mulatta*)? *Journal of Experimental Psychology: Animal Behavior Processes, 33*, 32–41.
Cantlon, J. F., Platt, M. L., & Brannon, E. M. (2009). Beyond the number domain. *Trends in Cognitive Sciences, 13*, 83–91.
Cantlon, J. F., Safford, K. E., & Brannon, E. M. (2010). Spontaneous analog number representations in 3-year-old children. *Developmental Science, 13*, 289–297.
Cordes, S., & Brannon, E. M. (2009). Crossing the divide: infants discriminate small from large numerosities. *Developmental Psychology, 45*, 1583–1594.
Cordes, S., Gelman, R., Gallistel, C. R., & Whalen, J. (2001). Variability signatures distinguish verbal from nonverbal counting for both large and small numbers. *Psychonomic Bulletin and Review, 8*, 698–707.
Dadda, M., Piffer, L., Agrillo, C., & Bisazza, A. (2009). Spontaneous number representation in mosquitofish. *Cognition, 112*, 343–348.
Davis, H., & Memmott, J. (1982). Counting behavior in animals: a critical evaluation. *Psychological Bulletin, 92*, 547–571.
Davis, H., & Perusse, R. (1988). Numerical competence in animals: definitional issues, current evidence, and a new research agenda. *Behavioral and Brain Sciences, 11*, 561–615.
Dehaene, S. (1997). *The number sense*. New York: Oxford University Press.
Dehaene, S. (2003). The neural basis of the Weber–Fechner law: a logarithmic mental number line. *Trends in Cognitive Sciences, 7*, 145–147.
Emmerton, J. (1998). Numerosity differences and effects of stimulus density on pigeons' discrimination performance. *Animal Learning and Behavior, 26*, 243–256.
Evans, T. A., Beran, M. J., Harris, E. H., & Rice, D. (2009). Quantity judgments of sequentially presented food items by capuchin monkeys (*Cebus apella*). *Animal Cognition, 12*, 97–105.
Feigenson, L., & Carey, S. (2005). On the limits of infants' quantification of small object arrays. *Cognition, 97*, 295–313.
Feigenson, L., Carey, S., & Hauser, M. D. (2002). The representations underlying infants' choice of more: object files versus analog magnitudes. *Psychological Science, 13*, 150–156.
Feigenson, L., Dehaene, S., & Spelke, E. (2004). Core systems of number. *Trends in Cognitive Sciences, 8*, 307–314.
Ferkin, M. H., Pierce, A. A., Sealand, R. O., & delBarco-Trillo, J. (2005). Meadow voles, *Microtus pennsylvanicus*, can distinguish more over-marks from fewer over-marks. *Animal Cognition, 8*, 182–189.

Gallistel, C. R., & Gelman, R. (1992). Preverbal and verbal counting and computation. *Cognition, 44*, 43–74.

Gallistel, C. R., & Gelman, R. (2000). Non-verbal numerical cognition: from reals to integers. *Trends in Cognitive Sciences, 4*, 59–65.

Garland, A., Low, J., & Burns, K. C. (2012). Large quantity discrimination by North Island robins (*Petroica longipes*). *Animal Cognition, 15*, 1129–1140.

Gebuis, T., Gevers, W., & Cohen Kadosh, R. (2014). Topographic representation of high-level cognition: numerosity or sensory processing? *Trends in Cognitive Sciences, 18*, 1–3.

Gomez-Laplaza, L. M., & Gerlai, R. (2011). Spontaneous discrimination of small quantities: shoaling preferences in angelfish (*Pterophyllum scalare*). *Animal Cognition, 14*, 565–574.

Gomez-Laplaza, L. M., & Gerlai, R. (2013). Quantification abilities in angelfish (*Pterophyllum scalare*): the influence of continuous variables. *Animal Cognition, 16*, 373–383.

Hanus, D., & Call, J. (2007). Discrete quantity judgments in the great apes (*Pan paniscus, Pan troglodytes, Gorilla gorilla, Pongo pygmaeus*): the effect of presenting whole sets versus item-by-item. *Journal of Comparative Psychology, 121*, 241–249.

Hauser, M. D., Carey, S., & Hauser, L. B. (2000). Spontaneous number representation in semi-free-ranging rhesus monkeys. *Proceedings of the Royal Society of London. Series B: Biological Sciences, 267*, 829–833.

Huntley-Fenner, G. (2001). Children's understanding of number is similar to adults' and rats': numerical estimation by 5–7-year-olds. *Cognition, 78*, B27–B40.

Huntley-Fenner, G., & Cannon, E. (2000). Preschoolers' magnitude comparisons are mediated by a preverbal analog mechanism. *Psychological Science, 11*, 147–152.

Irie-Sugimoto, N., Kobayashi, T., Sato, T., & Hasegawa, T. (2009). Relative quantity judgment by Asian elephants (*Elephas maximus*). *Animal Cognition, 12*, 193–199.

Jaakkola, K., Fellner, W., Erb, L., Rodriguez, M., & Guarino, E. (2005). Understanding of the concept of numerically "less" by bottlenose dolphins (*Tursiops truncatus*). *Journal of Comparative Psychology, 119*, 286–303.

Jones, S. M., Cantlon, J. F., Merritt, D. J., & Brannon, E. M. (2010). Context affects the numerical semantic congruity effect in rhesus monkeys (*Macaca mulatta*). *Behavioural Processes, 83*, 191–196.

Judge, P. G., Evans, T. A., & Vyas, D. K. (2005). Ordinal representation of numeric quantities by brown capuchin monkeys (*Cebus apella*). *Journal of Experimental Psychology: Animal Behavior Processes, 31*, 79–94.

Kilian, A., Yaman, S., Von Fersen, L., & Gunturkun, O. (2003). A bottlenose dolphin discriminates visual stimuli differing in numerosity. *Learning and Behavior, 31*, 133–142.

Krusche, P., Uller, C., & Dicke, U. (2010). Quantity discrimination in salamanders. *Journal of Experimental Biology, 213*, 1822–1828.

Leibovich, T., & Henik, A. (2014). Comparing performance in discrete and continuous comparison tasks. *Quarterly Journal of Experimental Psychology, 67*, 899–917.

Lewis, K. P., Jaffe, S., & Brannon, E. M. (2005). Analog number representations in mongoose lemurs (*Eulemur mongoz*): evidence from a search task. *Animal Cognition, 8*, 247–252.

McComb, K., Packer, C., & Pusey, A. (1994). Roaring and numerical assessment in contests between groups of female lions *Panthera leo*. *Animal Behaviour, 47*, 379–387.

Meck, W. H., & Church, R. M. (1983). A mode control model of counting and timing processes. *Journal of Experimental Psychology: Animal Behavior Processes, 9*, 320–324.

Menzel, E. W., Jr. (1960). Selection of food by size in the chimpanzee and comparison with human judgments. *Science, 131*, 1527–1528.

Menzel, E. W., Jr. (1961). Perception of food size in the chimpanzee. *Journal of Comparative and Physiological Psychology, 54*, 588–591.

Menzel, E. W., Jr., & Davenport, R. K., Jr. (1962). The effects of stimulus presentation variable upon chimpanzee's selection of food by size. *Journal of Comparative and Physiological Psychology, 55*, 235–239.

Merritt, D. J., MacLean, E. L., Crawford, J. C., & Brannon, E. M. (2011). Numerical rule-learning in ring-tailed lemurs (*Lemur catta*). *Frontiers in Comparative Psychology, 2*, Article 23.

Mix, K. S., Huttenlocher, J., & Levine, S. C. (2002). Multiple cues for quantification in infancy: is number one of them? *Psychological Bulletin, 128*, 278–294.

Parrish, A. E., & Beran, M. J. (2014a). When less is more: like humans, chimpanzees (*Pan troglodytes*) misperceive food amounts based on plate size. *Animal Cognition, 17*, 427–434.

Parrish, A. E., & Beran, M. J. (2014b). Chimpanzees sometimes see fuller as better: judgments of food quantities based on container size and fullness. *Behavioural Processes, 103*, 184–191.

Parrish, A. E., Evans, T. A., & Beran, M. J. (2015). Defining value through quantity and quality – chimpanzees;1; (*Pan troglodytes*) undervalue food quantities when items are broken. *Behavioural Processes, 111*, 118–126.

Pepperberg, I. M. (2006). Grey parrot numerical competence: a review. *Animal Cognition, 9*, 377–391.

Perdue, B. M., Talbot, C. G., Stone, A. M., & Beran, M. J. (2012). Putting the elephant back in the herd: elephant relative quantity judgments match those of other species. *Animal Cognition, 15*, 955–961.

Piffer, L., Agrillo, C., & Hyde, D. C. (2012). Small and large number discrimination in guppies. *Animal Cognition, 15*, 215–221.

Pisa, E. P., & Agrillo, C. (2009). Quantity discrimination in felines: a preliminary investigation of the domestic cat (*Felis silvestris catus*). *Journal of Ethology, 27*, 289–293.

Roberts, W. A. (2005). How do pigeons represent numbers? Studies of number scale bisection. *Behavioural Processes, 69*, 33–43.

Roberts, W. A. (2010). Distance and magnitude effects in sequential number discrimination by pigeons. *Journal of Experimental Psychology: Animal Behavior Processes, 36*, 206–216.

Rugani, R., Regolin, L., & Vallortigara, G. (2007). Rudimental numerical competence in 5-day-old domestic chicks (*Gallus gallus*): identification of ordinal position. *Journal of Experimental Psychology: Animal Behavior Processes, 33*, 21–31.

Rugani, R., Regolin, L., & Vallortigara, G. (2008). Discrimination of small numerosities in young chicks. *Journal of Experimental Psychology: Animal Behavior Processes, 34*, 388–399.

Rumbaugh, D. M., Savage-Rumbaugh, E. S., & Hegel, M. T. (1987). Summation in the chimpanzee (*Pan troglodytes*). *Journal of Experimental Psychology: Animal Behavior Processes, 13*, 107–115.

Sayers, K., & Menzel, C. R. (2012). Memory and foraging theory: chimpanzee utilization of optimality heuristics in the rank-order recovery of hidden foods. *Animal Behaviour, 84*, 795–803.

Scarf, D., Hayne, H., & Colombo, M. (2011). Pigeons on par with primates in numerical competence. *Science, 334*, 1664.

Siegler, R. S., & Opfer, J. E. (2003). The development of numerical estimation: evidence for multiple representations of numerical quantity. *Psychological Science, 14*, 237–243.

Silberberg, A., Widholm, J. J., Bresler, D., Fujita, K., & Anderson, J. (1998). Natural choice in nonhuman primates. *Journal of Experimental Psychology: Animal Behavior Processes, 24*, 215–228.

Simon, T. J. (1997). Reconceptualizing the origins of number knowledge: a "non-numerical" account. *Cognitive Development, 12*, 349–372.

Smith, B. R., Piel, A. K., & Candland, D. K. (2003). Numerity of a socially housed hamadryas baboon (*Papio hamadryas*) and a socially housed squirrel monkey (*Saimiri sciureus*). *Journal of Comparative Psychology, 117*, 217–225.

Terrell, D. F., & Thomas, R. K. (1990). Number-related discrimination and summation by squirrel monkeys (*Saimiri sciureus sciureus* and *S. boliviensus boliviensus*) on the basis of the number of sides of polygons. *Journal of Comparative Psychology, 104*, 238–247.

Thomas, R. K., & Chase, L. (1980). Relative numerousness judgments by squirrel monkeys. *Bulletin of the Psychonomic Society, 16*, 79–82.

Thomas, R. K., Fowlkes, D., & Vickery, J. D. (1980). Conceptual numerousness judgments by squirrel monkeys. *American Journal of Psychology, 93*, 247–257.

Uller, C., & Lewis, J. (2009). Horses (*Equus caballus*) select the greater of two quantities in small numerical contrasts. *Animal Cognition, 12*, 733–738.

Uller, C., Jaeger, R., Guidry, G., & Martin, C. (2003). Salamanders (*Plethodon cinereus*) go for more: rudiments of number in an amphibian. *Animal Cognition, 6*, 105–112.

Van Ittersum, K., & Wansink, B. (2012). Plate size and color suggestibility: the Delboeuf illusion's bias on serving and eating behavior. *Journal of Consumer Research, 39*, 215–228.

vanMarle, K., Aw, J., McCrink, K., & Santos, L. A. (2006). How capuchin monkeys (*Cebus apella*) quantify objects and substances. *Journal of Comparative Psychology, 120*, 416–426.

Vonk, J., & Beran, M. J. (2012). Bears "count" too: quantity estimation and comparison in black bears (*Ursus americanus*). *Animal Behaviour, 84*, 231–238.

Vonk, J., Torgerson-White, L., McGuire, M., Thueme, M., Thomas, J., & Beran, J. M. (2014). Quantity estimation and comparison in Western Lowland Gorillas (*Gorilla gorilla gorilla*). *Animal Cognition, 17*, 755–765.

Ward, C., & Smuts, B. B. (2007). Quantity-based judgments in the domestic dog (*Canis lupus familiaris*). *Animal Cognition, 10*, 71–80.

Whalen, J., Gallistel, C. R., & Gelman, R. (1999). Nonverbal counting in humans: the psychophysics of number representation. *Psychological Science, 10*, 130–137.

Wilson, M. L., Hauser, M. D., & Wrangham, R. W. (2001). Does participation in intergroup conflict depend on numerical assessment, range location, or rank for wild chimpanzees? *Animal Behaviour, 61*, 1203–1216.

Xu, F. (2003). Numerosity discrimination in infants: evidence for two systems of representations. *Cognition, 89*, B15–B25.

SECTION III

PROCESSES AND MECHANISMS

CHAPTER

9

"Number Sense": What's in a Name and Why Should We Bother?

Bert Reynvoet[*,**], *Karolien Smets*[*], *Delphine Sasanguie*[*,**]

[*]Brain & Cognition, Faculty of Psychology and Educational Sciences, KU Leuven, Leuven, Belgium; [**]Faculty of Psychology and Educational Sciences, KU Leuven, Kulak, Belgium

OUTLINE

Continuous Issues in Numerical Cognition. http://dx.doi.org/10.1016/B978-0-12-801637-4.00015-9

9.1 IMPORTANCE OF NUMBER SENSE

Mathematics is crucial in one's daily-life and professional activities (eg, Ancker & Kaufman, 2007; Finnie & Meng, 2001). For instance, if we have to compute how much we will have to pay for all our groceries or how many days we can still work before the next deadline, we rely heavily on mathematics. Because mathematics is everywhere, it has been claimed that numeracy and mathematics play an invaluable role in cognitive development and life success in contemporary society (Reyna, Nelson, Han, & Dieckmann, 2009). Both cognitive psychologists and mathematics educators agree on the fact that mathematics builds upon a "number sense" (eg, Dehaene, 2001; Verschaffel & De Corte, 1996). However, the way in which number sense is defined by cognitive psychologists and mathematics educators differs a lot.

In the field of *cognitive psychology*, the concept of number sense is often considered as the biologically determined ability to "understand, approximate and manipulate non-symbolic quantities" (Dehaene, 2001, p. 16). This means that number sense refers to a set of abilities consisting of the discrimination of sets of elements based on number (eg, 4 vs. 8 dots), estimating roughly the number of elements in a set and/or understanding number increases/decreases when elements are added/subtracted from a set. These abilities build on an adequate representation of number and therefore "number sense" and "representation of number" are often used as synonyms in this field. Research has suggested that humans (Halberda, Mazzocco, & Feigenson, 2008), but also nonhuman species (Agrillo, Piffer, & Bisazza, 2011; Meck & Church, 1983; Pahl, Si, & Zhang, 2013) dispose over such a number representation, enabling fast and accurate reasoning with nonsymbolic number (Feigenson, Dehaene, & Spelke, 2004). An important characteristic is that this system of number representations obeys Weber–Fechner's law (Fechner, 1890) and thus represents number in an approximate and compressed manner. Therefore, researchers refer to it as the "approximate number system" (ANS). Next to nonsymbolic numbers however, humans also learn symbolic numbers (eg, Arabic numerals) as part of their (in)formal education. Most researchers assume that these symbolic numbers acquire meaning by being associated with or mapped onto the congenital ANS (eg, Barth, La Mont, Lipton, & Spelke, 2005; Dehaene, 1992; Mundy & Gilmore, 2009). When cognitive psychologists refer to number sense, most of them implicitly refer to the system that represents number and is used to understand and manipulate *nonsymbolic* numbers, and to the later acquired *symbolic* numbers.

In contrast, *mathematics educators* consider number sense as a set of more complex abilities that are the product of mathematics education (Berch, 2005). For instance, on the website of the National Council of Teachers of Mathematics (NCTM) (http://www.nctm.org/), it can be found that "students with number sense naturally decompose numbers, use particular numbers as referents, solve problems using the relationships among operations and knowledge about the base-ten system, estimate a reasonable result for a problem, and have a disposition to make sense of numbers, problems, and results."

Although these definitions of number sense are very different, ranging from biologically determined quantity representations to more complex abilities like mastering the base-ten system, all concepts of number sense share the idea that number sense serves as a building block for higher mathematical abilities and success in later life, making it an extremely relevant and essential research topic.

In this chapter, we, as cognitive psychologists, will address number sense as *the ability to represent the number of both nonsymbolic and symbolic stimuli*. However, because of the fact that this ability is assumed to underlie higher mathematical abilities, we meet mathematics educators and will try to build bridges in order to contribute to a better educational practice.

9.2 ASSESSMENT OF NUMBER SENSE

In infants, psychologists traditionally assessed number sense with *implicit* paradigms like habituation (eg, Xu & Spelke, 2000) and change detection (eg, Libertus & Brannon, 2010). In a habituation paradigm, infants are continuously shown arrays of dots with the same numerosity, leading to a decrease of looking time toward the stimulus. Next, a deviant numerosity is shown and when looking times increase again, one can conclude that the infant has detected the difference between the initial and the deviant test numerosity. Infants improve in this with increasing age. At 6 months, they can discriminate numerosities differing by a ratio of 1:2. At 9–12 months, they are able to discriminate numerosities differing by a ratio of 2:3 (Xu & Spelke, 2000; Xu & Arriaga, 2007). In the change detection task, infants are presented with two streams of numerosities of which one remains constant in number (eg, 16-16-16...), while the other stream alternates between two different numerosities (eg, 8-16-8-...). Libertus and Brannon (2010) observed that infants looked longer at the changing stream when the difference or the ratio between the numerosities alternating in the changing stream was larger. This indicated that the infants were able to detect the numerical difference and were better at doing so when this difference was larger.

As opposed to the use of these implicit tasks, the way in which older children process nonsymbolic number and in a later stage also symbolic number, is in most studies assessed with *explicit* tasks. These children are for instance instructed to compare (ie, which one is more?) or to match (ie, are both stimuli numerically equivalent or not?) two nonsymbolic or symbolic stimuli. Typically, distance and size effects are reported in studies using these tasks. The distance effect (DE) refers to the observation that numbers are more easily compared when the numerical distance between them increases. The size effect (SE) is reflected by more difficult discrimination of two large numbers (eg, 8 and 9) compared to two smaller number (eg, 2 and 3), despite the fact that the distance between them is equal. These observations have led to the popular idea that number is represented on a "mental number line" (Dehaene, 1997; Gallistel & Gelman, 1992). Stimuli activate their corresponding number representations in an approximate manner on this line, resulting in overlapping representations for nearby numbers. This overlap makes it more difficult to discriminate between these numbers, hence evoking the DE. Furthermore, it is assumed that this number line is characterized by scalar variability or logarithmic compression, resulting in more overlap for larger numbers and consequently causing the SE. Because of this compressive nature of the number line, instead of (absolute) DEs, often also relative distances (ie, ratios) or Weber fractions (w) are reported, especially in studies assessing nonsymbolic number representations (eg, Halberda et al., 2008; Halberda & Feigenson, 2008). It has been repeatedly shown that children's overall performance on these explicit tasks improves with increasing age (eg, Halberda & Feigenson, 2008; Holloway & Ansari, 2009). This has led many authors to conclude that our representation of number becomes more precise over development.

Next to these discrimination tasks, converging evidence for this increasingly more precise representation with age has also been observed with other paradigms. The mental number line concept has for instance inspired Siegler and colleagues (Booth & Siegler, 2006; Opfer & Siegler, 2007; Siegler & Booth, 2004) to assess the number representation through number line tasks, an approach that was followed by many other researchers (Dehaene, Izard, Spelke, & Pica, 2008; Sasanguie, De Smedt, Defever, & Reynvoet, 2012; Schneider, Grabner, & Paetsch, 2009). In the number-to-position task, participants have to indicate the position of a nonsymbolic or symbolic number on an empty line that goes from 0 to 10, 100, or 1000, depending on the participants' age and development (eg, Booth & Siegler, 2006; Booth & Siegler, 2008; Sasanguie et al., 2012a). The rationale behind this task is that the positioning of numbers on the external line reflects the organization of the underlying mental number line. Typically, it is found that, with increasing age, the pattern to map numbers on the line shifts from a logarithmic (ie, larger numbers are put closer together than smaller numbers) toward a more precise linear pattern (eg, Booth & Siegler, 2006, 2008; Siegler & Booth, 2004), indicating that children represent number first in a compressed manner and gradually shift toward a more precise linear representation of number.

Although these tasks are by far the most frequently used tasks to assess number sense, we should note that also other explicit tasks have been proposed to assess how we represent number. Several authors have, for instance, used numerical estimation (ie, estimating the number of items in a set by providing an Arabic digit) as an indicator of number sense (eg, Mejias, Grégoire, & Noël, 2012; Sullivan & Barner, 2014). Others have proposed an approximate addition task, in which participants first have to add two numerical values and then compare the addition solution with a third value (eg, Gilmore, Attridge, & Inglis, 2011; Xenidou-Dervou, van der Schoot, & van Lieshout, 2014). These studies also fit well within the dominant picture that number is represented approximately and that the representation of number becomes more precise over development.

Whereas the majority of studies with children from kindergarten on and adults have used explicit designs, others have argued for *implicit tasks* to investigate the representation of number (Cohen Kadosh & Walsh, 2009; Defever, Gebuis, Sasanguie, & Reynvoet, 2011; Reynvoet, De Smedt, & Van den Bussche, 2009). Intentional or explicit tasks might encourage subjects to form explicit, task-relevant connections that may be different from the default connections observed under implicit conditions (Cohen Kadosh & Walsh, 2009). Two implicit tasks are commonly used to investigate the representation of number: the priming task and the numerical Stroop task. In a priming task, two stimuli are presented sequentially and responses on the second stimulus (ie, target) are analyzed as a function of the first stimulus (ie, prime). Because the relation between prime and target is irrelevant for the task instructions itself, it is assumed that one addresses the representation of number in an automatic way (eg, Reynvoet & Brysbaert, 2004). Typically, a priming DE is found reflected in faster reaction times when prime and target are numerically close. The priming DE (and its size) is assumed to reflect the representation of number. Crucially and in contrast with the findings of explicit tasks, the size of the priming DE does not change through development, indicating a stable representation of number even before formal education starts (eg, Defever et al., 2011; Reynvoet et al., 2009). In the numerical Stroop task, two numerical stimuli are presented that differ in number and physical size. In the crucial condition, participants have to ignore number, and decide which is the larger on the basis of physical size. Because number is irrelevant for the task, the pattern of results can be considered as evidence for automatic processing of number (eg, Henik & Tzelgov, 1982). Using this paradigm, it has been suggested that children who just acquired knowledge of Arabic digits have automatic access to the representation of number when arrays of dots are presented (Gebuis, Cohen Kadosh, de Haan, & Henik, 2009) but not when digits are presented. The automatic activation of the number representation by symbols develops gradually with increasing age (Girelli, Lucangeli, & Butterworth, 2000; Rubinsten, Henik, Berger, & Shahar-Shalev, 2002; Rubinsten & Henik, 2005). In sum, findings from implicit tasks have led to slightly different conclusions than explicit tasks. Results from priming and Stroop paradigms have shown that the representation of number is rather stable from kindergarten on and does not become more precise with increasing age. In addition, the automatic activation of this representation by symbolic stimuli needs to be established over time.

9.3 RELATION BETWEEN NUMBER SENSE AND MATHEMATICS ACHIEVEMENT

Many studies investigating typically developing children have shown that performance in explicit tasks like number discrimination (Bugden & Ansari, 2011; De Smedt, Verschaffel, & Ghesquière, 2009; Durand, Hulme, Larkin, & Snowling, 2005; Holloway & Ansari, 2009; Jordan, Kaplan, Ramineni, & Locuniak, 2009; Lyons, Price, Vaessen, Blomert, & Ansari, 2014; Rousselle & Noël, 2008) and number line estimation (Booth & Siegler, 2006, 2008; LeFevre et al., 2013; Link, Nuerk, & Moeller, 2014; Sasanguie, Van den Bussche, & Reynvoet, 2012) is related to individual differences in mathematical achievement. Furthermore, children with mathematical disabilities seem to have particular deficits in these basic number processing tasks too (Defever, De Smedt, & Reynvoet, 2013; Geary, Bailey, & Hoard, 2009; Geary, Hoard, Nugent, & Byrd-Craven, 2008; Iuculano, Tang, Hall, & Butterworth, 2008; Landerl & Kölle, 2009; Piazza et al., 2010). A less clear picture emerges from implicit tasks and their relation with mathematical achievement. Some studies reported differences in automatic measures between low and high mathematically performing children (Defever et al., 2011; Rubinsten & Henik, 2005). For instance, Rubinsten and Henik (2005) showed that, compared to controls, a numerical Stroop effect is absent in dyscalculic adults and Defever et al. (2011) showed that the size of the priming DE was related, although weakly, to individual differences on a mathematics test. In contrast, other studies failed to show evidence for a relation between implicit measures and mathematical achievement. For instance, Bugden and Ansari (2011) contrasted explicit (ie, number discrimination) and implicit (ie, numerical Stroop) measures and showed that only explicit measures are related to mathematical achievement. Furthermore, in a priming study comparing dyscalculic adults and controls, Defever, Göbel, Ghesquière and Reynvoet (2014) did observe similar priming effects in both groups (see also De Visscher & Noël, 2013, for similar findings in a case study).

Currently, researchers still disagree whether nonsymbolic or symbolic number processing is more related to later mathematical achievement. Several studies have shown that discrimination of nonsymbolic stimuli is related to the score on a mathematical achievement test (eg, Feigenson, Libertus, & Halberda, 2013; Halberda et al., 2008). In contrast, other studies found that the processing of symbolic number, but not nonsymbolic number, is related to mathematical achievement (Holloway & Ansari, 2009; Lyons et al., 2014; Sasanguie et al., 2012a). To date, there is thus no general consensus on what is *the best* predictor of mathematical achievement, but reviews and metaanalyses demonstrate that both nonsymbolic and symbolic number processing skills contribute to mathematical development (see De Smedt, Noël, Gilmore, & Ansari, 2013; Gebuis & Reynvoet, 2015 for discussions and Chen & Li, 2014; Fazio, Bailey, Thompson, & Siegler, 2014 for metaanalyses).

9.4 ISSUES WITH MEASURING NUMBER SENSE

Earlier in this chapter, we described how number sense or the representation of nonsymbolic and symbolic number is assessed with explicit and implicit tasks. When the representation of number is assessed explicitly, performance is related to mathematical achievement. In contrast, implicit measures are not (or only weakly) related to mathematical achievement. Here, we review evidence that cast doubt on the assumption that the explicit measures are valid ways of assessing number sense or the representation of number. To anticipate, we will argue that these explicit tasks additionally rely on other nonnumerical processes that may be responsible for the observed relation between explicit measures and mathematical achievement.

9.4.1 Number Line Estimation and Anchor Points

Although the performance in number discrimination and number line estimation is thought to reflect the characteristics of a common underlying "mental number line," recent evidence points toward a different origin to explain participants' performance on these tasks. Sasanguie and Reynvoet (2013), for instance, tested the assumption of concurrent validity of both tasks by examining the relation between the performance on number discrimination and number line estimation in typically developing children. They reasoned that children who demonstrate a linear representation in a number line task should have no or at least a smaller SE in the discrimination task, due to the absence of compression. In contrast, children who demonstrate a logarithmic representation in the number line estimation task are expected to show a (larger) SE because of the compression and thus more representational overlap on their mental number line for large numbers. The authors showed that no relation was present between the number line estimation pattern and the size of the SE in the number discrimination task. Children with a linear mapping had the same SE as children with a logarithmic mapping. These findings do not support the assumption of a common underlying representation in both tasks as proposed by many other authors (eg, Laski & Siegler, 2007). In line with these findings, recently, other explanations have been suggested to account for findings in number line estimation. For instance, Barth and Paladino (2011) demonstrated that children rely on the beginning and end points of the number line to estimate the correct position on a number line. By means of this strategy, children estimate the part (eg, 30) of a whole (eg, 100), or in other words, a proportion. According to these authors, the requirement of a proportion judgment in this task provides a better explanation than the idea of a logarithmic-to-linear representational shift does. Older children additionally rely on a midpoint strategy, in addition to using the beginning and end points of the number line: if the target position is closer to the midpoint of the line than to one of its extremes, older children start at the middle of the line in order to place the number, a finding that was confirmed by Schneider et al. (2008) using eye-tracking. Similarly, White and Szücs (2012) recently observed that first graders only use the beginning point and start to use both extremes only after 1 year of formal education. Second and third graders are capable of using a "halfway" internal midpoint, enabling them to be more accurate and linear on positioning the central numbers (eg, 11 and 13 in case of a 0–20 number range) on a number line. Recently, Sasanguie, Verschaffel, Reynvoet, and Luwel (2016) confirmed the idea of strategy use in a study in which they compared three developmental accounts used in previous literature to describe number line estimation performance [ie, (1) the logarithmic to linear representation shift: Siegler & Opfer, 2003; Siegler, Thompson, & Opfer, 2009; (2) the two linear-to-linear account: Ebersbach, Luwel, Frick, Onghena, & Verschaffel, 2008; Moeller, Pixner, Kaufmann, & Nuerk, 2009; and (3) the proportion judgment account: Barth & Paladino, 2011], using one single criterion to compare the accounts. To date, this study provides the strongest evidence that the development of children's symbolic and nonsymbolic number line estimations does not reflect a developmental change in their mental representations of number, but rather the extent to which they might be using (internal) anchor points.

9.4.2 Number Discrimination and Decisional Aspects

Common tasks to examine number discrimination are the already discussed comparison task (eg, Buckley & Gillman, 1974) and matching task (eg, Defever, Sasanguie, Vandewaetere, & Reynvoet, 2012). Both tasks lead to similar behavioral DEs and are therefore viewed as interchangeable tasks, despite the fact that these tasks differ in the specific instruction provided to the participants. The traditional view assumes that the DE originates because of overlap in the number representations on the "mental number line" (Libertus & Brannon, 2010; Restle, 1970). However, Verguts, Fias, and Stevens (2005) uncovered by means of computational modeling that the DE in the comparison task can alternatively be explained by a difference in weights between the relevant number and the response categories instead of assuming representational overlap. This is evidenced by studies in which a significant DE is obtained, but where representational overlap between the to-be-compared (nonnumerical) stimuli is not present (eg, Chiao, Bordeaux, & Ambady, 2001; Cohen Kadosh, Brodsky, Levin, & Henik, 2008). For instance, Chiao et al. (2001) observed a significant DE when participants were required to judge university ranks or social status of persons (eg, participants were faster to compare "president" versus "janitor" than they were to compare "assistant professor" versus "associate professor").

While overlap in the mental representations of the numerosities does not seem to be a necessary prerequisite for the emergence of a DE in the comparison task, the DE in the matching task can only be accounted for by assuming representational overlap (Van Opstal, Gevers, De Moor, & Verguts, 2008; Van Opstal & Verguts, 2011). For instance, Van Opstal and Verguts (2011) examined the presence of the DE with letter stimuli in both comparison and matching. Letters are robustly learned sequences, but they are not represented in the same manner as numbers are, as there is no overlap in the representations of neighboring letters. A significant DE with letters in comparison was observed, whereas a DE was absent in the letter matching task, indicating the necessity of representational overlap to obtain a DE in matching.

Further evidence for a distinct origin of the DE in comparison versus matching has been provided by a study of Smets, Gebuis, and Reynvoet (2013) contrasting the neural DEs of both tasks with electroencephalography (EEG). These authors showed that the comparison and matching task were similar with regard to (1) the behavioral DEs in accuracy and reaction time, and (2) the early neural components (ie, N1–P2p transition and P2p component) characterized by similar neural DEs in both tasks. However, the observation that participants performed significantly worse on the matching task (see also Gebuis & van der Smagt, 2011; Piazza, Izard, Pinel, Le Bihan, & Dehaene, 2004) in combination with the fact that the neural DE on the late P3 component was only present in the comparison task, led the authors to suggest an additional decisional stage for the comparison task, which may (partially) underlie the behavioral comparison DE (Verguts et al., 2005). Especially considering that the P3 component refers to a vast abundance of general cognitive abilities, such as attentional resources and working memory (Gray, Ambady, Lowenthal, & Deldin, 2004; Polich & Kok, 1995; Sutton, Braren, Zubin, & John, 1965), but also stimulus categorization (Gebuis, Kenemans, de Haan, & van der Smagt, 2010; Lansbergen & Kenemans, 2008) and response selection (Pritchard, Houlihan, & Robinson, 1999), this provides supplemental evidence for the existence of a decisional mechanism that is solely recruited in the comparison task. In another study, Smets, Gebuis, Defever, and Reynvoet (2014) contrasted the comparison and matching task within the same participants and observed that average accuracies and Weber fractions in these tasks were not correlated to each other, suggesting fundamentally different mechanisms to be at play in both tasks.

In sum, this range of behavioral, computational and imaging studies points out that different number discrimination tasks possibly assess different (additional) mechanisms. As made clear earlier, in particular the comparison task is most likely not a pure reflection of the underlying number representation.

9.4.3 Nonsymbolic Number Discrimination and the Influence of Continuous Visual Cues

Most theories assume that our number sense or ANS enables us to extract number from an array of dots *independently* from several visual cues, like total surface, dot size, and other continuous visual parameters (eg, Barth, Kanwisher, & Spelke, 2003; Piazza, Pinel, Le Bihan, & Dehaene, 2007). It is proposed that these sensory properties of numerosities are discounted in the initial stages of numerosity processing in order to arrive at an abstract sense of number (Stoianov & Zorzi, 2012).

In daily life, however, number and visual cues are highly correlated. For instance, if one compares two oranges to a more numerous set of ten oranges in reality, the total area covered by oranges will also be larger in the more numerous set: ten oranges take up more place than two oranges do. In order to study nonsymbolic numerosity processing while avoiding the confound of visual cues, researchers manipulate or control the visual cues of dot arrays, attempting to make them uninformative about numerosity (eg, Gebuis & Reynvoet, 2011a; Gilmore et al., 2013; Pica, Lemer, Izard, & Dehaene, 2004; Szücs, Nobes, Devine, Gabriel, & Gebuis, 2013). Despite all attempts to control these continuous visual cues, the importance of visual cues in nonsymbolic numerosity processing is underlined by the results of several studies. For instance, Clearfield and Mix (1999) showed that performance of infants aged 6–8 months in an ANS task was predominantly influenced by visual cues instead of numerosity: when visual cues and numerosity were separated, infants spontaneously attended to contour length rather than number. Also discrimination performance in children (Rousselle, Palmers, & Noël, 2004; Szücs et al., 2013) and adults (Gebuis & Reynvoet, 2011b; Szücs et al., 2013) has been found to be influenced by the visual cues of the nonsymbolic number stimuli. In a recent study, Smets, Sasanguie, Szücs, and Reynvoet (2015) contrasted two different methods of controlling for visual cues (ie, Dehaene, S., Izard, V., & Piazza, M. (2005). Control over non-numerical parameters in numerosity experiments. Unpublished manuscript. Available from: www. Unicog. Org.; Gebuis & Reynvoet, 2011a). Crucially, both methods build upon different assumptions on how to control for visual cues (for a discussion, see Smets et al., 2015). Strikingly, performance in both conditions differed and more importantly, the performance in one condition was not correlated with the performance in the other condition. Although the two methods thus attempted to control for the visual cues, these cues still might have affected the performance in both conditions and in a different manner, leading to unrelated results. The importance of visual cues is further illustrated by Gebuis and Reynvoet (2012), who found in a passive viewing task that when the entire set of stimuli was controlled for continuous visual cues, electrophysiological recordings revealed no number-related effects at all. When the data was reorganized according to continuous visual cues, however, traditional N1 and P2 effects emerged. In a follow-up study, this finding was replicated when subjects had to actively process number (Gebuis & Reynvoet, 2011b). These studies contradict the claim that number is processed independently from continuous visual cues. We suggest that when we derive number from dot collections or other sets of objects, the numerical product is at least partly based upon coexisting visual cues (Gebuis & Reynvoet, 2011b, 2012; Smets et al., 2015).

9.4.4 Number Discrimination and Domain-General Factors

The increasing awareness that visual cues influence nonsymbolic number processing has also led to the proposition that other confounding factors might be involved in the assessment of number sense as well. For instance, a natural consequence of controlling for continuous visual cues in nonsymbolic comparison tasks is the occurrence of congruent and incongruent trials. In congruent trials, the more numerous stimulus contains larger visual cues (eg, a larger total surface and larger dot size) and consequently, visual cues provide additional indications for the correct answer. In incongruent trials, however, the smaller numerosity has the larger visual cues and participants have to inhibit the visual cue information to reach the correct answer. Szücs et al. (2013) demonstrated that incongruent trials are more difficult than congruent trials. Additionally, they observed that whereas performance of children and adults on congruent trials is similar, children perform significantly worse than adults on incongruent trials, indicating that it is especially performance on incongruent trials that improves throughout development. This suggests that the development traditionally observed in numerosity comparison (eg, Halberda & Feigenson, 2008) is mainly due to improvement on incongruent trials and may be caused by better inhibition skills with increasing age instead of a better number sense. In line with this finding, Gilmore et al. (2013) suggested that the individual differences observed in nonsymbolic comparison, by some authors interpreted as individual differences in number sense (eg, Feigenson et al., 2004; Piazza et al., 2010), can also be caused by individual differences in inhibitory skills. These authors demonstrated that inhibitory control affects performance in a nonsymbolic comparison task and demonstrated that the relation between nonsymbolic comparison and mathematical achievement observed in many studies may be an artifact of inhibitory control skills.

Inhibition is merely one of many factors that may have confounded performance on number sense tasks. In the working memory (WM) model of Baddeley (2003), inhibition is one component of the central executive (CE), which controls, monitors, and regulates performance. The CE also supervises the processes of the other two systems: a visuospatial sketch pad (VSSP), which stores visuospatial information, and a phonological loop (PL), which stores phonological information and connects this information with our long-term memory. Recently, it was found that several of these WM components underlie performance in both nonsymbolic and symbolic number processing tasks (eg, Friso-van den Bos, Kroesbergen, & van Luit, 2014; Xenidou-Dervou, De Smedt, van der Schoot, & van Lieshout, 2013, 2014a; Xenidou-Dervou, van Lieshout, & van der Schoot, 2014). Moreover, developmental studies suggested that although nonsymbolic and symbolic number processing skills are related to mathematics achievement in the first year of formal schooling (Xenidou-Dervou, Ansari, Molenaar, van der Schoot, & van Lieshout, 2016), the unique contribution of number sense to mathematics achievement vanished in later years when the authors controlled for verbal and visuospatial WM components (Xenidou-Dervou et al., 2016). These studies suggest that, in addition to inhibition skills, other WM processes are also involved in number sense measures.

In sum, although more research is necessary to pin down the precise involvement of WM and cognitive control processes in number sense measures, it is clear that they are involved in these tasks under certain circumstances and may have confounded our view on number sense.

9.5 MOVING FORWARD: BUILDING A BRIDGE BETWEEN COGNITIVE PSYCHOLOGISTS AND MATHEMATICS EDUCATORS FOR THE SAKE OF EDUCATIONAL MATH PRACTICE

In this chapter, in line with the cognitive tradition, we defined number sense as the ability to represent nonsymbolic and symbolic number. This ability is assessed with different explicit measures, in particular number discrimination and number line estimation. The performance on these tasks is related to (later) mathematics achievement, suggesting that *the representation of number is an important building block for higher mathematical abilities.*

In the previous section, however, we summarized recent literature that has demonstrated that factors like the use of clever strategies, decisional aspects, continuous visual cues, and domain general abilities, such as inhibition and other aspects of working memory capacities, may have confounded the assessment of number sense. This overview results in the question we raised in the second part of the title of this chapter:

"should we bother about number sense?" Given these recent findings, should we, for instance, still put effort into developing training and intervention programs for number sense or a better representation of number (eg, Obersteiner, Reiss, & Ufer, 2013; Räsänen, Salminen, Wilson, Aunio, & Dehaene, 2009; Wilson, Dehaene, Dubois, & Fayol, 2009)? To give a motivated answer, we suggest to return to the definition of number sense we proposed in the beginning of this section and to dissect this definition in two parts: (1) number sense as the representation of number and (2) number sense as a building block for later mathematics.

If number sense refers to the way we mentally represent number, it is clear that previous studies using explicit measures are inconclusive. Most studies relied on number line estimation and number discrimination tasks to examine number sense and there is accumulating evidence that the performance in these tasks has been confounded. Therefore, the conclusion of many of these studies stating that we have an approximate system of numbers on which acquired symbols are mapped later on and which becomes more precise through development—as described in the introduction of this chapter—may thus be premature. If we want to address the issue of how number is represented, we believe that implicit paradigms like priming and numerical Stroop are better ways of assessing this. We already mentioned that the picture emerging from these studies is somewhat different than from explicit studies: the representation of number seems to be stable from kindergarten on. One disadvantage, however, is that these implicit measures result in small effects and are not very reliable (Sasanguie, Defever, Van den Bussche, & Reynvoet, 2011). To overcome the latter problem, we recently conducted some studies in our lab with the audiovisual matching paradigm (Sasanguie, De Smedt, & Reynvoet, in press; Sasanguie & Reynvoet, 2014) (Fig. 9.1; see also Barth et al., 2003).

In this paradigm, a visually presented number stimulus, either symbolic (digit) or nonsymbolic (set of dots), has to be matched with an auditory number stimulus, also in either a symbolic format (spoken number word) or a nonsymbolic format (set of sounds). This paradigm overcomes many

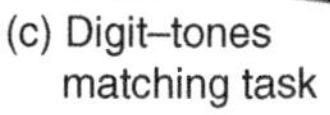

FIGURE 9.1 **Illustration of the audiovisual matching paradigm.** In all variants of the task, participants are first presented with an auditory stimulus (symbolic: number word or nonsymbolic: tone sequence) and afterward with a visual stimulus (symbolic: a digit or nonsymbolic: a dot pattern) and need to indicate whether what they saw matched with what they had heard or not. This resulted in four different conditions: a pure symbolic condition (a), two mixed symbolic-non-symbolic conditions (b and c) and a pure non-symbolic condition (d).

of the problems listed before because participants don't need to compare two visual stimuli and therefore cannot rely on continuous visual cues, which in turn makes that inhibition plays less of a role. The results of these studies disconfirm previous data by showing that a classic DE, traditionally interpreted as evidence for approximate representations, is only found in audiovisual matching tasks with at least one nonsymbolic stimulus and not in a pure symbolic audiovisual matching task. We believe this shows that symbolic number is represented in a separate, exact representational system that coexists with an approximate representation for nonsymbolic number (see also Lyons, Ansari, & Beilock, 2012; Sasanguie, Defever, Maertens, & Reynvoet, 2014 for similar ideas).

In sum, if we consider number sense as being able to represent symbolic and nonsymbolic number, many of these studies on number sense have probably revealed more about nonnumerical processes than we would like to believe. Further studies with implicit paradigms (eg, priming), explicit paradigms overcoming the problems raised here (eg, audiovisual matching) and passive paradigms in combination with neural measures (eg, Gebuis & Reynvoet, 2012) are required to further unravel our representation of number.

However, if we think of number sense as a building block for mathematical achievement, we adopt an approach that is pragmatic rather than fundamental and come close to the view of educational psychologists and math educators who are mainly interested in optimizing math education. In this regard, previous studies on number discrimination and number line estimation are extremely relevant as they have convincingly shown that these basic numerical abilities are unique predictors for later mathematical achievement. In addition, recently, a number of training and intervention studies showed that training these abilities resulted in better mathematics achievement, demonstrating the causal role of these abilities for math (eg, Kucian et al., 2011; Ramani & Siegler, 2008; Ramani, Siegler, & Hitti, 2012; Siegler & Ramani, 2009; Obersteiner et al., 2013; Räsänen et al., 2009; Wilson et al., 2009). Therefore, it can be concluded that, although it is unclear which cognitive processes exactly are responsible for the relation between the performance on these tasks and mathematical achievement, it is clear that number discrimination and number line estimation require subskills essential for mathematical achievement, making them extremely relevant for math educators, who want to gain insight into how math education can be optimized. In turn, cognitive psychologists may help math educators by systematically examining numerical and nonnumerical processes involved in these tasks and provide answers on how much "number" should be involved in their attempts of optimizing math education. Based on this review that demonstrates that only (confounded) explicit measures and not pure implicit measures are related to mathematics achievement, we tend to believe that the role of number representation will be limited compared to more general factors.

References

Agrillo, C., Piffer, L., & Bisazza, A. (2011). Number versus continuous quantity in numerosity judgments by fish. *Cognition, 119*(2), 281–287.

Ancker, J. S., & Kaufman, D. (2007). Rethinking health numeracy: a multidisciplinary literature review. *Journal of the American Medical Informatics Association, 14*(6), 713–721.

Baddeley, A. (2003). Working memory and language: an overview. *Journal of Communication Disorders, 36*(3), 189–208.

Barth, H. C., & Paladino, A. M. (2011). The development of numerical estimation: evidence against a representational shift. *Developmental Science, 14*, 125–135.

Barth, H., Kanwisher, N., & Spelke, E. (2003). The construction of large number representations in adults. *Cognition, 86*(3), 201–221.

Barth, H., La Mont, K., Lipton, J., & Spelke, E. S. (2005). Abstract number and arithmetic in preschool children. *Proceedings of the National Academy of Sciences, 102*(39), 14116–14121.

Berch, D. B. (2005). Making sense of number sense implications for children with mathematical disabilities. *Journal of Learning Disabilities, 38*(4), 333–339.

Booth, J. L., & Siegler, R. S. (2006). Developmental and individual differences in pure numerical estimation. *Developmental Psychology, 41*, 189–201.

Booth, J. L., & Siegler, R. S. (2008). Numerical magnitude representations influence arithmetic learning. *Child Development, 79*, 1016–1031.

Buckley, P. B., & Gillman, C. B. (1974). Comparisons of digits and dot patterns. *Journal of Experimental Psychology, 103*(6), 1131.

Bugden, S., & Ansari, D. (2011). Individual differences in children's mathematical competence are related to the intentional but not automatic processing of Arabic numerals. *Cognition, 118*, 35–47.

Chen, Q., & Li, J. (2014). Association between individual differences in non-symbolic number acuity and math performance: a meta-analysis. *Acta Psychologica, 148*, 163–172.

Chiao, J. Y., Bordeaux, A. R., & Ambady, N. (2001). Mental representations of social status. *Cognition, 93*(2), B49–B57.

Clearfield, M. W., & Mix, K. S. (1999). Number versus contour length in infants' discrimination of small visual sets. *Psychological Science, 10*(5), 408–411.

Cohen Kadosh, R., & Walsh, V. (2009). Numerical representation in the parietal lobes: abstract or not abstract? *Behavioral and Brain Sciences, 32*(3–4), 313–328.

Cohen Kadosh, R. C., Brodsky, W., Levin, M., & Henik, A. (2008). Mental representation: what can pitch tell us about the distance effect? *Cortex, 44*(4), 470–477.

De Smedt, B., Reynvoet, B., Swillen, A., Verschaffel, L., Boets, B., & Ghesquière, P. (2009a). Basic number processing and difficulties in single-digit arithmetic: evidence from Velo-Cardio-Facial Syndrome. *Cortex, 45*, 177–188.

De Smedt, B., Verschaffel, L., & Ghesquière, P. (2009b). The predictive value of numerical magnitude comparison for individual differences in mathematics achievement. *Journal of Experimental Child Psychology, 103*, 469–479.

De Smedt, B., Noël, M., Gilmore, C., & Ansari, D. (2013). The relationship between symbolic and non-symbolic numerical magnitude processing skills and the typical and atypical development of mathematics: a review of evidence from brain and behavior. *Trends in Neuroscience & Education, 2*, 48–55.

De Visscher, A., & Noël, M. (2013). A case study of arithmetic facts dyscalculia caused by a hypersensitivity-to-interference in memory. *Cortex, 49*(1), 50–70.

Defever, E., Gebuis, T., Sasanguie, D., & Reynvoet, B. (2011). Children's representation of symbolic and nonsymbolic magnitude examined with the priming paradigm. *Journal of Experimental Child Psychology, 109*(2), 174–186.

Defever, E., Sasanguie, D., Vandewaetere, M., & Reynvoet, B. (2012). What can the same–different task tell us about the development of magnitude representations? *Acta Psychologica, 140*(1), 35–42.

Defever, E., De Smedt, B., & Reynvoet, B. (2013). Numerical matching judgments in children with mathematical learning disabilities. *Research in Developmental Disabilities, 34*, 3128–3189.

Defever, E., Göbel, S., Ghesquière, P., & Reynvoet, B. (2014). Automatic number priming effects in adults with and without mathematical learning disabilities. *Frontiers in Psychology, 5*(4), 1–7.

Dehaene, S. (1992). Varieties of numerical abilities. *Cognition, 44*, 1–42.

Dehaene, S. (1997). *The number sense: How the mind creates mathematics*. New York: Oxford University Press.

Dehaene, S. (2001). Précis of the number sense. *Mind & Language, 16*(1), 16–36.

Dehaene, S., Izard, V., Spelke, E., & Pica, P. (2008). Log or linear? Distinct intuitions of the number scale in western and Amazonian indigene cultures. *Science, 320*(5880), 1217–1220.

Durand, M., Hulme, C., Larkin, R., & Snowling, M. (2005). The cognitive foundations of reading and arithmetic skills in 7- to 10-year-olds. *Journal of Experimental Child Psychology, 91*, 113–136.

Ebersbach, M., Luwel, K., Frick, A., Onghena, P., & Verschaffel, L. (2008). The relationship between the shape of the mental number line and familiarity with numbers in 5- to 9-year old children: evidence for a segmented linear model. *Journal of Experimental Child Psychology, 99*, 1–17.

Fazio, L. K., Bailey, D. H., Thompson, C. A., & Siegler, R. S. (2014). Relations of different types of numerical magnitude representations to each other and to mathematics achievement. *Journal of Experimental Child Psychology, 123*, 53–72.

Fechner, G. G. (1890). Elemente der psychophysik. Leipzig: Breitkopf Und Hartel. *Report in James, 2*, 50.

Feigenson, L., Dehaene, S., & Spelke, E. (2004). Core systems of number. *Trends in Cognitive Science, 8*(7), 307–314.

Feigenson, L., Libertus, M. E., & Halberda, J. (2013). Links between the intuitive sense of number and formal mathematics ability. *Child Development Perspectives, 7*(2), 74–79.

Finnie, R., & Meng, R. (2001). Cognitive skills and the youth labour market. *Applied Economics Letters, 8*(10), 675–679.

Friso-van den Bos, I., Kroesbergen, E. H., & van Luit, J. E. H. (2014). Number sense in kindergarten children: factor structure and working memory predictors. *Learning and Individual Differences, 33*, 23–29.

Gallistel, C. R., & Gelman, R. (1992). Preverbal and verbal counting and computation. *Cognition, 44*(1), 43–74.

Geary, D. C., Hoard, M. K., Nugent, L., & Byrd-Craven, J. (2008). Development of number line representations in children with mathematical learning disability. *Developmental Neuropsychology, 33*, 277–299.

Geary, D. C., Bailey, D. H., & Hoard, M. K. (2009). Predicting mathematical achievement and mathematical learning disability with a simple screening tool: the number sets test. *Journal of Psychoeducational Assessment, 27*, 265–279.

Gebuis, T., & Reynvoet, B. (2011a). Generating nonsymbolic number stimuli. *Behavior Research Methods, 43*(4), 981–986.

Gebuis, T., & Reynvoet, B. (2011b). The interplay between nonsymbolic number and its continuous visual properties. *Journal of Experimental Psychology: General, 141*(4), 642–648.

Gebuis, T., & Reynvoet, B. (2012). Continuous visual properties explain neural responses to non-symbolic number. *Psychophysiology, 49*(11), 1649–1659.

Gebuis, T., & Reynvoet, B. (2015). Number representations and their relation with mathematical ability. In A. Dowker, & R. Cohen-Kadosh (Eds.), *Oxford handbook of numerical*. Cognition: Oxford University Press.

Gebuis, T., & van der Smagt, M. J. (2011). False approximations of the approximate number system? *PloS One, 6*(10), e25405.

Gebuis, T., Cohen Kadosh, R., de Haan, E. H. F., & Henik, A. (2009). Automatic quantity processing in 5-year olds and adults. *Cognitive Processing, 10*(2), 133–142.

Gebuis, T., Kenemans, J. L., de Haan, E. H., & van der Smagt, M. J. (2010). Conflict processing of symbolic and non-symbolic numerosity. *Neuropsychologia*, *48*(2), 394–401.

Gilmore, C., Attridge, N., & Inglis, M. (2011). Measuring the approximate number system. *The Quarterly Journal of Experimental Psychology*, *64*(11), 2099–2109.

Gilmore, C., Attridge, N., Clayton, S., Cragg, L., Johnson, S., Marlow, N., Simms, V., & Inglis, M. (2013). Individual differences in inhibitory control, not non-verbal number acuity, correlate with mathematics achievement. *PLoS One*, *8*(6), e67374.

Girelli, L., Lucangeli, D., & Butterworth, B. (2000). The development of automaticity in accessing number magnitude. *Journal of Experimental Child Psychology*, *76*, 104–122.

Gray, H. M., Ambady, N., Lowenthal, W. T., & Deldin, P. (2004). P300 as an index of attention to self-relevant stimuli. *Journal of Experimental Social Psychology*, *40*(2), 216–224.

Halberda, J., & Feigenson, L. (2008). Developmental change in the acuity of the "Number Sense": the approximate number system in 3-, 4-, 5-, and 6-year-olds and adults. *Developmental Psychology*, *44*(5), 1457–1465.

Halberda, J., Mazzocco, M. M., & Feigenson, L. (2008). Individual differences in non-verbal number acuity correlate with maths achievement. *Nature*, *455*(7213), 665–668.

Henik, A., & Tzelgov, J. (1982). Is three greater than five: the relation between physical and semantic size in comparison tasks. *Memory & Cognition*, *10*(4), 389–395.

Holloway, I. D., & Ansari, D. (2009). Mapping numerical magnitudes onto symbols: the numerical distance effect and individual differences in children's mathematics achievement. *Journal of Experimental Child Psychology*, *103*, 17–29.

Iuculano, T., Tang, J., Hall, C., & Butterworth, B. (2008). Core information processing deficits in developmental dyscalculia and low numeracy. *Developmental Science*, *11*, 669–680.

Jordan, N. C., Kaplan, D., Ramineni, C., & Locuniak, M. N. (2009). Early math matters: kindergarten number competence and later mathematics outcomes. *Developmental Psychology*, *45*, 850–867.

Kucian, K., Grond, U., Rotzer, S., Henzi, B., Schönmann, D., Plangger, F., Gälli, M., Martin, E., & von Aster, M. (2011). Mental number line training in children with developmental dyscalculia. *NeuroImage*, *57*(3), 782–795.

Landerl, K., & Kölle, C. (2009). Typical and atypical development of basic numerical skills in elementary school. *Journal of Experimental Child Psychology*, *103*, 546–565.

Lansbergen, M. M., & Kenemans, J. L. (2008). Stroop interference and the timing of selective response activation. *Clinical Neurophysiology*, *119*(10), 2247–2254.

Laski, E. V., & Siegler, R. S. (2007). Is 27 a big number? Correlational and causal connections among numerical categorization, number line estimation, and numerical magnitude comparison. *Child Development*, *78*, 1723–1743.

LeFevre, J. A., Jimenez, L. C., Sowinski, C., Cankaya, O., Kamawar, D., & Skwarchuk, S. L. (2013). Charting the role of the number line in mathematical development. *Frontiers in Psychology*, *4*, 641.

Libertus, M. E., & Brannon, E. M. (2010). Stable individual differences in number discrimination in infancy. *Developmental Science*, *13*(6), 900–906.

Link, T., Nuerk, H. C., & Moeller, K. (2014). On the relation between the mental number line and arithmetic competencies. *The Quarterly Journal of Experimental Psychology*, *67*(8), 1597–1613.

Lyons, I. M., Ansari, D., & Beilock, S. L. (2012). Symbolic estrangement: evidence against a strong association between numerical symbols and the quantities they represent. *Journal of Experimental Psychology: General*, *141*(4), 635–641.

Lyons, I. M., Price, G. R., Vaessen, A., Blomert, L., & Ansari, D. (2014). Numerical predictors of arithmetic success in grades 1–6. *Developmental Science*, *17*(5), 714–726.

Meck, W. H., & Church, R. M. (1983). A mode control model of counting and timing processes. *Journal of Experimental Psychology: Animal Behavior Processes*, *9*(3), 320–334.

Mejias, S., Grégoire, J., & Noël, M. P. (2012). Numerical estimation in adults with and without developmental dyscalculia. *Learning and Individual Differences*, *22*(1), 164–170.

Moeller, K., Pixner, S., Kaufmann, L., & Nuerk, H. C. (2009). Children's early mental number line: logarithmic or rather decomposed linear? *Journal of Experimental Child Psychology, 103*, 503–515.

Mundy, E., & Gilmore, C. K. (2009). Children's mapping between symbolic and non-symbolic representations of number. *Journal of Experimental Child Psychology, 103*, 490–502.

Obersteiner, A., Reiss, K., & Ufer, S. (2013). How training on exact or approximate mental representations of number can enhance first-grade students' basic number processing and arithmetic skills. *Learning and Instruction, 23*, 125–135.

Opfer, J. E., & Siegler, R. S. (2007). Representational change and children's numerical estimation. *Cognitive Psychology, 55*, 169–195.

Pahl, M., Si, A., & Zhang, S. (2013). Numerical cognition in bees and other insects. *Frontiers in Psychology, 4*, 162.

Piazza, M., Izard, V., Pinel, P., Le Bihan, D., & Dehaene, S. (2004). Tuning curves for approximate numerosity in the human intraparietal sulcus. *Neuron, 44*(3), 547–555.

Piazza, M., Pinel, P., Le Bihan, D., & Dehaene, S. (2007). A magnitude code common to numerosities and number symbols in human intraparietal cortex. *Neuron, 53*(2), 293–305.

Piazza, M., Facoetti, A., Trussardi, A. N., Berteletti, I., Conte, S., Lucangeli, D., Dehaene, S., & Zorzi, M. (2010). Developmental trajectory of number acuity reveals a severe impairment in developmental dyscalculia. *Cognition, 116*(1), 33–41.

Pica, P., Lemer, C., Izard, V., & Dehaene, S. (2004). Exact and approximate arithmetic in an Amazonian indigene group. *Science, 306*(5695), 499–503.

Polich, J., & Kok, A. (1995). Cognitive and biological determinants of P300: an integrative review. *Biological Psychology, 41*(2), 103–146.

Pritchard, W. S., Houlihan, M. E., & Robinson, J. H. (1999). P300 and response selection: a new look using independent-components analysis. *Brain Topography, 12*(1), 31–37.

Ramani, G. B., & Siegler, R. S. (2008). Promoting broad and stable improvements in low-income children's numerical knowledge through playing number board games. *Child Development, 79*, 375–394.

Ramani, G. B., Siegler, R. S., & Hitti, A. (2012). Taking it to the classroom: number board games as a small group learning activity. *Journal of Educational Psychology, 104*(3), 661–672.

Räsänen, P., Salminen, J., Wilson, A., Aunio, P., & Dehaene, S. (2009). Computer-assisted intervention for children with low numeracy skills. *Cognitive Development, 24*, 450–472.

Restle, F. (1970). Speed of adding and comparing numbers. *Journal of Experimental Psychology, 83*(2), 274–278.

Reyna, V. F., Nelson, W. L., Han, P. K., & Dieckmann, N. F. (2009). How numeracy influences risk comprehension and medical decision making. *Psychological Bulletin, 135*(6), 943–973.

Reynvoet, B., & Brysbaert, M. (2004). Cross-notation number priming investigated at different stimulus onset asynchronies in parity and naming tasks. *Experimental Psychology, 51*(2), 81–90.

Reynvoet, B., De Smedt, B., & Van den Bussche, E. (2009). Children's representation of symbolic magnitude: the development of the priming distance effect. *Journal of Experimental Child Psychology, 103*, 480–489.

Rousselle, L., & Noël, M. P. (2008). The development of automatic numerosity processing in preschoolers: evidence for numerosity-perceptual interference. *Developmental Psychology, 44*(2), 544.

Rousselle, L., Palmers, E., & Noël, M. (2004). Magnitude comparison in preschoolers: what counts? Influence of perceptual variables. *Journal of Experimental Child Psychology, 87*(1), 57–84.

Rubinstein, O., & Henik, A. (2005). Automatic activation of internal magnitudes: a study of developmental dyscalculia. *Neuropsychology, 19*, 641–648.

Rubinsten, O., Henik, A., Berger, A., & Shahar-Shalev, S. (2002). The development of internal representations of magnitude and their association with Arabic numerals. *Journal of Experimental Child Psychology, 81*, 74–92.

Sasanguie, D., & Reynvoet, B. (2013). Number comparison and number line estimation rely on different mechanisms. *Psychologica Belgica, 53*(4), 17–35.

Sasanguie, D., & Reynvoet, B. (2014). Adults' arithmetic builds on fast and automatic processing of Arabic digits: evidence from an audiovisual matching paradigm. *PLoS One, 9*(2), e87739.

Sasanguie, D., Defever, E., Van den Bussche, E., & Reynvoet, B. (2011). The reliability of and the relation between non-symbolic numerical distance effects in comparison, same-different judgments and priming. *Acta Psychologica, 136*(1), 73–80.

Sasanguie, D., De Smedt, B., Defever, E., & Reynvoet, B. (2012a). Association between basic numerical abilities and mathematics achievement. *British Journal of Developmental Psychology, 30*(2), 344–357.

Sasanguie, D., Van den Bussche, E., & Reynvoet, B. (2012b). Predictors for mathematics achievement? Evidence from a longitudinal study. *Mind, Brain, and Education, 6*(3), 119–128.

Sasanguie, D., Defever, E., Maertens, B., & Reynvoet, B. (2014). The approximate number system is not predictive for symbolic number processing in kindergartners. *The Quarterly Journal of Experimental Psychology, 67*(2), 271–280.

Sasanguie, D., De Smedt, B., & Reynvoet, B. (in press). The exact symbolic magnitude system exists! Manuscript submitted for publication. *Psychological Research.*

Sasanguie, D., Verschaffel, L., Reynvoet, B., & Luwel, K. (2016). The development of symbolic and non-symbolic number line estimations: three developmental accounts contrasted within cross-sectional and longitudinal data. Manuscript under revision.

Schneider, M., Heine, A., Thaler, V., Torbeyns, J., De Smedt, B., Verschaffel, L., Jacobs, A. M., & Stern, E. A. (2008). Validation of eye movements as a measure of elementary school children's developing number sense. *Cognitive Development, 23*, 424–437.

Schneider, M., Grabner, R. H., & Paetsch, J. (2009). Mental number line, number line estimation, and mathematical achievement: their interrelations in grades 5 and 6. *Journal of Educational Psychology, 101*(2), 359–372.

Siegler, R. S., & Booth, J. L. (2004). Development of numerical estimation in young children. *Child Development, 75*, 428–444.

Siegler, R. S., & Opfer, J. E. (2003). The development of numerical estimation: evidence for multiple representations of numerical quantity. *Psychological Science, 14*, 237–243.

Siegler, R. S., & Ramani, G. B. (2009). Playing linear number board games—but not circular ones—improves low-income preschoolers' numerical understanding. *Journal of Educational Psychology, 101*, 545–560.

Siegler, R. S., Thompson, C. A., & Opfer, J. E. (2009). The logarithmic-to-linear shift: one learning sequence, many tasks, many time scales. *Mind, Brain, and Education, 3*(3), 143–150.

Smets, K., Gebuis, T., & Reynvoet, B. (2013). Comparing the neural distance effect derived from the non-symbolic comparison and the same-different task. *Frontiers in Human Neuroscience, 7*, 28.

Smets, K., Gebuis, T., Defever, E., & Reynvoet, B. (2014). Concurrent validity of approximate number sense tasks in adults and children. *Acta Psychologica, 150*, 120–128.

Smets, K., Sasanguie, D., Szücs, D., & Reynvoet, B. (2015). The effect of different methods to construct non-symbolic stimuli in numerosity estimation and comparison. *Journal of Cognitive Psychology, 27*(3), 310–325.

Stoianov, I., & Zorzi, M. (2012). Emergence of a "visual number sense" in hierarchical generative models. *Nature Neuroscience, 15*(2), 194–196.

Sullivan, J., & Barner, D. (2014). Inference and association in children's early numerical estimation. *Child Development, 85*, 1740–1755.

Sutton, S., Braren, M., Zubin, J., & John, E. R. (1965). Evoked-potential correlates of stimulus uncertainty. *Science, 150*(3700), 1187–1188.

Szücs, D., Nobes, A., Devine, A., Gabriel, F. C., & Gebuis, T. (2013). Visual stimulus parameters seriously compromise the measurement of approximate number system acuity and comparative effects between adults and children. *Frontiers in Psychology, 4*, 444.

Van Opstal, F., & Verguts, T. (2011). The origins of the numerical distance effect: the same–different task. *Journal of Cognitive Psychology, 23*(1), 112–120.

Van Opstal, F., Gevers, W., De Moor, W., & Verguts, T. (2008). Dissecting the symbolic distance effect: comparison and priming effects in numerical and nonnumerical orders. *Psychonomic Bulletin & Review, 15*(2), 419–425.

Verguts, T., Fias, W., & Stevens, M. (2005). A model of exact small-number representation. *Psychonomic Bulletin & Review, 12*(1), 66–80.

Verschaffel, L., & De Corte, E. (1996). Number and arithmetic. In A. Bishop, K. Clements, C. Keitel, & C. Laborde (Eds.), *International handbook of mathematics education. Part 1* (pp. 99–138). Dordrecht: Kluwer.

White, S., & Szücs, D. (2012). Representational change and strategy use in children's number line estimation during the first years of primary school. *Behavioral and Brain Functions, 8*(1), .

Wilson, A., Dehaene, S., Dubois, O., & Fayol, M. (2009). Effects of an adaptive game intervention on accessing number sense in low-socioeconomic-status kindergarten children. *Mind, Brain, and Education, 3*(4), 224–234.

Xenidou-Dervou, I., De Smedt, B., van der Schoot, M., & van Lieshout, E. C. D. M. (2013). Individual differences in kindergarten math achievement: the integrative roles of approximation skills and working memory. *Learning and Individual Differences, 28*, 119–129.

Xenidou-Dervou, I., van der Schoot, M., & van Lieshout, E. C. D. M. (2014a). Working memory and number line representations in single-digit addition: approximate versus exact, nonsymbolic versus symbolic. *Quarterly Journal of Experimental Psychology, 68*(6), 1–47.

Xenidou-Dervou, I., van Lieshout, E. C. D. M., & van der Schoot, M. (2014b). Working memory in nonsymbolic approximate arithmetic processing: a dual-task study with preschoolers. *Cognitive Science, 38*(1), 101–127.

Xenidou-Dervou, I., Ansari, D., Molenaar, D., van der Schoot, M., & van Lieshout, E.C.D.M. (2016). Longitudinal development of nonsymbolic and symbolic magnitude processing: contradictions, methodologies and moving forward. Manuscript submitted for publication.

Xu, F., & Arriaga, R. I. (2007). Number discrimination in 10-month-old infants. *British Journal of Developmental Psychology, 25*(1), 103–108.

Xu, F., & Spelke, E. S. (2000). Large number discrimination in 6-month-old infants. *Cognition, 74*, B1–B11.

CHAPTER

10

The Distribution Game: Evidence for Discrete Numerosity Coding in Preschool Children

Alain Content, Julie Nys

Laboratoire Cognition, Langage & Développement, Université Libre de Bruxelles, Brussels, Belgium

OUTLINE

10.1 AN INTERESTING IDEA

The idea that humans are endowed with a specific biological mechanism devoted to the extraction of number is a fascinating proposal. Among all dimensions of the environment, number is among the most abstract,

Continuous Issues in Numerical Cognition. http://dx.doi.org/10.1016/B978-0-12-801637-4.00010-X

most culture-related, and least immediate dimensions that one can think of. An additional reason we find the idea attractive is that since its earliest formulations, it was supported by empirical observations and simultaneously grounded on a simple and elegant processing mechanism. Converging behavioral, neuroimaging, and neurophysiological observations have been reported from animal, as well as human infant and adult studies (eg, Feigenson, Dehaene, & Spelke, 2004; Gallistel & Gelman, 2005; Mou & vanMarle, 2014; Nieder & Dehaene, 2009 for reviews). The hypothetical mechanism was described in functional terms by Meck and Church (1983) and later implemented in the form of an artificial neural network by Dehaene and Changeux (1993) (see also Verguts & Fias, 2004). The mechanism assumes an accumulator system that sums the approximately fixed quantity of activation generated by each item in the set. Hence, the state of the accumulator encodes the numerosity of the set, with some variation proportional to the number of activation increments. Therefore, the accumulator principle intrinsically entails that the standard deviation of the distribution of the encoded numerosity scales with its magnitude, a property termed "scalar variability" by Gallistel and Gelman (1992). One index of scalar variability is thus the observation that the coefficient of variation (CoV, the ratio of the standard deviation of estimates to their mean) is constant across target numerosities. If it is the case, the value of the CoV can be taken to characterize the acuity of the numerosity extraction mechanism.

Sometimes, appeals to hardwired constraints in developmental theories sound a bit like mystical beliefs, begging more questions than they solve. In the domain of number, by contrast, things seemed different. The simplicity of the accumulator principle and its prima facie neural plausibility made it worth considering as a potential foundation from which numerical intuition and arithmetic reasoning skills might later derive. Experimental studies with nonhuman animals and prelinguistic infants during the first year of life provided support for the idea that some capability to extract, represent, and reason about approximate quantity exists independently of and before any contact with cultural symbol systems. Furthermore, the idea fits with a conception of cognitive development in which innate cognitive constraints are viewed as a part of the basic equipment through which the organism interacts with its environment.

Yet not everybody shares this favorable a priori and there seems to be a split in the numerical cognition community at this time. Some researchers strongly believe in the existence of a *specific* mechanism that humans share with animals and which allows humans to extract approximate numerosity information from sensory stimuli. Challengers assume that general-purpose mechanisms which extract magnitude information from the senses are sufficient to account for early numerical competence. They assume that the abstract concept of number only emerges as the product of acquaintance with numerals, that is, the verbal and graphic symbols that cultures have invented to manipulate and communicate about quantities. Unfortunately the divide seems radical enough to sometimes hinder scientific debate. One of us was recently coauthoring a manuscript which showed no correlation across participants between numerosity comparison and numerosity estimation performance, and no correlation between nonsymbolic comparison and arithmetic performance. The authors argued that the absence of correlation between the Weber fraction computed from the comparison task and the coefficient of variation extracted from the estimation task was puzzling. As both tasks and measures are assumed to tap into the same feature, that is, the variability of the internal representations, a positive correlation would be expected if there are interindividual differences in the acuity of the mechanism that extracts numerosity information from visual stimuli. The manuscript received contrasting reviews. Some experts thought that the work was uninteresting and argued that the absence of correlation (as any null result) could be explained in various trivial ways. Others conversely encouraged the authors to focus on the absence of correlation between the two indices, considering that it was a serious blow against the approximate number sense (ANS) theory. The upshot, as it happens, was an unnecessary delay in the publication of a small empirical contribution of which the major aim was to appeal to a finer analysis of the tasks currently used to assess the acuity of adults' numerosity apprehension (Guillaume, Gevers, & Content, 2016; see also Smets, Sasanguie, Szücs, & Reynvoet, 2015 for a similar point). Although there is in our view abundant evidence in favor of the existence of a preverbal approximate numerosity perception mechanism, not all aspects and implications of the hypothesis have been satisfactorily established, and there is still room for discussion, as the present collection of papers demonstrates.

10.2 TWO ISSUES WITH ANS THEORY

In the recent years, the notion of a precocious (prelinguistic) perceptual system specifically dedicated to the extraction of approximate numerosity has been challenged on several grounds. A first reason is that number is naturally correlated with other characteristics, and it has proven extremely difficult to disentangle numerosity from other potentially salient perceptual dimensions. Thus, with visual collections, researchers have often pointed out that the number of elements correlates with the summed area occupied by the items, their summed contour, the overall color/luminance, the density (number of items per unit area), and the occupancy.[a] Attempts at controlling some of these potential confounds

[a]Note that we use "occupancy" in a strictly descriptive way to designate the exact proportion of the total area occupied by the items, and not in the specific sense of Allik's & Tuulmets's (1991) occupancy model.

would often create other covariations. For example, varying dot size to equate total area and occupancy only introduces yet another inverse relation between dot size and number, and equating density logically imposes to scale the available space in proportion of number. Similar necessary links occur with sequential series of auditory or visual events, since numerosity endures systematic relationships with pace and duration. These potential confounds are especially worth attention as they influence performance in nonsymbolic numerical comparison tasks despite efforts to control them (Gebuis & Reynvoet, 2012c; Szücs, Nobes, Devine, Gabriel, & Gebuis, 2013). For instance, Nys and Content (2012) manipulated number and summed area orthogonally by varying mean dot size and observed both interference and facilitation of the area dimension on numerosity comparison in adults. These findings have led some researchers to conclude that rather than being sensitive to the proper numerosity dimension, participants infer number by combining multiple other visual properties (Gebuis & Reynvoet, 2012a, 2012b, 2012c, 2013; Gebuis, Gevers, & Cohen Kadosh, 2014).

A second issue concerns interindividual differences. The seminal paper by Halberda, Mazzocco, and Feigenson (2008) indicating an association between ANS acuity and school achievements in mathematics launched a flurry of research investigating the relationship, with mixed results (see Chen & Li, 2014; Feigenson, Libertus, & Halberda, 2013 for recent reviews). The lack of clear and robust correlations has prompted some authors to cast doubt on the foundational role of the ANS for arithmetic and mathematical abilities, and to consider alternative developmental trajectories (De Smedt, Noël, Gilmore, & Ansari, 2013; Lyons & Ansari, 2015), for instance arguing for distinct approximate and exact representational systems, with the latter as the foundation of numeracy, arithmetic, and mathematical skill (Noël & Rousselle, 2011). While the issue is certainly important, both for developmental theory as well as in view of its potential educational consequences, it is logically distinct from the more basic claim that there exists a nonverbal approximate numerosity representation system which humans share with other nonhuman species. One can imagine several strands of explanations why the correlation would be weak or absent in spite of the existence of such a primitive system: To name just a few, it could be that the preverbal biological endowment does not vary in acuity or does so only minimally; or it could be that current assessment tools do not provide sufficiently accurate measures to capture the link; or that arithmetic and mathematical abilities are multidimensional constructs that do not lend themselves to single-factor associations. Be that as it may, the current evidence suggests the presence of a weak relationship, with bidirectional influences. On one hand, several studies with children (Mussolin, Nys, Content, & Leybaert, 2014) and adults (Nys et al., 2013; Piazza, Pica, Izard, Spelke, & Dehaene, 2013; see Mussolin, Nys, Leybaert

& Content, 2015, for a review) demonstrate that the degree of knowledge and experience with the numerals impacts on the acuity of the nonsymbolic numerosity perception system. On the other hand, while longitudinal studies indicate that preschool measures of acuity are predictive of primary school math achievements, only one to our knowledge produced evidence of a predictive link between acuity measures obtained well before language acquisition (at 6 months) and early mathematic abilities (around 3–6 years, Starr, Libertus, & Brannon, 2013).

10.3 ADULT'S ESTIMATION OF NUMEROSITY

The studies presented later in this chapter are aimed at examining how young children estimate the numerosity of sets of visual objects, and to what extent their estimation performance is related to their knowledge of the number-word sequence and their mastery of the counting procedure. The major question of these studies was whether children's estimation would provide evidence of a "nonverbal counting mechanism," as proposed by Whalen, Gallistel, and Gelman (1999). Whalen et al. (1999) used estimation to directly assess the hypothesis that mental representations of numerical magnitudes sustain scalar variability. They devised both a numerosity production task (emit a sequence of key-presses matching in number to an Arabic numeral) and a numerosity perception task (state verbally the number of randomly paced flashes presented) inspired from analogous animal studies. In both tasks, the responses increased in proportion to target values, and the variability in the responses was proportional to their mean, as demonstrated by stable coefficients of variations over the range of numerosities used. The authors concluded that their findings support the existence of a nonverbal system consistent with the accumulator principle, which provides mental representations of numerical magnitudes to map onto numerals.

To demonstrate that participants were not using silent verbal counting during the estimation task, Whalen et al. produced evidence that subvocal counting entails a slower response pace. In a later report, Cordes, Gelman, Gallistel, and Whalen (2001) compared performance in counting aloud conditions (either full counting or tens counting) with an articulatory suppression condition in which participants were instructed to estimate without counting and had to repeat the word *"the"* at every key-press. The number of presses was much closer to the numerical target in the counting condition than in the articulatory suppression condition. Furthermore, the CoVs were stable in the dual-task condition, whereas they decreased with numerosity in the full count and tens count conditions (see also Beran, Taglialatela, Flemming, James, & Washburn, 2006). Castronovo and Seron (2007) adapted Whalen et al.'s (1999) tasks to the auditory modality to compare blind and sighted adults' estimation ability. In the production experiment, targets (ie, verbal numerals), were presented orally and participants were requested to press a key as fast as possible without counting to approximate the target number. In the perception situation, participants heard sequences of randomly paced tones and were to report orally the corresponding number of events. Again, coefficients of variation appeared constant across numerosities, and blind participants were more accurate than sighted ones. Boisvert, Abroms, and Roberts (2003) also used a dual-task situation but with a different paradigm, inspired by their own previous work with pigeons. Participants saw a succession of visual events and they were instructed to either press a button when they felt that the target number of patterns was reached, or to report verbally the number of patterns presented. Thus, responses combined nonverbal behavioral production of target numerosities with verbal estimates. To avoid counting, participants were required to name the visual shapes as soon as they appeared on screen (eg, red square, blue rectangle, yellow circle, and so on). The results also indicated scalar variability. Interestingly, in one more complex experimental variant in which each successive event was composed of several identical shapes (1–10) participants had to count (ie, eight squares, three circles, and so on), only the verbal responses showed a larger underestimation effect, suggesting that concurrent counting impacts on verbal responses more than on manual responses. A discrepant result, however, was reported by Tan and Grace (2011), who had participants estimate the number of red objects in a sequential flow of 40 images, while simultaneously naming the objects. Three tasks were used: (1) decide whether there were many or few red objects, (2) produce the corresponding number of key-presses, or (3) type the corresponding number on the keyboard. Key-press and numeral responses produced similar underestimation effects for the larger numerosities. More importantly, the analysis of the CoVs suggested a bilinear function, with a negative slope up for the smaller numerosities up to six and a positive slope for larger values. The authors argued that previous studies may have used too few small numerosities to capture the bilinear trend in the CoVs.

In sum, although studies with human adults produced convergent evidence of an approximate estimation process obeying scalar variability, we believe that the findings would be unlikely to convince skeptics. First, the observations do not provide convincing evidence to fully discard the role of verbal tagging. It is doubtful that repetitive concurrent articulation completely suppresses any form of silent counting (see, eg, Logie & Baddeley, 1987, for a detailed analysis of the effect of concurrent articulation on counting). Participants could monitor the sequence of events or responses through the enumeration of phonological labels much faster than through subvocal articulation. Chunking processes might also contribute to diminish the verbal load (for instance, counting by five). Moreover, it is well known from the reading and word recognition literature that the use of articulatory suppression paradigms does not fully block access to phonology (Besner, Davies, & Daniels, 1981; Saito, 1997; Tree, Longmore, & Besner, 2011). Hence, stronger evidence would be obtained by showing that subjects can estimate the numerosity of collections that they simply *cannot* count. There is already some evidence to that effect from adults from cultures with limited numeral systems (Frank, Everett, Fedorenko, & Gibson, 2008; Pica, Lemer, Izard, & Dehaene, 2004; Piazza et al., 2013). Another approach, which we will explore, is to assess the estimation performance of young children whose mastery of the number–word sequence is still limited.

Second, most studies with adults investigated either the mapping of discrete quantity to Arabic numerals (the numerosity perception task) or the mapping of Arabic numerals to discrete responses (the numerosity production task). Thus, neither can provide direct, unambiguous evidence for a numerical magnitude representation system independent of numerical acculturation. Although it is plausible that both tasks involve a common intermediate magnitude representation, they entail further transcoding operations to map magnitudes with Arabic or verbal numerals (see Castronovo & Seron, 2007; Crollen, Castronovo, & Seron, 2011; Izard & Dehaene, 2008 for different views on this mapping process). In the experiments described later we focus on purely nonsymbolic estimation tasks, that is, situations in which participants have to map from one nonsymbolic format to another nonsymbolic format, so that they could in principle be realized without any knowledge of numerical symbol systems. For instance, Frank et al. (2008), Frank, Lai, Saxe, and Gibson (2012) used several variants of matching tasks to compare number processing in United States and Pirahã participants, an Amazonian tribe with a very limited number–word system. Subjects had to produce the same small quantity of objects that were presented to them, with the target sets either visible or not. The results supported the view that Pirahã and American adults used similar approximate estimation mechanisms when the latter were discouraged to count through concurrent articulation. Crollen et al. (2011) devised a computerized version of the numerosity matching task, with the advantage that larger numerosities could be tested. Participants saw dot collections on the display and were required to produce a set comprising approximately the same number of dots as the targets. They operated a rotating knob which controlled the number of dots flashed on screen, and had to press a button when they were satisfied by the generated numerosity. Crollen et al. found that estimates in the matching task were more accurate than in the perception and production conditions. They argued that underestimation (in the perception task) and overestimation (in the production task) were caused by the mapping between the internal nonsymbolic magnitude code and the Arabic numerals, implying that the matching task exclusively relies on the nonsymbolic magnitude code.

10.4 CHILDREN'S ESTIMATION OF NUMEROSITY

As with adults, most studies examining numerosity estimation in children used either perception or production tasks. For instance, Huntley-Fenner (2001) examined estimation in children between 5 and 7 years. Children saw briefly presented sets of black squares (5–11) and estimated the number of squares by pointing to one marking on a line with Arabic numerals from 1 to 20. Mean estimates were close to target numerosities, and a majority of children showed constant coefficients of variation. Moreover, younger children tended to have larger CoVs than older ones, and most of them had larger CoVs than adults in Whalen et al.'s experiments. Other studies have explored how children map Arabic numerals with spatial positions on a number line (Siegler & Opfer, 2003; see Siegler & Lortie-Forgues, 2014 for review).

To our knowledge, few studies have used purely nonsymbolic estimation tasks with preschoolers. Mejias and Schiltz (2013) used the numerosity matching technique devised by Crollen et al. (2011) with 5- and 6-year-old kindergarteners. To compare nonsymbolic and symbolic responses, the knob generated either dot collections or Arabic numerals. Children passed six trials with four target numbers, (8, 16, 34, and 64) in each condition. Ebersbach and Erz (2014) had kindergarteners, first-graders, third-graders, and adults pass a similar task with target magnitudes between 15 and 94. Children's knowledge of the number–word sequence was also assessed, and the mean level reached by the kindergarteners was 41. In both studies, nonsymbolic estimates as well as symbolic estimates increased monotonically with target numerosity. Unfortunately, however, neither study reported detailed results for the children with the lowest scores in the word-sequence task. Neither can thus inform on children's estimation performance outside their range of known numerals.

In the following sections, we describe two studies with children between 3 and 6 years of age, in which we explored a new experimental situation, the distribution game, loosely inspired from Whalen et al. (1999) but affording purely nonverbal estimation performance in children. The basic principle is quite simple: participants see collections of visual items, and they have to produce the corresponding number of key-presses. The major issue that we aimed at examining was whether the number of key-presses would show a systematic relation to the numerosity of the stimuli, and whether we would find evidence of scalar variability. We also assessed whether estimates would vary with the nonnumerical characteristics of the displays (ie, total area, dot size, density), and whether differences in acuity were related to knowledge of the numerical system.

One essential feature of this design is that, contrary to previous ones with the reproduction technique (Mejias, Mussolin, Rousselle, Grégoire, & Noël, 2012; Mejias & Schiltz, 2013; Ebersbach & Erz, 2014), there is no direct analogical mapping from the input in the visuospatial domain (the image of a set of randomly arranged objects) to the output, in the sequential, temporal modality. Thus, if children succeed in mapping the visual and the temporal magnitudes, it would imply that the information extracted and used is abstract enough to make that mapping possible.

10.5 STUDY 1: THE DISTRIBUTION GAME

Sixty-two children from 2;11 to 5;10 years of age participated in the first experiment. They were drawn from three preschool classes of a public middle-class school in a small town of Wallonia. Each child passed one 20- to 30-min session in a quiet room at school, with the estimation task followed by an assessment of numerical knowledge.

The estimation task was presented as a candy-distribution game. The experimenter told the story of a boy who wants to offer candies to all his friends in the playground. Playgrounds were represented on A4-sized plastified sheets with various numbers of items (drawings of identical animals, toys, puppets, or simple dots, from 3 to 19), arranged randomly. A device with a mechanical lever was handed to the child who was invited to press the lever to deliver candies to all kids in the playground. The lever was connected to an electronic counter that displayed the number of presses so that the experimenter could record it. The experimenter presented each plate until the child was ready to do the distribution. She then removed it from sight to avoid the use of a one-to-one correspondence strategy and started a timer to measure the duration of the distribution. When the child had finished, the experimenter recorded the total number of lever-presses and the duration of the distribution process. Eight training trials were presented first, followed by 36 experimental trials in a fixed random order (four different images with each odd numerosity from 3 to 19). After the estimation task, each participant was asked to produce their longest number-word sequence and to count collections of candies within the range attained. Children were divided in two knowledge-level groups, according to their knowledge of the number-word sequence and their performance in the *how many?* task (Table 10.1). They were identified as beginners if they could not produce the numeral sequence up to 10 nor successfully count collections above 10, and as advanced counters otherwise. Four children who had discrepant results on the two tests were discarded.

Fig. 10.1 presents the distribution of responses—number of presses—as a function of numerosity. Although some participants appear overgenerous, distributing largely more that the required number of sweets (sometimes more than 50!), on the average, the mean number of presses fits quite closely with the target numerosity and the range of variation clearly also increases with target numerosity.

TABLE 10.1 Characteristics of the Participants in Experiment 1, Study 1

Group		Beginners	Advanced
N		26	32
Age (months)	Mean (SD)	46 (8)	57 (9)
	Range	(35–67)	(36–70)
Number-word sequence	Mean (SD)	5.0 (2.5)	15.3 (2.2)
How many?	Mean (SD)	4.8 (2.7)	14.4 (2.2)

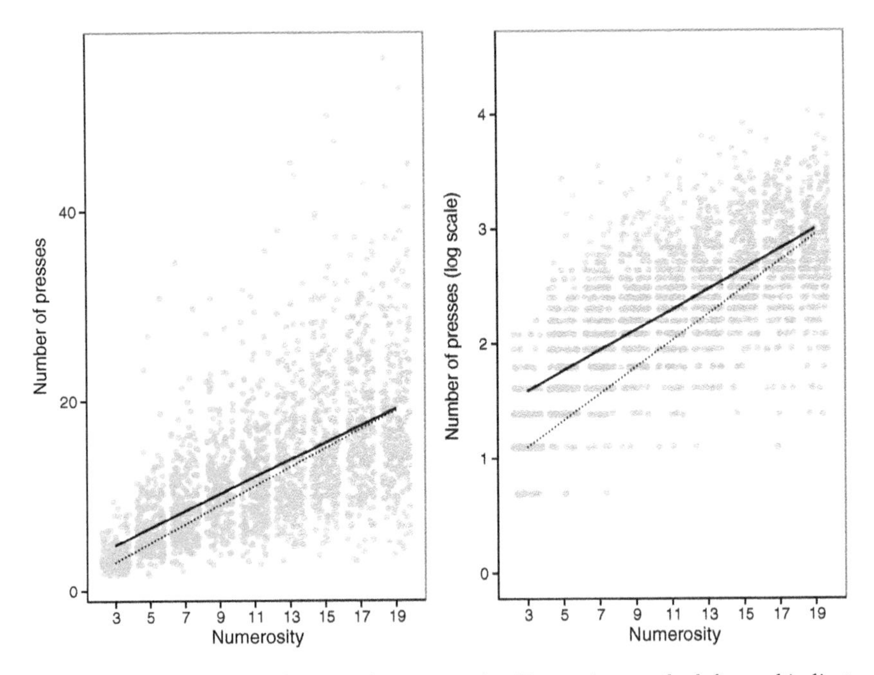

FIGURE 10.1 **Number of presses by numerosity.** Data points on the left panel indicate the increasing overall variability. The right panel displays the same data, showing that the range of variation appears constant under logarithmic compression. Data points at each numerosity are randomly jittered horizontally. The *dotted lines* indicate exact performance, and the *dashed lines* show the overall linear regression trend.

Interestingly, variability appears constant when the responses are presented on a logarithmic scale (Fig. 10.1, right panel), as would be expected if the variability is proportional to the magnitude. Fig. 10.2 displays the average number of presses, which thus corresponds to the mean numerosity estimate, as a function of numerosity and group. Both advanced counters and beginners produced monotonically increasing

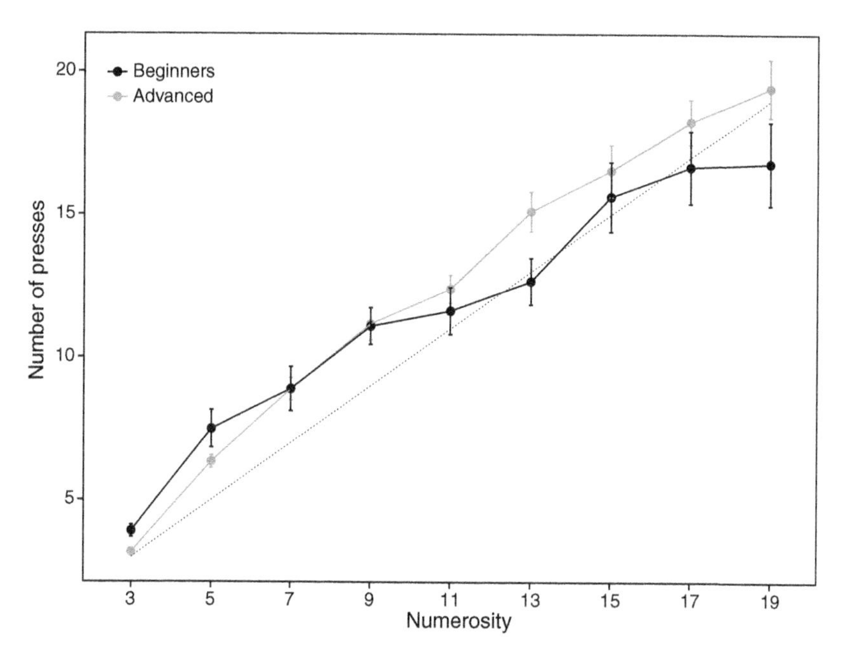

FIGURE 10.2 **Average of estimates per numerosity and knowledge level group.** Error bars show ±1 standard error. The *dotted line* indicates exact performance.

estimates. Because the variability of the estimates increases with numerosity, thus violating homoscedasticity, all analyses were performed on log-transformed variables. This has the advantage that if the relation between the estimates and the veridical numerosities follows a power law, as suggested by previous studies (eg, Cordes et al., 2001; Izard & Dehaene, 2008), the transformation should produce a linear relation, with a slope corresponding to the exponent of the power function. Thus, a slope of one corresponds to a linear relation, and slopes below one index, a negatively accelerated curve.

We fitted mixed models to the raw log-transformed estimates, with items and participants as random factors, and log-transformed numerosity and group as fixed factors. Both factors were centered to facilitate the interpretation of effects and interactions. The effect of numerosity was highly significant, $t(34) = 32.2$, $p < 0.0001$, as well as the interaction with group, $t(1994) = 7.15$, $p < 0.0001$, indicating that the slopes were different for the advanced counters and the beginners. Both groups obtained average slope values below 1, namely, 0.93 for the advanced group (closer to linearity) and 0.73 for the beginners. Thus, children overestimated small numerosities more than larger ones, although beginners appear to overestimate large numerosities less than advanced counters, as indicated by the numerosity by group interaction. Interestingly, an analysis restricted to

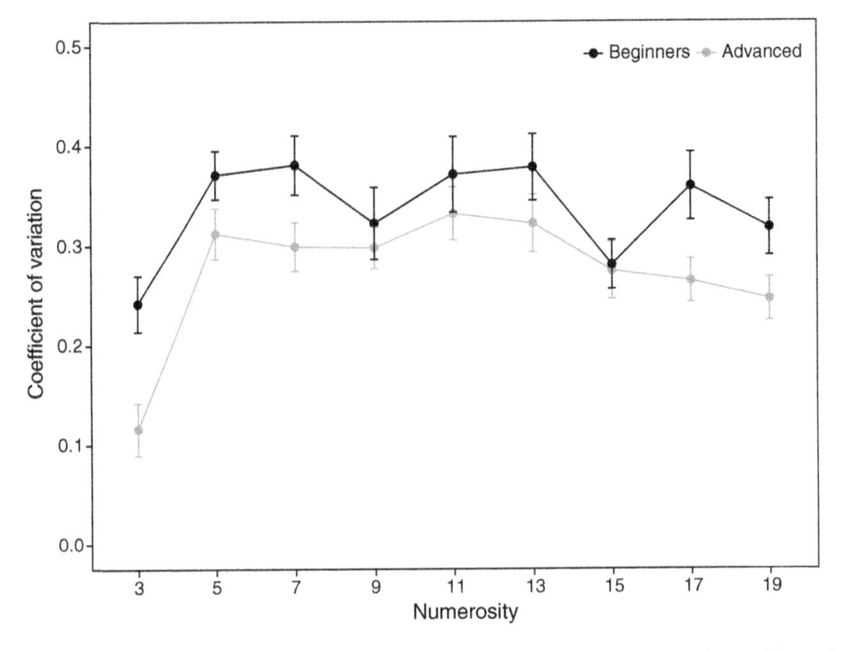

FIGURE 10.3 **Average coefficients of variation as a function of numerosity and knowledge level in Experiment 1.**

the numerosities above 10 confirmed the effect of numerosity, $t(34) = 8.9$, $p < 0.0001$, and indicated an overall group difference, $t(56) = -2.17$, $p < 0.05$ but no interaction. Both groups clearly show similar increases in the upper half of the target range, from 11 to 19, for which it is most unlikely that beginners could use verbal counting knowledge.

Besides the overall trends previously reported, we also wanted to assess performance on an individual basis. So we fitted a log–log linear regression to each child's raw estimates as a function of numerosity. All regressions were significant, indicating positive trends, but there were huge variations in the amount of unexplained variation (mean $R^2 = 0.65$, [0.17–0.92]) as well as in the individual slopes (mean slope, 0.84, range [0.30–1.33]).

Regarding the variability of estimates, Fig. 10.3 displays the average coefficients of variation for the two groups, as a function of numerosity. The average CoV appears smaller for advanced than for beginning counters. Moreover, in both groups, the CoV seems stable above 3, and much lower for 3 than for other numerosities. The smaller variability of responses on 3 is not surprising, as it is likely that many children could get the exact number of elements through either subitizing or counting.

We tested scalar variability in two ways. First, we fitted a linear regression model to each participant's CoVs as a function of numerosity (excluding 3). If the CoV is constant, the slope should be zero. Indeed, out of the 58 participants, only six obtained significant individual CoV slopes, five advanced counters, and one beginner, and even those were rather close to zero (−0.027 to 0.015). The mean CoV was well in the range of other measures of ANS acuity around the same age (Mussolin et al., 2015), and it was larger in the beginners group (CoV = 0.336), than in the advanced group, (CoV = 0.274), $t(53) = 3.71$, $p < 0.0005$. As expected, the CoV slopes were very close to zero in both the beginners group, −0.0036, $t(25) = -1.55$, $p = 0.13$, and the advanced group, −0.0043, although it was significantly negative in the latter case, $t(31) = -2.37$, $p < 0.05$, presumably due to a few children showing slight negative trends.

The previous analysis may, however, not be optimally sensitive because it tests for the *absence* of a linear relation. The obtained slopes are generally close to zero and very few children produce significant relations, but it is hard to know whether this is due to large variation or to the fact that the CoV is actually constant. A second, more stringent test assessed directly whether the standard deviation of the estimates for each target number is proportional to the mean of the corresponding estimates. To that end, for each participant, we fitted values derived from the scalar variability model, $s_i = k \times m_i$, with m_i and s_i representing the mean and standard deviation for target numerosity i. The constant k was estimated by averaging across the values of the participant's CoV and the predicted values of the standard deviation were calculated by multiplying k by the participant's mean estimates for each numerosity. Then, the scalar-variability fit was assessed by computing the summed residual squared error, which we compared to the fit of the least-square regression line, on one hand, and to that of a null model based on the assumption of a constant standard deviation, on the other hand. We reasoned that if the standard deviations are proportional to the mean of the estimates, the scalar-variability model should produce fits equivalent to the least-square regression model, and systematically better than the null model.

For 23 out of the 58 participants (40%), the least-square model did not reach significance ($\alpha = 0.05$), most likely due to noise (remember that there were only four data points per target value). Three subjects who showed slight increases of CoVs with numerosity obtained significantly better least-square than scalar variability fits. Among the 32 others, 31 (89%) showed no significant advantage of the least-square model over the scalar-variability model, and the scalar-variability model produced a significant or, for 5 participants, a marginally significant fit. A reanalysis of the raw estimation data for those 26 participants who obtained significant least-square and scalar-variability fits confirmed the previous conclusions, namely, a monotonic increase in estimates following a power function, slightly steeper for the advanced counters than for the beginners (Fig. 10.6).

To summarize the main findings of the first experiment, children's estimates increase quasilinearly with numerosity even in a range of

TABLE 10.2 Characteristics of the Participants in Experiment 2, Study 1

		A3		A4	
		Beginners	**Advanced**	**Beginners**	**Advanced**
N		12	5	9	9
Age (months)	Mean (SD)	45 (6)	53 (8)	44 (5)	52 (6)
	Range	(37–53)	(40–60)	(37–52)	(40–60)
Number-word sequence	Mean (SD)	5.8 (2.1)	16.0 (0.71)	5.3 (1.6)	14.1 (1.7)
How many?	Mean (SD)	5.3 (1.4)	15.4 (0.89)	4.3 (1.4)	13.4 (1.7)

magnitudes for which they do not yet master the number words and for numerosities above 10 that many of them cannot count. In addition, in spite of the very small number of data points per cell, the results fit quite well with the principle of scalar variability.

Experiment 2 used the same setting and procedure, but aimed in addition at assessing the influence of continuous variables on children's estimates. First, the size of the individual elements on the sheets decreased with numerosity to keep the aggregated area equal. If children use the cumulated area to evaluate the number of elements on each plate, their responses should remain constant in this condition. Second, to manipulate the density and display size, half the participants saw A4 (21 × 29 cm) sheets, while the other half received the same images in A3 format (42 × 29 cm), so the display area was doubled. Elements had the same size on both versions, so the density (number by unit area) and occupancy (total area occupied by unit area) were halved in the A3 format.

Forty children from the same school environment participated in the estimation task (5 were excluded because their knowledge level was not evaluated). None had been involved in the previous experiment. Unfortunately, the partition of the groups as a function of knowledge level was not even (Table 10.2), limiting the validity of group comparisons.

Yet, as in the previous experiment, mean estimates closely approximated stimulus numerosities and the results only suggested minimal effects of format (Fig. 10.4). We ran similar mixed model analyses as in Experiment 1, with the addition of the format factor. The effect of numerosity was highly significant, $t(38) = 25.76$, $p < 0.0001$, as well as the interaction with group, $t(1187) = 5.51$, $p < 0.0001$, and a second-order interaction with format, $t(1187) = -4.33$, $p < 0.0001$, indicating that the slopes were different as a function of group and format. An analysis restricted to the numerosities above 10 confirmed the effect of numerosity, $t(21) = 8.76$, $p < 0.0001$,

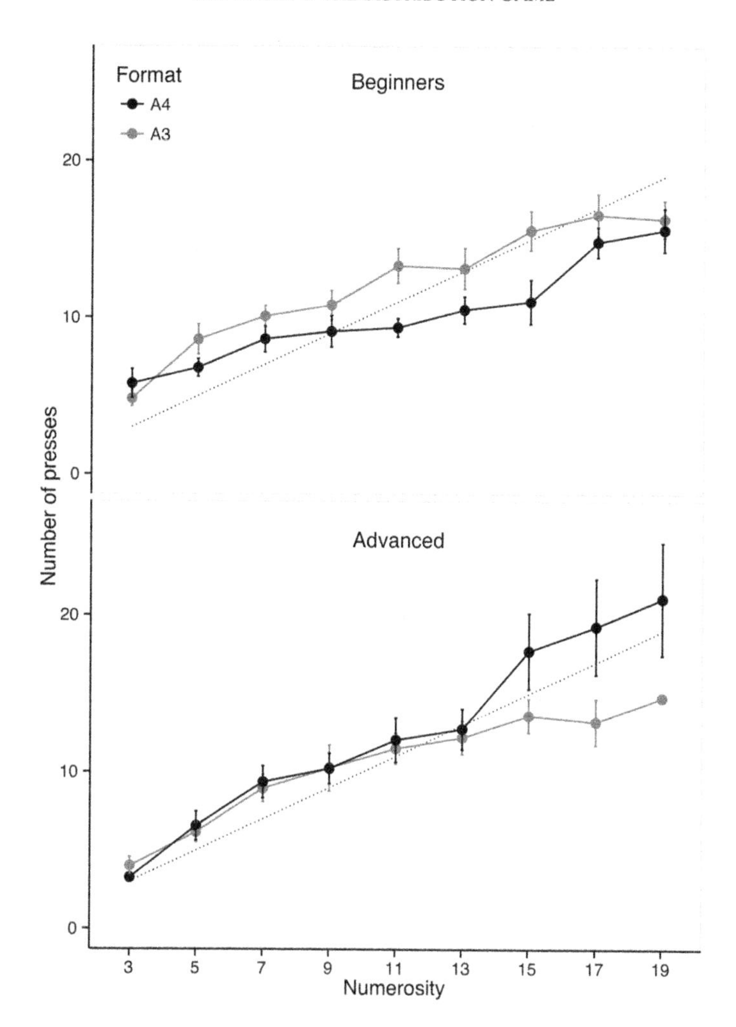

FIGURE 10.4 **Experiment 2: average of estimates per numerosity as a function of format and knowledge level group.** Error bars show ±1 standard error. The dotted lines indicate exact performance.

and revealed an overall format difference, $t(674) = 3.29$, $p = 0.001$ as well as an interaction between format and numerosity, $t(643) = -3.27$, $p = 0.001$, but no significant group difference.

Separate analyses for the beginners and advanced counters produced different patterns of results. For the beginners, there was an effect of numerosity, $t(35) = 18.50$, $p < 0.0001$, which also interacted with format, $t(699) = 2.45$, $p < 0.05$. The slope coefficient was smaller than for the advanced group, and slightly greater for the larger format (0.53 vs. 0.65). For the advanced group, there were main effects of format, $t(42) = 2.31$,

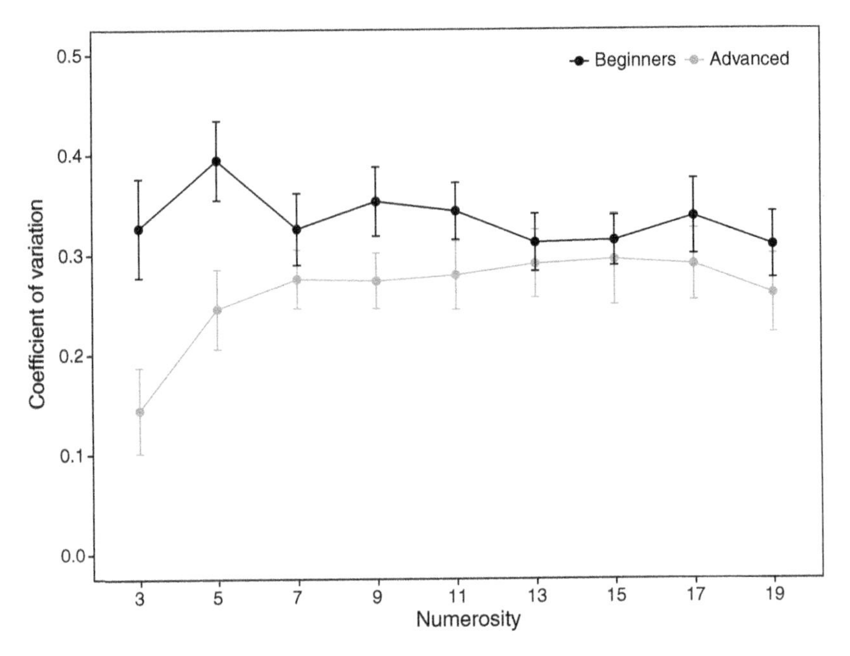

FIGURE 10.5 **Average coefficients of variation as a function of numerosity and knowledge level in Experiment 2.**

$p < 0.05$, and numerosity, $t(39) = 27.32$, $p < 0.0001$, as well as an interaction, $t(454) = -3.97$, $p < 0.0001$. As can be seen from Fig. 10.4, the slope was closer to 1 (ie, closer to linearity) for the smaller than for the larger format (0.91 vs. 0.70, respectively). Overall, log–log regressions on an individual basis were significant for all but one child, and there were again large variations both in the fit (mean $R^2 = 0.57$, [0.11–0.91]) and in the slopes (mean slope, 0.71, range [0.23–1.42]).

Analyses of variability produced results analog to those of Experiment 1 (Fig. 10.5). Only three children obtained significant CoV slopes with the log–log regression analysis, although with rather small values (−0.04 to 0.02). The mean CoV values were in the same range as in the previous experiment, and again larger for the beginners (0.335 vs. 0.261, $t(33) = 3.34$, $p < 0.005$). Contrary to Experiment 1 though, mean slopes did not significantly differ from zero in either group (beginners, slope = −0.0041, $t(20) = -1.36$, $p > 0.20$; advanced, slope = 0.0015, $t(13) = 0.574$, $p > 0.20$). The model comparison analysis led to convergent conclusions. Out of the 35 participants, 11 (31%) produced nonsignificant least-square fits. Among the 24 remaining, 20 had significant scalar-variability fits and showed no significant advantage of the least-square model over the scalar-variability model. As shown Fig. 10.6, the major features of the results are still present, and quite similar to those of Experiment 1.

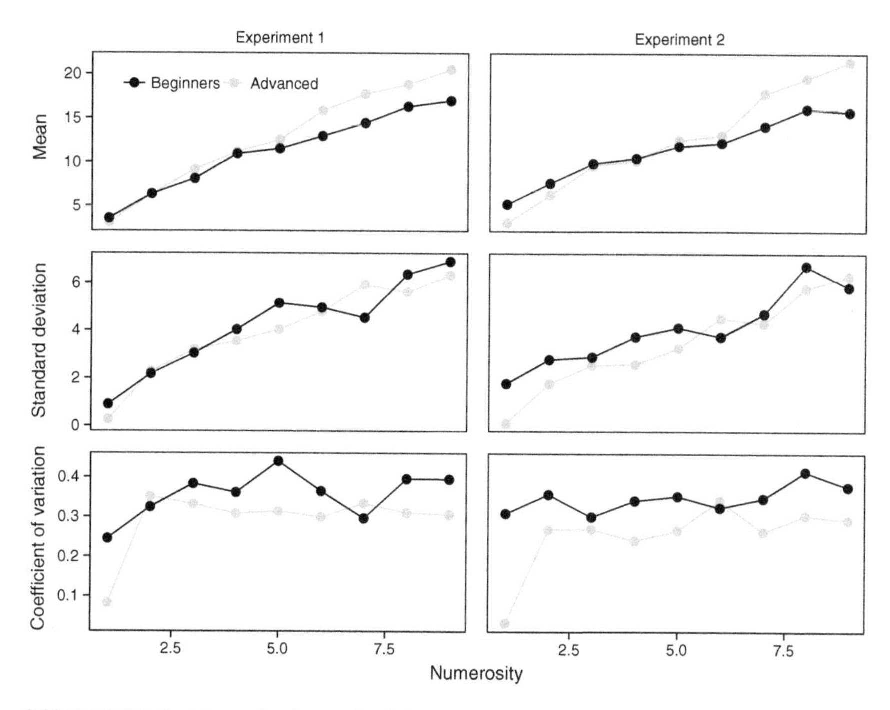

FIGURE 10.6 Means (top), standard deviations (mid), and coefficients of variation (bottom) for the subsets of children producing significant least-square and scalar-variability fits. Left: Experiment 1, $n = 12$ and 14 for the beginners and advanced group. Right: Experiment 2, $n = 12$ and 8.

In summary, these experiments convinced us that the distribution game was suitable for use with young children. The findings produced converging evidence that children aged between 3 and 5 are capable of estimating the numerosity of visual collections of entities, and the estimates for most of them showed scalar variability, suggestive of an accumulator mechanism. Estimates showed only minimal effects of display format, and the influence of format ran in opposite directions for beginners and more advanced counters, suggesting that format could only play a minor role in the process. Moreover, even children with very elementary knowledge of the number-word sequence and limited counting ability produced monotonically increasing number of presses for numerosities well above their counting range.

10.6 STUDY 2: A COMPUTER-CONTROLLED VARIANT

The second study was based on the same kind of distribution paradigm, but we switched to a computer-controlled setting for two main reasons. First, we thought it necessary in order to exclude potential alternative interpretations assuming experimenter-child implicit communication artifacts of the *Clever Hans* kind. In the previous study, as the experimenter was handing the sheets representing the collections to the child, she was aware of the numerosities and could involuntarily provide nonverbal bodily hints that may possibly have contributed to shape children's responses. This could be avoided if the spatial layout was organized so that the display would not be visible to the experimenter and the presentation of the collection and the recording of the button presses would be controlled by the computer. In addition, the use of the computer made it possible to precisely record the observation time, that is, the time during which the child examines the collection,[b] as well as the response duration, that is, the total time during which the child presses the key in response to a given collection. Conversely, children passed the whole test in a more autonomous way and their level of attention was thus less well controlled.

An additional change concerned the cover story. Because we wanted to present plain dot collections rather than pictures, the task was introduced in a different way, using Wall-E, an animation comedy figure well known to the children. In the film, Wall-E is a trash-compactor robot, whose belly is a cubic trash box, and we told the child that Wall-E very much likes to eat plastic balls. Unfortunately, he is a bit clumsy and often all the balls fall off the box. To help Wall-E recover all the balls, the child could press the magic button; each time the button is pressed, a ball would jump back into the box. Each trial began with the presentation of a collection of dots, which remained on screen until the child pressed the space bar. Then a short animation showed Wall-E coming in and opening his belly-box. The child could then press the mouse button with the display showing a fixed image of Wall-E surrounded with colorful balls. When the child pressed the space bar again, another animation would show Wall-E close his belly-box and leave.

We used six numerosities, 6, 9, 12, 15, 18, and 21. For each numerosity, except 12 (as will be explained later), we devised four dot collections. Half of those used a fixed dot size (65 pixels), so that total area covaried with numerosity (numerosity + area series, referred to as N + A). The other collections were constructed so that total area was kept equal, while dot size decreased with number (numerosity + dot size series, N + D). In addition, to assess whether estimates were influenced by dot size and total area, these trials were interspersed with 16 collections of 12 points with dot sizes and total area chosen to cover a similar range of values as the two other series (area + dot size series, A + D). Thus, if children respond on the basis of number rather than dot size or total area, they should produce similarly increasing numbers of presses for the N + A and N + D series, and stable

[b]Children had to press the space bar to launch a new trial, and to press it a second time to start the distribution. Hence, observation time is the time elapsed from the onset of the stimulus to the space bar key-press event.

TABLE 10.3 Characteristics of the Participants in Study 2

Condition		Count	Estimate	
			Beginners	**Advanced**
N		13	38*	17*
Age (months)	Mean (SD)	69 (5)	50 (10)	64 (8)
	Range	(62–78)	(38–72)	(49–75)
Sequence	Mean (SD)	29.8 (9.2)	9.3 (5.6)	22.6 (7.5)
How many?	Mean (SD)	12.1 (4.9)	4.8 (3.1)	14.4 (1.6)

* *Number knowledge was lost for five children in the estimation condition.*

numbers of presses in the A + D series.[c] The whole set of 36 collections was divided into two blocks, with equal proportions of each numerosity and series. Each block was preceded by two examples, either with small (two collections of three elements) or large collections (25 and 30), and the order of blocks was varied so that half the children had initial examples with large numbers and the other received examples with small numbers.

In a first control experiment (*count* situation), the instructions insisted on *putting exactly the same number of balls in the box* and no time pressure was exerted, so that children could count the dots before proceeding to amassing the balls. A small group of 13 children aged between 5;2 and 6;6 years passed this condition.

In the main experiment, the instructions insisted on speed in order to induce children to perform a rapid estimation of the dot collections without attempting to count. Sixty children between 3;2 and 6;3 years were tested. All were drawn from three public schools in small villages in a rural area of Wallonia (Table 10.3).

10.6.1 Count Condition

In the count situation, the mean number of presses closely approached the number of dots, and observation time increased regularly, with an average increment of approximately 0.25 s per dot, thus fully compatible with silent counting (Fig. 10.7).

A mixed model analysis on the log-transformed number of presses with numerosity and series (N + A vs. N + D) as factors revealed only a significant effect of number, $t(20) = 17.1$, $p < 0.0001$ and no difference between series, $t(20) = 0.43$, $p > 0.20$. Observation times also showed a clear effect

[c]Convex hull, the extent of the contour around the dots, was not controlled. Post hoc measurements indicated that it increased with numerosity and area, so it also increased in the fixed numerosity condition although in lower proportion.

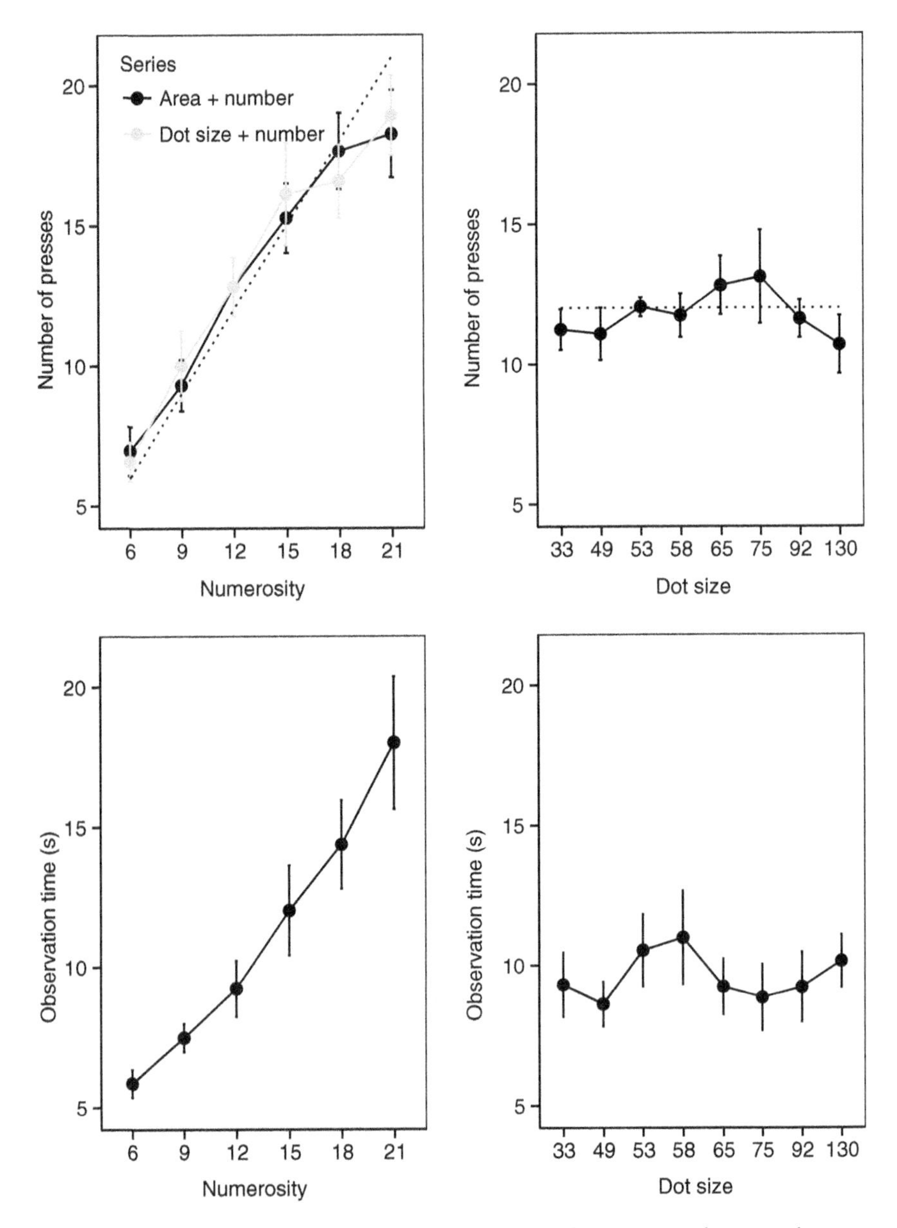

FIGURE 10.7 **Results for the *count* situation.** Left: performance as a function of numerosity and series for the N + A and N + D series (top: number of presses, bottom: observation time). Right: number of presses (top) and observation time (bottom) for the A + D series, as a function of dot size.

of numerosity, $t(20) = 7.63$, $p < 0.0001$. Coefficients of variation showed a tendency to increase over numerosity, as shown by a significant positive slope across individual regression fits, $t(12) = 4.11$, $p = 0.001$. Responses on the A + D series produced no evidence of an influence of dot size on number of presses, $t(14) = -0.74$, $p > 0.20$, or on observation time, $t(14) = 0.21$, $p > 0.20$. In brief, the results correspond to what would be expected given the instructions favoring counting.

10.6.2 Estimation Condition

In the estimation condition, some children were extraordinarily courageous and willing to help Wall-E and they produced huge numbers of presses. About 10% of responses exceeded 50, and more than 40 responses surpassed 100. (The record was 241 presses, for a total duration of 3 min and 16 s!) It is unclear whether these were due to a misunderstanding of the task by some participants, to the playfulness induced by the use of the computer, or to the clicking gesture required for the computer mouse rather than the more demanding lever-press. Lacking appropriate independent and nonarbitrary criteria to discard participants, we preferred to use all the data available, at the risk of increasing the noise.

Data were again analyzed with mixed models. The observation times were stable around 4 s, and there was no evidence of any influence of numerosity on observation times for the N + A and N + D series, $t(20) = -0.60$, $p > 0.20$, nor of dot size, $t(14) = 0.84$, $p > 0.20$, for the A + D series, confirming that children were not using counting. Number of presses were analyzed with series and numerosity as fixed factors (Fig. 10.8) for the stimuli varying in numerosity (N + A and N + D series).[d] The effect of numerosity was significant, $t(20) = 2.83$, $p = 0.01$. There was no significant difference between the two series, $t(20) = 0.84$, $p > 0.20$, and no interaction, $t(20) = 0.47$, $p > 0.20$. In the fixed numerosity stimuli (A + D series, Fig. 10.8), there was no indication of any effect of dot size, $t(14) = -0.23$, $p > 0.20$, suggesting that neither the increase in dot size, total area, or convex hull (as convex hull tended to covary with dot size) determine changes in number estimation.

Analysis of the variability was less conclusive than in the first study. Ignoring the block factor, each child produced four estimates by numerosity, two from the N + A and two from the N + D series of targets. The analysis of CoVs through linear regression by participant indicated no systematic variation of the CoV slope with numerosity in either group,

[d]Although for the sake of simplicity we did not include block and group in the analyses reported here, it is worth mentioning that the overestimation visible on Fig. 6.8 was mainly due to the subgroup of children who received large initial examples. It may be that those children interpreted the examples to suggest that they were expected to produce many presses.

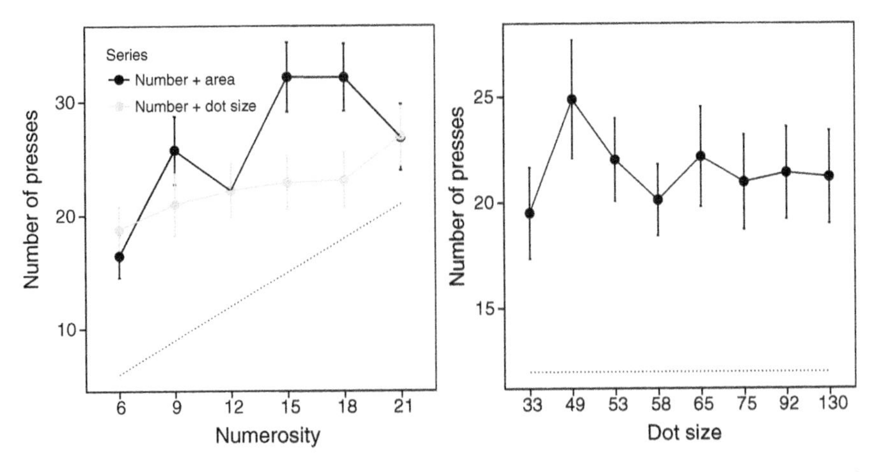

FIGURE 10.8 Mean estimates as a function of numerosity and series for the N + A and N + D series (left) and as a function of dot size for the fixed numerosity A + D series (right) The dotted line corresponds to exact performance.

$t(59) = 0.027$, $p > 0.20$. The mean CoV was larger than in the previous study, 0.593, reflecting the much larger range of responses observed in the current experiment than in the first study. In the model comparison analysis, the least-square regression of standard deviation on mean was significant in 29 of the 60 participants, and among those, only 13 produced similar fits of the least-square and scalar-variability models.

Finally, to examine the relation between estimation performance and knowledge of the numeral system, we split the participants into three groups. Out of the 55 children for whom counting and number-word sequence data were available, 38 had poor knowledge of the verbal numeral system, and failed in counting collections above 10. Among these, we distinguished between the 3-year-olds and the elder children ($n = 19$ in each group, mean age, 3;6 and 4;11 years). The third group comprised the 17 children who succeeded in counting above 10 (mean age, 5;5 years). As a group, the youngest children showed no sensitivity to numerosity, contrary to the two other groups that obtained similar patterns of results (Fig. 10.9). Individual log–log regression analyses indicated that only one of the younger children and half of those in each of the two other groups (10 out of 19 in the less advanced counters and 9 out of 17 in the more advanced counters) obtained significant effects of numerosity. Thus, the results provide additional evidence that children can estimate numerosities that they are unable to count.

Overall the results of this second study appear less convincing than those of the first one because of a general trend to overestimate and because the data are much noisier. Still, several elements are worth noting. The absence of any trace of an effect of number on observation times in the estimation condition confirms that children were not counting or even

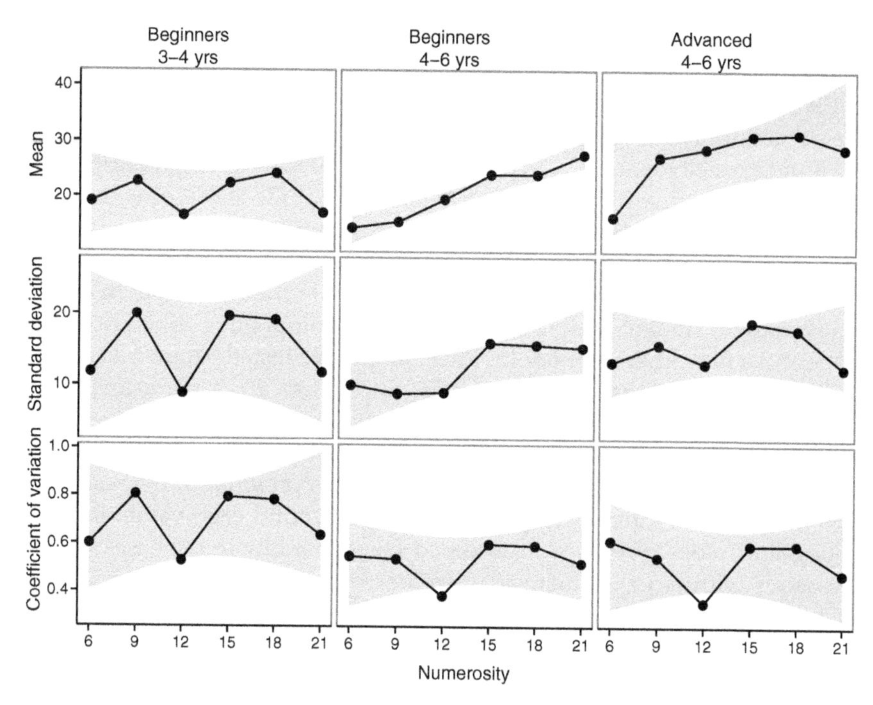

FIGURE 10.9 Means (top), standard deviations (mid), and coefficients of variation (bottom) by group. Left, 3- to 4-year-olds, $n = 19$, weak knowledge level; mid, $n = 19$, 4- to 6-year-olds, weak knowledge level; right, $n = 17$, 4- to 6-year-olds, high knowledge level.

enumerating the collections. The lack of any sign of an effect of dot size/cumulated area on number of presses confirms that children's estimates were not based on such nonnumerical visual cues. Some studies have produced evidence that convex hull, the extent of the contour surrounding the elements of the collection, could modulate performance in the comparison task (Gebuis & Reynvoet, 2012b). Although convex hull was not explicitly manipulated or controlled in the present study, posthoc measurements showed that it tended to covary with dot size. Hence, the absence of any effect of dot size in the fixed numerosity condition suggests that dot size, total area, and convex hull have little influence on children's estimation. Finally, although only little more than half the children between 4 and 6 years showed a significant effect of numerosity, they were equally distributed among less advanced and more advanced counters.

It is of course possible that the youngest children failed to understand what they were expected to do. As often occurs in developmental research, the findings do not permit us to discriminate between a failure to understand the nature of the task and a lack of sensitivity to numerosity. However, based on the first study as well as on more recent observations, we believe that the poor results of the younger children may be due to several unfortunate characteristics of this specific experiment. For one, children received only two examples before the experimental trials (rather than eight in the first study) and the examples were aimed at familiarizing them with the procedure rather than demonstrating the task or modeling the behavior, as no explicit feedback was provided. Moreover, the candy-distribution scenario may convey the notion of equal magnitude more directly than Wall-E's story. In addition, the fact that the computer displayed a picture of Wall-E with a large amount of colored balls around him during the response phase of the trials may have contributed to the confusion. In a new study using a similar computer-controlled paradigm but a different cover story and a few other methodological changes, we tested 3- and 5-year-old children. Children passed two sessions, and had six trials per numerosity at each session. Moreover, we added two examples in which the experimenter demonstrated the task, followed by four training trials (with numerosities 2, 9, 3, and 7) for which feedback was provided if the response was not within a range of approximately 2/3 around each value. In both young and older groups, we observed clear monotonic increases of the responses with target numerosity, together with proportional increases in standard deviations (Nys & Content, in preparation, see Fig. 10.10).

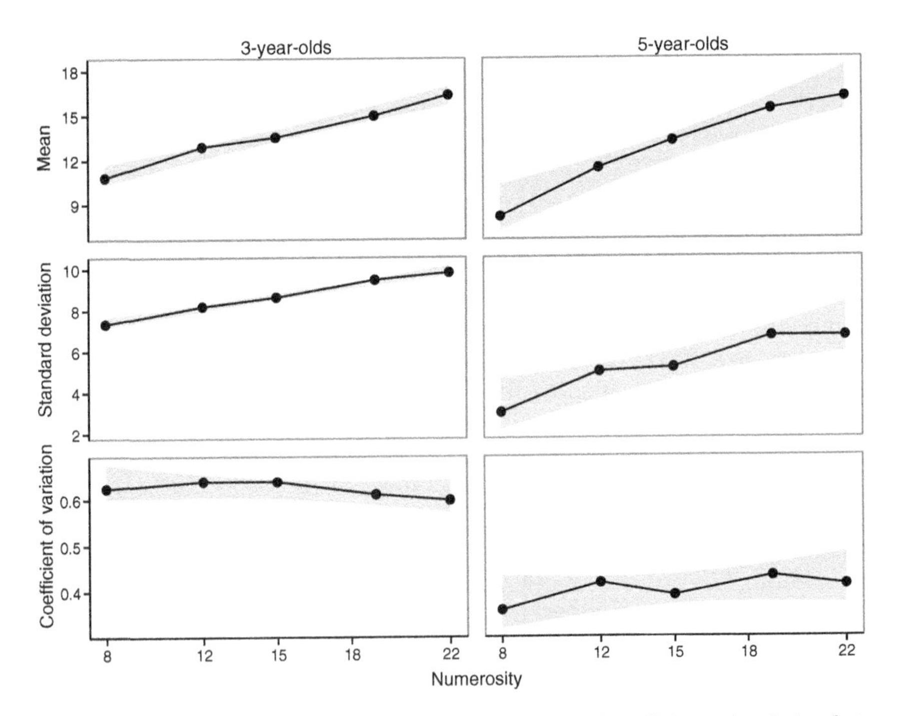

FIGURE 10.10 Means (top), standard deviations (mid), and coefficients of variation (bottom) by group. Left, 3- to 4-year-olds, $n = 19$; Right, $n = 17$, 5- to 6-year-olds. *From Nys and Content, in preparation.*

10.7 SUMMARY AND CONCLUSIONS

The experiments reported here converge to show that by the age of 3, children produce numerical estimates of discrete quantities that increase monotonically with the target magnitude. Although individual participants' estimates were often quite inaccurate and showed wide variability, average estimates approximated the target values quite closely in the first study. Crucially, this was also true when the analyses were restricted to children with extremely limited knowledge of the counting sequence and counting procedure, and to a range of value above those that they could enumerate exactly. Results were less clear in the second study, as many children tended to overestimate to a large extent. Despite this bias, the results again produced evidence of sensitivity to the numerosity of the stimulus collections.

The present experiments also demonstrate that nonnumerical visual cues play only a very minimal role, if any, in children's estimation. In the first study, the overall pattern of estimates was quite similar in Experiment 1 (in which element size was fixed and total area increased with numerosity) and in Experiment 2 (in which total area was kept constant by decreasing the size of individual elements). Moreover, the manipulation of display size and image density in Experiment 2 did not result in clear systematic differences. Although analyses indicated the presence of interactions between numerosity and format, the interactions ran in opposite directions in the beginners and advanced groups. The reverse interactions could perhaps be explained by supposing that display size constitutes a salient cue which causes interference for children with a limited understanding of the numeral system, whereas more numerically sophisticated children could discard display size and incorporate density or occupancy in their evaluation. This view is in line with reinterpretations of developmental patterns in number conservation studies in terms of inhibition (Houdé, 2000). However, even if the effect is taken at face value, display size can only be a modulating factor and cannot account for the increase in number of presses across numerosities.

In the second study, no marked difference was observed between numerosity estimates for stimuli in which aggregated area covaried with numerosity (the N + A series) and those in which dot size covaried with numerosity (the N + D series). Furthermore, stimuli varying in dot size and total area, with numerosity constant, did not engender any correlated variation in children's estimates. Thus, although we obviously cannot exclude the possibility that *some* still unknown nonnumerical perceptual cues contribute to the extraction of numerosity, the present findings provide clear evidence that *none* of the most salient cues, that is, dot size, aggregated area, density, and occupancy, are at play in children's numerical estimation.

Regarding the variations in children's estimates, despite the constraints of test duration and children's attention spans, which limited the number of trials presented and hence strongly restricted the reliability of the measurements, the results are globally consistent with the hypothesis of scalar variability. Excepted for responses on numerosity in the first study, which clearly showed less variation, all results converged to indicate that the variability of the estimates, measured by the standard deviation, increased in proportion of the mean value of the estimates. The lower variation for sets of three elements is not surprising, as the numerosity up to three or four could be rapidly extracted through subitizing, and virtually all children, even the younger ones, could recite the sequence and count up to three or four.

One essential feature of the nonverbal estimation task used in the present studies, which differs from previous reproduction paradigms (Crollen et al., 2011; Frank et al., 2012; Mejias et al., 2012), is that children have to map from one modality to another modality. Thus, here participants need to extract magnitude information from images of collections, in the visuospatial domain, and then map the encoded magnitude to a sequence of key-presses in the temporal dimension. There is no other feature common to the stimuli and the target responses than their numerosity. Hence, whatever magnitude cues may be extracted from the visual input would need to be transformed into some abstract number code in order to guide the response. How is such a mapping performed? We submit that the notion of an internal accumulator system serving as the interface or mediating representation device between the stimuli and responses provides an appropriate theoretical framework to account for the present estimation data. The accumulator system would extract an approximate internal representation of the numerosity of the stimulus collection. The present findings suggest that the internal magnitude is essentially determined by the numerosity of the stimulus collections and that nonnumerical cues bear minimal contributions, if any. The internal magnitude coding for the number of elements in the stimulus collections would in turn control the emission of presses, either on a discrete (number of presses) or continuous (duration of presses) manner. Whether responses are controlled by time or number is one issue that we have not explored in the present research.

In our view, one critical prediction of such an account is that the approximate internal magnitude should be *veridical*. With perception and production estimation tasks, the additional issue of the mapping between the internal code and the verbal or Arabic numerals (Castronovo & Seron, 2007; Izard & Dehaene, 2008) opens the possibility of calibration. As long as the tasks require to map symbols with numerosities, the exact value of the internal magnitude does not matter much. In fact, it could be arbitrary, as long as it increases monotonically with numerosity, since any monotonically set of internal states could be made to map to and from the numeral line. By contrast, in the case of nonverbal estimation, it seems natural to assume that the internal magnitude code controls action and that it should thus be without bias. What would be the adaptive advantage of a mechanism that massively overestimates or underestimates quantities? In that respect, the results of the second study seem puzzling (although it is plausible that extraneous factors such as expectations, compliance, or generosity are also at play). Further studies are needed to establish whether the overestimation constitutes a real phenomenon or an artifact in this particular study.

Finally, large individual differences in accuracy and acuity were observed. Whether these differences in estimation performance constitute a mark of differences in the precision of the mechanisms subserving the encoding of numerosity is uncertain, as they could as well be caused by general cognitive abilities, such as attention, memory, or perhaps visuospatial skill, as well as motivational disparities. Nevertheless, the differences in performance appeared to be related to some extent to the degree of familiarity with verbal numerals. Quite systematically, more advanced counters displayed estimation functions closer to a linear relation, and their estimates tended to vary less, as shown by smaller CoVs. Again, since the knowledge of numerals was partly confounded with age, we cannot exclude that the relation is mediated by general factors. Further studies using broader assessments would be required to determine whether there is a direct link between estimation acuity and numeral knowledge. Nor can the present research settle between cause and consequence. It is possible that early differences in ANS acuity influence the ease and speed of mapping numerals to internal magnitude representations. It is however tempting to speculate on the potential influence of counting and number-word knowledge on the precision of the magnitude code. The possibility that the numerosity apprehension mechanism improves during the course of development under the influence of command of cultural notations has been substantiated by several recent studies from our group. Nys et al. (2013) compared Western unschooled adults with matched subjects from the same areas and sociocultural background. In both symbolic and nonsymbolic comparison and numerosity matching tasks, unschooled adults were less precise than their pairs, suggesting that math education can contribute to enhance the precision of the ANS system. In a follow-up study, the same group of unschooled adults passed two perceptual estimation tasks with either number lines or dot collections as stimuli. In both, unschooled adults demonstrated larger underestimation and larger variability than their pairs. In a recent longitudinal study, Mussolin et al. (2014) tested a sample of 3- to 4-year-olds twice at an half-year interval and found asymetric relationships between counting and knowledge of the verbal numeral on the one hand and nonsymbolic comparison performance on the other hand. Accuracy in the *give-a-number* task and number-word knowledge predicted nonsymbolic comparison 7 months later, even when controlling for age, general intelligence, verbal and visuospatial short-term memory, and naming performance, whereas the reverse correlations were not significant. Those few studies are clearly not sufficient to settle the issue, and the relationships could as well be going both ways (as proposed for other aspects of cognitive development that entail intricate links with education and schooling, see for instance, Morais, Alegria, & Content, 1987). Yet, they imply that proper demonstrations of the causal influence of the ANS on numerical, arithmetic, and mathematical abilities need to use long-term longitudinal designs and start from an early age. Otherwise they will remain open to the eventuality that measures of numerosity perception are already influenced by the cultural and educational environment of the children. As it seems feasible with 3-year-olds, and perhaps even a bit earlier, the variant of the purely nonverbal estimation task that we have devised offers a promising technique to explore the ability to extract and manipulate numerical magnitude with children who are just beginning to get acquainted with number words.

References

Allik, J., & Tuulmets, T. (1991). Occupancy model of perceived numerosity. *Perception and Psychophysics*, *49*(4), 303–314.

Beran, M. J., Taglialatela, L. A., Flemming, T. M., James, F. M., & Washburn, D. A. (2006). Nonverbal estimation during numerosity judgements by adult humans. *The Quarterly Journal of Experimental Psychology*, *59*(12), 2065–2082.

Besner, D., Davies, J., & Daniels, S. (1981). Reading for meaning: the effects of concurrent articulation. *The Quarterly Journal of Experimental Psychology*, *33A*, 415–437.

Boisvert, M., Abroms, B., & Roberts, W. (2003). Human nonverbal counting estimated by response production and verbal report. *Psychonomic Bulletin and Review*, *10*(3), 683–690.

Castronovo, J., & Seron, X. (2007). Numerical estimation in blind subjects: evidence of the impact of blindness and its following experience. *Journal of Experimental Psychology*, *33*(5), 1089–1106.

Chen, Q., & Li, J. (2014). Association between individual differences in non-symbolic number acuity and math performance: a meta-analysis. *Acta Psychologica*, *148*, 163–172.

Cordes, S., Gelman, R., Gallistel, C., & Whalen, J. (2001). Variability signatures distinguish verbal from nonverbal counting for both large and small numbers. *Psychonomic Bulletin and Review*, *8*(4), 698–707.

Crollen, V., Castronovo, J., & Seron, X. (2011). Under- and over-estimation. *Experimental Psychology*, *58*(1), 39–49.

De Smedt, B., Noël, M. -P., Gilmore, C., & Ansari, D. (2013) How do symbolic and non-symbolic numerical magnitude processing skills relate to individual differences in children's mathematical skills? A review of evidence from brain and behavior. *Trends in Neuroscience and Education*, *2*(2), 48–55.

Dehaene, S., & Changeux, J. (1993). Development of elementary numerical abilities: a neuronal model. *Journal of Cognitive Neuroscience*, *5*(4), 390–407.

Ebersbach, M., & Erz, P. (2014). Symbolic versus non-symbolic magnitude estimations among children and adults. *Journal of Experimental Child Psychology*, *128*, 52–68.

Feigenson, L., Dehaene, S., & Spelke, E. S. (2004). Core systems of number. *Trends in Cognitive Sciences*, *8*(7), 307–314.

Feigenson, L., Libertus, M. E., & Halberda, J. (2013). Links between the intuitive sense of number and formal mathematics ability. *Child Development Perspectives, 7*(2), 74–79.

Frank, M., Everett, D., Fedorenko, E., & Gibson, E. (2008). Number as a cognitive technology: evidence from Pirahã language and cognition. *Cognition, 108*(3), 819–824.

Frank, M. C., Lai, P., Saxe, R., & Gibson, E. (2012). Verbal interference suppresses exact numerical representation. *Cognitive Psychology, 64*(1-2), 74–92.

Gallistel, C., & Gelman, R. (1992). Preverbal and verbal counting and computation. *Cognition, 44*(1-2), 43–74.

Gallistel, C. R., & Gelman, R. (2005). Mathematical cognition. In K. J. Holyoak, & R. G. Morrison (Eds.), *The Cambridge handbook of thinking and reasoning* (pp. 559–588). London: Cambridge University Press.

Gebuis, T., Gevers, W., & Cohen Kadosh, R. (2014). Topographic representation of high-level cognition: numerosity or sensory processing? *Trends in Cognitive Sciences, 18*, 1–3.

Gebuis, T., & Reynvoet, B. (2012a). Continuous visual properties explain neural responses to nonsymbolic number. *Psychophysiology, 49*(11), 1649–1659.

Gebuis, T., & Reynvoet, B. (2012b). The interplay between nonsymbolic number and its continuous visual properties. *Journal of Experimental Psychology, 141*(4), 642–648.

Gebuis, T., & Reynvoet, B. (2012c). The role of visual information in numerosity estimation. *PLoS One, 7*(5), e37426.

Gebuis, T., & Reynvoet, B. (2013). The neural mechanisms underlying passive and active processing of numerosity. *NeuroImage, 70*, 301–307.

Guillaume, M., & Gevers, W. (2016). Assessing the Approximate Number System: no relation between numerical comparison and estimation tasks. *Psychological Research, 80*(2), 248–258.

Halberda, J., Mazzocco, M. M., & Feigenson, L. (2008). Individual differences in non-verbal number acuity correlate with maths achievement. *Nature, 455*(7213), 665–668.

Houdé, O. (2000). Inhibition and cognitive development: object, number, categorization, and reasoning. *Cognitive Development, 15*(1), 63–73.

Huntley-Fenner, G. (2001). Children's understanding of number is similar to adults' and rats': numerical estimation by 5-7-year-olds. *Cognition, 78*(3), B27–B40.

Izard, V., & Dehaene, S. (2008). Calibrating the mental number line. *Cognition, 106*(3), 1221–1247.

Logie, R. H., & Baddeley, A. D. (1987). Cognitive processes in counting. *Journal of Experimental Psychology, 13*(2), 310–326.

Lyons, I. M., & Ansari, D. (2015). Foundations of children's numerical and mathematical skills: the roles of symbolic and nonsymbolic representations of numerical magnitude. *Advances in Child Development and Behavior, 48*, 93–116.

Meck, W. H., & Church, R. M. (1983). A mode control model of counting and timing processes. *Journal of Experimental Psychology, 9*(3), 320–334.

Mejias, S., & Schiltz, C. (2013). Estimation abilities of large numerosities in kindergartners. *Frontiers in Psychology, 4*, 518.

Mejias, S., Mussolin, C., Rousselle, L., Grégoire, J., & Noël, M. -P. (2012). Numerical and non-numerical estimation in children with and without mathematical learning disabilities. *Child Neuropsychology, 18*(6), 550–575.

Morais, J., Alegria, J., & Content, A. (1987). The relationships between segmental analysis and alphabetic literacy: an interactive view. *European Bulletin of Cognitive Psychology, 7*(5), 415–438.

Mou, Y., & vanMarle, K. (2014). Two core systems of numerical representation in infants. *Developmental Review, 34*(1), 1–25.

Mussolin, C., Nys, J., Content, A., & Leybaert, J. (2014). Symbolic number abilities predict later approximate number system acuity in preschool children. *PLoS One, 9*(3), e91839.

Mussolin, C., Nys, J., Leybaert, J., & Content, A. (2015). *How approximate and exact number skills are related to each other across development: A review*. Developmental Review. Available online 6 November 2015. doi:10.1016/j.dr.2014.11.001

Nieder, A., & Dehaene, S. (2009). Representation of number in the brain. *Annual Reviews of Neuroscience, 32*(1), 185–208.

Noël, M. P., & Rousselle, L. (2011). Developmental changes in the profiles of dyscalculia: an explanation based on a double exact-and-approximate number representation model. *Frontiers in Human Neuroscience, 5*, 165.

Nys, J., & Content, A. (2012). Judgement of discrete and continuous quantity in adults: number counts! *The Quarterly Journal of Experimental Psychology, 65*(4), 675–690.

Nys, J., & Content, A. (in preparation). Nonverbal numerical estimation in 3-5 year old children.

Nys, J., Ventura, P., Fernandes, T., Querido, L., Leybaert, J., & Content, A. (2013). Does math education modify the approximate number system? A comparison of schooled and unschooled adults. *Trends in Neuroscience and Education*, 2(1), 13–22.

Piazza, M., Pica, P., Izard, V., Spelke, E. S., & Dehaene, S. (2013). Education enhances the acuity of the nonverbal approximate number system. *Psychological Science, 24*(6), 1037–1043.

Pica, P., Lemer, C., Izard, V., & Dehaene, S. (2004). Exact and approximate arithmetic in an Amazonian indigene group. *Science, 306*(5695), 499–503.

Saito, S. (1997). When articulatory suppression does not suppress the activity of the phonological loop. *British Journal of Psychology, 88*(4), 565–578.

Siegler, R. S., & Lortie-Forgues, H. (2014). An integrative theory of numerical development. *Child Development Perspectives, 8*(3), 144–150.

Siegler, R. S., & Opfer, J. (2003). The development of numerical estimation: evidence for multiple representations of numerical quantity. *Psychological Science, 14*(3), 237–243.

Smets, K., Sasanguie, D., Szücs, D., & Reynvoet, B. (2015). The effect of different methods to construct non-symbolic stimuli in numerosity estimation and comparison. *Journal of Cognitive Psychology, 27*(3), 310–325.

Starr, A., Libertus, M. E., & Brannon, E. M. (2013). Number sense in infancy predicts mathematical abilities in childhood. *Proceedings of the National Academy of Sciences, 110*(45), 18116–18120.

Szucs, D., Nobes, A., Devine, A., Gabriel, F. C., & Gebuis, T. (2013). Visual stimulus parameters seriously compromise the measurement of approximate number system acuity and comparative effects between adults and children. *Frontiers in Psychology, 4*, 444.

Tan, L., & Grace, R. C. (2011). Human nonverbal discrimination of relative and absolute number. *Learning & Behavior, 40*(2), 170–179.

Tree, J. J., Longmore, C., & Besner, D. (2011). Orthography, phonology, short-term memory and the effects of concurrent articulation on rhyme and homophony judgements. *Acta Psychologica, 136*(1), 11–19.

Verguts, T., & Fias, W. (2004). Representation of number in animals and humans: a neural model. *Journal of Cognitive Neuroscience, 16*(9), 1493–1504.

Whalen, J., Gallistel, C., & Gelman, R. (1999). Nonverbal counting in humans: the psychophysics of number representation. *Psychological Science, 10*(2), 130–137.

CHAPTER

11

Magnitudes in the Coding of Visual Multitudes: Evidence From Adaptation

Frank H. Durgin

Swarthmore College, Department of Psychology, Swarthmore, PA, United States of America

OUTLINE

Continuous Issues in Numerical Cognition. http://dx.doi.org/10.1016/B978-0-12-801637-4.00011-1

Euclid's Elements (1956/300 BC), the classic treatise on geometry, is divided into 13 books, of which the seventh concerns number. Thus, Euclid's treatment of number is at the arithmetic center of his work on geometry. Before defining number itself, Euclid began with an almost psychological definition of the concept of a unit: "A unit is that by virtue of which each of the things that exist is called one" (p. 277). Euclid then proceeded to define number: "A number is a multitude composed of units." The crucial aspect of number that underlies all of the best work showing that animals represent number depends on this notion of the unit. For example, controlling for size, duration, and other variables is intended to provide evidence that the critical variable underlying an animal's responses maps onto *how many* things were presented rather than *how much*.

As Dehaene's (1997) popular review describes, the question of *what in the world* is being represented (eg, number) is separate from the question of *how in the brain* it is represented. Even if animals represent visual number, for example (are indeed sensitive to "how many," see Chapters 7 and 8), their representations of such multitudes might turn out to be supported by magnitude representations in the brain (or in the case of small numbers, by other kinds of tokening systems). Dehaene reviews the example of using reservoirs of water to maintain a record of numbers of things. In the study of visual number, yet another issue arises: could the visual information being used to estimate number itself be nonnumeric (ie, a continuous variable)? If so, how can it be made to serve as a proxy for number?

This chapter will concern itself with the encoding of spatially distinguished visual multitudes (eg, of dots) and consider the question of whether these multitudes are encoded by uniquely numeric information, or if they are parasitic on other sources of information. I will use adaptation data to argue for the latter.

11.1 UNITS AND THE SUBITIZING RANGE

Consider first the notion of subitizing (Kaufman, Lord, Reese, & Volkmann, 1949) as a different process from estimation (Taves, 1941). It was Jevons (1871) who first tossed small sets of beans from jars to see how accurately he could evaluate their number in an instant. Trick and Pylyshyn (1993, 1994) argued that subitizing corresponds to a small number of attentional markers, an idea that has been established as one of two "core" systems (Feigenson, Dehaene, & Spelke, 2004). If this theory is correct, then the basis for subitizing is a one-to-one correspondence between attentional markers and objects. In other words, this core system corresponds to processes relevant to Euclid's unit.

However, notice that even if this theory is correct, it requires that attentional markers can be enumerated, and it is not obvious why that would be particularly easy. Moreover, there is also evidence that even success at low numbers is aided by configural information (Mandler & Shebo, 1982). Number systems are opportunistic. But a more crucial fact about the subitizing range, from the perspective of magnitude estimation, is that success at subitizing is overdetermined: the Weber fraction for visual number is small enough (typically about 15%) that even if 3 and 4 are encoded by means of continuous variables (magnitudes), their values are not very confusable, being more than two JNDs apart. Evidence that 3 and 4 are hard for some animals to discriminate simply suggests a higher JND is relevant.

When it comes to discriminating among visual numbers beyond the subitizing range, there are two separate kinds of issue that arise. On the one hand, the ability to discriminate between visual numerosities typically shows a magnitude effect corresponding to a (more or less) logarithmic representation, and roughly consistent with the scaling of other psychophysical dimensions. On the other hand, the assignment of an actual numeric value becomes highly indefinite. Even if I can discriminate between two different numerosities with some precision, it need not follow that I know what either of those numerosities is relative to the ostensible unit that defines number in the Euclidean sense. To be able to say that one multitude is larger than another without knowing the value of either multitude is most consistent with the notion that multitudes are coded by visual magnitude representations, which also might be selected opportunistically (Durgin, 1995; Gebuis & Gevers, 2011; Gebuis & Reynvoet, 2012).

However, in the transition from the subitizing range to higher numbers, it remains possible that the units made cognitively accessible by the subitizable part of the range can be carried over into the higher part of the range. That is, in discriminating 5 from 4, a person can come to also be able to define the unit as 20% of the perceptual information that corresponds to 5. Therefore, we might expect that the subitizing range can serve to linearize some numerosities that extend beyond the subitizing range. This could occur if, in the course of evaluating numerosities that span values extending down into the subitizing range, it were possible to establish an absolute unit that can be applied even beyond the subitizing range. And if this were the case, there might remain a range of numerosites beyond the subitizing range that can nonetheless be coded with respect to a unit defined (perceptually) within the subitizing range.

For example, in their studies of numerosities of about 20 dots, Allik and Tuulmets (1991) seemed to identify a zone of numerosities where perceived numerosity was a function of "filled area" (*occupancy*)—a perfectly good sort of unit of magnitude that seemed to be used to define visual multitudes. A limitation of occupancy as a metric of numerosity was that it seemed not to work as a metric for higher numerosities because it reached a ceiling of occupancy with relatively sparse displays, although it provides an excellent account of clustering effects (ie, a reduction in perceived numerosity when subregions of dots are clustered together) in the medium range.

11.2 AFTEREFFECTS AND THE PERCEPTION OF TEXTURE ELEMENT DENSITY

Durgin (1995) showed that adapting to dense (high numerosity) textures of distinct elements seemed to affect higher visual numerosities more than it affected low numerosities. That is, at low numerosities (eg, 20) the effect of density adaptation seemed to primarily affect the distribution of texture rather than its apparent number. Durgin proposed that different kinds of information were used to estimate visual numerosity, and the information selected (occupancy, density) depended on specific stimuli used.

Studies of "texture density" adaptation had originally been reported by Walker (1966), but these were based on texture magnification and were thus confounded with size and spatial frequency. Anstis (1974) explicitly showed that perceived texture magnification (which he also called texture density) was affected by adaptation in a manner consistent with spatial frequency (ie, size) adaptation (Blakemore & Campbell, 1969; Blakemore & Sutton, 1969). But by controlling the spatial frequency of their textures, Durgin and Huk (1997) see also Durgin and Proffitt (1996) showed that the adaptation to texture element density observed by Durgin (1995) was distinct from the size adaptation studied by Anstis.

Whereas most studies of texture density adaptation do not assess the perceived numeric values (ie, absolute estimates rather than comparisons), one early study by Huk and Durgin (1996) investigated the effects of adaptation on the estimation of number qua number. Here I report a new analysis of the data from their study. Since the original study was published only as an abstract, I include the methods of the study here as well.

11.3 EXPERIMENT 1: MAGNITUDE ESTIMATION OF VISUAL NUMBER UNDER CONDITIONS OF ADAPTATION (HUK & DURGIN, 1996)

Whereas texture element density adaptation clearly affected the perception of relative number as assessed by direct visual comparison (Durgin, 1995, 2008), Huk and Durgin (1996) tested whether magnitude estimation of visual numbers could serve as a secondary measure of adaptation. Durgin (1995) had concluded that texture density adaptation produced a proportional decrease in perceived number for all numerosities above about 40, but never directly tested perceived number (ie, via magnitude estimation). The chief conclusion of their study was that over a large range of numerosities, magnitude estimation results were in accord with perceptual matching with respect to the numbers of dots that were perceptually equivalent in the adapted and unadapted fields. Here we consider the magnitude estimation data itself.

11.3.1 Method

11.3.1.1 *Participants*

Eight undergraduates at Swarthmore College participated in exchange for payment. All were naïve concerning the purpose of the experiment.

11.3.1.2 *Apparatus*

The experiment was run using an Apple Macintosh computer using custom programming based on VideoToolbox (Pelli, 1997), and an Eyelink eye-tracker to monitor fixation using custom software.

11.3.1.3 *Displays*

White dots, 3.4 × 3.4 arcmin square (2 × 2 pixels) were presented against a medium gray background to the left or right of fixation. The adaptation and test regions to the left and right of fixation were 7 degrees wide (250 pixels) and 14 degrees high (500 pixels), and were separated by 1.4 degrees (50 pixels).

11.3.1.4 *Design*

Participants were adapted to dense texture (high numerosity) on one side of fixation and to a very low numerosity on the other side. Participants then made 240 magnitude estimates of the numbers of dots presented in a region to the left or to the right of fixation. Three estimates were made on each side for each of 40 numerosities ranging logarithmically from 5 to 1152 dots (the numbers 5, 6, 7, 8, 9 were multiplied by 2^x where x ranged from 0 to 7). These estimates were interspersed with 120 judgments of relative numerosity (which side has more dots). A staircase method was used to measure points of subjective equality for numerosities of 5, 10, 20, 40, 80, 160, 320, and 640 dots presented in the region adapted to very dense visual textures.

11.3.1.5 *Procedure*

One side of fixation (randomized between subjects) was first adapted with a very high density/numerosity of dots (1500 dots in a region that was 7 degrees wide and 14 degrees tall; that is, 15 dots/deg^2). The other side was exposed, during adaptation, to fairly small numbers of dots (15; ie, 0.15 dots/deg^2). There was an initial 3-min period of adaptation consisting of 180 newly randomized texture pairs that were each presented for 400 ms, separated by an ISI of 600 ms. The remainder of the experiment consisted of individual trials that included a top-up adaptation of two more texture pairs, followed, after 600 ms, either by the presentation of a single texture to the left or right of fixation or by the presentation of two visual textures. The texture(s) was/were presented for 400 ms and the computer then waited for the subject to type in an estimate of the number of dots presented (if one field) or to indicate which field has more dots (if two fields). There were a total of 240 estimation trials (following 20 implicit practice trials) and about 160 comparison trials. The entire procedure took about an hour.

11.3.2 Results and Discussion

Mean numerosity estimates are plotted in log/log space in Fig. 11.1 for numerosities presented in the adapted side (white plot markers) and in the unadapted side (black plot markers). Estimates are consistently lower in the adapted region than in the unadapted region for the same numbers of dots. At the upper range of numerosities, these two curves appear roughly parallel, which is consistent with a constant proportional reduction in perceived number.

However, the power functions fit to each of these curves that are shown in Fig. 11.1 fail to capture an important feature of the data: there seems to be an elbow in the function between 20 and 40 dots. Fig. 11.2 shows the data split into two parts: numerosities less than 20 dots and numerosities of at least 40 dots. In Fig. 11.2 we can see that numerosity estimates in the range from 5 to 18 dots appear to be scaled very differently from those from 40 to 1152.

Consider first the effect of density adaptation on numerosity estimation for 40 dots or more (Fig. 11.2, upper panel). In this range the exponent of

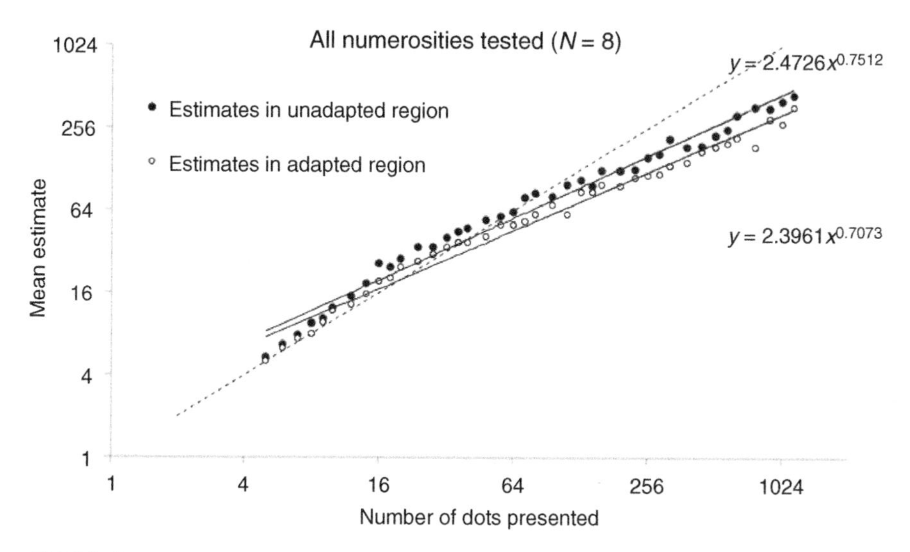

FIGURE 11.1 **Numerosity estimates from Experiment 1 (Huk & Durgin, 1996).** At most numerosities, estimates are lower for dot displays presented in a region adapted to dense texture (1500 dots) rather than sparse texture (15 dots).

the power functions are quite similar for estimates in both the adapted and the unadapted regions. The average exponent for power function fits to individual participant data were 0.62 (SE = 0.04) in the unadapted region and 0.60 (SE = 0.02) in the adapted region. These clearly did not differ from each other, $t(7) = 0.7$, n.s. This replicates the common finding that numerosity estimation is roughly linear in log–log space with an exponent less than 1 (Krueger, 1972, 1982). The fact that the two lines are parallel is consistent with the finding that density adaptation produces a proportional reduction in perceived number across a broad range of numerosities (Durgin, 1995).

In this upper range, the mean ratio of adapted to unadapted estimates was 0.77, but because the exponent of the power function in this range is about 0.61, this means that 1.5 times as many dots (ie, 1/0.65) had to be presented in the adapted region as in the unadapted region to produce the same numeric estimate. In other words, this predicts a matching ratio of about 0.65 following adaptation. For numerosities of 40–640, the matching data of Huk and Durgin (1996) showed a mean matching proportion of 0.65 following adaptation, indicating overall consistency between matching and estimation. However, the ratio varied from 0.82 to 0.54 as the standard numerosity increased from 40 to 640, showing some evidence that comparison and estimation were constrained somewhat differently.

Of course, at the very lowest number tested (which is near the subitizing range) we can see (Fig. 11.2; lower panel) that there is no effect of adaptation on estimation. Observers could reliably estimate 5 dots even

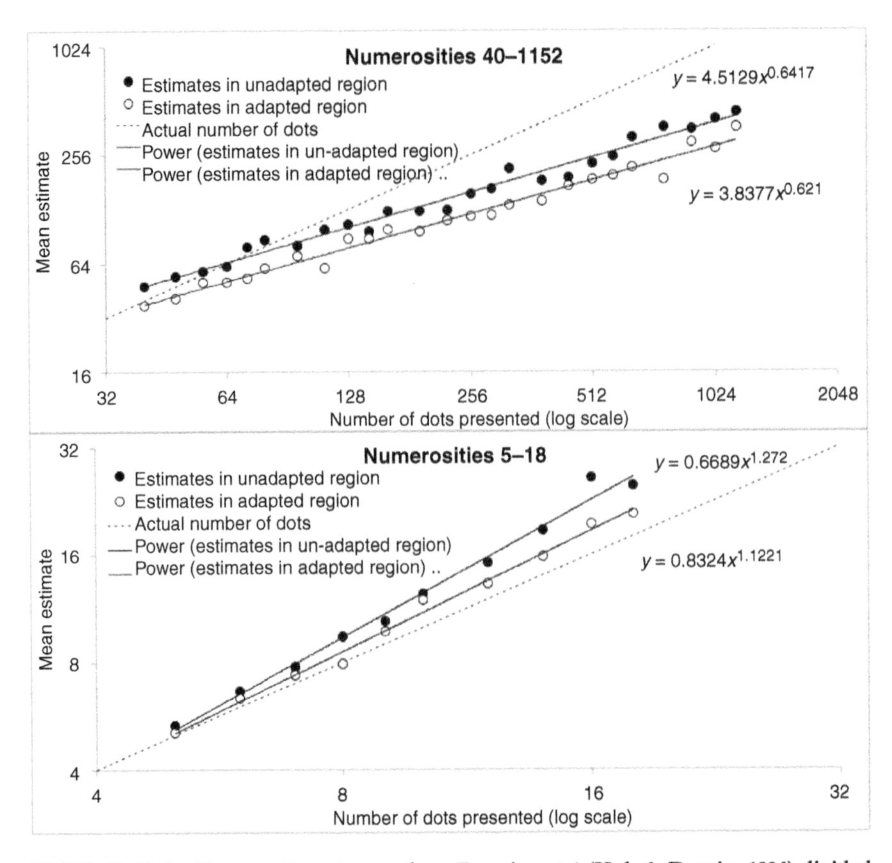

FIGURE 11.2 **Numerosity estimates from Experiment 1 (Huk & Durgin, 1996) divided into an upper range (top) and a lower range (bottom).** In the upper range the exponent of the power function is about 0.6, and estimates are proportionally lower in a region adapted to dense texture. For lower numerosities, the effect of adaptation increases with numerosity, and the exponents of the power functions are nearly 1, reflecting a more linear coding regime.

in the adapted region, whereas there is a growing separation between the adapted and unadapted curves as the number of dots increases from 5 to 18. This is consistent with the idea that density information has greater influence as numerosity increases well past the subitizing range. The mean exponent of power functions (fit to individual observer data) in the unadapted field in this range was 1.29 (SE = 0.13), which did not differ reliably from 1.0 [$t(7) = 1.92$, $p = 0.0963$], nor did it differ reliably from the mean exponent in the adapted region, 1.11 (SE = 0.06), $t(7) = 1.13$, $p = 0.296$. In other words, for the range of numerosities extending from 5 up to 18, the average exponents did not differ from 1 (the median exponent was 1.09 in the unadapted region), indicating a nearly linear relationship between presented number and estimated number. This near linearity is consistent with the participants having access to a useful unit of number. This is likely related to their relatively accurate estimation of the lowest number, 5, which is at the edge of the classic subitizing range. It may also have been supported by the presence of subitizable subgroups within the dots presented at other relatively low numerosities. Note that some prior fractionation studies (Taves, 1941) have seen evidence of exponents greater than 1, which is also a feature of overconstancy findings (eg, in size perception: Carlson, 1962) associated with cognitive attempts to compensate for perceptual compression.

11.3.3 Conclusions

The magnitude estimation data of Huk and Durgin (1996) were reanalyzed to show three theoretically important patterns:

1. At high numerosities (>40), estimates made in densely adapted regions were proportionally lower than those in unadapted regions. The data suggests a matching value of about 0.65 following adaptation, roughly consistent with actual matching data following adaptation.
2. Near the subitizing range, there was no effect of density adaptation on numeric estimation. However, the effect of density adaptation seemed to gain more weight as numerosity increased. This is consistent with the notion that perceived density is one of the perceptual magnitudes that is used to evaluate numerosity, and that it carries greater weight for larger numbers (more recent evidence consistent with this is reported both by Anobile, Cicchini, & Burr, 2013 and by Dakin, Tibber, Greenwood, Kingdom, & Morgan, 2011).
   These first two conclusions are consistent with the claims of Durgin (1995) regarding the relationship between density adaptation and numerosity perception. However, the third conclusion is a novel observation concerning the scaling of number:
3. For low numerosities beyond the subitizing range, the scaling of magnitude estimation was nearly linear in this experiment (exponent near 1.0), consistent with the idea that evaluating numbers near the subitizing range may implicitly provide a unit that can be extended beyond the subitizing range. That is, in the context of evaluating numbers of dots, participants can learn something about how to convert visual magnitudes into explicit verbal multitudes (numerals) by virtue of being given a magnitude value of the unit itself when processing in the subitizing range where individual units are larger than the JND. This unit scale only applies to relatively low numerosities because of changes in the way visual information can be processed as higher numbers of dots produce more crowding, and thus more need for summarization at higher densities.

The fact that numerosity estimation typically has a low exponent for higher numbers clearly suggests that the underlying visual variables involved are magnitudes that roughly correspond to multitudes. But these variables are not perceptually represented by a quantized or digital code. The argument being proposed here is that for lower magnitudes a meaningful unit can be more easily *established* near the subitizing range (ie, the perceptual "unit" must be established for any given multitude) and this can provide a basis for maintaining a relatively linear representation of number over a range of numerosities extending somewhat beyond the subitizing range.

This ability to represent numbers digitally depends on cognitive tools that are certainly provided by language. Magnitude estimation tells us something about perceptual experience, but it also depends on participants having fairly sophisticated knowledge about numbers. For the Pirahã, whose language concerning numeric quantity seems to be limited to "one," "a couple," and "many," even numbers of items as low as 5 do not get matched exactly unless presented in spatially matched arrays—so that the concept of number is made irrelevant by being replaced by spatial symmetry (Gordon, 2004).

11.3.4 Interlude: Visual Unitization for Large Multitudes

For large numbers of elements or for abstract elements (Durgin & Huk, 1997), the problem of unitization has not received much study. If each unit is made up three small lines, then is 100 units really 300 things? How can the visual system properly interpret such a situation? Based on a model appealing to divisive normalization (Heeger, 1992), it would be reasonable for the visual system to identify a unit that exists at one scale of visual analysis (eg, the three lines), and then use the computed energy of the unit to scale the energy of the collection. The visual system is readily poised to accomplish this using fairly low-level visual analyses (Carandini & Heeger, 2012). Burr and Ross (2008) have jested that this method is like counting one's sheep by first counting the legs and dividing by 4. But if what one had was a map of sheep legs on one's retina, what might one do?

Abdul-Malak and Durgin (2009) directly tested whether literal numerosity of legs or of sheep were more relevant in an adaptation experiment. They first asked whether 100 disconnected legs would really appear equally as numerous as 100 sheep with 4 legs each? They further asked whether a perceptually equivalent number of sheep legs (which turns out to be about 200 legs per 100 sheep) would be as effective as 100 sheep at producing adaptation relevant to the estimation of numerosity? Finally they asked: is the effectiveness of 100 sheep as an adapting display any different than the effect of their disembodied legs? The conclusion of these studies, which will be described later, seem to support the theoretical claims that (1) relative number and apparent relative number dissociate when the units are sufficiently different in scale (sheep vs. legs), and (2) relative apparent number dissociates from effective density for purposes of adaptation (see also Durgin, 2008).

11.4 EXPERIMENT 2: WHAT IS THE UNIT IN NUMBER ADAPTATION? (ABDUL-MALAK & DURGIN, 2009)

Perceptual systems represent number insofar as they respect something like the Euclidean concept of number—a multitude of units. Texture density adaptation suggests that the level at which density is represented might not be at all the same as the level at which number is cognitively available. This kind of dissociation happens all the time in perception: early visual processes take advantage of information that is not preserved at higher (eg, conscious) levels of cognition. Thus, while Durgin and Huk (1997) argued that texture density involved discriminating units, they successfully treated texture density as being present even when discrete units were not (eg, in cloud-like textures). Evidently the "units" over which the visual analysis of texture density could be computed need not correspond to units that would be recognized in conscious awareness, although they might normally so correspond.

If the unit for adaptation of visual number is the same as the unit for the perception of visual number, then pairs of adapting displays matched in (perceived) visual number should produce no differential adaptation. In Experiment 2, participants were tested in two kinds of task. In the first task, displays composed of "sheep" and of "legs" were compared for their apparent visual number in order to find the relative numbers of sheep and legs that would appear to have the same number of items. It was stressed to participants that they were to compare the absolute number of items (sheep in one case, legs in the other) in the displays. Based on pilot studies we anticipated that nearly twice as many legs as sheep were necessary to make the displays seem similar in numerosity. Of course, because each sheep had four legs, the relative density of lines in each texture was still such that a display of 100 sheep was much denser in texture features than a display of 200 legs (Fig. 11.3).

In the second part of the experiment, this perceptually matched 100:200 ratio was used as competing adapting stimuli. That is, the same participants were adapted to displays in which 100 sheep were presented to one side of fixation and 200 legs were presented to the other side. Since the adapting displays were expected to be approximately matched in apparent numerosity, an apparent-numerosity account of number adaptation should predict no different adaptation. While adapted, participants made relative numerosity judgments concerning textures composed of similar texture elements. For example, in one block of trials, textures of sheep

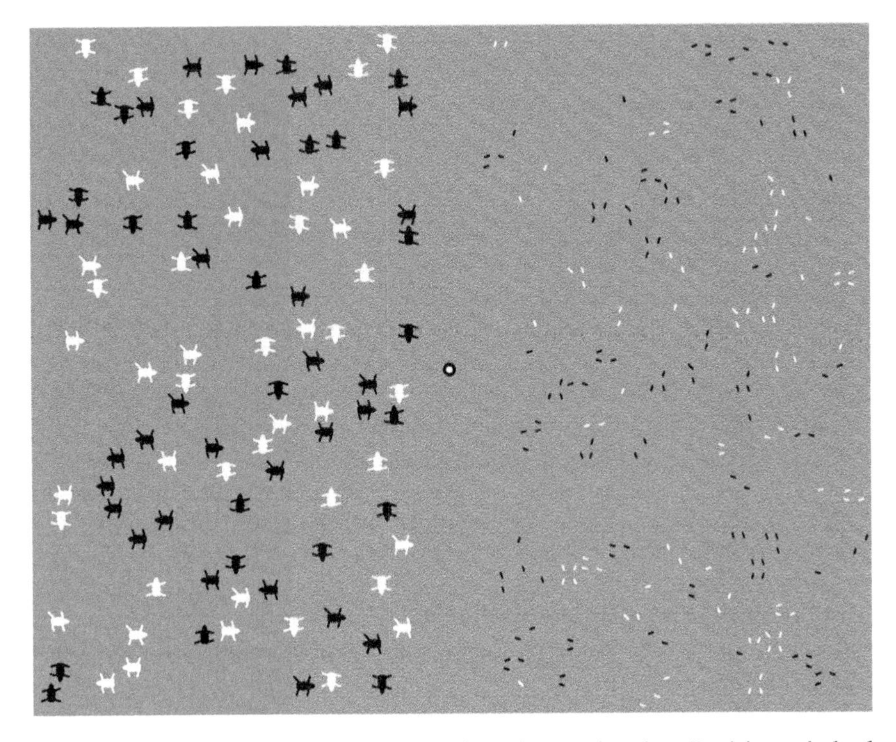

FIGURE 11.3 **A sample adapting display of 100 sheep and 200 legs. Participants judged such displays to be approximately equal in numerosity.**

presented in the sheep-adapted region were compared to textures of sheep in the legs-adapted region. In a separate block of trials, textures of legs presented in the sheep-adapted region of the visual field were compared to textures of legs presented in the legs-adapted portion of the visual field. Based on a texture density account, we expected that because each sheep contained several visual texture elements, the sheep-adapted side would be more adapted, and that this would be true whether the adaptation was measured with sheep or with legs (Durgin & Huk, 1997; Durgin, 2008).

11.4.1 Methods

11.4.1.1 Participants

Twenty-four Swarthmore College undergraduate students participated for pay or for course credit. None were informed of the hypothesis.

11.4.1.2 Design and Procedure

Participants performed three tasks. All did the sheep/leg numerosity matching task first (based on 80 trials) and then were adapted for each of the remaining tasks to displays matched in apparent numerosity of sheep and legs. Half the participants were adapted with sheep on the left and half with sheep on the right. Half of each group were tested first for matches between displays of sheep (60 trials) and then tested for matches between displays of legs (60 trials). The other half were tested first with displays of legs.

In each of the two adaptation tasks there was an initial period of repeated exposure to 30 newly generated displays of 100 sheep on one side and 200 legs on the other (display duration: 200 ms, ISI: 1250 ms). Subsequent trials were preceded by three more adapting flashes (all textures were generated on the fly). Each test trial was preceded by a warning beep. Responses (which side had appeared to have more elements) were recorded by key-press.

11.4.1.3 *Displays*

A sample display is shown in Fig. 11.3. Each "sheep" is composed of a central blob and four lines forming the legs. Sheep were colored black or white at random. To partly retain the clumpiness of the sheep texture, the legs were scattered in clumps of 1–4 legs in the absence of sheep bodies. The entire display was 1280 × 1024 pixels, and subtended approximately 36 × 27 degrees from the viewing distance of 57 cm. Sheep legs were about 9 pixels in length (0.25 degrees) and sheep bodies about 27 pixels in length (0.75 degrees). The central 2 degrees of the display was not textured but contained a fixation circle, as shown. The displays each appeared for 200 ms, with the fixation mark and background gray remaining for 1250 ms between presentations.

11.4.1.4 *Measurement*

A staircase method was used to estimate the point of subjective equality (PSE) between "sheep" and "legs" textures. In the first part of the experiment, 100 sheep were presented in one field on each trial and variable numbers of legs were presented in the other field (side was randomized). The number presented was varied according to a staircase procedure. One staircase started at 220 legs, the other started at 50 legs. Each value was stepped (up or down) by 20 legs in accord with participant responses. In the second part of the experiment, the sheep-adapted side was tested with 100 legs or 50 sheep. The legs-adapted side had a variable number, with parameters in proportion to those used in the staircase of part 1.

A logistic function was fit to the response probabilities for each task for each participant as a function of presented number, so as to estimate the PSE as the point of 50% likelihood of choosing either side as more numerous. As anticipated from pilot investigations, the mean number of legs required to match 100 sheep was about 200 ($M = 198$; $SE = 14$) in the initial comparison task. Thus, the visual numerosities used in the adaptation tasks (100 sheep on one side and 200 legs on the other) were ostensibly equal.

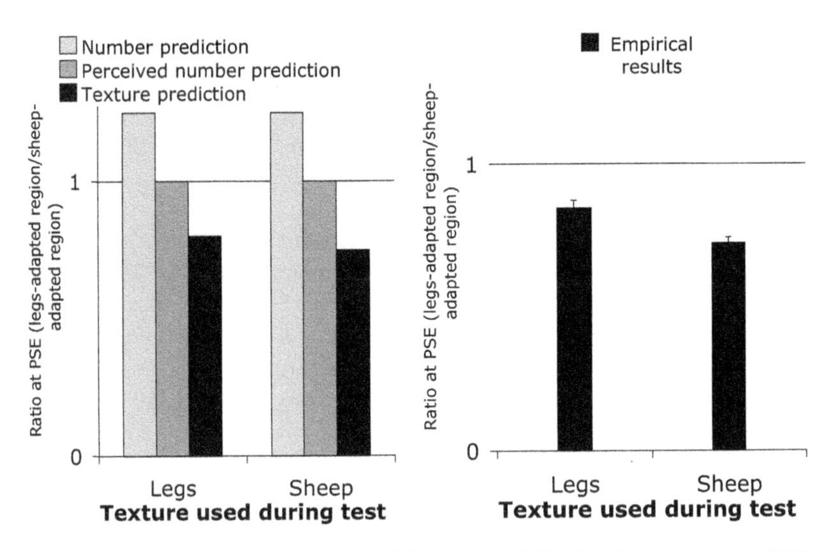

FIGURE 11.4 **Predictions and results of Experiment 2.** Predicted postadaptation PSE ratios based on number, perceived number, and texture density (left). Actual PSE ratios (right). Standard errors of the means are shown.

The PSEs following adaptation were expressed as ratios between the numbers in the legs-adapted side and in the sheep-adapted side at PSE. If actual number controlled adaptation, these ratios should be greater than 1. If apparent number controlled adaptation, they should not differ from 1. If texture density controlled adaptation, they should be less than 1 (and differ slightly according to texture similarity as demonstrated by Durgin & Huk, 1997). These predictions are shown in the left panel of Fig. 11.4.

11.4.2 Results

11.4.2.1 Matching

Overall, the average number of legs required to match 100 sheep was 198 ± 69 (SD), or about 2 legs for each sheep, which meant that our adapting displays of 100 sheep and 200 legs were, as intended, of approximately equivalent apparent numerosity.

11.4.2.2 Aftereffects

If the 100 sheep produced more adaptation than the 200 legs, we would expect the PSE ratios to be less than 1. Indeed, as shown in the right panel of Fig. 11.4, the ratios at PSE were less than 1.0, both when measured with textures of sheep, $t(23) = 14.6$, $p < 0.0001$, and when measured with textures of legs, $t(23) = 4.97$, $p < 0.0001$. Consistent with the partial texture specificity reported by Durgin and Huk (1997), the aftereffects measured with sheep were stronger (0.72) than those measured with legs (0.85), $t(23) = 3.98$, $p = 0.0006$.

11.4.3 Discussion

Direct matches between textures of sheep (with each sheep including four legs/lines) and textures of legs (single lines) showed that twice as many legs as sheep were needed for the two sets to appear visually similar in number. This sharp mismatch in perceived number is consistent with the idea that numerosity judgments are cognitive assessments based on perceptual information, but are not always particularly accurate at comparing different texture types (at least for numerosities as high as 100).

Moreover, the greater density of texture features on the "sheep" side produced larger aftereffects even for leg displays. Thus, the (consciously) obvious unitization of the sheep legs by the sheep bodies did not reduce their effectiveness as adapting elements. Although fewer in number, the sheep were more powerful adapters of relative number as measured by adaptation, even when matched for apparent numerosity.

11.5 EXPERIMENT 2B: THE EFFECTS OF CLUMPING

It is well known that elements appear less numerous when they form clumps than when they are evenly spread apart from one another within the same area. This observation was the basis for developing the occupancy model of numerosity (Allik & Tuulmets, 1991). Could the numerosity matches between sheep and legs in Experiment 2 reflect the partial clumping of the legs? To test the effect of partial clumping on perceived number, we replaced the 100 sheep with 150 individual legs that were prevented from overlapping and repeated the matching task of Experiment 2 with 19 new participants. The procedure was otherwise the same as in the matching task of Experiment 2.

11.5.1 Results and Discussion

A sample stimulus is shown in Fig. 11.5 that represents the average PSE. As expected based on prior findings of element distribution (eg, Allik & Tuulmets, 1991), when the legs were clumped as they had been in Experiment 2, slightly more were required to match 150 unclumped legs. The average PSE value in this case was 167 (SD 23), which was reliably greater than 150, $t(18) = 3.31$, $p = 0.0039$. Thus, the apparent reduction in numerosity due to leg clumping was by only 10%. Evidently the mismatches between apparent numerosities of legs and sheep in the matching task of Experiment 2 were not due simply to differences in the distribution of the elements.

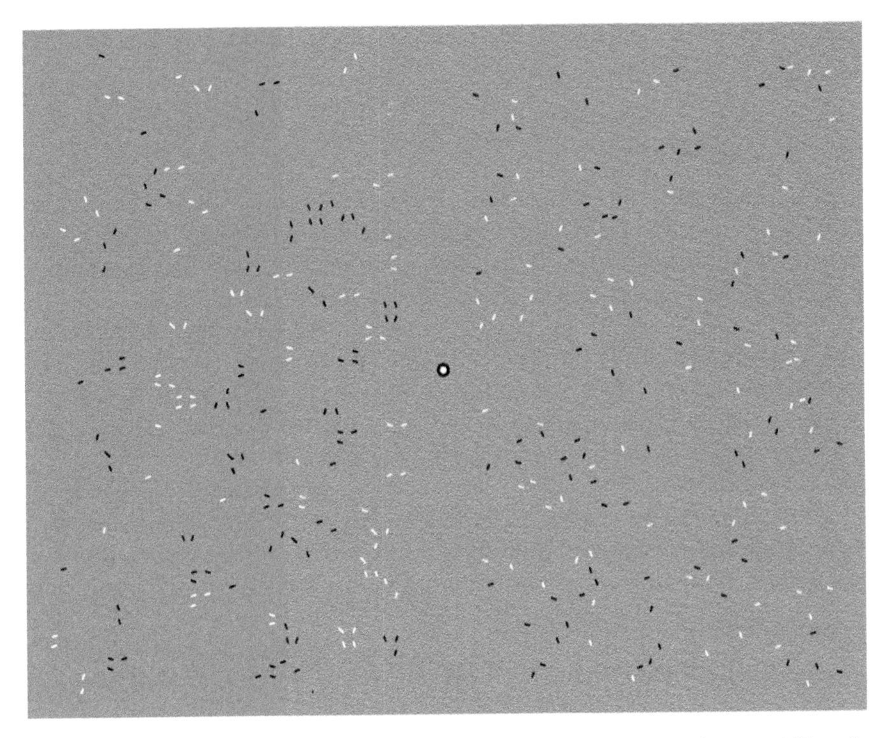

FIGURE 11.5 **Effects of clumping of legs on perceived numerosity.** There are 167 partly clumped legs on the left side and 150 unclumped legs on the right. Participants judged the numerosities of displays like these to be perceptually equal in number.

11.6 EXPERIMENT 3: DISEMBODIED SHEEP LEGS

In Experiment 2, unitization (sheep bodies) failed to fully reduce both the conscious estimation of the sheep/leg numerosity and their effective texture density adaptation. Twice as many legs as sheep were required to perceptually match the apparent numerosity of the sheep as judged by direct visual comparison. But even when twice as many legs were used during adaptation, the sheep still had a bigger influence (if both had been equally adapting, we should have measured no difference in perceived number at all following adaptation). To assess whether the sheep bodies (unitization) had affected adaptation at all, we simply removed the sheep bodies during adaptation so that participants were adapted to the legs of 100 sheep (ie, 400 legs in clumps of 4) on one side and 200 legs (partly clumped, as in Experiment 1) on the other as shown in Fig. 11.6.

If unitization by means of sheep bodies failed to block texture density adaptation at all, then removing the sheep bodies might make very little difference. That is, the amount of "leg texture" is unchanged, and the results of Experiment 2 suggest that unitization by means of sheep bodies

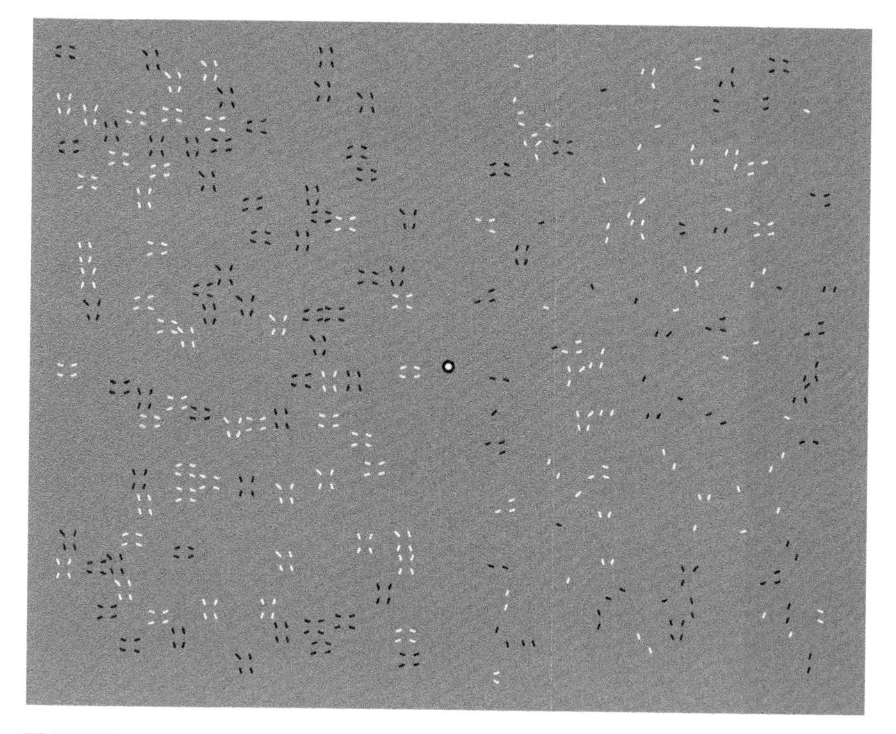

FIGURE 11.6 Sample adapting display for Experiment 4 in which 100 sets of sheep legs (ie, 400 legs) have replaced the sheep on the left. The sheep legs appear far more numerous than did the sheep. Adaptation resulting from them is quantitatively the same as when they were connected by sheep bodies in Experiment 1.

had even less effect on adaptation than the partial effect it had on numerosity comparison.

11.6.1 Methods

11.6.1.1 Participants

Twelve new participants were drawn from the same population. (Additional data were collected from the 19 participants from Experiment 2b using different adaptation parameters.)

11.6.1.2 Procedure

Apart from the adaptation stimuli, the design was identical to the adaptation task of Experiment 2 in which comparisons were made between two fields of legs following adaptation. For all participants, the adapting textures of 100 sheep were replaced by textures in which only the legs of 100 sheep were drawn (ie, 400 legs), as shown in Fig. 11.6. For the 12 new participants, the number of legs in the opposing field during adaptation remained at 200. For the 19 participants from Experiment 2b, the number of partly clumped legs on the opposite side during adaptation was reduced from 200 to 140. This was done to show that the matches following adaptation were indeed sensitive to relative density.

11.6.2 Results and Discussion

Adaptation to 100 sets of 4 sheep legs opposite 200 partly clumped legs, produced average aftereffects (0.84) when evaluating the relative numerosities of two regions of legs that were indistinguishable in magnitude from those produced in Experiment 2. The comparable aftereffect after adapting to 100 unitized sheep opposite 200 partly clumped legs was 0.85. Thus, removing the unitizing sheep bodies neither increased adaptation (as would be predicted if perceived numerosity were the source of adaptation), nor did removing the sheep bodies decrease adaptation as might be expected under a pure "texture energy" model. Rather, the amount of adaptation remained unchanged. It is somewhat surprising that the sheep bodies seem to have had essentially no effect whatsoever. Indeed, Emilie Shepherd and I have since replicated this observation using random assignment to conditions within the same experiment because it seemed so striking. The results were the same as reported here.

Conversely, when the number of legs in the opposite side was reduced during adaptation (by 30%), the measured difference in adaptation of the two sides at PSE (0.74) was increased (ie, its deviation from a ratio of 1.0 increased) accordingly, $t(41) = 2.54$, $p = 0.0149$.

Note that aftereffects of texture density/number are normally fairly large in magnitude. Typically, about twice as many elements are required in a strongly adapted region to match a weakly adapted region (Durgin, 1995; Durgin & Proffitt, 1996). In Experiment 2, the sheep textures produced greater adaptation despite being matched for numerosity with the legs, but the magnitude of the effect was fairly small (though not insubstantial) consistent with the fact that both sides of fixation were adapted. What was measured was differential adaptation. Here we have shown that the same level of differential adaptation is produced even without the sheep bodies, but that differential adaptation is (of course) increased if the other field is reduced in density by just 30%.

11.7 EXPERIMENT 4: EVIDENCE FOR THE SUCCESS OF UNITIZATION

Experiments 2 and 3 show that unitization of the sheep stimuli failed in two different respects. First, using such visually different units revealed that the comparison of numerosity was dependent on visual information that was not terribly well correlated with number: 200 legs were judged equal in numerosity with 100 sheep in Experiment 2. Second, even with subjective numerosity matched, 100 sheep acted, for purposes of adaptation, like 400 legs (as shown in Experiment 3) rather than merely 200. Based on these observations, it seems that the visual system did indeed "count" all the legs (when adapting), but only divided by two when matching. The unit for texture element density adaptation did not correspond to the unit for number comparison.

This finding is not a problem just for direct visual number theorists. The idea that texture density underlies numerosity discrimination at high numerosities still should involve the idea that the visual system can be successful at unitization across different object types. This seemed supported by earlier observations of the transfer of adaptation from one type to another (Durgin & Huk, 1997). Texture density was construed by Durgin (1995; Durgin & Huk, 1997) to be a local metric of multitude/area, even though it was assumed that continuous variables (magnitudes) were being used to represent visual multitudes. True "density" coding (rather than mere texture energy) requires sensitivity to texture units. Durgin (1999) proposed that this depended on divisive normalization: global texture energy divided by local texture energy (ie, local to the unit) can provide a magnitude to represent a multitude (ie, number or density) by virtue of having a well-defined unit. But Experiments 2 and 3 showed that the blob-like sheep bodies failed to produce integrated units with respect to adaptation processes (see also Durgin, 2008). Thus, the four legs of the sheep were what counted, as it were, in producing the level of adaptation observed in Experiments 2 and 3. This suggests that effective density (for adaptation) was based on features rather than conjoined units.

It seems surprising that the removal of sheep bodies did not seem to change the amount of adaptation at all. It could be argued that the loss of texture "energy" caused by removing the bodies was compensated for by the doubled number of line terminators that were exposed. Indeed, four-legged "sheep" have only twice as many terminators as do isolated legs, so it even seemed possible that the primary active variable in Experiments 2 and 3 was the number of terminators visible.

In Experiment 4, this terminator idea was tested more directly by creating elements that were similar in many properties, such as luminance and size, but differed in the number of terminators. In this experiment, the methods were similar to the adaptation portion of Experiment 2 but the competing texture fields during adaptation were constructed of elements made of four lines that could either be configured as a square, with four vertices and an enclosed area or a rectilinear hash-tag symbol (#) with 8 terminators (Fig. 11.7). The test textures were composed of either plus signs (+) or circles (o) similar in size to the adaptation elements. These texture types were selected because they contained distinct features of

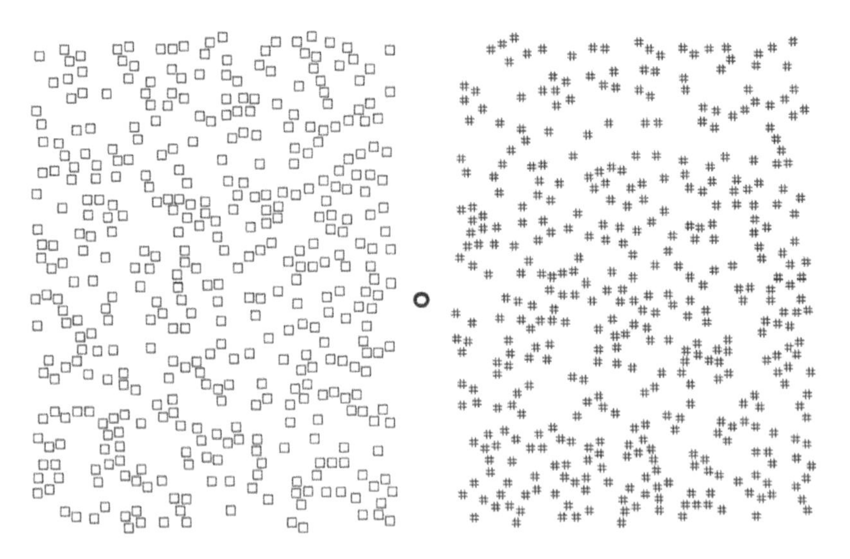

FIGURE 11.7 **Squares and hash-tags adapting display like those used in Experiment 4.**

the adapting stimuli—multiple terminators, or a largeish enclosed area. A pure feature-based analysis would suggest we would see strong feature texture specificity from adaptation, and probably that hash-tags would be much more powerful adapters overall.

11.7.1 Method

11.7.1.1 Participants

Twenty-four Swarthmore College undergraduate students participated for pay or for course credit. None were informed of the hypothesis.

11.7.1.2 Stimuli

In this experiment, the adaptation fields were twice as big in area (12 × 16 degrees) as the test fields (8.5 × 11.3 degrees) to better ensure that small fixation errors would not affect alignment of adapted and tested regions. During adaptation, 400 texture elements were presented to each side of fixation. In one field (whether left or right was alternated across participants), the adapting elements were squares composed of four black lines (2 vertical, 2 horizontal, with a length of 11 pixels each, or about 0.3 degrees in length). Each line was 1 pixel wide. In the other field the same lines were arranged so that they crossed each other at 1/3 intervals (ie, at the 4th and 8th pixels) to form a hash-tag symbol with all 8 of the line terminations exposed. Each texture unit was at least 2 pixels from every other unit.

On each comparison trial, both of the texture fields contained the same types of element. On half the test trials both fields contained circles (an enclosed region, 11 pixels in diameter) and on half the test trials both fields contained plus symbols (a vertical and horizontal line that crossed at their centers). The standard field contained 100 elements.

11.7.1.3 Design

Half the participants were adapted with the pound symbols on the left (squares on right) and half to pound symbols on the right. All were tested with both circles and with plus symbols (with continued top-up adaptation between trials). Unbeknown to the participants, the variable texture was always on the right side of fixation.

11.7.1.4 Procedure

The procedure was similar to that of the adaptation part of Experiments 2 and 3, except for the specific stimuli types. Adaptation textures are represented in Fig. 11.7.

11.7.2 Results and Discussion

If texture density adaptation were based on a system that is unable to unitize (normalize according to local energy), we should expect very different adaptation effects for the two texture types: the extra terminators on the hash-tags should tend to produce greater texture energy and therefore greater adaptation than the square, especially when tested with plus symbols. Conversely, the test circles might be affected more by the squares because both are enclosed shapes. In fact, matches differed reliably between the two test textures in the predicted direction, $t(23) = 5.01, p < 0.0001$, but the differential effect was quite small. For circle stimuli, the overall effect resulted in a reliable 6% reduction in the square-adapted field, $t(23) = 2.51$, $p = 0.0196$ (ie, 6% more circles were present in the square-adapted field at PSE). For plus signs, the overall effect resulted in a (nonreliable) 4% reduction in the hash-tag-adapted field, $t(23) = 1.52$, $p = 0.1412$. These results show minor texture-specific adaptation, but they do not support the texture energy hypothesis. Rather, in the present experiment, density adaptation (where unit density was equated in both adaptation fields) appears to be surprisingly robust to unit type.

11.8 GENERAL DISCUSSION AND CONCLUSIONS

Humans and other animals do seem to have nonsymbolic representations that reflect relative number, but these systems are neither magical (ie, direct conduits to true number) nor error-free. Based on Euclid's definition of number (a multitude of units), we have considered the human process of assigning numbers to small and large collections—both with and without adaptation to dense visual texture (Experiment 1). We have also considered the process of comparing and adapting to collections made up of different kinds of units (Experiments 2, 3, and 4). What have we learned?

11.8.1 Magnitudes All the Way Down?

In Experiment 1 we observed that for numbers below 20, numerosity estimates were fairly linear with actual number, whether presented in an adapted region or in an unadapted region, but that number estimates were scaled logarithmically beyond about 40 dots (see also Kaufman et al., 1949). The linear range may reflect the ability of human observers to extract a unit from the low end of a range and then to extend that unit well past the subitizing range. In support of this idea, it is worth noting that the performance of observers in our experiment contrasts with those asked to make single trial estimates of number (Krueger, 1982) or to make estimates only of numerosities well above 20 (Krueger, 1972). Without the context of seeing numbers in the subitizing range, Krueger (1972, 1982) found marked underestimation even for 25 dots, whereas we observed slight overestimation in this range in Experiment 1. Moreover, Kaufman et al. (1949), who used a range from 2 to 200 dots for numerosity estimation, also observed excellent scaling of estimates up to about 20 dots.

Overall, this lends support to the notion that even in the lower range (eg, 20 dots and below), numerosity estimates depend on magnitude representations that have to be scaled by a unit magnitude. The paradigm used in Experiment 1 allowed participants to establish a unit (by intermittent exposure to the subitizing range) that could have served to calibrate their estimates in this lower range. This was true of the design used by Kaufman et al. (1949) as well (whose estimation data also show the elbow shown in our data), but it was not true of the designs used by Krueger (1972). This perceptually established unit (that we hypothesize is afforded by designs that dip into the subitizing range) even affected estimates far beyond 20 dots, with the result that for actual numerosities of approximately 100 dots, Kaufman et al. and our participants (in the unadapted field) each gave mean estimates very close to 75% of the actual number, whereas Krueger's (1972) participants gave estimates that were only about 60% of the actual number: the difference in these estimates corresponds to the differential success of the different groups at 20 dots, which in turn seems to have depended on their exposure to the subitizing range.

Of course, other explanations of this pattern are possible, but this is exactly the pattern one would expect if numerosity estimation normally depends on measuring a total magnitude and then dividing it by a unit magnitude: to the extent that the range of stimuli presented allows one to better measure the unit magnitude, the scaling of all values will be improved. However, the type of magnitude information available probably changes as displays become more numerous, rendering the unit magnitude less effective as a divisor. For example, if the unit were measured in "occupancy" units (Allik & Tuulmets, 1991), it would be effective in the occupancy range (ie, 20 dots), but fairly useless in at higher ranges where density information seems to dominate numerosity comparisons (Durgin, 1995). This does suggest that numerosities of about 20 are not primarily measured by density (a point where both Anobile et al., 2013, and Durgin, 1995, agree), but it does not show that numerosity in this range is directly perceived as metric number. Rather it tends to suggest that number is typically perceived as a scaled magnitude—scaled by a unit measure that may or may not be well calibrated.

11.8.2 What is the Unit?

The series of studies reported here using different units (eg, sheep vs. sheep legs: Experiments 2–4) used large numerosities (closer to 100 than to 20), and in this range, at least, we saw quite clearly that the numerosity comparisons were not very accurate across extremely different units. One hundred sheep were judged about equal in apparent numerosity to 200 legs. Moreover 100 sheep produced much greater adaptation than did 200 legs. Thus, the adapting magnitude of 100 sheep was greater than the 200 legs. (It was, in fact, equal to the adapting power of 400 legs!) This suggests that the numerosity comparison process that participants used to compare collections of sheep with collections of legs did show evidence of compensating for the different units (otherwise 100 sheep should have appeared equal to 400 legs), but was only partly successful. The adaptation process, on the other hand, seems to have either ignored the unit altogether or defined it rather differently than we would. Perhaps for adaptation, the unit was the density of lines.

Thinking that perhaps texture features (eg, terminators) were what mattered, we created very different appearing elements in Experiment 4 that were both composed of 4 lines (squares and hash-tags), but these proved to have nearly identical adapting power. This is consistent with the "density of lines" theory, but further systematic investigation is needed. Clearly sheep bodies failed to unitize the sheep legs as observed by early visual analyzers. Perhaps the presence of clearly articulated "parts" is important for the adaptation process.

11.8.3 Relative Number System

Based on the observations of Experiment 1, it would be interesting to look at numerosity estimation processes in a paradigm in which the subitizing range was included for both sheep and for legs. It seems likely that participants could be self-trained to be better calibrated at estimating either legs or sheep. But even if estimation for each type of texture improved, it would be interesting to ask whether that would be sufficient to overcome the bias we observed in the direct comparison of the apparent numerosities of sheep and of isolated legs. In other domains, perceptual biases of comparison can persist even when numeric scaling of magnitudes shows marked improvement (eg, Durgin, Leonard-Solis, Masters, Schmelz, & Li, 2012).

It is evidently not trivial to establish a unit ratio with which to make comparisons across apples and oranges (or artichokes and bananas). If numerosity perception were based on a representation of number (with a natural unit), comparing across numbers of different units should probably not be so difficult. The fact that it is easier to accurately compare numerosities across similar types of elements than across different types of elements implies that no division by a unit is ever made in these comparisons. If the unit can be assumed to be the same in both displays, then the numerosity-relevant magnitude of one display can be directly compared with the numerosity-relevant magnitude of the other. Note that sensitivity to numerosity-relevant aspects of a display (ie, compensating for density differences, or size differences or what have you) is an important property of a system hypothesized to be a number system. But if the imputed number system does not or cannot reliably apply a unit to scale the number, then it is probably better to construe it as a magnitude system that is tuned to compare relative number (of like things) rather than as a system designed to measure absolute number. For many tasks concerning choice of food or a choice between fighting or fleeing (eg, Chapter 8), sensitivity to relative number is typically sufficient to make a decision.

11.8.4 Conclusions Regarding Adaptation

In Experiment 1 we saw that adaptation to dense or numerous displays affected both numerosity comparison processes and numerosity estimation processes. The proposal here is that adaptation can affect the computation of the numerosity correlate (often density and/or area), prior to any attempt to scale numerosity. Thus, the 100 sheep, which possessed 400 legs, were much more powerful adapters than the 200 legs (which were matched in apparent numerosity) in Experiment 2. But even when we required participants to scale perceived numerosities (ie, to give estimates in Experiment 1), which requires applying a unit, we saw that their estimates did indeed shift in a way that could be roughly (though not perfectly) predicted by the postadaptation comparison data: estimates in the dense-adapted region were proportionally lower than estimates in the nonadapted region. The only exception to this was at the edge of the subitizing range (ie, 5 dots), where density information is almost certainly irrelevant.

Thus, the present data seem to support the idea that it is not number that was adapted but something more magnitude-like, like density.

11.8.5 Conclusions Regarding Visual Number

In many evolutionary contexts, discriminating relative number is probably essential and thus sensitivity to relative number is important. In these contexts, it is no small achievement to isolate relative number from other features like relative size or relative density, and it would be a mistake to downplay the importance of relative number skills. But absolute number is a different thing altogether than relative number. It requires not only an external unit (one sheep or one leg), but also an internal means of computing total number based on that unit. For large numbers, at least, humans are biased to underestimate visual numerosity (eg, Kaufman et al., 1949; Krueger, 1972), and here we have shown that they are quite biased when, for example, comparing the numerosities of depicted sheep and legs. We have sought to emphasize that number cannot be understood without the concept of the unit, but in contexts where only a relative number decision must be made, representing the magnitude of the unit is unnecessary. Our results suggest that properly calibrated unit magnitudes are not readily available humans for numerosities beyond the subitizing range, but that exposure to the subitizing range can help to establish a calibrated unit somewhat beyond that range. Instead of speaking of an approximate number system (rather than an exact number system), perhaps we should be emphasizing that most studies involve comparison rather than estimation and thus examine a relative number system in which unit magnitudes can remain entirely unspecified. It is in this sense that the nonsymbolic representations of numbers in our heads might normally be considered to be (holistic) magnitudes rather than multitudes (of units).

References

Abdul-Malak, D., & Durgin, F. H. (2009). Dividing the legs of sheep: does Burr's Australian stockman strategy work? *Journal of Vision, 9*(8), 980.

Allik, J., & Tuulmets, T. (1991). Occupancy model of perceived numerosity. *Perception & Psychophysics, 49*, 303–314.

Anobile, G., Cicchini, G. M., & Burr, D. M. (2013). Separate mechanisms for perception of numerosity and density. *Psychological Science, 25*, 265–270.

Anstis, S. M. (1974). Size adaptation to visual texture and print: evidence for spatial-frequency analysis. *American Journal of Psychology, 87*, 261–267.

Blakemore, C. T., & Campbell, F. W. (1969). On the existence of neurones in the human visual system selectively sensitive to the orientation and size of retinal images. *The Journal of Physiology, 203*(1), 237–260.

Blakemore, C. T., & Sutton, P. (1969). Size adaptation: a new aftereffect. *Science, 166*, 245–247.

Burr, D., & Ross, J. (2008). A visual sense of number. *Current Biology, 18*(6), 425–428.

Carandini, M., & Heeger, D. J. (2012). Normalization as a canonical neural computation. *Nature Reviews Neuroscience, 13*(1), 51–62.

Carlson, V. R. (1962). Size-constancy judgments and perceptual compromise. *Journal of Experimental Psychology, 63*(1), 68–73.
Dakin, S. C., Tibber, M. S., Greenwood, J. A., Kingdom, F. A. A., & Morgan, M. J. (2011). A common visual metric for approximate number and density. *Proceedings of the National Academy of Sciences, 108*(49), 19552–19557.
Dehaene, S. (1997). *The number sense*. Oxford: Oxford University Press.
Durgin, F. H. (1995). Texture density adaptation and the perceived numerosity and distribution of texture. *Journal of Experimental Psychology, 21*, 149–169.
Durgin, F. H. (1999). A model of texture density encoding. *Investigative Ophthalmology & Visual Science, 40*, S200.
Durgin, F. H. (2008). Texture density adaptation and visual number revisited. *Current Biology, 18*, R855–R856.
Durgin, F. H., & Huk, A. C. (1997). Texture density aftereffects in the perception of artificial and natural textures. *Vision Research, 37*, 3273–3282.
Durgin, F. H., & Proffitt, D. R. (1996). Visual learning in the perception of texture: simple and contingent aftereffects of texture density. *Spatial Vision, 9*, 423–474.
Durgin, F. H., Leonard-Solis, K., Masters, O., Schmelz, B., & Li, Z. (2012). Expert performance by athletes in the verbal estimation of spatial extents does not alter their perceptual metric of space. *i-Perception, 3*(5), 357–367.
Euclid (1956). *The elements* (T.L. Heath, Trans.). New York: Dover. (Original manuscript, 300 BC)
Feigenson, L., Dehaene, S., & Spelke, E. (2004). Core systems of number. *Trends in Cognitive Sciences, 8*(7), 307–314.
Gebuis, T., & Gevers, W. (2011). Numerosities and space; indeed a cognitive illusion! A reply to de Hevia and Spelke (2009). *Cognition, 121*(2), 248–252.
Gebuis, T., & Reynvoet, B. (2012). The interplay between nonsymbolic number and its continuous visual properties. *Journal of Experimental Psychology, 141*(4), 642.
Gordon, P. (2004). Numerical cognition without words: evidence from Amazonia. *Science, 306*, 496–499.
Heeger, D. J. (1992). Normalization of cell responses in cat striate cortex. *Visual Neuroscience, 9*(02), 181–197.
Huk, A. C., & Durgin, F. H. (1996). Concordance of numerosity comparison and numerosity estimation: evidence from adaptation. *Investigative Ophthalmology & Visual Science, 37*(3), 1341.
Jevons, W. S. (1871). The power of numerical discrimination. *Nature, 3*, 363–372.
Kaufman, E. L., Lord, M. W., Reese, T. W., & Volkmann, J. (1949). The discrimination of visual number. *The American Journal of Psychology, 62*(4), 498–525.
Krueger, L. E. (1972). Perceived numerosity. *Perception & Psychophysics, 11*(1), 5–9.
Krueger, L. E. (1982). Single judgments of numerosity. *Perception & Psychophysics, 31*(2), 175–182.
Mandler, G., & Shebo, B. J. (1982). Subitizing: an analysis of its component processes. *Journal of Experimental Psychology, 111*(1), 1–22.
Pelli, D. G. (1997). The VideoToolbox software for visual psychophysics: transforming numbers into movies. *Spatial Vision, 10*(4), 437–442.
Taves, E. H. (1941). Two mechanisms for the perception of visual numerousness. *Archives of Psychology, 37*, 1–47.
Trick, L. M., & Pylyshyn, Z. W. (1993). What enumeration studies can show us about spatial attention: evidence for limited capacity preattentive processing. *Journal of Experimental Psychology, 19*(2), 331.
Trick, L. M., & Pylyshyn, Z. W. (1994). Why are small and large numbers enumerated differently? A limited-capacity preattentive stage in vision. *Psychological review, 101*(1), 80.
Walker, J. (1966). Textural aftereffects: tactual and visual. Unpublished doctoral dissertation, University of Colorado, Boulder.

CHAPTER

12

Ordinal Instinct: A Neurocognitive Perspective and Methodological Issues

Orly Rubinsten

Edmond J. Safra Brain Research Center for the Study of Learning Disabilities, Department of Learning Disabilities, University of Haifa, Haifa, Israel; Center for the Neurocognitive Basis of Numerical Cognition, Israel Science Foundation, Israel

OUTLINE

As early as 450 BCE, the Greek philosopher Anaxagoras of Clazomenae argued for the existence of an ordering principle in the universe and introduced the concept of Mind as an ordering tool (Burnet, 1908). Interestingly, contemporary scientific and clinical approaches have not availed themselves of this insight. Here, I propose an innovative outlook that argues for the existence of a neurocognitive system designed to evaluate ordinal relationships. This neurocognitive system, termed here "the

Continuous Issues in Numerical Cognition. http://dx.doi.org/10.1016/B978-0-12-801637-4.00012-3

ordinal instinct," is tuned to optimally pick up ordinal information by *implicitly* and *unconsciously* capturing sequential regularities in the surrounding world. Importantly, I review data that may suggest that human numerical intelligence significantly relies on the ordinal instinct through biological development and experience.

12.1 SCIENTIFIC KNOWLEDGE AND DEVELOPMENTS

Numerical cognition is important because mathematical skills are essential for productive functioning in our progressively complex, technological society. In addition, numerical development has been a focus of the continuing theoretical debate concerning the origins of cognition and how it develops throughout life. Approximately 10% of the population has low numeracy skills, a condition called developmental dyscalculia (DD) or mathematical learning disability (MLD) (Kaufmann et al., 2013; Rubinsten & Henik, 2009; Szűcs & Goswami, 2013). Numerical difficulties result in reduced educational and employment achievements, as well as increased costs to physical and mental health (eg, Parsons & Bynner, 2005; Reyna, Nelson, Han, & Dieckmann, 2009). Some argue that in Western societies, poor numeracy is a more significant handicap than poor literacy (eg, Estrada, Martin-Hryniewicz, Peek, Collins, & Byrd, 2004; Parsons & Bynner, 2005). Hence, mathematical skills may have an impact on social mobility and poverty levels. Despite its manifest importance, paradoxically, and compared to other cognitive abilities such as reading and attention, the neurocognitive development of numerical abilities has been neglected by both clinicians and researchers.

12.2 NEURAL AND COGNITIVE FOUNDATIONS OF NUMERICAL KNOWLEDGE

Kant argued that human beings are endowed with innate knowledge that is not related to experience, as part of their rational nature (Kant, 1908). Following such ideas, modern research has argued that mathematics rests on mental representations that developed in the course of evolution (Cantlon, Platt, & Brannon, 2009; Feigenson, Dehaene, & Spelke, 2004). These core representations include a nonsymbolic numerical system that represents the approximate numerical value of a collection of objects (ie, *numerosity*). Nonsymbolic approximation is a key evolutionary ability, which is believed to develop without formal teaching. Indeed, studies have found that infants from birth (eg, Izard, Sann, Spelke, & Streri, 2009) and even nonhuman animals (eg, fish—Piffer, Petrazzini, & Agrillo, 2013) display the ability to discriminate quantities. In contrast, symbolic and exact numerical representations (eg, Arabic numerals such as 6) vary across cultures, are influenced by language, and arise late in human development (as opposed to nonsymbolic approximates).

Both exact symbolic and approximated nonsymbolic numerical knowledge is most commonly measured by comparison tasks, in which participants are asked to identify the larger among two arrays of dots. One major signature of nonsymbolic core numerical representations that is present in human and nonhuman animals is that comparisons are subject to a ratio (minimum/maximum) limit: Accuracy decreases while reaction time (RT) increases as the ratio of the compared numbers approaches one (ie, the ratio effect; Cantlon et al., 2009). This ratio effect is the result of both the distance between two quantities and their location on the mental number line. Similarly, the larger the distance between two numbers to be compared the faster the response (ie, the distance effect; eg, Ansari, 2008). The distance effect was first reported by Moyer and Landauer (1967) and has since been replicated by many researchers under a variety of conditions (eg, Dehaene, 1989; Kucian, Loenneker, Martin, & von Aster, 2011). The numerical distance effect has been found in children (eg, Holloway & Ansari, 2009; Landerl & Kölle, 2009; Sekular & Mierkiewicz, 1997) and in primates (Nieder, Freedman, & Miller, 2002). Similarly, the ratio effect has been demonstrated in infants (Xu & Spelke, 2000; Xu, Spelke, & Goddard, 2005), in young children (Barth, La Mont, Lipton, & Spelke, 2005), and in animals (eg, Cantlon & Brannon, 2007; Hauser, Tsao, Garcia, & Spelke, 2003). Both of these signature effects (ratio and distance) are compromised in developmental dyscalculia (DD) (for the distance effect, see Heine et al., 2013; Mussolin et al., 2010; Price, Holloway, Vesterinen, Rasanen, & Ansari, 2007; Soltesz, Szucs, Dekany, Markus, & Csepe, 2007) (for the ratio effect, see Kovas et al., 2009; Libertus, Feigenson, & Halberda, 2011). Further research identified the neural tissues involved in ratio and distance effects. Specifically, the parietal lobes and in particular the intraparietal sulcus (IPS) serve the mental operations involved in these effects (eg, Castelli, Glaser, & Butterworth, 2006; Cohen Kadosh, Bien, & Sack, 2012; Cohen Kadosh, Cohen Kadosh, Kaas, Henik, & Goebel, 2007; Fias, Lammertyn, & Reynvoet, 2003; Piazza, Izard, Pinel, Le Bihan, & Dehaene, 2004).

This accumulated body of results has led to an accepted view of the approximate numerical system (numerosity) acting as an innate foundation of numerical knowledge (eg, Butterworth, 2010). Moreover, it has been suggested that DD involves a deficit in this foundation of knowledge (eg, Butterworth, 2010) (but see Kaufmann et al., 2013; Rubinsten & Henik, 2009).

However, several findings and methodological issues suggest that this wide agreement should be reexamined. (1) A major obstacle to the study of cognitive and neural correlates of numerical knowledge is encompassed in the difficulty of teasing apart, at the cognitive level of analysis, processes which are involved—to varying degrees—in both ordinal and quantity processes. Under normal conditions, order and quantity are inseparably bound. To be able to estimate which of two quantities is larger, one needs to know the position of the specific quantity in relation to other quantities, that is, ordinal information. Specifically, the ratio and distance effects, which are considered signatures of a phylogenetically ancient core quantitative system, are a result of the ability to assess ordinal relationships and not only quantities.

(2) Occasional reports show that the ability to process ordinality was found in primates (eg, Brannon, Cantlon, & Terrace, 2006; Cantlon & Brannon, 2006) and in other animal species phylogenetically distant from humans, such as rats (Suzuki & Kobayashi, 2000) and parrots (Pepperberg, 2006). Infants' ability to detect ordinality was found with size-based sequences (De Hevia & Spelke, 2010; Macchi Cassia, Picozzi, Girelli, & de Hevia, 2012), arbitrary sequences of different items (eg, Lewkowicz & Berent, 2009), and with quantities (eg, Brannon, 2002; De Hevia & Spelke, 2010; Feigenson, Carey, & Hauser, 2002; VanMarle, 2013) (Gevers, Reynvoet, & Fias, 2003). These findings provide clues to the existence of an evolutionary-based ordinal input analyzer. However, these findings are typically intepreted such that participants are able to actively compute ordinal information or to actively learn how to place stimuli based on their relative ordinal position and that this skill plays a role in numerical cognition. For example, it has been shown that nonhuman primates can learn to use ordering rules similar to those that emerge during human development (McGonigle & Chalmers, 2006).

In contrast to the argument that ordinal perception is learned during development, an alternative interpretation to previous findings and arguments may be suggested. That is, it may be argued that humans have a neurocognitive platform of order, which is termed here the "ordinal instinct." This ordinal instinct is at the heart of the preverbal ability to *implicitly*, *inattentively*, and *immediately* perceive ordinal relationships (as opposed to active, attentive, and exact computation) between any types of stimuli in the visual scene. In current opinion, numerosity (not order) is agreed to be at the heart of the numerical core system. In addition, previous research claimed that ordinality is at work in solving unique numerical problems such as counting (Gallistel & Gelman, 1992), but, to the best of our knowledge, the ability to process ordinal information is not currently recognized as a system that evolved under selection pressures for ordinal representations. Thus, the fact that at later stages of development ordinality served other purposes, including numerical representations, is not acknowledged by contemporary science. Despite the fact that much of our perceptual abilities, which are highly important for survival, these perceptual abilities are all based on lower-order geometric properties, such as ordinal relations. Examples include perceiving objects in three-dimensional space (Koenderink, van Doorn, Kappers, & Todd, 2002; Norman & Todd, 1998), face recognition (Gilad, Meng, & Sinha, 2009), and movement perception. That is, even highly complex tasks such as tossing an object to a target, could be performed simply based on an ordinal representation of depth and a corresponding ordinal representation of limb force (Todd & Norman, 2003). Moreover, possible neural constructs have been suggested for such an ordinal mechanism. Neurons in the early stages of the mammalian visual pathway provide a plausible substrate for extracting ordinal image relationships (Albrecht & Hamilton, 1982). In addition, Sawamura, Shima, and Tanji (2002) found single-cell neurons located in the superior lobule in the parietal cortex of monkeys that are selective for ordinal position in sequences of arm movements. Hence, it may be suggested that the evolution and development of brain systems critically involved in the perception of action, space, and numerical cognition (Walsh, 2003) depend on an initial ability to evaluate order. That is, it may be speculated and suggested as a theoretical framework that there are two separate coexisting systems: a developmentally older ordinal system and a separate numerosity system. It may be proposed that the neurocognitive system that manages the ordinal instinct might have had a major role in the development of the occipitoparietal dorsal brain system, known to be involved in the perception of action, spatial locations, and numbers. As mentioned earlier in this chapter, ordinal information is indeed highly linked with perceiving action (eg, Sawamura et al., 2002), space (eg, Gilad et al., 2009), and numbers. Accordingly, the ordinal instinct might be also involved with setting up neural and cognitive networks of the numerosity system.

Indeed several recent findings point to specific brain areas that are involved with abstract representation of ordinal information that is not necessarily number specific. Two recent papers (Fias, Lammertyn, Caessens, & Orban, 2007; Ischebeck et al., 2008) compared the neural bases of symbolic ordinality and numerical processing by using functional magnetic resonance imaging (fMRI). Fias et al. (2007) contrasted brain activation during symbolic numerical comparisons (eg, which comes first—3 or 9) with brain activation during comparisons of nonnumerical stimuli that carry ordinal information. Ischebeck et al. (2008) compared word generation of numbers, months, both denotes ordinal information, with word generation of animals which do not have ordinal meaning. These studies found that the anterior intraparietal sulcus [IPS; specifically, the horizontal segment of the IPS (hIPS) in Fias's study and in Ischebeck study, it was the hIPS2 which is the most anterior and lateral part of the IPS] responds equally to both numerical and nonnumerical ordinal information. Similarly, Kaufmann, Vogel, Starke, Kremser, and Schocke (2009) found that in children the processing of numerical (symbolic) and nonnumerical (sizes) ordinality is supported by the IPS. Collectively, these findings suggest that the IPS and, specifically, the anterior region of the IPS may be involved in the abstract representation of ordinal information that is not number specific. Hence, there are domain-general representations of ordinal information that are involved with any type of stimulus that embodies ordinal information such as numerical, magnitude, and alphabetical stimuli. However, Zorzi, Di Bono, and Fias (2011) used support vector machines to reanalyze the data of Fias et al. (2007), who found a clear dissociation between processing numerical versus alphabetical orders in bilateral horizontal IPS. These findings support previous neuropsychological studies with brain-damaged patients (eg, Delazer & Butterworth, 1997; Turconi & Seron, 2002), and suggest, in contrast to the Fias or Ischebeck arguments (Fias et al., 2007; Ischebeck et al., 2008) that ordinal and quantity processing dissociate at both the behavioral and biological levels.

Accordingly, scientific evidence is inconclusive; some evidence suggests that a single numerical magnitude system operates over both quantity and ordinality information, whereas other evidence shows signatures of separate cognitive systems of ordinality and quantity. One reason for the lack of clarity in previous results (eg, Fias et al., 2007 vs. Ischebeck et al., 2008) could be the use of different methodologies in the study of ordinality. I turn to further discuss this issue.

12.3 METHODOLOGICAL ISSUES IN STUDYING ORDINALITY

Ordinality is considered here as a segment of the visual scene at a given location that is perceived by an observer to have a beginning and an end of sequential relationships between countable or noncountable items. I call the process by which observers identify these beginnings and endings and their relationships—order perception, ordinality, or *the ordinal instinct*. This concept suggests that although ordinality is in the mind of the beholder, it is tied to visual scenes (Zacks & Tversky, 2001). Hence, effects of both environment, biology, and cognition should be investigated.

Beside quantities, digit symbols have their own ordinal features that may modulate their effect (eg, when they produce a conflict and are presented in a nonordered fashion). These ordinal features could be employed to ask questions about digit or quantitative representations.

We (eg, Rubinsten & Sury, 2011; see Fig. 12.1) and others (eg, Brannon & Terrace, 1998; Franklin, Jonides, & Smith, 2009; Kaufmann et al., 2009; Lyons & Beilock, 2011, 2013; Macchi Cassia et al., 2012; Merritt, MacLean, Jaffe, & Brannon, 2007; Terrace, Chen, & Jaswal, 1996) have been using similar situations to study ordinality. The basic paradigm presents participants with n (more than 2) stimuli (eg, groups of dots—Fig. 12.2; or Arabic numerals, shades of gray, size, etc.) in an ordered or nonordered form (eg, groups of dots—see Fig. 12.1; Arabic numeral, shades of gray, different

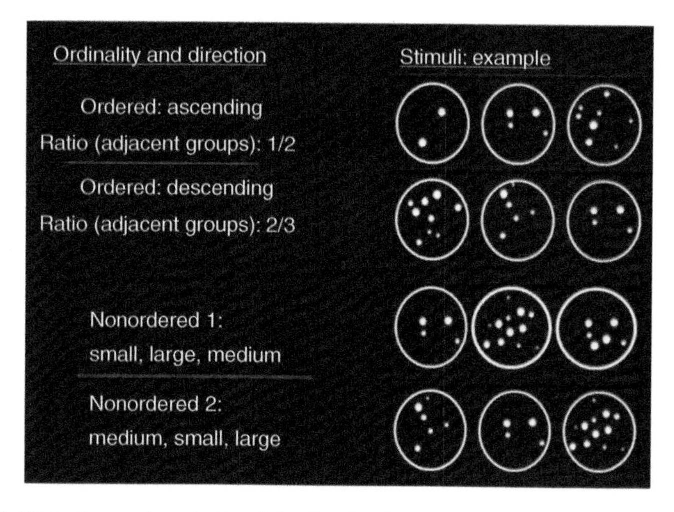

FIGURE 12.1 Examples of stimuli (each stimulus is represented by one *black rectangle* that includes three simultaneously presented *circles with dots*). The task is to decide whether the three circles include dots that are presented in an ordered or nonordered fashion (in the current example, based on quantity).

physical sizes of arbitrary stimuli, etc.). Participants' task is to indicate whether the stimuli are ordered or not.

Counting is not possible due to the brief presentation. Specifically, the ordinal instinct should be studied similar to investigations of numerical estimations. That is, estimation of numbers relates to the strategy employed when a stimulus configuration is comprised of a large number of items and is presented briefly (Pavese & Umiltà, 1998). It is an intuition

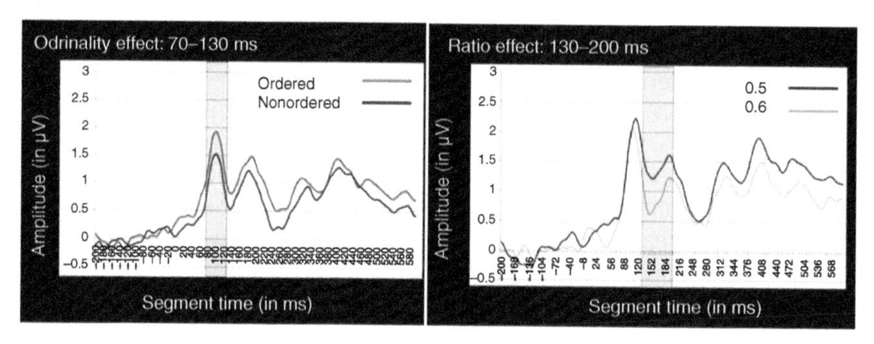

FIGURE 12.2 An ERP study showed that the P1 component, between 80 and 130 ms after stimulus onset (left), was modulated by ordinal information regardless of ratio. A later positive component (right), separated from the ordinal effect, ranging roughly from 130 and 200 ms, was sensitive to the "ratio" manipulation (larger amplitudes in response to smaller ratios as compared to larger ratios).

available to humans regardless of language and education; hence, numerical estimation is considered to be part of the core numerical system (Dehaene, 2009) that is innately available to humans and nonhuman animals (Cantlon et al., 2009). Therefore, to study the ordinal instinct as a core ordinal system, critical variables should include nonsymbolic and approximations of ordinality (as opposed to linguistic/symbolic counting).

In addition, in most of the studies in the field of ordinal processing, participants are presented with pairs of items (eg, numbers, letters, months, etc.) and are asked to decide whether these pairs are presented in an ascending or descending order (eg, Fias et al., 2007; Turconi, Campbell, & Seron, 2006) or to decide which one of the items appears before/after in a sequence (eg, Brannon & Terrace, 1998; Brannon & Van de Walle, 2001). In contrast, presenting three or more items as a single stimulus, and forcing participants to pay attention to all the numbers or quantities as a triad or more (eg, Franklin et al., 2009; Kaufmann et al., 2009; Rubinsten, Dana, Lavro, & Berger, 2013; Rubinsten & Sury, 2011) can better emphasize the ordinal relationship of a sequence. The study of ordinality can be compared to statistical methods for measuring and analyzing change (eg, change such as ascending, descending, ordered, nonordered). Such statistical methods for change include individual growth modeling (eg, Rogosa & Zimowski, 1982), hierarchical linear modeling (Bryk & Raudenbush, 1992), random coefficient regression (Hedeker, 2015), and multilevel modeling (Goldstein, 2011). These statistical methods require at least three points of measurement. This suggests that change (including order perception) cannot be viewed as the difference between "before" and "after" or "smaller" and "larger." That is, two points, which are typically used by modern science in investigations of ordinality, may not be sufficient to reveal the features of the trajectory. Hence, to better separate ordinal from quantity processing, the basic paradigm should present participants with n (but more than 2) stimuli (eg, groups of dots—Fig. 12.1) that are displayed briefly and simultaneously.

Why simultaneously? Order and timing are dissociate (as was shown, eg, by Schubotz & von Cramon, 2001). For example, it had been shown that timing and ordering are closely related but act as independent processes in the programming of movement (eg, MacKay, 1985). Also, when presenting stimuli as a dynamic sequence along a timeline, additional cognitive processes are being imposed by the task structure, such as short-term memory (or working memory). Short-term memory is capacity limited, and encoding into explicit short-term memory requires attention (Rensink, 2000; Sperling, 1960). Hence, in the investigation of ordinality, when presenting variables in a dynamic presentation, there is a need to actively initiate order encoding (Holcombe, Treisman, & Kanwisher, 2001; Potter, 1976) (as opposed to implicit nonattentive approximation in ordinal instinct). Indeed, in the numerical cognition field, it had been shown that when the elements of a set are presented one after the other, they are enumerated consecutively (Cordes, Gelman, Gallistel, & Whalen, 2001), a process that requires symbolic counting, which is based on learned number symbols. In contrast, when presented simultaneously as in multiple-item patterns, numerosity can be estimated at a single glance (eg, Barth, Kanwisher, & Spelke, 2003) and show equal numbers of scanning eye movements (Nieder & Miller, 2004).

Examples for studies that used such a methodological paradigm have been given by Brannon and Terrace (1998) who presented four items of different sizes and shapes to monkeys; Terrace et al. (1996), who presented three-colored fields of light, achromatic geometric shapes, and lines' orientation to pigeons; Merritt et al. (2007) who used three and four digitized color photographs to *Lemur catta*; Kaufmann et al. (2009) who presented three Arabic numerals or symbols varying in font sizes to children with dyscalculia; Lyons & Beilock (2013) who presented three groups of dots, Arabic numerals, and luminance to intact developing university students and three Arabic numerals in university students (Lyons & Beilock, 2011); Macchi Cassia et al., 2012, who presented three sizes to 4-month-old infants; also by Rubinsten and Sury who presented three groups of dots to university students with dyscalculia (Rubinsten & Sury, 2011) and Franklin et al. (2009), who used three double-digit numbers or names of months.

In addition, it should be noted that controlling or manipulating multiple visual parameters is not an easy task since there is a considerable correlation between numerosity and many other different visual cues (eg, Gebuis & Reynvoet, 2012; Leibovich & Henik, 2013). Accordingly, density and aggregate surface (area) are typically being controlled within the previously mentioned methodology, but this aspect (ie, visual cues) should be further investigated. Specifically, multiple visual parameters should be better controlled (ie, by being kept constant or randomized) or manipulate simultaneously. This may allow investigators to study whether participants process ordinal information of the visual cues even though they are uninformative about numerosity. In such case, it may be concluded that ordinality acts, as an instinct that enables implicit analysis of the visual scene.

Finally, and as can be seen in Fig. 12.1, to dissociate ordinality from numerosity, we systematically manipulate the ratio between adjacent stimuli. We have shown that such a methodology (eg, Fig. 12.1) is capable of teasing apart numerosity and ordinality, at both the cognitive and biological levels (Rubinsten et al., 2013). That is, in an event-related potentials (ERP) experiment, we (Rubinsten et al., 2013) found that regardless of the ratio between the three groups of dots, the amplitude of the ERP wave, between 80 and 130 ms after stimulus onset (known as the P1 component), was modulated by the ordinal information (ie, ordered vs. nonordered stimuli) and showed a restricted posterior occipitoparietal distribution. Ratio effect, on the other hand, began and ended later. That is, a later positive component, separated from the ordinal effect, ranging roughly from 130 to 200 ms, was sensitive to the "ratio" manipulation: it showed larger amplitudes in response to smaller ratio between the three circles, compare to larger ratios.

To note, the P1 component has been associated primarily as an early sensory/perceptual responses. It also should be noted that the current ordinal effect has been found to resist extensive manipulation of the nonnumerical parameters of the display, thus evading simple explanations in terms of density, or area. That is, no visual analysis (eg, density or area) was possible other than ordinal information. Accordingly, and taking together the behavioral as well as the ERP data, it was argued that P1 component is also associated with an early visual mechanism dedicated to analyzing ordinal information and providing a sensory representation of order to higher-level perceptual systems (eg, systems that process quantity or direction). Hence, it may be suggested that ordinal relationships are perceived earlier than numerosity, independent of the ratio between quantities. In addition, ordinality is processed very early after stimulus presentation, too early to be attentively or explicitly processed, supporting the claim that ordinality is implicitly perceived (Fig. 12.2).

Yet, to best capture the neural and cognitive signature of the ordinal instinct, within the basic paradigm described earlier (Fig. 12.1), the issue of nonsymbolic versus symbolic presentations should also be taken into account. I elaborate later.

12.4 SYMBOLIC VERSUS NONSYMBOLIC REPRESENTATION

Symbolic numerical representations are distinct, accurate, verbally based, culturally dependent, and require explicit learning (eg, Arabic numerals such as "6" or number words such as "SIX").

Specifically, very recent evidence, show that active symbolic number-ordering acts as an important factor in the aquisition of symbolic numerals (Lyons & Beilock, 2009). For example, children's performance in a symbolic ordering task, in which they are asked to decide if three written Arabic numerals are in increasing order or not, better predicts their symbolic arithmetic performance in grades 1 to 6, than number comparison (Lyons, Price, Vaessen, Blomert, & Ansari, 2014). Such findings support the role of active symbolic ordering in the acquisition of symbolic numerals.

In contrast, the ordinal instinct targets the nonsymbolic, preverbal, and approximate aspect of ordinality (ordinal instinct) as the basis of higher-order mathematics. Even Charles Ransom Gallistel and Rochel Gelman, the influential pioneers in the field of numerical neurocognition, argued that "The preverbal process for comparing magnitudes (preverbal ordination) renders the ordering of the verbal numbers intelligible" (Gallistel & Gelman, 1992).

It is clear that education enables the shift from an informal approximate representations of ordinality to a formal, symbolic, exact representation of ordinal relationships (eg, of Arabic numerals in the counting system; eg, Lyons & Beilock, 2009). Lyons and Beilock (2009) showed that in learning settings, working memory capacity is highly important for linking the ordinal representations of symbolic numerical with nonsymbolic numerical information. Specifically, in their first experiment, Lyons and Beilock (2009) trained adult participants to associate dot quantities with novel symbols and tested, among other things, the ordinal perception of the novel symbols. Findings showed that participants who were higher in working memory capacity learned ordinal information about the symbols that lower working memory individuals did not. In addition, different languages, which are symbolic by definition, have an important role. Accordingly, the direction of the writing system, for example, may have a significant impact on the processing of ordinal information. Indeed, we (eg, Hochman Cohen, Berger, Rubinsten, & Henik, 2014; Prior, Katz, Mahajna, & Rubinsten, 2015), have shown significant differences between Hebrew and Arabic native speakers in processing numerical information (not ordinality) due to the different directions of the numerical writing systems (ie, in contrast to Arabic numerals, Indian numerals, used by many Arabic-speaking peoples, are written from right to left). In addition, we have reported (Sury, Rubinsten, De Visscher, & Noël, submitted) a cross-linguistic experiment of ordinal processing exploring the relationship between ordinal and numerical information. Two groups of Hebrew-speaking Israeli and French-speaking Belgian children (6–9-year-olds) completed ordinal judging tasks with quantities (ie, groups of dots) while the direction of the sequence (ascending/descending) was manipulated. Findings indicate that both groups estimate order from 6 years of age (as indicated by the ordinal effect, which is measured by differences in performances to ordered vs. nonordered stimuli). However, responses to ordinal sequences were modulated by direction; Hebrew-speaking children processed ordinal descending direction more accurately (compatible with the right-to-left Hebrew-writing system). In contrast, French-speaking children were more accurate in processing the ascending sequences. That is, the ability to estimate order (ie, the ordinal instinct) exists from an early age in both languages, but linguistic information (eg, direction of the writing system) modulates the development of the ability to estimate ordinal information.

To note, the aforementioned findings, about differences and similarities between ordinal and numerical perception, are all related to intact developing populations. Accordingly and as will be discussed in what follows, it may be very interesting to ask questions about developmental patterns of ordinality and numerosity of cases with numerical deficiencies (DD) and math anxiety.

12.5 ORDINAL INSTINCT AND DEVELOPMENTAL DYSCALCULIA

There are several different lines of evidence for the biological and cognitive precursors of DD. Specifically, specialized neural circuits for numerical processing have been identified in the parietal lobes of the brain, especially the left and right intraparietal sulci (Cohen Kadosh et al., 2007b) and the left angular gyrus (Ansari, 2008; Fias & Fischer, 2005). People with DD show strikingly poor performance on very simple tasks such as difficulty in retrieving arithmetical facts, using arithmetical procedures (Shalev & Gross-Tsur, 2001), and the use of immature problem-solving strategies, even in adulthood (eg, using finger counting—Jordan, Hanich, & Kaplan, 2003). DD is also reflected in deficiencies in basic numerical functions such as comparing nonsymbolic numerical quantities (eg, Halberda, Ly, Wilmer, Naiman, & Germine, 2012; Piazza et al., 2010), processing numbers symbolically (eg, in Arabic notation—Stock, Desoete, & Roeyers, 2010), recognizing patterns and subitizing (Ashkenazi, Mark-Zigdon, & Henik, 2013) and visuospatial working memory (for review, see our paper Kaufmann et al., 2013; Rotzer et al., 2009). For example, Price et al. (2007) found a weak IPS activation in DD children compared to controls when comparing nonsymbolic numerical stimuli. Moreover, Landerl and Kölle (2009) found that 8–10-year-old DD children were slower than controls in both symbolic and nonsymbolic number comparisons. Mussolin et al. (2010) found that 10- and 11-year-old children with DD showed larger distance effect in both symbolic and nonsymbolic numerical comparisons. Also, using the event-related potentials (ERPs) methodology, Soltesz et al. (2007) found that, compared to controls, adolescents with DD show no late event-related brain potentials (ERPs) distance effect between 400 and 440 ms on right parietal electrodes when comparing Arabic numerals. Such findings suggest that the processing of numerical information and representations of the distances between them is abnormal in DD.

Collectively, the behavioral and electrophysiological evidence points to a deficient mental number line, in which quantities are spatially mapped based on distances, as one of the possible outcomes or sources of DD. However, all of these cognitive skills may be also significantly linked with ordinal processing. Accordingly, and as previously suggested, it is possible that the precursor of DD (Rubinsten & Henik, 2009) is a difficulty in the processing and evaluation of ordinal relationships (the ordinal instinct), which may lead to deficiencies in higher mathematical abilities. Indeed, we (Rubinsten & Sury, 2011) showed that adult DD participants exhibit deficiencies in processing ordinal information. Specifically, DD participants had a normal ratio effect in the nonsymbolic ordinal task (Fig. 12.1), regardless of the perceptual condition (ie, constant area or constant density conditions but also in randomized area and density condition). In addition, in this nonsymbolic task, DD participants did not show the ordinality effect in constant area or in density conditions. However, when both cues (ie, area and density) were not controlled (ie, randomized), DD participants showed the ordinality effect, but only in the descending direction. In the symbolic task, the ordinality effect was modulated by ratio and direction, which are more natural to symbolic representations. These findings suggest DD may be deficient in their ability to process ordinal information but also that there might be two separate cognitive representations of ordinal and quantity information and that linguistic abilities may facilitate estimation of ordinal information.

12.6 CONCLUSIONS

The suggested ordinal instinct may have a number of implications for the study of ordinality and numerical cognition. First, it may be suggested that quantities and ordinal information, at least at a certain stage of cognitive processing, are distinct and are not necessarily "two sides of the same coin" (Jacob & Neider, 2008). Second, if human beings are indeed able to estimate order as part of their core cognitive system, it will mean that among some, such as those with developmental dyscalculia, this ability might be deficient. In such a case, those people are partially "blind" to ordinal information and not only to quantity. This hypothesis needs to be investigated because (1) it will help uncover the neurocognitive mechanisms that are basic to DD, and (2) it will shed light on the important issue of how numerical cognition, and more specifically the ordinal instinct, can be promoted by classroom practice in typically as well as atypically developing children.

References

Albrecht, D. G., & Hamilton, D. B. (1982). Striate cortex of monkey and cat: contrast response function. *Journal of Neurophysiology, 48*(1), 217–237.

Ansari, D. (2008). Effects of development and enculturation on number representation in the brain. *Nature Reviews Neuroscience, 9*, 278–291.

Ashkenazi, S., Mark-Zigdon, N., & Henik, A. (2013). Do subitizing deficits in developmental dyscalculia involve pattern recognition weakness? *Developmental Science, 16*(1), 35–46.

Barth, H., Kanwisher, N., & Spelke, E. (2003). The construction of large number representations in adults. *Cognition, 86*(3), 201–221.

Barth, H., La Mont, K., Lipton, J., & Spelke, E. S. (2005). Abstract number and arithmetic in preschool children. *Proceedings of the National Academy of Sciences USA, 102*(39), 14116–14121.

Brannon, E. M. (2002). The development of ordinal knowledge in infancy. *Cognition, 836*, 223–240.

Brannon, E. M., & Terrace, H. S. (1998). Ordering of the numerosities 1 to 9 by monkeys. *Science, 282*(5389), 746–749.

Brannon, E. M., & Van de Walle, G. A. (2001). The development of ordinal numerical competence in young children. *Cognitive Psychology, 43*, 53–81.

Brannon, E. M., Cantlon, J. F., & Terrace, H. S. (2006). The role of reference points in ordinal numerical comparisons by Rhesus macaques (*Macaca mulatta*). *Journal of Experimental Psychology: Animal Behavior Processes, 32*, 120–134.

Bryk, A. S., & Raudenbush, S. W. (1992). *Hierarchical linear models in social and behavioral research: applications and data analysis methods*. Newbury Park, CA: Sage.

Burnet, J. (1908). *Early greek philosophy* (2nd ed.). London: A. & C. Black.

Butterworth, B. (2010). Foundational numerical capacities and the origins of dyscalculia. *Trends in Cognitive Sciences, 14*, 534–541.

Cantlon, J. F., & Brannon, E. M. (2006). Shared system for ordering small and large numbers in monkeys and humans. *Psychological Science, 17*, 401–406.

Cantlon, J. F., & Brannon, E. M. (2007). How much does number matter to a monkey (*Macaca mulatta*)? *Journal of Experimental Psychology: Animal Behavior Processes, 33*(1), 32.

Cantlon, J. F., Platt, M. L., & Brannon, E. M. (2009). Beyond the number domain. *Trends in Cognitive Sciences, 13*, 83–91.

Castelli, F., Glaser, D. E., & Butterworth, B. (2006). Discrete and analogue quantity processing in the parietal lobe: a functional MRI study. *Proceedings of the National Academy of Sciences USA, 103*(12), 4693–4698.

Cohen Kadosh, R., Cohen Kadosh, C., Kaas, A., Henik, A., & Goebel, R. (2007a). Notation-dependent and -independent representations of numbers in the parietal lobes. *Neuron, 53*, 307–314.

Cohen Kadosh, R., Cohen Kadosh, K., Schuhmann, T., Kaas, A., Goebel, R., Henik, A., & Sack, A. T. (2007b). Virtual dyscalculia after TMS to the right parietal lobe: a combined fMRI and neuronavigated TMS study. *Current Biology, 17*, 689–693.

Cohen Kadosh, R., Bien, N., & Sack, A. T. (2012). Automatic and intentional number processing both rely on intact right parietal cortex: a combined fMRI and neuronavigated TMS study. *Frontiers in Human Neuroscience, 6*, 2.

Cordes, S., Gelman, R., Gallistel, C. R., & Whalen, J. (2001). Variability signatures distinguish verbal from nonverbal counting for both large and small numbers. *Psychonomic Bulletin & Review, 8*(4), 698–707.

De Hevia, M. D., & Spelke, E. S. (2010). Number-space mapping in human infants. *Psychological Science, 21*(5), 653–660.

Dehaene, S. (1989). The psychophysics of numerical comparison: a reexamination of apparently incompatible data. *Perception & Psychophysics, 45*, 557–566.

Dehaene, S. (2009). Origins of mathematical intuitions: the case of arithmetic. *Annals of the New York Academy, 1156*, 232–259.

Delazer, M., & Butterworth, B. (1997). A dissociation of number meanings. *Cognitive Neuropsychology, 14*, 613–636.

Estrada, C. A., Martin-Hryniewicz, M., Peek, B. T., Collins, C., & Byrd, J. C. (2004). Literacy and numeracy skills and anticoagulation control. *The American Journal of the Medical Sciences, 328*, 88–93.

Feigenson, L., Carey, S., & Hauser, M. (2002). The representations underlying infants' choice of more: object files versus analog magnitudes. *Psychological Science, 13*(2), 150–156.

Feigenson, L., Dehaene, S., & Spelke, E. S. (2004). Core systems of number. *Trend in Cognitive Science, 8*, 307–314.

Fias, W., & Fischer, M. H. (2005). Spatial representation of numbers. In J. I. D. Campbell (Ed.), *Handbook of mathematical cognition* (pp. 43–54). New York: Psychology Press.

Fias, W., Lammertyn, J., & Reynvoet, B. (2003). Parietal representation of symbolic and non-symbolic magnitude. *Journal of Cognitive Neuroscience, 15*, 47–56.

Fias, W., Lammertyn, J., Caessens, B., & Orban, G. A. (2007). Processing of abstract ordinal knowledge in the horizontal segment of the intraparietal sulcus. *Journal of Neuroscience, 27*, 8952–8956.

Franklin, M. S., Jonides, J., & Smith, E. E. (2009). Processing of order information for numbers and months. *Memory & Cognition, 37*, 644–654.

Gallistel, C. R., & Gelman, R. (1992). Preverbal and verbal counting and computation. *Cognition, 44*, 43–74.

Gebuis, T., & Reynvoet, B. (2012). The interplay between nonsymbolic number and its continuous visual properties. *Journal of Experimental Psychology: General, 141*(4), 642.

Gevers, W., Reynvoet, B., & Fias, W. (2003). The mental representation of ordinal sequences is spatially organized. *Cognition, 87*(3), B87–B95.

Gilad, S., Meng, M., & Sinha, P. (2009). Role of ordinal contrast relationships in face encoding. *Proceedings of the National Academy of Sciences, 106*(13), 5353–5358.

Goldstein, H. (2011). *Multilevel statistical models* (Vol. 922). Chichester: John Wiley & Sons.

Halberda, J., Ly, R., Wilmer, J. B., Naiman, D. Q., & Germine, L. (2012). Number sense across the lifespan as revealed by a massive Internet-based sample. *Proceedings of the National Academy of Sciences, 109*(28), 11116–11120.

Hauser, M. D., Tsao, F., Garcia, P., & Spelke, E. S. (2003). Evolutionary foundations of number: spontaneous representation of numerical magnitudes by cotton-top tamarins. *Proceedings of the Royal Society of London Series B, 270*(1523), 1441–1446.

Hedeker, D. (2015). Methods for multilevel ordinal data in prevention research. *Prevention Science, 16*(7), 997–1006.

Heine, A., Wissmann, J., Tamm, S., De Smedt, B., Schneider, M., Stern, E., & Jacobs, A. M. (2013). An electrophysiological investigation of non-symbolic magnitude processing: numerical distance effects in children with and without mathematical learning disabilities. *Cortex, 49*(8), 2162–2177.

Hochman Cohen, H., Berger, A., Rubinsten, O., & Henik, A. (2014). Does the learning of two symbolic sets of numbers affect the automaticity of number processing in children? *Journal of Experimental Child Psychology, 121*, 96–110.

Holcombe, A. O., Treisman, A., & Kanwisher, N. (2001). The midstream order deficit. *Perception & Psychophysics, 63*(2), 322–329.

Holloway, I. D., & Ansari, D. (2009). Mapping numerical magnitudes onto symbols: the numerical distance effect and individual differences in children's mathematics achievement. *Journal of Experimental Child Psychology, 103*(1), 17–29.

Ischebeck, A., Heim, S., Siedentopf, C., Zamarian, L., Schocke, M., Kremser, C., & Delazer, M. (2008). Are numbers special? Comparing the generation of verbal materials from ordered categories (months) to numbers and other categories (animals) in an fMRI study. *Human Brain Mapping, 29*, 894–909.

Izard, V., Sann, C., Spelke, E. S., & Streri, A. (2009). Newborn infants perceive abstract numbers. *Proceedings of the National Academy of Sciences USA, 106*, 10382–10385.

Jacob, N. S., & Neider, A. (2008). The ABC of cardinal and ordinal number representations. *Trends in Cognitive Sciences, 12*, 41–43.

Jordan, N., Hanich, L. B., & Kaplan, D. (2003). A longitudinal study of mathematical competencies in children with specific mathematics difficulties versus children with co-morbid mathematics and reading difficulties. *Child Development, 74*(3), 834–850.

Kant, I. (1908). *Critique of pure reason* (N. K. Smith, Trans.). London: Macmillan. (Original work published 1781)

Kaufmann, L., Vogel, S., Starke, M., Kremser, C., & Schocke, M. (2009). Numerical and non-numerical ordinality processing in children with and without developmental dyscalculia: evidence from fMRI. *Cognitive Development, 24*(4), 486–494.

Kaufmann, L., Mazzocco, M. M., Dowker, A., von Aster, M., Göbel, S. M., Grabner, R. H., & Hans-Christoph, N. (2013). Dyscalculia from a developmental and differential perspective. *Frontiers in Psychology, 4*, 516.

Koenderink, J. J., van Doorn, A. J., Kappers, A. M., & Todd, J. T. (2002). Pappus in optical space. *Perception & Psychophysics, 64*(3), 380–391.

Kovas, Y., Giampietro, V., Viding, E., Ng, V., Brammer, M., Barker, G. J., & Plomin, R. (2009). Brain correlates of non-symbolic numerosity estimation in low and high mathematical ability children. *PLoS One, 4*(2), e4587.

Kucian, K., Loenneker, T., Martin, E., & von Aster, M. (2011). Non-symbolic numerical distance effect in children with and without developmental dyscalculia: a parametric fMRI study. *Developmental Neuropsychology, 36*(6), 741–762.

Landerl, K., & Kölle, C. (2009). Typical and atypical development of basic numerical skills in elementary school. *Journal of Experimental Child Psychology, 103*, 546–565.

Leibovich, T., & Henik, A. (2013). Magnitude processing in non-symbolic stimuli. *Frontiers in Psychology, 4*, 375.

Lewkowicz, D. J., & Berent, I. (2009). Sequence learning in 4-month-old infants: do infants represent ordinal information? *Child Development, 80*(6), 1811–1823.

Libertus, M. E., Feigenson, L., & Halberda, J. (2011). Preschool acuity of the approximate number system correlates with school math ability. *Developmental Science, 14*(6), 1292–1300.

Lyons, I. M., & Beilock, S. L. (2009). Beyond quantity: individual differences in working memory and the ordinal understanding of numerical symbols. *Cognition, 113*(2), 189–204.

Lyons, I. M., & Beilock, S. L. (2011). Numerical ordering ability mediates the relation between number-sense and arithmetic competence. *Cognition, 121*(2), 256–261.

Lyons, I. M., & Beilock, S. L. (2013). Ordinality and the nature of symbolic numbers. *The Journal of Neuroscience, 33*(43), 17052–17061.

Lyons, I. M., Price, G. R., Vaessen, A., Blomert, L., & Ansari, D. (2014). Numerical predictors of arithmetic success in grades 1–6. *Developmental Science, 17*(5), 714–726.

Macchi Cassia, V., Picozzi, M., Girelli, L., & de Hevia, M. D. (2012). Increasing magnitude counts more: asymmetrical processing of ordinality in 4-month-old infants. *Cognition, 124*(2), 183–193.

MacKay, D. G. (1985). A theory of the representation, organization and timing of action with implications for sequencing disorders. *Advances in psychology, 23*, 267–308.

McGonigle, B., & Chalmers, M. (2006). Ordering and executive functioning as a window on the evolution and development of cognitive systems. *International Journal of Comparative Psychology, 19*(2), .

Merritt, D., MacLean, E. L., Jaffe, S., & Brannon, E. M. (2007). A comparative analysis of serial ordering in ring-tailed lemurs (*Lemur catta*). *Journal of Comparative Psychology, 121*(4), 363.

Moyer, R. S., & Landauer, T. K. (1967). Time required for judgement of numerical inequality. *Nature, 215*, 1519–1520.

Mussolin, C., Mejias, S., & Noël, M. (2010). Symbolic and nonsymbolic number comparison in children with and without dyscalculia. *Cognition, 115*, 10–25.

Nieder, A., & Miller, E. K. (2004). Analog numerical representations in rhesus monkeys: evidence for parallel processing. *Journal of Cognitive Neuroscience, 16*(5), 889–901.

Nieder, A., Freedman, D. J., & Miller, E. K. (2002). Representation of the quantity of visual items in the primate prefrontal cortex. *Science, 297*(5587), 1708–1711.

Norman, J. F., & Todd, J. T. (1998). Stereoscopic discrimination of interval and ordinal depth relations on smooth surfaces and in empty space. *Perception, 27*(3), 257–272.

Parsons, S., & Bynner, J. (2005). *Does numeracy matter more?* London: National Research and Development Centre for adult literacy and numeracy.

Pavese, A., & Umiltà, C. (1998). Symbolic distance between numerosity and identity modulates Stroop interference. *Journal of Experimental Psychology, 24*, 1535–1545.

Pepperberg, I. M. (2006). Cognitive and communicative abilities of Grey parrots. *Applied Animal Behaviour Science, 100*(1), 77–86.
Piazza, M., Izard, V., Pinel, P., Le Bihan, D., & Dehaene, S. (2004). Tuning curves for approximate numerosity in the human intraparietal sulcus. *Neuron, 44*(3), 547–555.
Piazza, M., Facoetti, A., Trussardi, A. N., Berteletti, I., Conte, S., Lucangeli, D., & Zorzi, M. (2010). Developmental trajectory of number acuity reveals a severe impairment in developmental dyscalculia. *Cognition, 116*, 33–41.
Piffer, L., Petrazzini, M. E. M., & Agrillo, C. (2013). Large number discrimination in newborn fish. *PLoS One, 8*(4), e62466.
Potter, M. C. (1976). Short-term conceptual memory for pictures. *Journal of Experimental Psychology, 2*(5), 509.
Price, G. R., Holloway, I., Vesterinen, M., Rasanen, P., & Ansari, D. (2007). Impaired parietal magnitude processing in developmental dyscalculia. *Current Biology, 17*, R1042–R1043.
Prior, A., Katz, M., Mahajna, I., & Rubinsten, O. (2015). Number word structure in first and second language influences arithmetic skills. *Frontiers in Psychology, 6*.
Rensink, R. A. (2000). The dynamic representation of scenes. *Visual Cognition, 7*(1–3), 17–42.
Reyna, V. F., Nelson, W. L., Han, P. K., & Dieckmann, N. F. (2009). How numeracy influences risk comprehension and medical decision making. *Psychological Bulletin, 135*(6), 943.
Rogosa, D. R. D. B., & Zimowski, M. (1982). A growth curve approach to the measurement of change. *Psychological Bulletin, 90*, 726–748.
Rotzer, S., Loenneker, T., Kucian, K., Martin, E., Klaver, P., & von Aster, M. (2009). Dysfunctional neural network of spatial working memory contributes to developmental dyscalculia. *Neuropsychologia, 47*, 2859–2865.
Rubinsten, O., & Henik, A. (2009). Developmental dyscalculia: heterogeneity may not mean different mechanisms. *Trends in Cognitive Sciences, 13*, 92–99.
Rubinsten, O., & Sury, D. (2011). Processing ordinality and quantity: the case of developmental dyscalculia. *PLoS One, 6*, e24079.
Rubinsten, O., Dana, S., Lavro, D., & Berger, A. (2013). Processing ordinality and quantity: ERP evidence of separate mechanisms. *Brain and Cognition, 82*(2), 201–212.
Sawamura, H., Shima, K., & Tanji, J. (2002). Numerical representation for action in the parietal cortex of the monkey. *Nature, 415*(6874), 918–922.
Schubotz, R. I., & von Cramon, D. Y. (2001). Interval and ordinal properties of sequences are associated with distinct premotor areas. *Cerebral Cortex, 11*(3), 210–222.
Sekular, R., & Mierkiewicz, D. (1997). Children's judgement of numerical inequality. *Child Development, 48*, 630–633.
Shalev, R. S., & Gross-Tsur, V. (2001). Developmental dyscalculia. *Pediatric Neurology, 24*(5), 337–342.
Soltesz, F., Szucs, D., Dekany, J., Markus, A., & Csepe, V. (2007). A combined event-related potential and neuropsychological investigation of developmental dyscalculia. *Neuroscience Letters, 417*, 181–186.
Sperling, G. (1960). The information available in brief visual presentations. *Psychological Monographs, 74*(11), 1.
Stock, P., Desoete, A., & Roeyers, H. (2010). Detecting children with arithmetic disabilities from kindergarten: evidence from a 3-year longitudinal study on the role of preparatory arithmetic abilities. *Journal of Learning Disabilities, 43*(3), 250–268.
Sury, D., Rubinsten, O., De Visscher, A., & Noël, M.-P. (submitted). The effect of language on numerical processing in six to nine years old children: a cross-cultural study.
Suzuki, K., & Kobayashi, T. (2000). Numerical competence in rats (*Rattus norvegicus*): Davis and Bradford (1986) extended. *Journal of Comparative Psychology, 114*(1), 73–85.
Szűcs, D., & Goswami, U. (2013). Developmental dyscalculia: fresh perspectives. *Trends in Neuroscience and Education, 2*(2), 33–37.
Terrace, H., Chen, S., & Jaswal, V. (1996). Recall of three-item sequences by pigeons. *Animal Learning & Behavior, 24*(2), 193–205.

Todd, J. T., & Norman, J. F. (2003). The visual perception of 3-D shape from multiple cues: are observers capable of perceiving metric structure? *Perception & Psychophysics*, *65*(1), 31–47.
Turconi, E., & Seron, X. (2002). Dissociation between order and quantity meaning in a patient with Gerstmann syndrome. *Cortex*, *38*, 911–914.
Turconi, E., Campbell, J. I. D., & Seron, X. (2006). Numerical order and quantity processing in number comparison. *Cognition*, *98*, 273–285.
VanMarle, K. (2013). Infants use different mechanisms to make small and large number ordinal judgments. *Journal of Experimental Child Psychology*, *114*(1), 102–110.
Walsh, V. N. (2003). A theory of magnitude: common cortical metrics of time, space and quantity. *Trends in Cognitive Sciences*, *7*, 483–488.
Xu, F., & Spelke, E. S. (2000). Large number discrimination in 6-month-old infants. *Cognition*, *74*(1), B1–B11.
Xu, F., Spelke, E. S., & Goddard, S. (2005). Number sense in human infants. *Developmental Science*, *8*(1), 88–101.
Zacks, J. M., & Tversky, B. (2001). Event structure in perception and conception. *Psychological Bulletin*, *127*(1), 3.
Zorzi, M., Di Bono, M. G., & Fias, W. (2011). Distinct representations of numerical and non-numerical order in the human intraparietal sulcus revealed by multivariate pattern recognition. *Neuroimage*, *56*(2), 674–680.

CHAPTER

13

Discrete and Continuous Presentation of Quantities in Science and Mathematics Education

Ruth Stavy, Reuven Babai

Department of Mathematics, Science and Technology Education, The Constantiner School of Education and The Sagol School of Neuroscience, Tel Aviv University, Tel Aviv, Israel

OUTLINE

13.1 INTRODUCTION

Through our senses we are able to estimate and compare discrete and continuous quantities. Discrete quantities include nonsymbolic numerosities, and continuous quantities encompass length, area, volume, weight, temperature, and so forth. This ability to estimate and compare

Continuous Issues in Numerical Cognition. http://dx.doi.org/10.1016/B978-0-12-801637-4.00013-5

quantities is limited by our senses and cognitive system. In order to overcome this limitation, humankind developed the formal number system for exact measurement, comparison, and calculation of discrete quantities and the measuring units that allow exact measurement (each with its own measurement error), comparison, and calculations for continuous quantities. The measuring units enable one to treat continuous quantities as discrete ones, using the number system for measuring, comparing, and calculating them.

Many quantities studied in school science and mathematics, such as length, area, volume, weight, and temperature, are continuous. It is well known that many students encounter difficulties in dealing with such quantities (Erickson, 1979; Driver, 1989). In our view these difficulties could stem from interference of salient irrelevant variables (automatically processed) with formal/logical reasoning about the relevant variable. It appears that certain variables of the task are so salient that they are automatically processed and thus may interfere with correct reasoning (a detailed description of salience can be found in Babai, Nattiv, & Stavy, 2016). This interference is reflected in students' erroneous responses to numerous tasks in science and mathematics education, even when students have the knowledge and skills to solve these tasks correctly (Babai & Stavy, 2015; Stavy & Tirosh, 2000; Strauss & Stavy, 1982).

In this chapter we will focus on the question whether changing the mode of presentation of continuous and discrete quantities will affect students' performance, for example, their ability to overcome the interference and correctly compare them.

We examine this question through three studies. Study 1 focuses on a comparison of perimeters task in discrete and continuous modes of presentation. Study 2 focuses on a comparison of ratios task (color intensity of mixtures of red and white paint quantities) in discrete and continuous modes of presentation. Study 3 focuses on comparison of area and number in discrete and continuous modes of presentation.

13.2 STUDY 1: COMPARISON OF PERIMETERS

This study focuses on a well-known difficulty that students of different ages experience in comparing perimeters of geometrical shapes. Many students claim that shapes with a larger area have a larger perimeter. It has been suggested that students' difficulties in this task stem from the interference caused by the salient, irrelevant variable, area, when comparing the shapes' perimeters (D'Amore & Fandiño Pinilla, 2006; Marchett, Medici, Vighi, & Zaccomer, 2005; Stavy & Tirosh, 2000).

In previous studies, participants were asked to compare the perimeters of two geometrical shapes (ie, to decide whether the perimeter of the

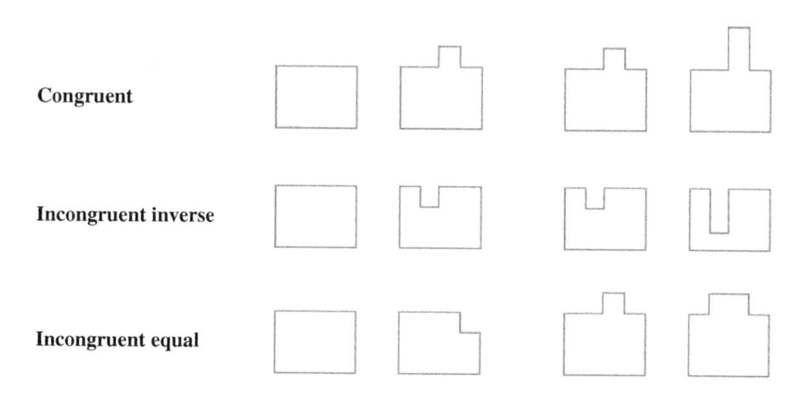

FIGURE 13.1 **Examples of congruent and incongruent comparison of perimeters task conditions.**

left/right shape was larger or both perimeters were equal) in congruent and incongruent conditions (Fig. 13.1).

1. Congruent—in this case, there is no interference of the salient, irrelevant variable, area, as one shape has a larger area and a larger perimeter than does the other shape.
2. Incongruent—in this case, there is interference of the salient, irrelevant variable, area, with perimeter comparison, as one shape has a larger area but a smaller perimeter (incongruent inverse) than does the other shape, or an equal perimeter (incongruent equal).

Among schoolchildren, adolescents, and adults, it has been consistently found that in the congruent condition accuracy was higher and reaction time for correct responses was faster than in the incongruent conditions. When participants were asked to compare the areas of the shapes, almost all of the responses were correct and relatively quick (faster than for perimeter comparison) in all conditions. These findings suggest that area is indeed the salient variable in this task and that participants have difficulty in ignoring it when comparing perimeters (Babai, Levyadun, Stavy, & Tirosh, 2006; Babai, Zilber, Stavy, & Tirosh, 2010; Stavy & Babai, 2008, 2010). These studies suggested that when processing the salient variable, area, and processing the variable, perimeter, result in the same conclusion (congruent condition), no interference is created and participants respond correctly and rapidly. If, however, the processing of area and perimeter result in two different conclusions (as in incongruent conditions), a conflict is created. Overcoming this conflict and answering correctly is a demanding and time-consuming process.

A brain-imaging study employing this task revealed that reasoning in the congruent condition was associated with specific activation of parietal brain regions that are known to be involved in perceptual and spatial

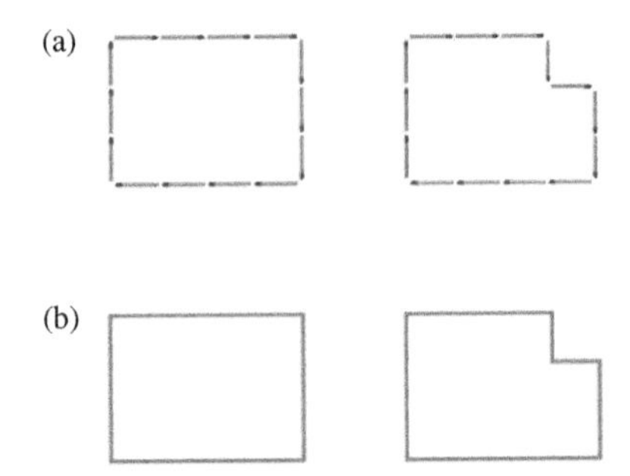

FIGURE 13.2 **(a) Discrete (drawn as built from matchsticks) and (b) continuous (drawn as a solid line) modes of presentation of a comparison of perimeters incongruent equal trial.**

processing (Stavy, Goel, Critchley, & Dolan, 2006). Reasoning in the incongruent condition was associated with specific activation of regions in the prefrontal cortex when answering correctly, suggesting that inhibition was required in order to overcome the interference, as these brain regions are known for their executive inhibitory control over other brain regions (Stavy et al., 2006). This fMRI study further indicated that increasing the level of salience of the irrelevant variable, area, is associated with lower accuracy and enhanced activation of parietal brain regions. These findings suggested that lowering the level of salience of the irrelevant variable area and/or increasing the level of salience of the relevant variable, perimeter, would increase students rate of correct responses. Drawing the perimeters of the shapes as built from separate units, matchsticks (Fig. 13.2a), rather than drawing them continuously as a solid line (Fig. 13.2b), is likely to increase the salience level of the relevant variable perimeter and somewhat decrease the level of salience of the irrelevant variable, area. Such presentation may allow appropriate solution strategies such as moving of segments, counting them, or both. A discrete mode of presentation of the perimeter could, therefore, enhance students' performance.

We explored whether changing the mode of presentation of the continuous variable, perimeter, into a discrete mode, would improve students' success, that is, their ability to correctly compare perimeters (Babai et al., 2016). Fifty-eight fifth and sixth graders participated in this study. We used two modes of presentation of the perimeters of the shapes (Fig. 13.2). In the discrete mode of presentation the perimeters were drawn as built from separate units—matchsticks (Fig. 13.2a). In the continuous mode of presentation, the perimeters were drawn as solid lines (Fig. 13.2b).

Table 13.1 shows rate of success and reaction time for correct responses and their standard error of the mean (SEM), for the continuous and discrete comparison of perimeters task, and for the three task conditions (congruent, incongruent inverse, and incongruent equal).

TABLE 13.1 Rate and Reaction Time of Correct Responses for the Comparison of Perimeters Task in Continuous ($n = 29$) and Discrete ($n = 29$) Modes of Presentation

	Correct (%) (SEM)		Reaction time (ms) (SEM)	
Congruity	**Continuous**	**Discrete**	**Continuous**	**Discrete**
Congruent	81.3 (4.5)	90.1 (3.8)	1835 (203)	1716 (133)
Incongruent inverse	50.0 (7.0)	72.8 (6.8)	1946 (213)	2415 (230)
Incongruent equal	21.3 (4.4)	32.8 (5.0)	1729 (635)	3750 (618)

It was found that success rate in the discrete mode of presentation was significantly higher than in the continuous mode and significantly higher in congruent trials than in incongruent ones (in both modes of presentation).

Because the success rate of incongruent equal trials was very low, there was an insufficient number of students for statistical analysis of reaction time, and analysis of reaction time of correct responses was carried out on congruent and incongruent inverse conditions only. Analysis of reaction time revealed no effect of mode of presentation. As expected, reaction time in incongruent inverse trials was significantly longer than that in congruent ones.

It is interesting to note that using this discrete mode of presentation but asking the students to compare the number of matchsticks between the two shapes resulted in very similar accuracy and reaction time to that of the comparison of perimeters task in this mode of presentation (Babai et al., 2015).

The findings of Study 1 suggest that the discrete mode of presentation strongly enhances the salience of the relevant variable, perimeter, and somewhat decreases the salience of the irrelevant variable, area. This enhancement specifically relates to the salience of the perimeter's segments that should be mentally moved when solving the task. These changes in saliency suggest appropriate solution strategies. In the continuous mode of presentation, however, no hint of such possibility of mentally "breaking" the solid line into relevant segments is given.

A possible educational implication of these findings related to the comparison of perimeters task is to use the discrete mode of presentation as an "anchoring task." In the "teaching by analogy" educational approach, an anchoring task is a task known to elicit a correct response and supports appropriate solution strategies in a subsequent "target task" (Clement, 1993; Stavy, 1991; Tsamir, 2003).

In a recent study we explored whether an educational intervention of first performing the discrete mode of presentation (as an anchoring task) would improve students' success in a subsequent continuous mode (as a target task) (Babai et al., 2016). It was found that, indeed, success in the continuous mode of presentation significantly increased as a result of this educational intervention (ie, when performed after discrete mode). Presenting the tasks in a reversed order had no effect on success rate.

13.3 STUDY 2: COMPARISON OF RATIOS

Comparison of ratios is commonplace in science, mathematics, and everyday life. Research has shown that it presents difficulties to many students and adults (for a recent review, see Ni & Zhou, 2005; Siegler, Fazio, Bailey, & Zhou, 2013).

The common difficulty in comparison of ratios is explained by the interference of the salient irrelevant whole-number comparison, that is, the whole-number bias. For example, in a large survey of adolescents' concept of probability (Green, 1983), students were presented with two jars (Jar J containing 3 red balls and 1 white ball, and Jar K containing 6 red balls and 2 white balls), and asked to decide if they were more likely to pick a red ball from one of the two jars or if the chance was equal. The most common response, Jar K, was incorrect, in line with the whole-number bias.

In previous comparison of ratios studies (Babai & Stavy, 2015; Stavy & Babai, 2011), participants were asked to compare the intensity of color of two mixtures of red and white paint drops (Fig. 13.3) and were asked to judge whether the right or left mixture had a darker shade, or if the two mixtures had an equal shade.

The intensity of color is determined by the ratio, number of red paint drops divided by the number of white paint drops. In this task, the number of red paint drops is the salient variable when the difference between the

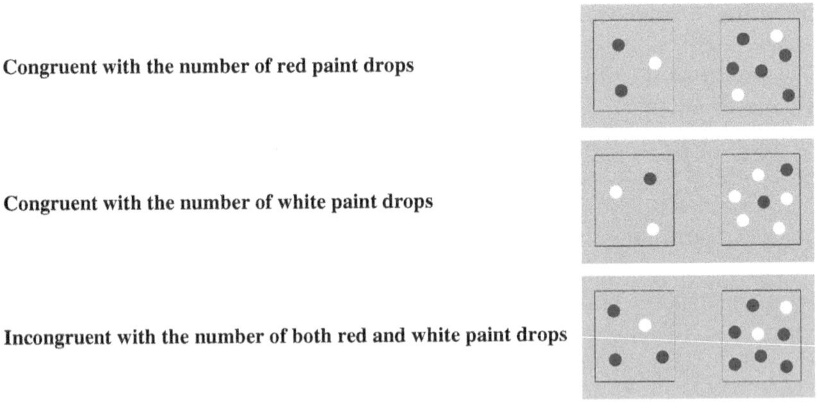

Congruent with the number of red paint drops

Congruent with the number of white paint drops

Incongruent with the number of both red and white paint drops

FIGURE 13.3 **Examples of congruent and incongruent comparison of ratios task conditions.**

number of red drops is larger than that of the difference between the number of white drops (between the two mixtures). When the difference between the number of white drops is larger than that of the difference between the number of red drops, the number of white paint drops is the salient variable.

Fig. 13.3 shows that in some of the conditions there was no interference of the salient variable (congruent with the number of red paint drops or congruent with the number of white paint drops). In these trials one of the mixtures had a larger number of red/white paint drops and a darker/lighter shade. In the other condition, there was interference of the salient variable (incongruent with the number of both red and white paint drops). In this condition, one mixture had a larger number of red/white paint drops and the shades (ratios) of both mixtures were equal.

We found that accuracy was significantly higher and reaction time for correct responses was significantly shorter in congruent as compared to incongruent comparison of ratios trials (Babai & Stavy, 2015; Stavy & Babai, 2011). Similar findings were found in the context of probability (Babai, Brecher, Stavy, & Tirosh, 2006).

We suggested that the number of red/white paint drops, the salient variable in the task, is probably automatically (intuitively) processed. Its processing runs parallel to the comparison of ratios. In the congruent condition, the processing of the number of red/white paint drops and that of the ratios produce the same conclusion and participants respond correctly and quickly. In the incongruent condition, however, the processing of the number of red/white paint drops and that of the ratios produce conflicting conclusions. The created conflict can be resolved either by overcoming the intuitive interference (a time-consuming process) or by giving an incorrect response.

We explored whether changing the mode of presentation of the red/white paint variable from discrete to continuous will affect participants' performance, that is, their ability to correctly compare the ratios. Sixty-four university students participated in this study. We used two modes of presentation of the comparison of ratios task (Fig. 13.4). In the discrete mode of presentation, the amounts of red/white paint was drawn as equal-sized

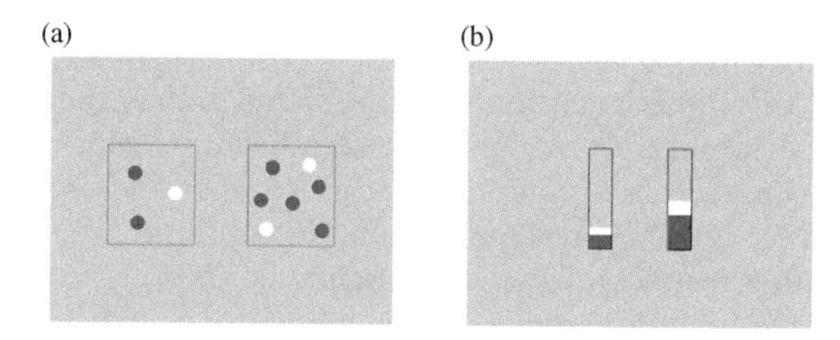

FIGURE 13.4 **Modes of presentation used in the comparison of ratios task.** (a) Discrete representation and (b) continuous representation.

TABLE 13.2 Rate of and Reaction Time of Correct Responses for the Comparison of Ratios Tasks in Continuous (n = 34) and Discrete (n = 30) Modes of Presentation

	Correct (%) (SEM)		Reaction time (ms) (SEM)	
Congruity	**Continuous**	**Discrete**	**Continuous**	**Discrete**
Congruent with quantity of red paint	59.4 (4.3)	88.01 (2.3)	4065 (364)	5478 (445)
Congruent with quantity of white paint	62.6 (3.5)	87.7 (1.9)	5296 (3793)	6379 (488)
Incongruent with both	55.4 (5.2)	71.9 (4.9)	5538 (325)	6000 (323)

circles (representing paint drops, see Fig. 13.4a). In the continuous mode of presentation, the amounts of red/white paint were drawn as red/white filled areas (Fig. 13.4b).

Table 13.2 shows rate of success and reaction time for correct responses and their SEM, for the continuous and discrete comparison of ratios tasks, for the three task conditions (congruent with the amount of red paint, congruent with the amount of white paint, incongruent with amounts of both red and white paint).

Findings show that the success rate in the discrete mode of presentation was significantly higher than in continuous mode and significantly lower in incongruent trials than in the other two congruent trial conditions.

Analysis of reaction time of correct responses revealed no significant difference between the two modes of presentation. Reaction time for correct responses in the congruent with the amount of red paint condition was significantly shorter than that for the other two conditions.

The findings of Study 2 suggest that the discrete mode of presentation enables solvers to exactly measure (count) the amounts of red and white paint drops in each mixture and therefore to calculate the ratios for comparison. Continuous mode of presentation enables only the estimation of these amounts and therefore would result in a lower success rate, yet, it enables correct responses quite well, way above chance level even in the incongruent condition.

Numerical presentation of the number of red and white paint drops resulted in very similar accuracies and a significantly faster reaction time to that of the comparison of ratios in the discrete mode of presentation.

To recap, the discrete presentation of the continuous quantities described in Studies 1 and 2 improved participants' ability to compare them and to overcome the interference. It seems that the discrete mode of presentation changes the perceptual information in such a way that it encourages the use of relevant solution strategies.

In each of these studies one variable was salient: the area in the comparison of perimeters task and the amount of red/white paint in the comparison of ratios task. The other variable was not salient and required reasoning processes: mentally manipulating the shapes in the comparison of perimeters task and calculating the ratios in the comparison of ratios task. Indeed, in the comparison of the perimeters task, for example, the interference was not symmetrical, as the area interfered with the perimeter, but the perimeter did not interfere with the area. Accuracy in comparison of areas was significantly higher and reaction time significantly shorter in all task conditions.

13.4 STUDY 3: COMPARISON OF AREAS AND NUMBERS IN CONTINUOUS AND DISCRETE PRESENTATION MODES

In Study 3 we focus on the interference of two salient variables, the continuous variable, area and the discrete variable, number, presented in two modes of presentation, continuous and discrete. One of the basic properties of natural numbers is its being discrete, whereas one the basic properties of area is its being continuous. We explore whether the mode of presentation of these variables will have an effect on the participants' performance. The aim of this study is to find out whether the processing of a variable is determined by its basic properties (ie, its being discrete or continuous) or whether it also depends on the presentation mode.

Previous studies have shown that an Arabic numeral's physical size (its area) interferes with the comparison of its numerical magnitude, and vice versa (Besner & Coltheart, 1979; Henik & Tzelgov, 1982). Recently it has been shown that participants could perform both the comparison of continuous quantities (area, length) and discrete quantities (numerosity) accurately and fast. Their performance was better in terms of both accuracy and reaction time for congruent than for incongruent trials (Barth, 2008; Dormal & Pesenti, 2007; Hurewitz, Gelman, & Schnitzer, 2006; Nys & Content, 2012). These studies suggested that discrete quantity information interfered with continuous quantity judgments and vice versa. In all these studies the stimuli were presented discretely.

In Study 3 we explored, in a Stroop-like paradigm, the effect of presenting the continuous and discrete variables, area and number, each in a continuous and a discrete mode of presentation. For the discrete variable, number, the discrete mode of presentation is compatible (with its basic property) and the continuous mode is incompatible. For the continuous variable, area, the continuous mode of presentation is compatible (with its basic property) and the discrete mode is incompatible.

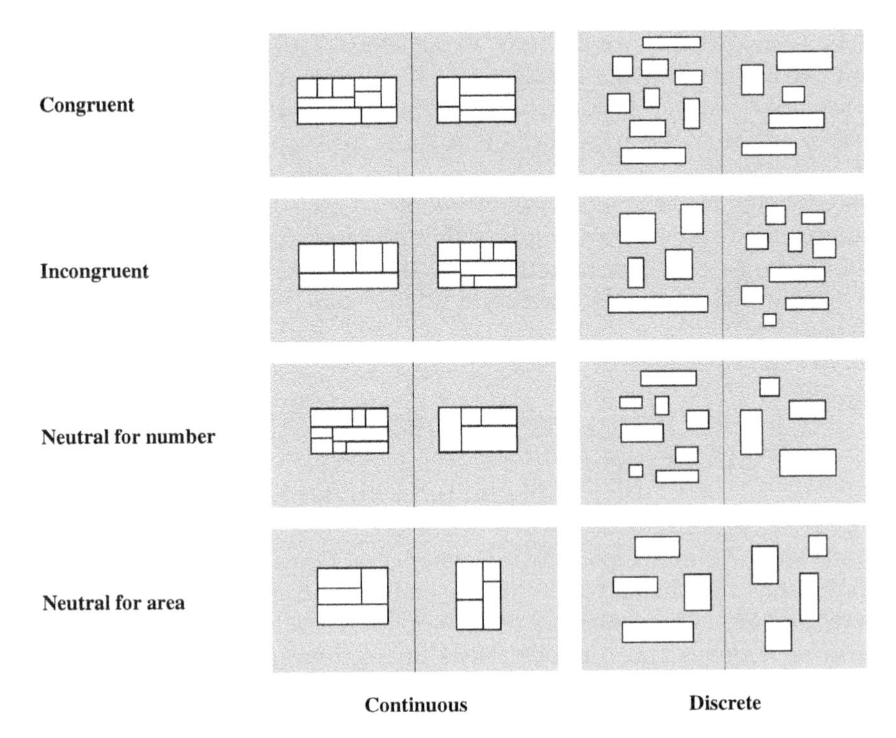

FIGURE 13.5 **Examples of congruent, incongruent, and neutral trials used in comparison of numbers and areas tasks in discrete and continuous modes of presentation.**

We used the following conditions (Fig. 13.5): congruent in which there was larger numerosity and larger total area in the same field; incongruent in which there was larger numerosity and smaller total area in the same field; neutral for number, in which there was same total area and larger numerosity in one field; and neutral for area, in which there was same numerosity and larger total area in one field. We used two modes of presentation of the area and the number tasks (Fig. 13.5). In the continuous mode of presentation, the areas were drawn as adjacent rectangles. In the discrete mode of presentation, the area was drawn as separate rectangles. The ratios between the areas (in the comparison of areas tasks) and the ratios between the numerosities (in the comparison of numbers tasks) were identical. The trials that were presented in the number tasks and the area tasks had the same numerosities and the same total areas and were shown in the same order in both tests. Eighteen university students participated in this study. In the number tasks, participants were asked to decide which field (left or right) had larger number of rectangles. In the area tasks, participants were asked to decide which field (left or right) had larger total area (of rectangles).

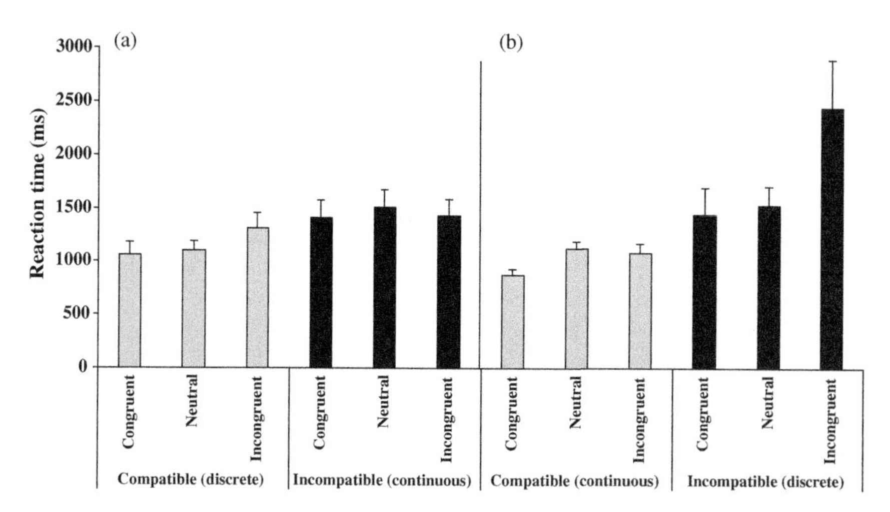

FIGURE 13.6 **Reaction times and their SEM, for the number (a) and area (b) tasks, each in compatible and incompatible presentation modes.**

Fig. 13.6 shows reaction times and their SEM for the number (a) and area (b) tasks, each in compatible and incompatible presentation modes, for the three task conditions (congruent, neutral, and incongruent).

In the number tasks, analysis of success rates revealed no significant difference between the two modes of presentation. Accuracy was very high in all conditions and ranged from 97 to 100%. Reaction time was significantly shorter in the compatible (discrete) mode of presentation as compared with the incompatible (continuous) one. In the compatible mode, the incongruent condition yielded a significantly longer reaction time than did the other two conditions. In the incompatible mode, no significant differences were found among the different conditions.

In the area tasks, it was found that the success rate in the compatible (continuous) mode of presentation was significantly higher than in the incompatible (discrete) mode (100 vs. 95%, $p < 0.001$). The success rate was significantly higher in the congruent and the neutral conditions (99%) than in the incongruent one in the incompatible mode of presentation (89%), $p = 0.001$ for both. Reaction time was significantly shorter in the compatible mode of presentation as compared with the incompatible one ($p = 0.002$). In the compatible mode, the congruent condition yielded significantly shorter reaction time than did the incongruent condition ($p = 0.045$) and the neutral one ($p < 0.001$). In the incompatible mode of presentation, the incongruent condition yielded significantly longer reaction time than did the congruent and neutral conditions ($p = 0.033$ and $p = 0.037$, respectively).

With regard to the neutral conditions, when there was neither facilitation nor interference, in the compatible mode of presentation, no significant differences were found between success rates and reaction times of the area and the number (area, 100% and 1121 ms; number, 100% and 1101 ms). In addition, in the incompatible mode of presentation, no significant differences were found between success rates and reaction times of the area and the number (area, 99% and 1528 ms; number, 97% and 1505 ms). However, when comparing the neutral conditions between compatible and incompatible modes of presentation, in each variable (number and area), the incompatible mode of presentation yielded a significantly lower success rate in the number task ($p = 0.007$) and significantly longer reaction times in both the number ($p = 0.006$) and area ($p = 0.024$) tasks.

To summarize, the findings of Study 3 show that the presentation mode affects the participants' performance. In the compatible mode of presentation, the comparison of congruent and incongruent conditions reveals that the variable, area, interferes with that of number and vice versa, whereas in the incompatible mode of presentation, the variable, number, interferes with that of the area with no reciprocal interference.

It seems that in the compatible mode of presentation, the relevant variable is processed more automatically (more salient). The incompatible mode of presentation probably demands additional processes (counting, adding, etc.). The incompatible mode of presentation, that is, the continuous presentation for the discrete variable, number, and the discrete presentation for the continuous variable, area, seems to decrease the salience of the relevant variable. Thus, the interference of the irrelevant variable, which is presented in its compatible mode, increases. This is clearly seen in the area tasks; in the incompatible mode of presentation, the interference of the variable, number, is significantly larger than in the compatible mode, in terms of both accuracy and reaction time. This effect is probably due to the fact that one cannot immediately perceive the size of the area, but one has to estimate the cumulative area by mentally adding all area pieces. In the incompatible mode of the number task, however, participants can count the number of all pieces and therefore can overcome the interference of the variable, area.

13.5 DISCUSSION

Continuous quantities are part of everyday science and mathematics education at all levels and are difficult for many students. These difficulties may arise from the interference of salient irrelevant variables, which are automatically/intuitively processed. In this chapter we have focused on the effect of mode of presentation of continuous and discrete quantities that affects the level of salience of the variables in the task. We demonstrate how continuous and discrete modes of presentation affect students' ability to compare these variables.

Study 1 suggests that the discrete mode of presentation strongly enhances the salience of the relevant variable, perimeter. This enhancement specifically relates to the salience of the perimeter's segments that should be mentally moved when solving the task. These changes in saliency suggest relevant solution strategies such as moving or counting the perimeter segments. Such a presentation was shown to have an educational benefit, as students who performed the discrete mode of presentation improved their success rate in a subsequent continuous mode of presentation (Babai et al., 2016).

Study 2 suggests that the discrete mode of presentation enables the solvers to measure exactly (to count) the amounts of red and white paint drops in each mixture and therefore to correctly compare the intensity of colors in the mixtures.

Both Studies 1 and 2 show that the discrete mode of presentation (in comparison to the continuous one) improves the participants' ability to correctly compare the relevant variable. It seems that in both studies, this mode of presentation changes the perceptual information in a way that encourages the use of relevant solution strategies.

Study 3 gives a different pattern of behavior. It suggests that a discrete presentation of a continuous quantity (ie, area) decreases participants' ability to correctly compare it.

In Studies 1 and 2, in order to solve the task, the participant's have to compare the relevant variable by formal/logical reasoning processes and to overcome intuitive interference of the irrelevant variables. The discrete mode of presentation, in these studies, supported the formal/logical reasoning processes needed in order to solve the task correctly. In Study 3, however, when the relevant variable (ie, area) is presented continuously, it is intuitively/automatically processed. Presenting it discreetly demands additional processes (counting, adding, etc.) and therefore results in reduced participants' performance.

To recap, the current chapter shows that changing the presentation mode changes the perceptual information, leading to a change in salience level. Such a change could encourage or discourage the use of relevant solution strategies and thus improve or reduce the participants' success. It appears that the processing of a quantity and the level of its salience and automaticity is determined by: (1) its basic properties (discrete, continuous, intensive, extensive, etc.); (2) presentation mode (discrete, continuous, etc.); (3) the participants' formal knowledge; and (4) the mental processes needed to solve the task (counting, calculating, moving segments, etc.). These conclusions are important for understanding the reasoning processes related to quantities and have practical implications in science and mathematics education.

Acknowledgment

The comparison of ratios study (Study 2) was supported by the Israel Science Foundation [Grant no. 464/07].

References

Babai, R., & Stavy, R. (2015). Intuitive interference in science and mathematics education. In Z. Smyrnaiou, M. Riopel, & M. Sotiriou (Eds.), *Recent advances in science and technology education, ranging from modern pedagogies to neuroeducation and assessment*. Newcastle upon Tyne: Cambridge Scholars Publishing.

Babai, R., Brecher, T., Stavy, R., & Tirosh, D. (2006a). Intuitive interference in probabilistic reasoning. *International Journal of Science and Mathematics Education, 4*, 627–639.

Babai, R., Levyadun, T., Stavy, R., & Tirosh, D. (2006b). Intuitive rules in science and mathematics: a reaction time study. *International Journal of Mathematical Education in Science and Technology, 37*, 913–924.

Babai, R., Zilber, H., Stavy, R., & Tirosh, D. (2010). The effect of intervention on accuracy of students' responses and reaction times to geometry problems. *International Journal of Science and Mathematics Education, 8*, 185–201.

Babai, R., Nattiv, L., & Stavy, R. (2016). Comparison of perimeters: improving students' performance by increasing the salience of the relevant variable. *ZDM Mathematics Education.*

Barth, H. C. (2008). Judgments of discrete and continuous quantity: an illusory Stroop effect. *Cognition, 109*, 251–266.

Besner, D., & Coltheart, M. (1979). Ideographic and alphabetic processing in skilled reading of English. *Neuropsychologia, 17*, 467–472.

Clement, J. (1993). Using bridging analogies and anchoring intuitions to deal with students' preconceptions about physics. *Journal of Research in Science Teaching, 30*, 1241–1257.

D'Amore, B., & Fandiño Pinilla, M. I. (2006). Relationships between area and perimeter: beliefs of teachers and students. *Mediterranean Journal for Research in Mathematics Education, 5*, 1–29.

Dormal, V., & Pesenti, M. (2007). Numerosity-length interference: a Stroop experiment. *Experimental Psychology, 54*, 289–297.

Driver, R. (1989). Students' conceptions and the learning of science. *International Journal of Science Education, 11*, 481–490.

Erickson, G. L. (1979). Children's conceptions of heat and temperature. *Science Education, 63*, 221–230.

Green, D. R. (1983). A survey of probabilistic concepts in 3000 students aged 11–16 years. In D. R. Grey, P. Holmes, V. Barnett, & G. M. Constable (Eds.), *Proceedings of the first International Conference on Teaching Statistics* (pp. 766–783). Sheffield, UK: Teaching Statistics Trust.

Henik, A., & Tzelgov, J. (1982). Is three greater than five: the relation between physical and semantic size in comparison tasks. *Memory & Cognition, 10*, 389–395.

Hurewitz, F., Gelman, R., & Schnitzer, B. (2006). Sometimes area counts more than number. *Proceedings of the National Academy of Sciences, 103*, 19599–19604.

Marchett, P., Medici, D., Vighi, P., & Zaccomer, E. (2005). Comparing perimeters and area children's pre-conceptions and spontaneous procedures. *Proceedings CERME, 4*, 766–776.

Ni, Y., & Zhou, Y. D. (2005). Teaching and learning fraction and rational numbers: the origins and implications of whole number bias. *Educational Psychologist, 40*, 27–52.

Nys, J., & Content, A. (2012). Judgement of discrete and continuous quantity in adults: number counts! *The Quarterly Journal of Experimental Psychology, 65*, 675–690.

Siegler, R. S., Fazio, L. K., Bailey, D. H., & Zhou, X. (2013). Fractions: the new frontier for theories of numerical development. *Trends in Cognitive Sciences, 17*, 13–19.

Stavy, R. (1991). Using analogy to overcome misconceptions about conservation of matter. *Journal of Research in Science Teaching, 28*, 305–313.

Stavy, R., & Babai, R. (2008). Complexity of shapes and quantitative reasoning in geometry. *Mind, Brain, and Education, 2*, 170–176.

Stavy, R., & Babai, R. (2010). Overcoming intuitive interference in mathematics: insights from behavioral, brain imaging and intervention studies. *ZDM Mathematics Education, 42*, 621–633.

Stavy, R., & Babai, R. (2011). *Development of proportional reasoning: a reaction time study*. Paper presented at the 41st Annual Meeting of the Jean Piaget Society, Berkeley, CA.

Stavy, R., & Tirosh, D. (2000). *How students (mis-)understand science and mathematics.* New York: Teachers College Press.

Stavy, R., Goel, V., Critchley, H., & Dolan, R. (2006). Intuitive interference in quantitative reasoning. *Brain Research, 1073–1074*, 383–388.

Strauss, S., & Stavy, R. (1982). U-shaped behavioral growth: implications for theories of development. In W. W. Hartup (Ed.), *Review of child development research* (pp. 547–599). Chicago: University of Chicago Press.

Tsamir, P. (2003). From "easy" to "difficult" or vice versa: the case of infinite sets. *Focus on Learning Problems in Mathematics, 25*, 1–17.

CHAPTER

14

Interaction of Numerical and Nonnumerical Parameters in Magnitude Comparison Tasks With Children and Their Relation to Arithmetic Performance

Swiya Nath, Dénes Szűcs

Department of Psychology, University of Cambridge, Cambridge, United Kingdom

OUTLINE

Continuous Issues in Numerical Cognition. http://dx.doi.org/10.1016/B978-0-12-801637-4.00018-4

14.1 INTRODUCTION

Determining and measuring the underlying mechanisms of mathematics development allow for earlier detection with appropriate and timely intervention in children with mathematics learning disabilities. However, the tasks used to assess the supposed underlying mechanisms of mathematics have perceptual processing confounds that override numerical processing. In this chapter, we critically analyze nonsymbolic magnitude comparison tasks used to assess the so-called approximate number system (ANS), and question whether the ANS is an underlying mechanism of mathematics development and performance.

Acuity of a putative ANS is thought to be the underlying mechanism of mathematics development and ergo learning disabilities specific to mathematics. Acuity of the ANS, also known as the "internal Weber fraction" (w), is generally tested with a simple magnitude comparison task (Piazza, Izard, Pinel, Le Bihan, & Dehaene, 2004; Piazza et al., 2010; Halberda & Feigenson, 2008; Halberda, Mazzocco, & Feigenson, 2008; Mazzocco, Feigenson, & Halberda, 2011). Participants are asked to compare two numerosities (the number of presented items) and select the side that they perceive has more items. By fitting the participant's responses to a sigmoid function, the participant's w is computed. Hence, w is a direct function of accuracy of the participant's responses in the nonsymbolic magnitude discrimination task.

The theory behind the ANS is that since infants (Xu & Spelke, 2000), animals (Agrillo, Piffer, Bisazza, & Butterworth, 2012), and societies without formal education (Pica, Lemer, Izard, & Dehaene, 2004) are able to discriminate between numerosities, this must be the evolutionary underlying mechanism of numerical processing of magnitude representation (Halberda et al., 2008). However, experiments on magnitude discrimination tasks have found that infants rely on total surface area more than the numerosity in judging which stimuli has "more" (Mix, Levine, & Huttenlocher, 1997). This also makes evolutionary sense. For example, in nature 6 bananas is not only more in number, but also physically larger than 3 bananas. This applies to most things in nature, where *more* means *bigger*. This brings into question whether nonsymbolic magnitude discrimination tasks are influenced primarily by perceptual properties, or whether they solely test numerical processing.

In this chapter we discuss the problems faced with nonsymbolic magnitude comparison tasks used to measure supposed underlying mechanisms of mathematics development. The visual cues that participants can rely on to compare magnitudes in the nonsymbolic magnitude comparison task essentially makes it a Stroop task, better suited to measure inhibition control than the supposed ANS (Henik & Tzelgov, 1982; Szűcs, Nobes, Devine, Gabriel, & Gebuis, 2013). Children with developmental dyscalculia were similar to typically developing children on their acuity of the ANS; however, they were impaired on inhibition control and working memory (Szűcs, Devine, Soltesz, Nobes, & Gabriel, 2013). Research has accordingly turned the focus toward the relationship between working memory and mathematics development. Visuospatial memory was a significant predictor of mathematics performance in children even when verbal memory, IQ, and executive functions were included in the linear regression models (Szűcs, Devine, Soltesz, Nobes, & Gabriel, 2014). The issues of perceptual processing confounds in the nonsymbolic magnitude comparison tasks attempting to test the acuity of the supposed ANS, and the subsequent switch in research focus on the relationship between mathematics development and working memory are discussed in this chapter.

14.2 APPROXIMATE NUMBER SYSTEM AND MATHEMATICS DEVELOPMENT

While some argue that the ANS is predictive of later mathematics performance, others argue that formal education in mathematics further refine acuity of the ANS. Verbal counting abilities enable the understanding that numerical quantities are independent of physical properties like size and other perceptual properties. The developmental relationships in children (4- to 7-year-olds) between magnitude comparison, number knowledge, and counting knowledge was tested and compared to general cognitive abilities like short-term memory (Soltész, Szucs, & Szucs, 2010). Magnitude comparison was not correlated to verbal counting knowledge or to mathematics performance on simple arithmetic tests. While younger children (4-year-olds) were able to detect magnitudes, they were unable to discriminate between the magnitudes of numerosities. Older children showed responses on the nonsymbolic magnitude comparison task that reflected ANS (Soltész et al., 2010). This suggests that younger children were influenced by task-irrelevant perceptual properties of the nonsymbolic magnitude comparison task that had congruent (larger size for larger numerosity, and smaller size for smaller numerosity), and incongruent (smaller size for larger numerosity, and larger size for smaller numerosity) trials. Another study also finds that performance on magnitude detection is better and more correlated to mathematics performance than magnitude comparison (Gebuis & Van der Smagt, 2011a). Performance suggesting sensitivity to task-irrelevant perceptual properties was correlated to short-term memory and age (Soltész et al., 2010). Developmental improvement (4- and 5-year-olds) on the incongruent conditions of the nonsymbolic magnitude comparison task was not correlated developmental change in mathematics performance when children (5- and 6-year-olds) start formal schooling. Therefore, development of general cognitive abilities may have had a larger role in magnitude discrimination tasks than an innate ANS (Soltész et al., 2010). Likewise, another study testing the development of performance on nonsymbolic magnitude comparison tasks and counting knowledge found that the two developmental trajectories were not related (Rousselle & Noël, 2008). They tested nonsymbolic magnitude comparison tasks with dot arrays and bars, and children (3- to 6-year-olds) were asked to either judge the numerosity (discrete quantification), or the surface area filled (continuous quantification). While sensitivity to irrelevant numerical cues increased with age, the influence of irrelevant perceptual cues remained stable throughout development (Rousselle & Noël, 2008). The increased sensitivity to numerical properties may reflect children's exposure to numeracy and formal education. However, the automatic perceptual processing that is already in place in 3-year-old children (Rousselle & Noël, 2008) may override numerical processing in nonsymbolic magnitude discrimination tasks (Gebuis & Reynvoet, 2013; Gebuis, Kenemans, de Haan, & Van der Smagt, 2010).

14.3 NONSYMBOLIC MAGNITUDE DISCRIMINATION TASK AND VISUAL PARAMETERS

In the nonsymbolic magnitude discrimination task, participants can rely on visual cues to compare the numerosities (Gebuis & Reynvoet, 2011, 2012b; Gebuis & Van der Smagt, 2011b; Gebuis & Gevers, 2011; Defever, Reynvoet, & Gebuis, 2013). Participants are asked to compare two numerosities (the number of presented items) and select the side that they perceive has more items. By fitting the participant's responses to a sigmoid function, the participant's acuity of the ANS, or the "w" is computed. Hence, w is a direct function of accuracy of the participant's responses in the nonsymbolic magnitude discrimination task. However, the visual parameters (eg, surface, density) of the stimuli allow participants to rely on visual cues to compare numerosities (Gebuis & Reynvoet, 2012b). Therefore, the accuracy on the nonsymbolic magnitude discrimination task computed as w may be driven by visual cues instead of numerical processes. In fact, visual parameters are so heavily linked to the stimuli that the nonnumerical parameters cannot be controlled in such a way that participants cannot rely on visual cues in each individual trial (Gebuis & Reynvoet, 2012a, 2012b). To counteract participants' reliance on visual cues, visual parameters of the stimuli are manipulated and counterbalanced across trials. In congruent trials, the size of the stimuli (typically dots) are larger for the larger of the two numerosities. In incongruent trials, the size of the stimuli is smaller for the larger of the two numerosities. However, this conventional method only accounts for a single visual cue across the trials, and does not include any neutral conditions. Participants are still able to rely on different visual cues, or integrate multiple visual cues to compare numerosities without numerical processing (Gebuis & Reynvoet, 2011). Added to this, the visual parameters of the stimuli can be directly related to the distance ratio of the numerosities being compared such that a larger difference in visual stimulus properties reflects a larger distance between numerosities. Therefore, the w computed by the ratio effect can be the result of participants relying on visual cues, instead of comparing the number of items. Most importantly, a nonsymbolic discrimination task with congruent and incongruent trials where visual parameters of the stimuli correlate positively or negatively with numerosity essentially becomes a Stroop task, testing inhibition (Henik & Tzelgov, 1982), instead of numerical processing captured by the acuity of the supposed ANS (Soltész et al., 2010).

When visual parameters were controlled in a more stringent manner across the trials, w was strongly correlated to the congruency effect (Soltész et al., 2010) and inhibition (Fuhs & McNeil, 2013). Congruency effects occur when participants are unable to inhibit task-irrelevant visual information in the incongruent condition because they cannot focus their attention only on task-relevant numerical parameters. Visual parameters were controlled on five sensory properties: (1) surface: total surface of the dots in one dot array; (2) diameter: the average diameter of the dots in one dot array; (3) contour length: the total contour length of all dots in one dot array; (4) convex hull: the smallest contour that can be drawn around the dots on one dot array; and (5) density: surface divided by convex hull. Surface, diameter, and contour length change in relation with each other; however, the convex hull can be manipulated without changing other visual parameters. Counterbalancing the convex hull with other visual parameters allowed for trials that were partly congruent and partly incongruent. Doing so disabled participants from relying on a single visual cue across all trials. Adults and children performed equally well in the congruent condition. Performance was below chance levels in the incongruent condition in children. Congruency effects heavily influenced the fit of the responses to the sigmoid model used to compute w (Szűcs et al., 2013b). This shows evidence that visual parameters compromise the assumption that nonsymbolic magnitude discrimination tasks test acuity of the ANS (Gebuis & Reynvoet, 2011, 2012b; Defever et al., 2013). Instead, the nonsymbolic magnitude discrimination task may simply be a reflection of inhibition control as congruency effects rely on interference suppression. Likewise, another study (Clayton & Gilmore, 2015) examined whether nonsymbolic magnitude discrimination tasks test ANS acuity (revealed by significant effects of numerosity ratio), or inhibitory control (revealed by congruency effects of visual parameters of the task). The results show that in larger set sizes, the convex hull influenced the participant's discrimination of numerosities, while in smaller set sizes, the size of the dot influenced the participant's discrimination of numerosities. The authors take this as evidence that nonsymbolic magnitude discrimination tasks test inhibitory control, and studies using the same to test ANS acuity may be problematic given the confounds of visual parameters (Clayton & Gilmore, 2015). In a study with children (5-year-olds) from low socioeconomic status, inhibition, and vocabulary were analyzed along with ANS that was tested by comparing two arrays of stars and selecting the array with more number of stars (Fuhs & McNeil, 2013). Correlation analysis revealed that ANS, particularly the incongruent condition was more strongly related to inhibition control than to mathematics performance. Hierarchical regression analysis revealed that ANS, which was marginally related to mathematics performance, became nonsignificant when vocabulary was entered into the model. Only the incongruent conditions remained significant when vocabulary was controlled; however, this relationship became nonsignificant when inhibition control was entered into the model (Fuhs & McNeil, 2013). Furthermore, in another study, children from mid- to high-socioeconomic backgrounds were tested on nonsymbolic magnitude comparison task, inhibition, and general intelligence (Gilmore et al., 2013). Mathematics performance was only correlated to the incongruent conditions of the nonsymbolic magnitude comparison task. Children with better inhibition control had a smaller reaction time difference between congruent and incongruent conditions. Performance on the nonsymbolic magnitude comparison task was not a significant predictor of mathematics performance when inhibition control was entered into the model (Gilmore et al., 2013). These studies on the ANS and inhibition reveal that inhibition control has a stronger relationship with mathematics performance than measures of ANS.

The confounds of visual parameters were further tested by comparing tasks that do and do not control for visual cues across trials (Smets, Sasanguie, Szücs, & Reynvoet, 2014). The nonsymbolic magnitude discrimination task that controls only for a single visual cue (dot size) and a task that controls for five sensory properties mentioned previously were compared to a task that only had congruent trials without incongruent trials. Although magnitude estimation of numerosities were equal in all three task conditions, magnitude comparisons of numerosities were influenced by visual cues task condition. Accuracy of magnitude comparison on the congruent task was correlated to accuracy on the task controlling for a single visual cue, and performance was high. However, accuracy of magnitude comparison on the task controlling for multiple visual cues was not correlated to the other two tasks, and performance was significantly lower (Smets et al., 2014). These results suggest that the nonsymbolic magnitude discrimination task is heavily influenced by visual cues. A supposed underlying ANS system would not result in variances of accuracy in judging numerosities between tasks with different controls of visual cues.

14.4 DEVELOPMENTAL DYSCALCULIA AND THE STATE OF THE ANS

Further evidence for the hypothesis that inhibition control, rather than w as the factor correlated to mathematics performance, comes from literature on children with developmental dyscalculia (Szűcs et al., 2013a). Developmental dyscalculia (DD) is a specific impairment of mathematics ability that may affect 3–6% of the population. A dominant magnitude representation theory of DD assumes that DD is related to the impairment of the ANS (Piazza et al., 2010) or a "number module" (Landerl, Bevan, & Butterworth, 2004), tested with nonsymbolic magnitude discrimination tasks to compute w. Another magnitude representation theory suggests that the link between w and numerical symbols are impaired in DD (Rousselle & Noël, 2008; De Smedt & Gilmore, 2011), and is tested with symbolic numerosity comparison (eg, "Which is larger; 3 or 4?"). Both magnitude representation theories cite correlates in the intraparietal sulcus (IPS), however, the IPS is also associated with other cognitive abilities required for mathematics performance like attention and working memory. Other studies have found correlations between DD and inhibition (Bull & Scerif, 2001; Bull, Johnston, & Roy, 1999; Passolunghi, Cornoldi, & De Liberto, 1999; Passolunghi & Siegel, 2004; McKenzie, Bull, & Gray, 2003; Espy et al., 2004; Blair & Razza, 2007; Swanson, 2011), spatial processing (Rourke & Conway, 1997; Rourke, 1993), and working memory (Hitch & McAuley, 1991; Passolunghi & Siegel, 2001, 2004; Keeler & Swanson, 2001; Bull, Espy, & Wiebe, 2008; Swanson, 2006; Geary, 2004). The contrasting five alternative theories (magnitude representation, working memory, inhibition, attention, and spatial processing) of DD were tested in 9- to 10-year-old primary school children (Szűcs et al., 2013a). Results show evidence for the theories of inhibition and working memory. DD participants had higher congruency effects than control participants in the nonsymbolic magnitude discrimination task and an animal Stroop task. This suggests that participants with DD were unable to resist the intrusion of task-irrelevant stimulus dimensions as efficiently as controls (Szűcs et al., 2013a). In the Animal Stroop task, participants had to select the side on which the animal was larger in real life. In congruent trials, the relative sizes of the two animals to be compared were realistic on the monitor screen. In incongruent trials, the larger animal in real life was smaller than the other animal on the monitor screen. Apart from inhibition impairment, DD participants also had lower visuospatial short-term memory and working memory performance than controls. There were no differences between participants with DD and controls on measures of w, rebutting the magnitude representation theory that DD is related to an impairment of the ANS (Szűcs et al., 2013a). These results are in line with previous studies that also did not detect any difference in w between DD participants and controls (Landerl et al., 2004; Kucian et al., 2006, 2011;

Rousselle & Noël, 2008; Soltész, Szűcs, Dékány, Márkus, & Csépe, 2007; Landerl & Kolle, 2009; Mussolin et al., 2010b; Kovas et al., 2009), while other studies reported such a difference (Mussolin et al., 2010a; Piazza et al., 2010; Mazzocco et al., 2011). However these studies used nonsymbolic magnitude discrimination tasks that are confounded with visual parameter cues discussed previously (Gebuis & Reynvoet, 2012b). Given the lack of evidence for magnitude representation theory, we focus on the other alternative theories of DD (working memory, executive functions, attention, and spatial processing) and their correlation to mathematics performance.

In summary, the nonsymbolic magnitude discrimination task that is conventionally used to test acuity on the supposed approximate number system poses visual parameters confounds wherein participants can rely on visual cues to make comparison judgments not based on numerical processing (Gebuis & Gevers, 2011; Gebuis & Reynvoet, 2011, 2012b; Gebuis & Van der Smagt, 2011b). Performance on the nonsymbolic magnitude discrimination task has not been found to be related to verbal counting in young children (Soltész et al., 2010; Rousselle & Noël, 2008). Younger children (4-year-olds) were unable to discriminate magnitude numerosities (Soltész et al., 2010), and while sensitivity to numerical cues increases with development, the influence of irrelevant perceptual cues remains constant (Rousselle & Noël, 2008). This suggests that automatic perceptual processing overrides numerical processing in nonsymbolic magnitude discrimination tasks in children (Gebuis & Reynvoet, 2013). Performance on the nonsymbolic magnitude comparison task was influenced by varying levels of control of visual parameters (Smets et al., 2014). On nonsymbolic magnitude discrimination tasks with more stringent control on visual parameters disabling participants from relying on visual cues to make magnitude comparison judgments, congruency effects and inhibition was correlated to performance on the task in both adults and children, while acuity of the ANS was not (Soltész et al., 2010; Clayton & Gilmore, 2015). Furthermore, children with developmental dyscalculia also showed stronger congruency effects and impairment in inhibition than control participants, while they were comparable on measures of w (Szűcs et al., 2013a). With this evidence, we conclude that the nonsymbolic magnitude comparison task is better suited to measure inhibition control than the acuity of ANS it is conventionally supposed to measure.

14.5 WORKING MEMORY, MATHEMATICS PERFORMANCE, AND THE ANS TASK CONTEXT

Research has found confounds in the nonsymbolic magnitude comparison tasks conventionally used to test the development of numerical processing and the acuity of the ANS (w), which is thought to be the underlying mechanism of numerical processing. However, when acuity of the ANS was compared to other cognitive measures such as inhibition (Soltész et al., 2010; Clayton & Gilmore, 2015) and working memory (Soltész et al., 2010), *w* was not correlated to performance on the magnitude discrimination tasks (Soltész et al., 2010), or mathematics performance (Szűcs et al., 2013a). Therefore, research on mathematics development has recently shifted focus from the ANS toward working memory and inhibition.

Baddeley's influential tripartite memory model differentiates between verbal and visuospatial memory slave systems, under an overarching umbrella of central executive functions (Baddeley, 1986) with an episodic buffer linking these memory systems to long-term memory (Baddeley, 2000). Verbal memory is storage of phonological information, and visuospatial memory is storage of color, shape, movement, and location information. Verbal memory is further divided into short-term memory, and working memory. Short-term memory is simple recall, whereas working memory (recall and processing) is thought to involve executive functions such as controlled attention, in addition to short-term memory (Engle, Tuholski, Laughlin, & Conway, 1999). In regards to the theoretical model of visuospatial memory, models test the fractionation of visuospatial memory divided into short-term memory and working memory (Alloway, Gathercole, & Pickering, 2006), or into visual and spatial memory (Gathercole & Pickering, 2000; Cornoldi & Vecchi, 2004; Miyake, Friedman, Rettinger, Shah, & Hegarty, 2001; Logie & Pearson, 1997).

Studies have tested ANS alongside working memory in relation to mathematics performance (Szűcs et al., 2013a; Xenidou-Dervou, De Smedt, van der Schoot, & van Lieshout, 2013; Halberda et al., 2008). Using structural equation modeling, the relationship between working memory and ANS tested with both symbolic and nonsymbolic addition and magnitude comparison tasks was analyzed (Xenidou-Dervou et al., 2013). In the nonsymbolic magnitude addition task, blue dots fell into a gray box followed by red dots. Children had to respond whether there were more red or blue dots. In the nonsymbolic magnitude comparison task, dots fell into Peter's and Sarah's gray boxes, and the children had to respond whether Peter or Sarah had more number of dots. The symbolic tasks followed the same paradigm using Arabic numerals instead of dots. The structural equation modeling revealed that working memory predicted mathematics performance beyond that of ANS measures. The nonsymbolic magnitude tasks were only indirectly related to mathematics performance, and symbolic magnitude tasks mediated the relationship (Xenidou-Dervou et al., 2013). The same research group conducted a dual task study wherein the nonsymbolic magnitude addition task was carried out during the delay phase between stimulus presentation and stimulus recall of verbal memory, visual memory, and spatial memory tasks (Xenidou-Dervou, Lieshout, & Schoot, 2014). In the central executive dual task, kindergarten children said "A" for tones of high pitches, and "O" for low pitches while performing the nonsymbolic magnitude addition task. In the spatial memory task, performance was at chance level even during the standalone condition, suggesting that the spatial memory task similar to the Corsi block task was too difficult for the age (6-year-old) range tested. Results show that performance on the nonsymbolic magnitude addition task decreased during central executive function interference (Xenidou-Dervou et al., 2014). In a slightly different nonsymbolic magnitude discrimination task, participants saw a ratio of dots of two different colors intermingled within one dot array, and had to discriminate which color of dots were more than the other. Only dot size was controlled across trials. The acuity of the ANS was found to be correlated to past mathematics performance in 5- to 11-year-old children, even when working memory, executive functions, and intelligence were controlled (Halberda et al., 2008). Other than the confounds of controlling for only one visual parameter across trials, the study used measures of visual working memory that has not been found to be correlated to mathematics performance previously (Passolunghi & Mammarella, 2012). In 9-year-old children with mathematics learning disability perform poorly on complex spatial tasks compared to typically developing children, but were indistinguishable on their performance on visual memory tasks (Passolunghi & Mammarella, 2012). This suggests that mathematics performance is correlated to spatial memory, more than visual memory. The data from studies testing the relationship between ANS, working memory, and mathematics are seemingly contradictory. One study found only central executive functions interfere with nonsymbolic magnitude addition task performance (Xenidou-Dervou et al., 2014). Another found working memory predicts mathematics performance beyond the measures of ANS through structural equation modeling (Xenidou-Dervou et al., 2013). Yet another found ANS was correlated to mathematics performance retrospectively, even when working memory and intelligence were controlled (Halberda et al., 2008). However, the studies that did not find the relationship between working memory and mathematics to be stronger than the relationship between ANS and mathematics used visuospatial memory measures that either had floor effects (Xenidou-Dervou et al., 2014), or are known not to be correlated with mathematics performance (Halberda et al., 2008).

The relationship between working memory and mathematics performance is further supported by a study comparing and contrasting underlying cognitive skills associated with mathematics development, along with magnitude comparison tasks (Szűcs et al., 2014). Verbal and visuospatial short-term memory and working memory were tested in 9-year-old children, along with other executive functions of task switching, attention, and inhibition, as well as nonverbal intelligence, spatial ability, and phonological decoding. The relationship between these variables were tested against symbolic and nonsymbolic magnitude comparison measures, subitizing, and mathematics performance tests. The only magnitude comparison task related to mathematics performance was the symbolic magnitude comparison task in which participants had to respond whether the number presented was greater than 5, or less than 5. Linear regression analysis revealed that visuospatial short-term memory and visuospatial working memory significantly predicted mathematics performance along with general cognitive abilities of phonological decoding, spatial ability, vocabulary, and task switching. These results suggest an executive function's centric model of mathematics development, rather than models based on magnitude comparison tasks (Szűcs et al., 2014).

14.6 MATHEMATICS PERFORMANCE, WORKING MEMORY, AND INHIBITION

Stronger evidence for the link between working memory, inhibition, and mathematics performance has been accumulating over the recent years. Executive functions of inhibition, task switching, and updating was tested on children as predictors of mathematics performance. In 7-year-old children, each of the executive functions significantly predicted mathematics performance, while the dual task did not (Bull & Scerif, 2001). In the dual task, children were asked to recall digits while crossing out boxes on a grid linking the boxes to make a path. Such a dual task is unusual since the recall task tests verbal memory, whereas the interference task is visuospatial in nature. A similar study with two tasks for each of the executive functions of updating, shifting, and inhibition was run on children (11-year-olds), using domain specific working memory tasks (St Clair-Thompson & Gathercole, 2006). Shifting failed to load on as a distinct factor. Updating and domain-specific working memory measures loaded onto one composite working memory factor, while inhibition loaded onto another factor. Working memory and inhibition predicted unique variance of English and mathematics performance. When the data was analyzed by domain-specific working memory measures, verbal working memory was correlated to English performance, while visuospatial memory was correlated to English and mathematics performance (St Clair-Thompson & Gathercole, 2006). A longitudinal study spanning a year, tested updating, shifting, and inhibition in 6-year-old children (Van der Ven, Kroesbergen, Boom, & Leseman, 2012). They found inhibition and shifting loaded onto one factor, but was not correlated to mathematics performance. Updating, which was tested with domain-specific verbal and visuospatial memory, was a separate factor and was strongly related with the development of mathematics performance (Van der Ven et al., 2012). School entry-level (4-year-olds) performance on executive functions of updating, shifting, and inhibition along with measures of verbal and visuospatial memory were assessed to study the longitudinal predictive value of academic attainment (Bull et al., 2008). Short-term verbal memory was correlated to reading and mathematics performance, whereas short-term visuospatial memory was correlated to mathematics performance in preschool. Executive functions and working memory were correlated to mathematics performance in the third year, with the strongest correlation being visuospatial working memory. Inhibition was related to both reading and mathematics performance (Bull et al., 2008). In all four studies, working memory was categorized under the executive function of updating, and was consistently found to be related to mathematics performance (Bull et al., 2008; Van der Ven et al., 2012; St Clair-Thompson & Gathercole, 2006; Bull & Scerif, 2001). In studies further distinguishing the relationship between mathematics performance and domain-specific working memory tasks, they found visuospatial memory uniquely related to mathematics performance (Bull et al., 2008; St Clair-Thompson & Gathercole, 2006).

In summary, the research on mathematics development has shifted focus toward working memory. Operational definitions and task purity have led to contradictory findings in the developmental trajectory of the relationship between mathematics performance and working memory. For example, while one study comparing *w* and working memory found only working memory correlated to mathematics performance (Szűcs et al., 2013a), another found the opposite results (Halberda et al., 2008). This controversy is because the latter study tested visual working memory that is not correlated to mathematics performance (Passolunghi & Mammarella, 2012), whereas the former tested visuospatial working memory. Linear regression modeling finds that visuospatial memory, and not verbal memory, is a significant predictor of mathematics performance in 9-year-old children (Szűcs et al., 2014). Evidence suggests that the relationship between visuospatial memory and mathematics performance is robust, and may be an underlying cognitive mechanism of numerical processing.

14.7 VISUOSPATIAL MEMORY, MATHEMATICS, AND CONSTRUCTION PLAY

The developmental precursor of the relationship between visuospatial memory and mathematics performance may be construction play. Construction play is thought to develop logico-mathematical skills, and is defined by Piaget as activities producing symbolic products (Wolfgang & Phelps, 1983). The relationship between construction play and mathematics performance is mediated by visuospatial memory (Nath & Szűcs, 2014). Although mediational analysis is not causal and requires future studies, construction play may be a promising intervention tool to facilitate visuospatial memory, and in turn develop mathematics performance.

Studies on construction play have primarily focused the relationship between construction play and spatial abilities. In children (3- to 5-year-olds), boys playing with masculine activities (blocks and large motor toys) was correlated to spatial ability measured by the Wechsler Preschool and Primary Scale of Intelligence Block Design and Preschool Embedded Figures Test (EFT). However, for girls, there were no significant relationships between cognitive development and play preferences (Connor & Serbin, 1977). In a study with older children (9-year-olds), although preference for playing with Legos was not correlated to spatial ability, the number of errors and time taken to accurately construct a Lego model was negatively correlated to mental rotation ability (Brosnan, 1998). Likewise, rather than preference for construction play, accuracy on structured block play was correlated with Block Design and Copying Blocks, but not with the Children's EFT (Caldera et al., 1999). In a study of 9- and 12-year-old children, construction play was significantly related to Water Level tasks and Block Design but not EFT (Robert & Héroux, 2004). All the studies found construction play was correlated to Block Design. It is important to note that in the studies comparing preference versus accuracy on construction play, only accuracy was correlated to spatial ability (Brosnan, 1998; Caldera et al., 1999).

Studies have found construction play is correlated to mathematics performance across development. In adolescents, the structural balance of construction play was correlated to mathematics performance (Casey, Pezaris, & Bassi, 2012). In elementary school children (7- to 8-year-old, 10- to 11-year-old, and 13- to 14-year-old), construction ability was correlated with mathematics performance, and spatial ability (tested with Surface Development Test) mediated this relationship in the older age groups (Richardson, Hunt, & Richardson, 2014). The correlation between construction play and mathematics performance has been tested in preschool children as well. Children's (3-year-olds) construction ability to build a model according to given instructions was correlated to early mathematics performance (Verdine et al., 2014). Another study observed the adaptivity and complexity of construction play in a preschool (3- to 4-year-old children) child care program, and analyzed the relationship with the participants' mathematics performance until secondary school (Stannard, Wolfgang, Jones, & Phelps, 2001). Although construction play was not correlated to mathematics performance at younger years, construction play in preschool was correlated to mathematics performance in grade 7 (12-year-olds) and beyond. This could be evidence that construction play can be the developmental precursor to more complex mathematical processing only taught in higher years of schooling (Stannard et al., 2001). However, given the lapse of time and development between the observations of construction ability in preschool, and the measures of mathematics performance several years later, it is difficult to assess whether the findings relate specifically to construction play or as a result of general cognitive development, which was not tested.

The relationship between construction play and mathematics performance has been established, and evidence shows the underlying mechanism of this relationship is visuospatial memory (Nath & Szűcs, 2014). Children's (7-year-olds) construction ability was tested using a Lego construction task paradigm at increasing task difficulty determined by a mathematical formula derived through regression analysis (Richardson, Jones, & Torrance, 2004):

$$\text{Task difficulty} = 10^{[(0.020\ \text{components})+(-0.117\ \text{symmetrical planes})+(0.047\ \text{novel assemblies})+(0.028\ \text{selections})+1.464]}$$

The four task variables found to systematically increase construction task difficulty are (1) symmetrical planes (number of orientations required to find correct placement), (2) novel assemblies (number of unique assemblies), (3) components (number of pieces to create structure), and (4) selections (total number of components to select required components from) (Richardson et al., 2004). Children were also tested on Automated Working Memory Assessment measures of visuospatial and verbal short-term memory and working memory. Mathematical performance was measured through the WIAT-II Numerical Operations, and the Word Reading subtest was used as a control variable. Cognitive abilities of updating, shifting, and attention, along with IQ measures of WISC-III Block Design and Vocabulary were tested. Lego construction ability was significantly related to visuospatial working memory, Raven's CPM, Block Design and mathematics performance, but not to verbal memory, vocabulary, reading performance, or any of the executive function measures. This suggests that Lego construction ability is uniquely correlated with mathematics performance, and that this relationship is not driven by general cognitive ability or intelligence. Through mediation analysis following the recommendations of Baron and Kenny (1986) and Preacher and Hayes (2008), we found that the relationship between Lego construction ability and mathematics performance was uniquely mediated by visuospatial working memory and Raven's CPM. Spatial reasoning of Block Design did not mediate the relationship between Lego construction ability and mathematics performance.

In summary, the relationship between mathematics performance and construction play is well established. Although previous research finds correlations with construction play and spatial abilities, we found the relationship between construction play and mathematics performance was mediated by visuospatial memory, independent of spatial ability in 7-year-old children. Although mediation analysis does not provide causal data, construction play may be a promising intervention tool for impairments in mathematics related to visuospatial memory. Using the Lego construction paradigm (Richardson et al., 2004) as an intervention tool allows task difficulty to be systematically increased in increments. Such intervention training raises the possibility of using Lego construction tasks to facilitate visuospatial memory, and mathematics performance. Future work with an intervention study will be required to test this hypothesis.

14.8 CONCLUSIONS

The symbolic and nonsymbolic magnitude comparison tasks used to measure the approximate number system have perceptual processing confounds, which brings up the question of task validity (Gebuis & Gevers, 2011; Gebuis & Reynvoet, 2011, 2012b; Gebuis & van der Smagt, 2011b). The control of visual cues based on a single visual parameter in the nonsymbolic magnitude comparison task essentially make it a Stroop task, better suited to measure inhibitory control rather than ANS. In fact, studies with more stringent control of the visual parameters find responses on the nonsymbolic magnitude comparison task correlated to inhibition control and working memory, but not the supposed w underlying numerical processing (Szűcs et al., 2013b; Xenidou-Dervou et al., 2013). Data from children with developmental dyscalculia show impairment in inhibition control and working memory, but are comparable to controls on measures of w (Szűcs et al., 2013a). Further delving into the research on the relationship between working memory and mathematics development reveals that visuospatial memory is a significant predictor of mathematics performance even when IQ and executive functions are included in the linear regression models (Szűcs et al., 2014; Passolunghi, Mammarella, & Altoè, 2008). Mediational analysis reveals that visuospatial memory mediates the relationship between construction ability and mathematics performance (Nath & Szűcs, 2014), which may lead to promising research on construction play as the developmental precursor of the relationship between mathematics performance and visuospatial memory. Given the robust data, evidence suggests visuospatial memory and inhibition control as an underlying cognitive mechanism of numerical processing.

References

Agrillo, C., Piffer, L., Bisazza, A., & Butterworth, B. (2012). Evidence for two numerical systems that are similar in humans and guppies. *PLoS One*, *7*(2), e31923.

Alloway, T. P., Gathercole, S. E., & Pickering, S. J. (2006). Verbal and visuospatial short-term and working memory in children: are they separable? *Child Development*, *77*(6), 1698–1716.

Baddeley, A. D. (1986). *Working memory*. Oxford: Oxford University Press.
Baddeley, A. D. (2000). The episodic buffer: a new component of working memory? *Trends in Cognitive Sciences*, *4*(11), 417e422.
Baron, R. M., & Kenny, D. A. (1986). The moderator–mediator variable distinction in social psychological research: conceptual, strategic, and statistical considerations. *Journal of Personality and Social Psychology*, *51*(6), 1173–1182.
Blair, C., & Razza, R. P. (2007). Relating effortful control, executive function, and false belief understanding to emerging math and literacy ability in kindergarten. *Child Development*, *78*(2), 647–663.
Bull, R., & Scerif, G. (2001). Executive functioning as a predictor of children's mathematics ability: inhibition, switching, and working memory. *Developmental Neuropsychology*, *19*(3), 273–293.
Bull, R., Espy, K. A., & Wiebe, S. A. (2008). Short-term memory, working memory, and executive functioning in preschoolers: longitudinal predictors of mathematical achievement at age 7 years. *Developmental Neuropsychology*, *33*(3), 205–228.
Bull, R., Johnston, R. S., & Roy, J. A. (1999). Exploring the roles of the visual-spatial sketch pad and central executive in children's arithmetical skills: views from cognition and developmental neuropsychology. *Developmental Neuropsychology*, *15*(3), 421–442.
Brosnan, M. J. (1998). Spatial ability in children's play with Lego blocks. *Perceptual and Motor Skills*, *87*(1), 19–28.
Caldera, Y. M., Mc Culp, A., O'Brien, M., Truglio, R. T., Alvarez, M., & Huston, A. C. (1999). Children's play preferences, construction play with blocks, and visual-spatial skills: are they related? *International Journal of Behavioral Development*, *23*(4), 855–872.
Casey, B. M., Pezaris, E. E., & Bassi, J. (2012). Adolescent boys' and girls' block constructions differ in structural balance: a block-building characteristic related to math achievement. *Learning and Individual Differences*, *22*(1), 25–36.
Clayton, S., & Gilmore, C. (2015). Inhibition in dot comparison tasks. *ZDM Mathematics Education*, *47*, 759–770.
Connor, J. M., & Serbin, L. A. (1977). Behaviorally based masculine-and feminine-activity-preference scales for preschoolers: correlates with other classroom behaviors and cognitive tests. *Child Development*, *48*, 1411–1416.
Cornoldi, C., & Vecchi, T. (2004). *Visuo-spatial working memory and individual differences*. Hove, UK: Psychology Press.
De Smedt, B., & Gilmore, C. K. (2011). Defective number module or impaired access? Numerical magnitude processing in first graders with mathematical difficulties. *Journal of Experimental Child Psychology*, *108*(2), 278–292.
Defever, E., Reynvoet, B., & Gebuis, T. (2013). Task and age dependent effects of visual stimulus properties on children's explicit numerosity judgments. *Journal of Experimental Child Psychology*, *116*(2), 216–233.
Engle, R. W., Tuholski, S. W., Laughlin, J. E., & Conway, A. R. (1999). Working memory, short-term memory, and general fluid intelligence: a latent-variable approach. *Journal of Experimental Psychology: General*, *128*(3), 309–331.
Espy, K. A., McDiarmid, M. M., Cwik, M. F., Stalets, M. M., Hamby, A., & Senn, T. E. (2004). The contribution of executive functions to emergent mathematic skills in preschool children. *Developmental Neuropsychology*, *26*(1), 465–486.
Fuhs, M. W., & McNeil, N. M. (2013). ANS acuity and mathematics ability in preschoolers from low-income homes: contributions of inhibitory control. *Developmental Science*, *16*(1), 136–148.
Gathercole, S. E., & Pickering, S. J. (2000). Assessment of working memory in six-and seven-year-old children. *Journal of Educational Psychology*, *92*(2), 377–390.
Geary, D. C. (2004). Mathematics and learning disabilities. *Journal of Learning Disabilities*, *37*(1), 4–15.
Gebuis, T., & Gevers, W. (2011). Numerosities and space, indeed a cognitive illusion! A reply to de Hevia and Spelke (2009). *Cognition*, *121*(2), 253–255.

Gebuis, T., & Reynvoet, B. (2011). The interplay between non-symbolic number and its continuous visual properties. *Journal of Experimental Psychology: General, 141*, 642–648.

Gebuis, T., & Reynvoet, B. (2012a). Continuous visual properties explain neural responses to non-symbolic number. *Psychophysiology, 49*(11), 1481–1491.

Gebuis, T., & Reynvoet, B. (2012b). The role of visual information in numerosity estimation. *PLoS One, 7*(5).

Gebuis, T., & Reynvoet, B. (2013). The neural mechanisms underlying passive and active processing of numerosity. *Neuroimage, 15*(70), 301–307.

Gebuis, T., & Van der Smagt, M. J. (2011a). False approximations of the approximate number system. *PLoS One, 6*(10), e25405.

Gebuis, T., & Van der Smagt, M. J. (2011b). Incongruence in number-luminance congruency effects. *Attention Perception & Psychophysics, 37*(1), 259–265.

Gebuis, T., Kenemans, J. L., de Haan, E. H. F., & Van der Smagt, M. J. (2010). Conflict processing of symbolic and non-symbolic numerosity. *Neuropsychologia, 48*(2), 394–401.

Gilmore, C., Attridge, N., Clayton, S., Cragg, L., Johnson, S., Marlow, N., ... & Inglis, M. (2013). Individual differences in inhibitory control, not non-verbal number acuity, correlate with mathematics achievement. *PLoS One, 8*(6), e67374.

Halberda, J., & Feigenson, L. (2008). Developmental change in the acuity of the "number sense": The approximate number system in 3-, 4-, 5-, and 6-year-olds and adults. *Developmental Psychology, 44*(5), 1457–1465.

Halberda, J., Mazzocco, M. M., & Feigenson, L. (2008). Individual differences in non-verbal number acuity correlate with maths achievement. *Nature, 455*(7213), 665–668.

Henik, A., & Tzelgov, J. (1982). Is three greater than five: The relation between physical and semantic size in comparison tasks. *Memory & Cognition, 10*(4), 389–395.

Hitch, G. J., & McAuley, E. (1991). Working memory in children with specific arithmetical learning difficulties. *British Journal of Psychology, 82*(3), 375–386.

Keeler, M. L., & Swanson, H. L. (2001). Does strategy knowledge influence working memory in children with mathematical disabilities? *Journal of Learning Disabilities, 34*(5), 418–434.

Kovas, Y., Giampietro, V., Viding, E., Ng, V., Brammer, M., Barker, G. J., ... & Plomin, R. (2009). Brain correlates of non-symbolic numerosity estimation in low and high mathematical ability children. *PLoS One, 4*(2), e4587.

Kucian, K., Loenneker, T., Dietrich, T., Dosch, M., Martin, E., & Von Aster, M. (2006). Impaired neural networks for approximate calculation in dyscalculic children: a functional MRI study. *Behavioral and Brain Functions, 2*(1), 1.

Kucian, K., Loenneker, T., Martin, E., & von Aster, M. (2011). Non-symbolic numerical distance effect in children with and without developmental dyscalculia: a parametric fMRI study. *Developmental Neuropsychology, 36*(6), 741–762.

Landerl, K., & Kölle, C. (2009). Typical and atypical development of basic numerical skills in elementary school. *Journal of Experimental Child Psychology, 103*(4), 546–565.

Landerl, K., Bevan, A., & Butterworth, B. (2004). Developmental dyscalculia and basic numerical capacities: a study of 8–9-year-old students. *Cognition, 93*(2), 99–125.

Logie, R. H., & Pearson, D. G. (1997). The inner eye and the inner scribe of visuo-spatial working memory: evidence from developmental fractionation. *European Journal of Cognitive Psychology, 9*(3), 241–257.

Mazzocco, M. M., Feigenson, L., & Halberda, J. (2011). Impaired acuity of the approximate number system underlies mathematical learning disability (dyscalculia). *Child Development, 82*(4), 1224–1237.

McKenzie, B., Bull, R., & Gray, C. (2003). The effects of phonological and visual-spatial interference on children's arithmetical performance. *Educational and Child Psychology, 20*(3), 93–108.

Mix, K. S., Levine, S. C., & Huttenlocher, J. (1997). Numerical abstraction in infants: another look. *Developmental Psychology, 33*(3), 423–428.

Miyake, A., Friedman, N. P., Rettinger, D. A., Shah, P., & Hegarty, M. (2001). How are visuo-spatial working memory, executive functioning, and spatial abilities related? A latent-variable analysis. *Journal of Experimental Psychology: General*, *130*(4), 621–640.

Mussolin, C., De Volder, A., Grandin, C., Schlögel, X., Nassogne, M. C., & Noël, M. P. (2010b). Neural correlates of symbolic number comparison in developmental dyscalculia. *Journal of Cognitive Neuroscience*, *22*(5), 860–874.

Mussolin, C., Mejias, S., & Noël, M. P. (2010a). Symbolic and nonsymbolic number comparison in children with and without dyscalculia. *Cognition*, *115*(1), 10–25.

Nath, S., & Szűcs, D. (2014). Construction play and cognitive skills associated with the development of mathematical abilities in 7-year-old children. *Learning and Instruction*, *32*, 73–80.

Passolunghi, M. C., Cornoldi, C., & De Liberto, S. (1999). Working memory and intrusions of irrelevant information in a group of specific poor problem solvers. *Memory & Cognition*, *27*(5), 779–790.

Passolunghi, M. C., & Mammarella, I. C. (2012). Selective spatial working memory impairment in a group of children with mathematics learning disabilities and poor problem-solving skills. *Journal of Learning Disabilities*, *45*(4), 341–350.

Passolunghi, M. C., Mammarella, I. C., & Altoè, G. (2008). Cognitive abilities as precursors of the early acquisition of mathematical skills during first through second grades. *Developmental Neuropsychology*, *33*(3), 229–250.

Passolunghi, M. C., & Siegel, L. S. (2001). Short-term memory, working memory, and inhibitory control in children with difficulties in arithmetic problem solving. *Journal of Experimental Child Psychology*, *80*(1), 44–57.

Passolunghi, M. C., & Siegel, L. S. (2004). Working memory and access to numerical information in children with disability in mathematics. *Journal of Experimental Child Psychology*, *88*(4), 348–367.

Piazza, M., Facoetti, A., Trussardi, A. N., Berteletti, I., Conte, S., Lucangeli, D., & Zorzi, M. (2010). Developmental trajectory of number acuity reveals a severe impairment in developmental dyscalculia. *Cognition*, *116*(1), 33–41.

Piazza, M., Izard, V., Pinel, P., Le Bihan, D., & Dehaene, S. (2004). Tuning curves for approximate numerosity in the human intraparietal sulcus. *Neuron*, *44*(3), 547–555.

Pica, P., Lemer, C., Izard, V., & Dehaene, S. (2004). Exact and approximate arithmetic in an Amazonian indigene group. *Science*, *306*(5695), 499–503.

Preacher, K. J., & Hayes, A. F. (2008). Asymptotic and resampling strategies for assessing and comparing indirect effects in multiple mediator models. *Behavior Research Methods*, *40*(3), 879–891.

Richardson, M., Hunt, T. E., & Richardson, C. (2014). Children's construction task performance and spatial ability: controlling task complexity and predicting mathematics performance. *Perceptual & Motor Skills*, *119*(3), 741–757.

Richardson, M., Jones, G., & Torrance, M. (2004). Identifying the task variables that influence perceived object assembly complexity. *Ergonomics*, *47*(9), 945–964.

Robert, M., & Héroux, G. (2004). Visuo-spatial play experience: forerunner of visuo-spatial achievement in preadolescent and adolescent boys and girls? *Infant and Child Development*, *13*(1), 49–78.

Rourke, B. P. (1993). Arithmetic disabilities, specific and otherwise: a neuropsychological perspective. *Journal of Learning Disabilities*, *26*(4), 214–226.

Rourke, B. P., & Conway, J. A. (1997). Disabilities of arithmetic and mathematical reasoning perspectives from neurology and neuropsychology. *Journal of Learning Disabilities*, *30*(1), 34–46.

Rousselle, L., & Noël, M. P. (2008). The development of automatic numerosity processing in preschoolers: evidence for numerosity-perceptual interference. *Developmental Psychology*, *44*(2), 544–560.

Smets, K., Sasanguie, D., Szücs, D., & Reynvoet, B. (2014). The effect of different methods to construct non-symbolic stimuli in numerosity estimation and comparison. *Journal of Cognitive Psychology*, *27*, 1–16.

Soltész, F., Szűcs, D., Dékány, J., Márkus, A., & Csépe, V. (2007). A combined event-related potential and neuropsychological investigation of developmental dyscalculia. *Neuroscience Letters*, *417*(2), 181–186.

Soltész, F., Szucs, D., & Szucs, L. (2010). Relationships between magnitude representation, counting and memory in 4-to 7-year-old children: a developmental study. *Behavioral and Brain Functions*, *6*(13), 1–14.

Stannard, L., Wolfgang, C. H., Jones, I., & Phelps, P. (2001). A longitudinal study of the predictive relations among construction play and mathematical achievement. *Early Child Development and Care*, *167*(1), 115–125.

St Clair-Thompson, H. L., & Gathercole, S. E. (2006). Executive functions and achievements in school: shifting, updating, inhibition, and working memory. *The Quarterly Journal of Experimental Psychology*, *59*(4), 745–759.

Swanson, H. L. (2006). Cognitive processes that underlie mathematical precociousness in young children. *Journal of Experimental Child Psychology*, *93*(3), 239–264.

Swanson, H. L. (2011). Working memory, attention, and mathematical problem solving: a longitudinal study of elementary school children. *Journal of Educational Psychology*, *103*(4), 821.

Szűcs, D., Devine, A., Soltesz, F., Nobes, A., & Gabriel, F. (2013a). Developmental dyscalculia is related to visuo-spatial memory and inhibition impairment. *Cortex*, *49*(10), 2674–2688.

Szűcs, D., Nobes, A., Devine, A., Gabriel, F. C., & Gebuis, T. (2013b). Visual stimulus parameters seriously compromise the measurement of approximate number system acuity and comparative effects between adults and children. *Frontiers in Psychology*, *4*, 444.

Szűcs, D., Devine, A., Soltesz, F., Nobes, A., & Gabriel, F. (2014). Cognitive components of a mathematical processing network in 9-year-old children. *Developmental Science*, *17*(4), 506–524.

Van der Ven, S. H., Kroesbergen, E. H., Boom, J., & Leseman, P. P. (2012). The development of executive functions and early mathematics: a dynamic relationship. *British Journal of Educational Psychology*, *82*(1), 100–119.

Verdine, B. N., Golinkoff, R. M., Hirsh-Pasek, K., Newcombe, N. S., Filipowicz, A. T., & Chang, A. (2014). Deconstructing building blocks: preschoolers' spatial assembly performance relates to early mathematical skills. *Child Development*, *85*(3), 1062–1076.

Wolfgang, C. H., & Phelps, P. (1983). Preschool play materials preference inventory. *Early Child Development and Care*, *12*(2), 127–141.

Xenidou-Dervou, I., De Smedt, B., van der Schoot, M., & van Lieshout, E. C. (2013). Individual differences in kindergarten math achievement: the integrative roles of approximation skills and working memory. *Learning and Individual Differences*, *28*, 119–129.

Xenidou-Dervou, I., Lieshout, E. C., & Schoot, M. (2014). Working memory in nonsymbolic approximate arithmetic processing: a dual-task study with preschoolers. *Cognitive Science*, *38*(1), 101–127.

Xu, F., & Spelke, E. S. (2000). Large number discrimination in 6-month-old infants. *Cognition*, *74*(1), B1–B11.

SECTION IV

MODELS

CHAPTER

15

Symbolic and Nonsymbolic Representation of Number in the Human Parietal Cortex: A Review of the State-of-the-Art, Outstanding Questions and Future Directions

H. Moriah Sokolowski, Daniel Ansari

Numerical Cognition Laboratory, Department of Psychology; Brain & Mind Institute, University of Western Ontario, Ontario, Canada

OUTLINE

Continuous Issues in Numerical Cognition. http://dx.doi.org/10.1016/B978-0-12-801637-4.00015-9

Without numbers, modern-day society could not exist. We could not build cities, calculate budgets, travel across the world, or democratically elect a government. Since the ability to understand numbers is a major pillar of our contemporary existence, it is critical to study how humans represent numbers. Humans have the unique ability to represent numbers either nonsymbolically or symbolically. For example, a set of six items can be represented as "••••••" (nonsymbolic), the Arabic digit "6," or the word "six" (symbolic). A prominent view in the field of numerical cognition is that nonsymbolic numbers are represented by an evolutionarily ancient and innate approximate number system (ANS), used specifically to represent nonsymbolic numerical magnitudes (Dehaene, Dehaene-Lambertz, & Cohen, 1998; Nieder & Dehaene, 2009; Pica, Lemer, Izard, & Dehaene, 2004). In contrast, symbolic numerical representations are thought to be learned over the course of development through cultural transmission (Coolidge & Overmann, 2012).

The overarching aim of this chapter is to review what is known about how nonsymbolic and symbolic numbers are represented in the human brain. In order to do this, the chapter will first provide a detailed and critical review of research examining neural activation during nonsymbolic and symbolic number processing. We will address the question of whether numbers are processed abstractly by distinguishing brain regions that respond to numbers across formats from brain regions that show format-specific processing of numbers. Throughout the chapter, we will discuss how the neural underpinnings of symbolic and nonsymbolic number processing change across development. We will close with a discussion of the association between the brain regions that respond to numerical (both symbolic and nonsymbolic) compared to nonnumerical magnitudes such as space, time, and luminance. Because the focus of this chapter is on neural representations underlying symbolic and nonsymbolic number processing in the parietal lobes, we will primarily discuss human neuroimaging studies.

15.1 NUMBER IN THE BRAIN

The first insights into what brain regions support numerical processing arose from case studies of patients with brain damage that resulted in number processing deficits. These studies reported that acquired calculation deficits appeared in patients with brain damage near the parieto-occipito-temporal junction (Gerstmann, 1940; Henschen, 1919). Patients with lesions in left fronto-parietal regions also showed a specific deficit for numerical processing (Cipolotti, Butterworth, & Denes, 1991; Dehaene & Cohen, 1997). This early indication that damage to the parietal cortex impairs numerical abilities has driven researchers to canvass the brain in search of the neural underpinnings of number processing.

15.2 NONSYMBOLIC NUMBER IN THE BRAIN

An intuitive sense of magnitude has been shown to have phylogenetic and ontogenetic continuity (Dehaene, 2007). Specifically, the ability to distinguish between nonsymbolic numerical magnitudes is thought to rely on an ANS (Feigenson, Dehaene, & Spelke, 2004; Nieder & Dehaene, 2009). Two basic properties of the ANS are as follows: (1) The ANS is used to process magnitudes greater than three or four items. It is thought that an exact number processing system is used to estimate fewer than three or four items (Demeyere, Rotshtein, & Humphreys, 2014; Le Corre & Carey, 2007; Piazza, Mechelli, Butterworth, & Price, 2002). (2) Accuracy and reaction times on numerical discrimination tasks are sensitive to the numerical ratio or the distance of the numerical quantities being compared. For example, the ratio of the numbers 2 and 8 is 0.25 and the distance of the numbers 2 and 8 is 6. Performance on numerical discrimination tasks (eg, judging which of two numbers is numerically larger) decreases as the ratio between the two quantities being compared increases. Similarly, performance on numerical discrimination tasks decreases as the distance between the two numbers being compared decreases (ie, the numerical distance effect). Typicially, participants are faster and more accurate at discriminating between the numbers 2 and 8 compared to discriminating between the numbers 2 and 3 (Buckley & Gillman, 1974; Halberda & Feigenson, 2008; Moyer & Landauer, 1967). In view of these effects, it has been argued that numerical quantities are analog and approximate rather than exact, since distance and ratio effects would not be present if each number was represented fully independently of adjacent numbers. Consistent with this interpretation, it is assumed that close numbers (such as 6 and 7) share overlapping representations and this overlap increases with the relative size of the numbers (their numerical ratio).

15.2.1 Adult Studies

Neuroimaging studies examining nonsymbolic number processing in humans have yielded a set of reproducible brain areas in the frontal and parietal lobes that are consistently activated by nonsymbolic number stimuli (Ansari & Dhital, 2006; Haist, Wazny, Toomarian, & Adamo, 2014; Lyons & Ansari, 2009; Piazza, Izard, Pinel, Le Bihan, & Dehaene, 2004; Piazza, Mechelli, Price, & Butterworth, 2006; Piazza, Pinel, Le Bihan, & Dehaene, 2007). These studies have focused on the parietal cortex and specifically the intraparietal sulcus (IPS) (depicted in Fig. 15.1) as supporting nonsymbolic numerical representations. Several neuroimaging experimental paradigms, including numerical discrimination tasks, ordering tasks, and functional magnetic resonance imaging (fMRI) adaptation studies, have been used to unpack the role of the IPS in nonsymbolic number processing (Castelli, Glaser, &

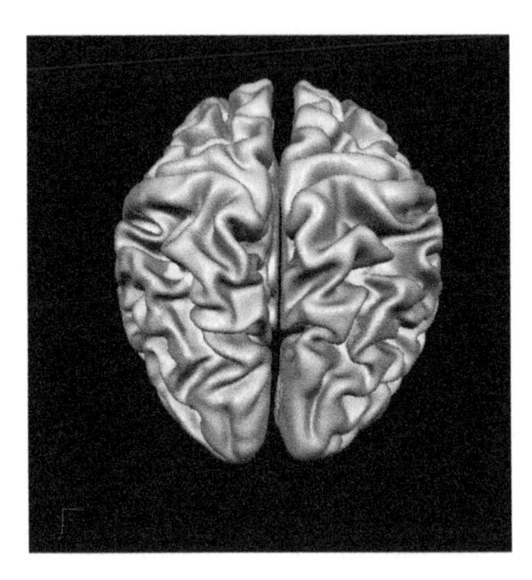

FIGURE 15.1 **Functional neuroimaging studies have implicated the bilateral intraparietal sulcus (hIPS) for the processing of numerical magnitudes.** The hIPS is shown in yellow. Generated using Brain Voyager Tutor Version 2.5.

Butterworth, 2006; Crollen, Grade, Pesenti, & Dormal, 2013; Dormal, Dormal, Joassin, & Pesenti, 2012; Roggeman, Santens, Fias, & Verguts, 2011).

Many neuroimaging studies have used a nonsymbolic number discrimination task to study brain regions that support the ANS. In this task, the participant is presented with two arrays of dots and must choose the array that has more dots (ie, the larger numerical quantity) (Fig. 15.2a). Such studies have demonstrated that regions in the frontal and parietal cortex, specifically the right IPS, are activated during nonsymbolic number discrimination tasks (Holloway, Price, & Ansari, 2010; Jacob & Nieder, 2009; Lyons & Beilock, 2013; Piazza et al., 2004, 2007; Roggeman et al., 2011). Moreover these investigations have revealed that parietal neural activation during the nonsymbolic number discrimination task is dependent on the ratio or distance between the two quantities being compared (Ansari, Dhital, & Siong, 2006; Ansari & Dhital, 2006; Cantlon, Libertus, Pinel, Dehaene, Brannon, & Pelphrey, 2009; Castelli et al., 2006; Holloway et al., 2010; Piazza et al., 2004, 2007). This effect is typically referred to as the neural ratio effect or the neural distance effect. Holloway and Ansari (2010) demonstrated this effect by revealing a decrease of activation in parietal regions as a function of nonsymbolic (and symbolic) distance, during numerical discrimination tasks. Specifically, the IPS showed greater neural activation in response to small numerical distances compared to large numerical distances (Holloway & Ansari, 2010). Taken together, there is a wealth of research that has implicated the IPS as important for ratio-dependent processing of nonsymbolic numerical magnitudes during nonsymbolic number processing.

Although comparison tasks are most commonly used to study the neural underpinnings of nonsymbolic number processing, it is important to

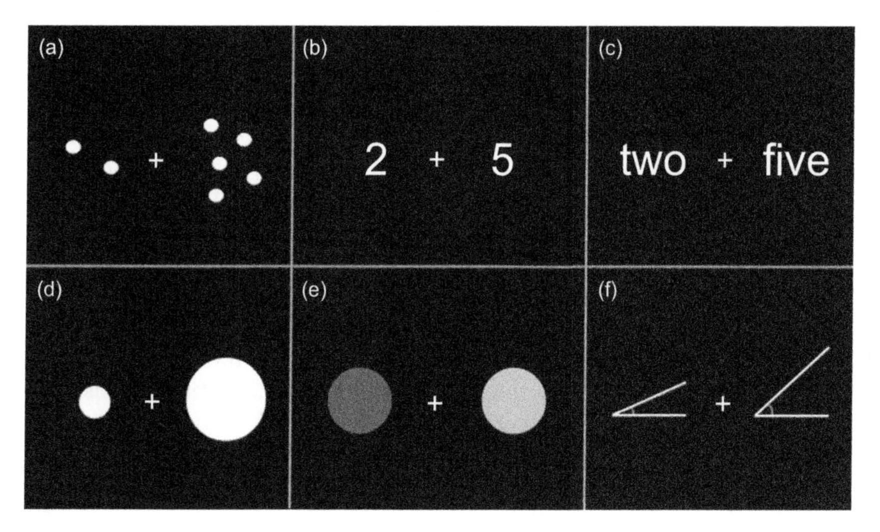

FIGURE 15.2 **Examples of numerical and nonnumerical magnitude stimuli used in discrimination tasks.** Three numerical magnitudes include: (a) nonsymbolic dot arrays; (b) Arabic digits; (c) number words. Three nonnumerical magnitudes include: (d) physical size; (e) luminance; (f) angles.

note that the IPS is also activated when participants select a response, regardless of whether the task is related to numerical processing (Culham & Kanwisher, 2001; Göbel, Johansen-Berg, Behrens, & Rushworth, 2004; Göbel & Rushworth, 2004). This means that neuroimaging studies that have implicated the IPS as important for number processing may be confounded by response selection. In support of this, Göbel and colleagues found that IPS activation did not differ when participants made numerical judgments (whether a number is larger or smaller than a reference number) compared to nonnumerical judgments (whether a symbol contains a vertical line). The absence of number-specific activation in the IPS has driven researchers to question whether the IPS responds to numerical stimuli or whether this activation reflects the neural correlates of general response selection.

In order to overcome the confound of response selection, researchers have used fMRI adaptation [a neuroimaging tool that can measure brain response to a stimuli in the absence of an active task (Grill-Spector, Henson, & Martin, 2006)] to reveal whether the IPS is activated during number processing in the absence of response selection. fMRI adaptation is a neuroimaging technique that measures brain changes that are the result of repeated exposure to a stimulus. The neuronal populations that are sensitive to a stimulus will "adapt" (ie, reduce in their response) as a function of stimulus repetition. When the stimulus changes the adapted regions recover from the repetition-induced reduction in activity. Piazza et al. (2004) were first to use fMRI adaptation methodology to study the neural correlates of nonsymbolic number processing (Piazza et al., 2004). In this study, neural populations were adapted to nonsymbolic arrays (with 16 or 32 dots). Deviant stimuli (with varying ratios between the adapted number and the deviant number) were presented during one in every eight repeats of the adapted number. Deviant numbers presented after habituation to 16 were 8, 10, 13, 16, 20, 24, and 32. Deviant numbers presented after habituation to 32 were 16, 21, 26, 32, 40, 48, and 64. In this study, deviant trials were correlated with neural activation in the IPS that increased as a function of the ratio between the adapted and deviant number (Piazza et al., 2004). Therefore, through the use of fMRI adaptation, these authors revealed that the IPS is activated by the passive viewing of nonsymbolic numerical magnitudes. This evidence implies that the IPS activation, reported in studies that use numerical discrimination tasks, cannot be solely explained by response selection related processing in the IPS. Therefore, converging evidence implicates the parietal cortex and specifically the IPS as an important brain region used to process nonsymbolic numerical magnitudes in human adults.

15.2.2 Developmental Studies

An important empirical question is whether children process nonsymbolic numerical magnitudes in the same way as adults. Neuroimaging has been used to address this question through the study of brain activation during nonsymbolic number processing in infants and children (Ansari & Dhital, 2006; Cantlon, Brannon, Carter, & Pelphrey, 2006; Cantlon, Libertus et al., 2009; Holloway & Ansari, 2010; Kaufmann et al., 2008). The findings of these neuroimaging studies have been disparate. While some studies have found little or no differences in the neural correlates of nonsymbolic number processing as a function of age, others have revealed age-related differences. A study by Cantlon et al. (2006) used an fMRI adaptation paradigm to examine neural responses to visual arrays where the number of elements presented deviated from a standard dot array stimulus by a ratio of 2:1 in 4-year-old children and adults. Results revealed that the IPS responded similarly in adults and 4-year-old children during passive viewing of nonsymbolic numbers. These findings indicate that the neural underpinnings of nonsymbolic numerical processing show continuity across development (Cantlon et al., 2006). In a similar vein, Izard, Dehaene-Lambertz, and Dehaene (2008) compared visual event-related potentials in 3-month-old infants, evoked by changes in the cardinal properties of a set. These data revealed that 3-month-old infants produced activation similar to adult participants in a right frontoparietal network during nonsymbolic number processing (Izard et al., 2008). These results provide further evidence to support the notion that there is developmental continuity in the neural correlates underlying nonsymbolic number processing. However, several studies have highlighted age-related differences in brain activation patterns supporting the ratio effect during nonsymbolic number discriminations (Ansari & Dhital, 2006; Holloway & Ansari, 2010). One such study by Ansari and Dhital (2006) used fMRI to examine the neural correlates that support nonsymbolic magnitude judgments in children compared to adults. Although behavioral performance was approximately equivalent across children and adults, adult participants exhibited a larger neural distance effect than children in the left IPS. This study revealed age-related differences in nonsymbolic number processing (Ansari & Dhital, 2006). The finding that the neural underpinnings of nonsymbolic number processing change across developmental time has been replicated across several neuroimaging studies (Ansari & Dhital, 2006; Holloway & Ansari, 2010; Kaufmann et al., 2008). There is a critical difference between the study by Cantlon et al. that demonstrated developmental continuity for the processing of nonsymbolic number compared to studies that find change in neural activation across development. Namely, Cantlon et al. utilized a passive design. In contrast, studies highlighting age-related changes (Ansari & Dhital, 2006; Holloway et al., 2010) used active paradigms. Therefore, these conflicting results may be due to differences between implicit compared to explicit processing of nonsymbolic numerical magnitudes. Specifically, implicit processing of nonsymbolic numbers may not change much across development, while explicit processing of nonsymbolic numbers may change greatly as a function of age. The distinct developmental trajectories for explicit compared to implicit processing may be due to the fact that unlike implicit tasks, explicit tasks require the individual to differentiate between various aspects of the stimuli in order to focus on a specific aspect of the stimuli and make a response. For example, during implicit processing of nonsymbolic stimuli the participant may observe numerosity as well as other continuous visual properties of the dot arrays (such as the area of the dots). However, the explicit processing of nonsymbolic stimuli requires focusing on numerosity and therefore ignoring or inhibiting other visual properties of the stimuli. Importantly, this proposed distinction between the developmental trajectories of implicit and explicit processing of nonsymbolic numbers has never been empirically tested within a single sample. Together, these pediatric neuroimaging studies suggest that the system used to approximate numerosities develops early in human ontogeny, but also report age-related differences in the neural processing of nonsymbolic numbers across development.

15.3 SYMBOLIC NUMBER IN THE BRAIN

Unlike nonhuman animals, humans have the unique ability to represent numbers both nonsymbolically and symbolically. Nonsymbolic representations of numerical magnitudes are iconic representations. This means that the quantity of a nonsymbolic number can be determined through counting or estimation. Symbolic numbers are culturally acquired representations of numerical magnitudes. For example, the number word "two" or the Arabic digit "2" represents the numerical magnitude two. The relationship between symbols and numerical magnitudes are noniconic, meaning that the sound or shape of the symbol is not directly related to the numerical magnitude that it represents. Thus, the numerical meaning of number symbols must be learned over the course of development. Research has investigated which brain regions or network of brain regions are important for the acquisition of symbolic representations of numerical magnitudes (Ansari, Fugelsang, Dhital, & Venkatraman, 2006; Ansari, Garcia, Lucas, Hamon, & Dhital, 2005; Bugden, Price, McLean, & Ansari, 2012; Holloway, Battista, Vogel, & Ansari, 2013; Pinel, Dehaene, Rivière, & LeBihan, 2001). The following section will review these studies and discuss what they have revealed about symbolic representations of numerical magnitudes.

15.3.1 Adult Studies

Functional neuroimaging methods have been used to determine brain regions that are activated during symbolic number processing in human adults and children. In adults, a variety of neuroimaging experimental paradigms have been applied. These include symbolic number discrimination paradigms (Holloway et al., 2010; Pinel et al., 1999, 2001), ordinal processing tasks (Lyons & Beilock, 2013), and fMRI adaptation passive viewing paradigms (Holloway et al., 2013; Notebaert, Pesenti, & Reynvoet, 2010). The finding that the bilateral IPS supports symbolic number processing has been predominantly driven by studies that examined neural activation during cardinal processing (Ansari & Dhital, 2006; Ansari, Fugelsang et al., 2006; Kaufmann et al., 2005; Piazza et al., 2007; Pinel et al., 1999, 2001). Cardinality is the numerical quantity of a set of items. Specifically, it refers to how many items are in a set. Typically, neuroimaging studies used symbolic number discrimination tasks, which tap into the neural correlates associated with processing the relative cardinality of symbolic numbers (Ansari, Fugelsang et al., 2006; Holloway et al., 2010; Pinel et al., 1999, 2001). These neuroimaging studies have revealed consistent activation in regions of the bilateral IPS (Fig. 15.1) during symbolic numerical discrimination tasks.

Moreover, similarly to nonsymbolic numbers, neural activation during symbolic discriminations is modulated by the distance of the symbols being compared (Holloway & Ansari, 2010). Therefore, in general, the neural underpinnings of symbolic numerical discrimination tasks, namely ratio- and distance-dependent IPS activation, parallel brain activation during nonsymbolic discrimination tasks. There have also been studies investigating symbolic ordinality, which is the position of numbers within a sequence (Franklin & Jonides, 2008; Lyons & Beilock, 2013). To date, it is unclear whether number discrimination tasks and ordering tasks rely on similar or different networks. However, considering ordinality may shed light on the flexibility and precision of symbolic representations of numerical magnitudes.

The results of the symbolic studies that use active paradigms, such as a numerical discrimination task, must be interpreted with caution as they are confounded by response selection. In an effort to study the neuronal correlates of symbolic number processing in the absence of response selection, several studies have used fMRI adaptation (Notebaert, Nelis, & Reynvoet, 2011; Piazza et al., 2007). One such study by Notebaert and colleagues used fMRI adaptation to investigate which brain regions were activated during the passive viewing of small (single digit, such as "6") and large (double digit, such as "32") numbers (Notebaert et al., 2011). Six numerical deviants, with varying ratios between the adapted number and the deviant number, were presented during the repetition of both the small and large number adaptation conditions. The left IPS responded to numerical deviants in both the small and large condition. The neural response in the left IPS increased as the ratio between the adapted and deviant number increased (Notebaert et al., 2011). Holloway et al. (2013) used a similar fMRI adaptation paradigm as implemented by Notebaert et al. (2011) and replicated these findings (Holloway et al., 2013). Together, these studies demonstrate that the left IPS responds to the passive viewing of symbolic numbers in a ratio-dependent manner. Therefore, as was found to be case for nonsymbolic numerical magnitudes, discussed previously, symbolic number processing in the IPS can be demonstrated even when the task does not require response selection.

15.3.2 Developmental Studies

Symbolic number knowledge is a culturally acquired ability rather than a biological primitive, and thus must be learned over the course of development. Therefore, cognitive neuroscientists have predicted that the brain response to number symbols will change over development. This prediction has been tested in a small but growing body of research that has studied the neural underpinnings of symbolic number processing across developmental time (Cantlon, Libertus et al., 2009; Holloway & Ansari, 2010; Kaufmann et al., 2006). These studies show that in children prefrontal regions tend to be more strongly activated during symbolic number comparisons compared to adults. Importantly, IPS activation during symbolic number comparison increased across development (Ansari et al., 2005; Kaufmann et al., 2006). This frontoparietal shift from childhood to adulthood may reflect a decrease in the need for domain-general cognitive resources such as working memory and attention as children begin to process number symbols automatically.

Response selection confounds in number discrimination studies are more profound in developmental studies because response time and accuracy differ in children compared to adults. Therefore, the neural activity differences observed in children compared to adults may be driven by differences in response performance rather than symbolic number processing. An fMRI adaptation study with a similar design as Notebaert et al. (2011) and Holloway et al. (2013) was used to investigate the neural underpinning's of symbolic number processing in children aged 6–14 years (Vogel, Goffin, & Ansari, 2014). In this study, Vogel et al. examined the neural ratio-dependent response to symbolic number deviants across a group of children of varying ages. The right IPS was found to exhibit a ratio-dependent response to numerical deviants across children of all ages. In contrast, activation in the left IPS was correlated with age, revealing that older children activated the left IPS more than younger children when passively viewing numerical symbols (Fig. 15.3) (Vogel et al., 2014). This suggests that the left and right parietal lobes have distinct developmental

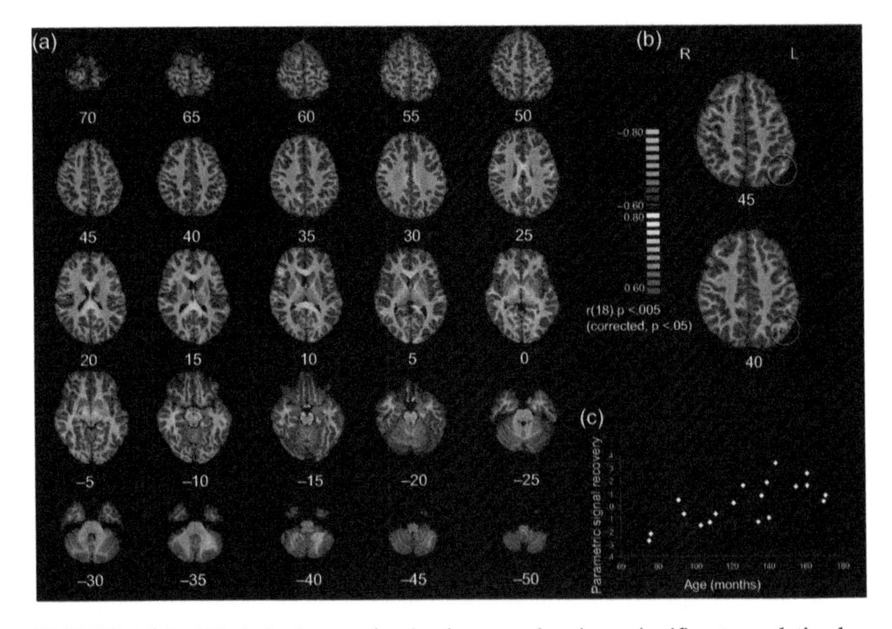

FIGURE 15.3 **Whole brain neural activation map showing a significant correlation between age and the ratio-dependent signal recovery in the left IPS.** (a) Axial brain slices labeled according to Talairach coordinates on the *z*-axis covering the whole brain. Slices are ordered from superior to inferior. (b) Two axial brain slices showing the correlation between ratio dependent activation in the left IPS and age. (c) A graphical representation of the correlation between age and the neural ratio effect in the left IPS. Scatterplot between age and ratio dependent signal recovery in the left IPS. *From Vogel et al. (2014), Fig. 4.*

trajectories during the acquisition of symbolic numerical processing. Importantly, in contrast to other studies examining the neural correlates of symbolic number representations across development (Cantlon, Libertus et al., 2009; Kaufmann et al., 2006), Vogel et al. (2014) did not find a decrease in activation of frontal regions across development for the passive viewing of symbolic numbers. This suggests that it is possible that the decrease in activation in the frontal cortex across development is related to response selection rather than basic number processing. Therefore, this study highlights the importance of the IPS for processing number across development.

In order to determine whether this age-related specialization in the parietal cortex is related to improvement in math ability across development, Bugden et al. (2012) investigated the relationship between the neural correlates that support symbolic number discrimination and mathematical abilities in children. Results revealed that children who scored higher on a mathematics fluency test exhibited more ratio-dependent activation in the left IPS during symbolic number discrimination tasks (Bugden et al., 2012). This finding shows that children with more mature neural patterns during symbolic discriminations exhibit higher mathematical competence. The developmental neuroimaging research, discussed previously, illustrates hemispheric asymmetries in the neural development underlying the acquisition of symbolic number comprehension. Specifically, the left IPS seems to become increasingly specialized for automatic symbolic number processing as a function of children's chronological and mathematical competence.

15.4 ABSTRACT REPRESENTATION OF NUMERICAL MAGNITUDES

Thus far, empirical evidence concerning the representation of nonsymbolic and symbolic number in the brain has been reviewed separately. In addition to this, a large body of research has evaluated the extent to which nonsymbolic and symbolic numerical representations are qualitatively similar or different from one another at the neural level (Cantlon, Libertus et al., 2009; Cohen Kadosh, Cohen Kadosh, Kaas, Henik, & Goebel, 2007; Holloway et al., 2010; Piazza et al., 2007). This work has explored whether the brain represents numbers using (1) an abstract (format-independent) system, (2) a format-dependent system, or (3) both a format-independent and format-dependent system for different levels of processing.

Researchers have hypothesized that symbolic numbers are rooted in the previously discussed primitive nonsymbolic ANS (Dehaene & Cohen, 1997; Eger, Sterzer, Russ, Giraud, & Kleinschmidt, 2003; Nieder & Dehaene, 2009; Piazza et al., 2007). Studies that have investigated the neural correlates of a single format of numerical representation have implicated the IPS during number discrimination paradigms when the numerical stimuli were presented symbolically, as Arabic digits (Ansari & Dhital, 2006; Venkatraman, Ansari, & Chee, 2005) and number words (Ansari, Fugelsang et al., 2006; Cohen Kadosh et al., 2007; Eger et al., 2003), or as nonsymbolic dot arrays (Ansari & Dhital, 2006; He, Zuo, Chen, & Humphreys, 2013; Holloway et al., 2010). The fact that independent studies with heterogeneous number formats all report activation in the bilateral IPS during numerical discrimination paradigms suggests that regions in the IPS are used to process format-independent quantities. However, such evidence is qualitative in nature and there is a need to address the notion of an abstract representation of number quantitatively. In view of this, several neuroimaging techniques have been used to directly test whether different number formats share common neural representations.

fMRI conjunction analyses have been used to determine which brain regions process numbers abstractly by revealing which brain regions are activated by both symbolic and nonsymbolic numbers (Cantlon, Libertus et al., 2009; Holloway & Ansari, 2010; Holloway et al., 2010). fMRI conjunction analysis is a statistical tool that is used to reveal whether two or more conditions activate the same brain regions. The results of these studies revealed that regions in the parietal cortex, specifically the left superior parietal lobule (SPL) (Cantlon, Libertus et al., 2009) and the right inferior parietal lobule (IPL) (Holloway & Ansari, 2010; Holloway et al., 2010), are activated by the conjunction of symbolic and nonsymbolic numbers. Therefore, although conjunction analyses consistently find overlapping activation for symbolic and nonsymbolic numerical magnitudes, the laterality of the region of parietal cortex thought to underlie abstract numerical representations remains unclear.

Another method used to further illuminate which parietal regions support abstract numerical processing is to examine whether processing symbolic numbers also activates regions that relate to nonsymbolic representations of number and vice versa. This method is referred to as cross-format activation. In an fMRI adaptation study that examined cross-format activation for symbolic and nonsymbolic numbers (Piazza et al., 2007), participants were adapted to either symbolic or nonsymbolic numerical quantities. Participants were subsequently presented with numerical deviants in either the same or different format as the adapted number (either symbolic or nonsymbolic). Results revealed that a bilateral frontoparietal network responded to cross-format numerical deviants. Specifically, the numerical distance between the adapted and deviant number correlated with bilateral IPS activity. These data support the idea that the IPS hosts an abstract representation of numerical quantity. However, contrary recent findings show that symbolic and nonsymbolic numbers may be represented more distinctly than was previously assumed (Bulthé, De Smedt, & Op de

Beeck, 2014; Cohen Kadosh & Walsh, 2009; Damarla & Just, 2013; Eger et al., 2009).

Cohen Kadosh and Walsh mounted a significant challenge to the commonly held view that number is represented abstractly in the brain (Cohen Kadosh & Walsh, 2009). First, they highlighted how findings supporting the idea that number is represented abstractly are often based on null results. Specifically, researchers have claimed that different notations of number have similar representations because researchers found no differences in brain activation patterns between different notations of number (Shuman & Kanwisher, 2004). Second, they argued that similar brain activations may be driven by distinct neuronal populations that cannot be distinguished using traditional neuroimaging techniques (Damarla & Just, 2013; Raizada & Poldrack, 2007). Last, they suggested that shared neural representations may be driven by specific task requirements rather than abstract numerical processing (Ansari, 2007; Göbel et al., 2004).

There is an important distinction between the format-dependent processing hypothesis presented by Cohen Kadosh and Walsh (2009) and ongoing work examining the neural correlates of symbolic and nonsymbolic numerical magnitude processing (Holloway & Ansari, 2010; Lyons, Ansari, & Beilock, 2014; Shuman & Kanwisher, 2004; Venkatraman et al., 2005). In the format-dependent processing hypothesis, number words, Arabic digits, and nonsymbolic dot arrays (Fig. 15.2a–c) are conceptualized orthogonally from one another. This hypothesis focuses more on differences between symbolic number formats (such as number words compared to Arabic digits) than differences between these symbolic number formats and nonsymbolic numbers (Cohen Kadosh et al., 2007). As already alluded to previously, current theories and computational models suggest that the acquisition of symbolic numerical representations may be grounded in a nonsymbolic number processing system (Coolidge & Overmann, 2012). Consequently, an empirical evaluation of differences and similarities in neural activation during symbolic and nonsymbolic numerical processing is necessary to determine how symbolic numbers are constructed. Because a primary focus of the format-dependent processing hypothesis (Cohen Kadosh & Walsh, 2009) is comparing brain representations of different symbolic representations, it does not address whether symbolic numbers are mapped onto nonsymbolic representations of number. Therefore, although the format-dependent processing hypothesis provides valuable insights on the neural underpinnings of different number formats, comparing symbolic and nonsymbolic number is critical to determine whether symbols are rooted in an approximate nonsymbolic number system.

Indeed, recent empirical research has highlighted striking differences between brain activation patterns of symbolic and nonsymbolic numerical stimuli (Cantlon, Libertus et al., 2009; Holloway et al., 2010; Piazza et al., 2007; Venkatraman et al., 2005). For instance, Holloway et al. (2010)

used a number discrimination paradigm to test whether the functional neuroanatomy underlying symbolic and nonsymbolic processing is distinct. These authors found that the right IPS was active for both symbolic and nonsymbolic stimuli. In contrast to the format-general activation in the right IPS, symbolic numbers produced specific activation in the left angular and left superior temporal gyri, while nonsymbolic number processing recruited regions in the right posterior SPL (Holloway et al., 2010). Additional differences between the neural underpinnings of symbolic and nonsymbolic numbers were found through an investigation of differences between neural activation during ordinal processing compared to number comparison (Lyons & Beilock, 2013). Moreover, activation during the number comparison task was found in the bilateral IPS for both symbolic and nonsymbolic number processing. Importantly, although ordinal processing of nonsymbolic numbers also correlated with activation the bilateral IPS, IPS activation was not observed during the symbolic ordering task. Instead, the left premotor cortex was activated during symbolic ordering. These results highlight striking differences between the neural correlates of symbolic compared to nonsymbolic numbers during ordinal processing (Lyons & Beilock, 2013). In a different study, brain activation also differed during symbolic compared to nonsymbolic calculation tasks (Venkatraman et al., 2005). Specifically, nonsymbolic addition produced greater activation in brain regions in the right parietal and frontal cortex compared to symbolic addition. Together, these studies suggest that distinct regions in the parietal cortex are used to process symbolic and nonsymbolic numbers. Moreover, these data imply format-dependent hemispheric specialization in the parietal cortex.

Consistent with the previous findings, fMRI adaptation studies have revealed distinct activation during the passive viewing of symbolic compared to nonsymbolic numbers (Cohen Kadosh et al., 2011; Piazza et al., 2007). Although Piazza et al. concluded that the IPS processes numbers abstractly, the IPS exhibited an asymmetry in the neural responses to symbolic compared to nonsymbolic numbers. More specifically, recovery in the left IPS was modulated by numerical distance when the left IPS was adapted to dots and the deviant stimulus was a digit. However, when deviant dots were presented among adapted digits, the recovery in the IPS did not relate to numerical distance. Therefore, although Piazza et al. support the idea that number is represented abstractly, the data revealed asymmetry in the neural responses in the IPS for symbolic compared to nonsymbolic numbers. Hence, this study provides some support for format-dependent processing of symbolic and nonsymbolic numbers. In another fMRI adaptation study, the IPS was adapted to a specific number format (eg, "2") and then the participant was presented with one of the following deviants: (1) a different number format that represented the same numerical quantity (eg, "••"), (2) a different numerical quantity (eg, "••••"), and (3) the same number format that represented a different numerical quantity (eg, "4") (Cohen Kadosh et al., 2011). Results revealed that number format changes led to greater IPS activation compared to quantity changes (Cohen Kadosh et al., 2011). This finding speaks against a cross-format coding of quantity, but rather suggests differences in format processing that are independent of quantity change.

The studies discussed previously used a univariate fMRI approach to examine distinct and overlapping regions of neural activation during symbolic and nonsymbolic number processing. This approach has methodological limitations which impact the inferences that can be drawn. In particular, the univariate fMRI approach has a relatively coarse spatial resolution because it averages activation across voxels (three-dimensional pixels in the brain). Therefore, although a specific brain region was activated by two different experimental conditions (ie, symbolic and nonsymbolic), the signal, giving rise to the activation associated with the two conditions could be different. In other words, spatial overlap in univariate analysis cannot be used to infer representational similarity. To address this problem, researchers have implemented novel multivariate analytic techniques often referred to as multi-voxel pattern analysis (MVPA). Rather than averaging activation across voxels, MVPA analyzes patterns of activation across groups of voxels. The utilization of this technique has provided a more fine-grained understanding of IPS activation during symbolic and nonsymbolic number processing (Damarla & Just, 2013; Eger et al., 2009; Lyons et al., 2014). Eger et al. (2009) were the first to use multivariate pattern recognition to decode patterns that represented specific numerical quantities in symbolic and nonsymbolic formats. The results revealed distinct patterns of activation in the IPS for symbolic compared to nonsymbolic number formats. The activation patterns in the parietal cortex for a specific quantity represented by a dot array could be predicted above chance (57%) by the brain activation patterns in the IPS evoked by digits. However, the activation pattern in the IPS elicited by the dots could not predict activation during processing of Arabic digits above chance (51%) (Eger et al., 2009). In a similar study, Damarla and Just (2013) examined the utility of using multi-voxel patterns in the IPS to distinguish between (1) two different quantities in the same format (such as "3" from "5") and (2) two different formats representing the same quantity (such as "3" from "•••"). Results from this study revealed that multi-voxel patterns in the IPS can be used to classify different quantities, represented in the same format (" 3" vs. "5") but not the same quantity presented in different formats ("3" vs. "•••"). These findings were replicated at the whole brain level (Bulthé et al., 2014). A related investigation used representational similarity analysis (RSA) to examine the correlation between multi-voxel patterns for symbolic compared to nonsymbolic numbers (Lyons et al., 2014). The results show correlations between patterns of activation for two numbers

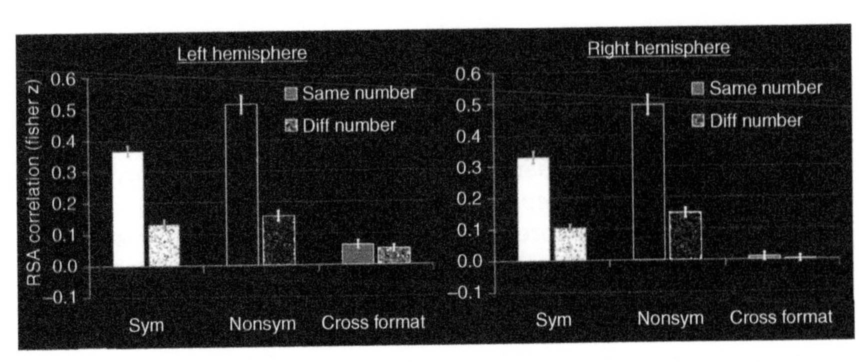

FIGURE 15.4 **Correlations between brain activations for the same numbers (eg, 2 and 2) and different numbers represented symbolically are shown in white.** Correlations between brain activations for the same numbers and different numbers represented nonsymbolically are shown in black. The gray bars represent correlations across symbolic and nonsymbolic number formats for the same number and different number. Correlation values were transformed using Fisher-z. Error bars represent standard error of the mean. *From Lyons et al. (2014), Fig. 7.*

that were presented in the same format (either both symbolically or both nonsymbolically). There was no correlation between patterns of activation for numbers presented in different formats (one symbolically and the other nonsymbolically) (Fig. 15.4) (Lyons et al., 2014). Together, these multivariate analyses support the idea that the representations for symbolic and nonsymbolic numbers in the human IPS are not as tightly linked as was previously thought and thereby challenge the notion that the IPS represents symbolic and nonsymbolic representations of number in an abstract format.

15.4.1 Developmental Studies

It remains unclear whether symbolic numerical representations are scaffolded by nonsymbolic representation during development. Researchers have hypothesized that the meaning of number symbols is directly linked to the corresponding nonsymbolic magnitude (Dehaene & Cohen, 1997; Dehaene et al., 1998; Feigenson et al., 2004). A small but growing body of research has compared symbolic and nonsymbolic number processing in children. fMRI conjunction analyses revealed regions in the parietal cortex that respond to both symbolic and nonsymbolic numbers in children (Cantlon, Libertus et al., 2009; Holloway & Ansari, 2010). Specifically, Cantlon and colleagues (2009) used an fMRI numerical discrimination paradigm to examine the brain mechanisms that 6- and 7-year-old children and adults use to solve both symbolic and nonsymbolic numerical comparison tasks across formats. The data revealed that 6- and 7-year-old children and adults produced activation in the bilateral occipitotemporal and left SPL during symbolic and nonsymbolic number discrimination paradigms. Children also recruited the inferior frontal cortex more than adults on the number discrimination task (Cantlon, Libertus et al., 2009). A related study compared brain activation in 7- and 8-year-old children compared to adults during symbolic and nonsymbolic number discrimination paradigms (Holloway & Ansari, 2010). The data revealed age-related increases in right IPS activation related to the conjunction of symbolic and nonsymbolic number comparison. Adults exhibited a neural ratio effect in the right IPS for both symbolic and nonsymbolic stimuli. However, children exhibited the same neural ratio effect for nonsymbolic but not for symbolic stimuli (Holloway & Ansari, 2010). This finding illuminates the neural underpinnings that may support the acquisition of symbolic knowledge, namely, that the processing of symbolic numerical magnitudes in the right IPS is a result of developmental specialization. A metaanalytic synthesis of eight studies of typically developing children revealed that children produce consistent frontoparietal activation during number processing (Kaufmann, Wood, Rubinsten, & Henik, 2011). This study provided a qualitative comparison of activation across symbolic and nonsymbolic developmental neuroimaging studies and revealed minimal overlap in the neural response to symbolic compared to nonsymbolic stimuli. On the one hand, symbolic number processing produced activation in the bilateral parietal lobes, including the left SPL and the right IPS. On the other hand, nonsymbolic number processing was associated with activation in the right anterior IPS (Kaufmann et al., 2011). These developmental data do not support the idea that numbers are processed using an abstract number representation system. Instead, this research suggests that children have format-specific representations of symbolic and nonsymbolic number in the parietal cortex and that these vary as a function of hemisphere.

15.5 NONNUMERICAL MAGNITUDES

As the previous review of studies illustrates, regions in the parietal cortex, specifically along the bilateral IPS, have been consistently implicated as critical for the processing of numerical magnitudes (Ansari, Dhital et al., 2006; Nieder & Dehaene, 2009). The consensus in the field of numerical cognition has been that number operates within its own domain and that the IPS hosts a specific number processing system (Dehaene, Piazza, Pinel, & Cohen, 2003; Piazza et al., 2007). However, behavioral and neuroimaging research has led researchers to question whether the IPS hosts a domain-specific number processing system or whether it indeed houses a more general system used to process both numerical and nonnumerical magnitudes. In what follows, we outline research that suggests that number is processed using a general magnitude system.

Since the late 1960s, research has shown behavioral parallels between numerical and nonnumerical magnitude processing (Moyer & Landauer, 1967). A nonnumerical magnitude refers to the size or extent of a continuous dimension such as space, time, or luminance (Fig. 15.2d–f). The striking behavioral similarities between numerical and nonnumerical magnitude processing have been empirically replicated (for reviews see Cantlon, Platt, & Brannon, 2009b; Cohen Kadosh, Lammertyn, & Izard, 2008).

This accumulation of evidence led to the proposal of a theory of magnitude (ATOM) (Walsh, 2003). ATOM was the first theoretical framework to suggest that we have a generalized magnitude system used to represent number, space, and time. Since the generation of ATOM, researchers have fiercely debated whether the human brain represents number using a number-specific representational system or a general system used to process numerical and nonnumerical magnitudes (Cantlon et al., 2009b; Cohen Kadosh et al., 2008; Simon, 1999; Walsh, 2003).

15.5.1 Symbolic Numbers and Nonnumerical Magnitudes

Researchers have used neuroimaging techniques to examine neural activation during the processing of numerical compared to nonnumerical magnitudes (Dormal, Andres, & Pesenti, 2012; Dormal & Pesenti, 2009; Fias, Lammertyn, Reynvoet, Dupont, & Orban, 2003; Pinel, Piazza, Le Bihan, & Dehaene, 2004). These studies have reported both distinct and overlapping brain regions for numerical and nonnumerical magnitude processing (Cohen Kadosh et al., 2008). The first of these studies by Fias et al. (2003) used positron emission tomography (PET) to examine neural activity during magnitude discriminations of nonnumerical (line lengths and angle sizes) and numerical (two-digit Arabic number symbols) stimuli. The left IPS was activated during numerical and nonnumerical magnitude processing, supporting the notion that a single general magnitude processing system is used to process numerical and nonnumerical magnitudes. However, the anterior left IPS showed greater activation during numerical compared to nonnumerical judgments (Fias et al., 2003). Subsequent studies used fMRI to examine activation during numerical, physical size, and luminance discrimination tasks (Cohen Kadosh et al., 2005; Pinel et al., 2004). Specifically, results from Pinel et al. (2004) revealed that number and size comparison tasks both activated a parietal network including regions in the IPS, while size and luminance comparison tasks activated regions in the occipitotemporal cortex (Pinel et al., 2004). Consistent with this finding, Cohen Kadosh et al. (2005) reported IPS activation during number, physical size, and luminance comparison tasks. However, the left IPS and right temporal regions were specifically activated by the number comparison task (Cohen Kadosh et al., 2005). A recent study extended this line of research by examining neural activation during a

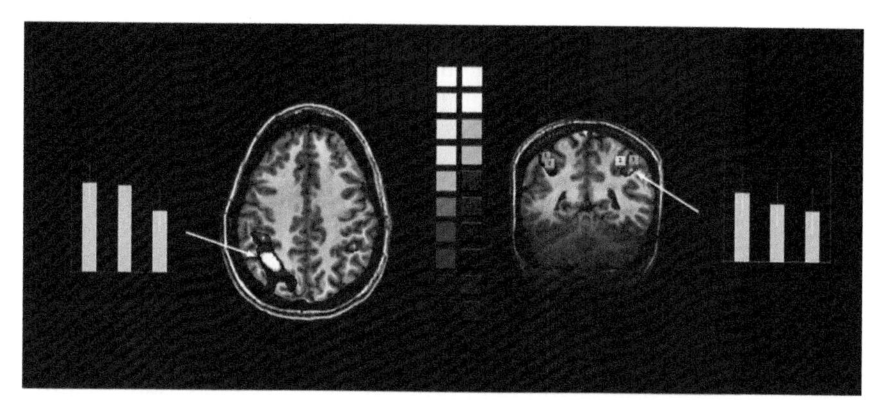

FIGURE 15.5 **Neural activation during the spatial estimation of number and brightness.** Areas in *blue* represent areas that were activated during the spatial estimation of both number and brightness. Areas in *red* represent regions that were activated only when the position of numbers was estimated. *From Vogel et al. (2013), Fig. 3.*

number line estimation task, where participants mapped numerical (Arabic digits) and nonnumerical (luminance) stimuli onto a number line (Vogel, Grabner, Schneider, Siegler, & Ansari, 2013). The results of this study revealed right IPS activation during numerical and nonnumerical magnitude estimations. In this study, symbolic number estimation was specifically correlated with additional activation in the bilateral IPS (Fig. 15.5) (Vogel et al., 2013). Together, these studies reveal that both overlapping and distinct neural populations are used to process numerical and nonnumerical magnitudes. Specifically, the right IPS shows overlap between numerical and nonnumerical magnitude processing. In addition, these studies have consistently revealed involvement of the left IPS for symbolic number processing. These findings parallel results of studies that compared brain regions associated with symbolic and nonsymbolic numerical representations.

15.5.2 Nonsymbolic Numbers and Nonnumerical Magnitudes

All of the studies with nonnumerical stimuli discussed compared symbolic numerical stimuli with nonnumerical magnitudes. Converging and diverging activation was also found for number and nonnumerical magnitudes when number was represented nonsymbolically, as a discrete array of dots or as a series of tones (Dormal, Dormal et al., 2012; Dormal & Pesenti, 2009). One such study by Castelli et al. (2006) revealed that processing nonsymbolic numerical magnitudes (estimation of "how many things") compared to processing of nonnumerical magnitudes (estimation of "how much" or "how long") produced greater activation in the bilateral IPS. In a similar study, Dormal and Pesenti (2009) examined neural activation in response to nonsymbolic magnitudes (lines made up of discrete dots) and nonnumerical magnitudes (continuous line lengths). While nonsymbolic numerical magnitudes and nonnumerical magnitudes activated the right IPS, the left IPS was specifically activated during nonsymbolic number processing but not during nonnumerical processing. This study provides additional support for the idea that the right IPS hosts a general system used to process all magnitudes and indicates that the left IPS may be specialized for processing numerical quantities (Dormal & Pesenti, 2009). This idea was further supported in a study that showed that the right IPS was commonly activated by nonsymbolic number and the nonnumerical magnitude, duration (Dormal, Dormal et al., 2012). Lateralization of general magnitude processing and symbolic number processing was reported by Chassy and Grodd (2012). This study was first to examine both symbolic (positive and negative integers) and nonsymbolic numbers (dot arrays) as well as nonnumerical magnitudes (disk size) within a single experimental design. In this study, nonsymbolic dot arrays were grouped with nonnumerical disk size in their analyses. Brain activation patterns during processing of dot arrays and disk sizes were compared to brain activation patterns during processing of positive and negative Arabic digits. The results demonstrated that dots, disks and Arabic digits correlated with activation in the right IPS. However, only symbolic number processing correlated with activation in the left IPS (Chassy & Grodd, 2012). Overall, many of these studies indicate that the right IPS houses a general magnitude system used to process both numerical and nonnumerical magnitudes, while activation in the left IPS seems to be related to the processing of symbolic numbers. However, some of the studies discussed in this review concluded that nonsymbolic numerical processing is stronger in the left IPS when compared with nonnumerical magnitudes. Therefore, the notion of left lateralization for symbols is not unequivocal across studies. In view of this, quantitative metaanalytic tools must be used to systematically evaluate trends in the hemispheric asymmetry of symbolic and nonsymbolic number processing across empirical studies.

15.5.3 Continuous Visual Properties in Nonsymbolic Dot Arrays

Continuity between animals, infants, and human adults in the ability to discriminate between nonsymbolic numerical magnitudes provided early evidence for the theory that humans possess a number-specific neural system such as the "approximate number system" rather than a general magnitude system. Importantly, this research has most commonly used dot arrays to measure nonsymbolic numerical representations. Therefore, it is critical for researchers to recognize that estimating the numerical magnitude of a nonsymbolic dot array is inherently confounded by the continuous properties (such as area and density) associated with that dot array. This means that if a quantity changes, the continuous properties associated with that quantity also change. As it is not possible to completely control the effect of these continuous properties (Gebuis & Reynvoet, 2012; Leibovich & Henik, 2013), measuring nonsymbolic numerical magnitude processing is inherently confounded with measuring nonnumerical magnitude processing. Evidence showing that the IPS is responsive to numerical and nonnumerical stimuli (Dormal, Dormal et al., 2012; Dormal & Pesenti, 2009; Pinel et al., 2004) implies that nonsymbolic number and continuous visual cues are processed by the same underlying neural mechanism. However, the number of dots (often referred to as numerosity) in nonsymbolic dot arrays is always correlated with continuous visual properties (such as surface area or density of dots). For example, as the number of dots increases, the surface area of dots will also increase. It is therefore conceivable that coactivation of numerical and nonnumerical cues in neuroimaging research is due to the continuous visual properties present in both nonsymbolic numerical and nonnumerical processing.

The hypothesis that common behavioral and neural effects in numerical and nonnumerical processing are driven by continuous visual properties (visual cues) has been empirically tested using several methodologies (Gebuis & Reynvoet, 2012, 2014). More specifically, Gebuis and Reynvoet (2012) used electroencephalography (EEG) to study how number and visual cues affect N1 and P2 waveform components (Gebuis & Reynvoet, 2012). N1 and P2 components are lateral occipital electrode sites that have been shown to relate to nonsymbolic number processing (Temple & Posner, 1998). Specifically, Gebuis and Reynvoet (2012) conducted two experiments using electroencephalography to look at whether number and three visual cues, namely convex hull, aggregate surface area, and diameter, affect these components of the EEG time-course differentially. They did not specifically control for visual cues. Instead, to study number, this experiment averaged across the visual cue conditions in order to cancel out visual cue-related effects. Results revealed that visual cues affected the N1 waveform and both number and visual cues affected the P2 waveform component. Specifically, the N1 component became more negative with increasing convex hull and less negative with increasing aggregate surface area and increasing diameter. P2 visual cue effects were also in opposite directions for convex hull compared to aggregate surface area and increasing diameter. In the second experiment, Gebuis and Reynvoet (2012) controlled for visual cues by manipulating the three most prominent visual cues (convex hull, aggregate surface area, and diameter) in a way that ensured that each visual cue did not correlate with number across trials. This technique was used to disentangle the effect of number and visual cues. In other words, the main difference between experiment 1 and experiment

2 is that in experiment 2, visual cues did not increase or decrease with increasing number. Results from experiment 2 revealed that when visual cues were properly controlled, no significant increase or decrease of the N1 or P2 component was related to increasing or decreasing number. This study suggests that number-specific effects on brain activation are likely the result of improperly controlled visual cues rather than the specific processing of nonsymbolic numbers (Gebuis & Reynvoet, 2012). Therefore, these findings significantly challenge the theory that humans possess an ANS that automatically processes nonsymbolic numbers independently of correlated visual cues. The same authors extended this finding to the study of nonsymbolic ordering (Gebuis & Reynvoet, 2014). In particular, EEG was used to investigate the relationship between numerosity and visual cues during a passive viewing ordinality task. In this task, participants were presented with five nonsymbolic dot arrays. Both the numerosity and visual cues increased across the first four stimuli that were presented. The numerosity and visual cues of the fifth stimulus were manipulated in one of four ways (both increased, both decreased, numerosity increased while visual cues decreased, visual cues increased while numerosity increased). The results showed different neural responses when numerosity and visual processing cues changed in the same direction (eg, numerosity increased when surface area increased) compared to different directions (eg, numerosity increased when surface area decreased) (Gebuis & Reynvoet, 2014). Together, these studies support the idea that visual cues are linked to numerosity at the neuronal level across tasks. These insights into how sensory cues relate to the processing of quantities may be fundamental for furthering our understanding of how we represent nonsymbolic numbers and subsequently how symbols are mapped onto these representations.

15.6 CONCLUSIONS

Humans have the unique ability to process quantity both nonsymbolically and symbolically. Brain representations that underlie nonsymbolic numbers are thought to be innate and have a long evolutionary history (Cantlon, 2012). Contrastingly, brain representations of symbolic numbers must be constructed across development through learning and education (Ansari, 2008). The current chapter provides an overview of research that has examined (1) the neural circuitry that underlies both nonsymbolic and symbolic number processing; (2) the ongoing debate regarding whether number is represented abstractly in the brain; and (3) how symbolic and nonsymbolic numerical representations relate to the neural representations of nonnumerical continuous stimuli. Taken together, relatively recent advances in noninvasive neuroimaging techniques have led to a remarkable expansion in our knowledge of how the brain processes number representations.

Early research supported the notion that number representations in the brain are similar across formats and therefore abstract. This hypothesis was supported by empirical evidence that consistently implicated the bilateral IPS in the processing of both nonsymbolic and symbolic numbers. In contrast with this view, a growing body of research has unveiled brain activation patterns that indicate that format specific processing of symbolic and nonsymbolic number occurs in the IPS. These data suggest that the right IPS may process numbers abstractly as both symbolic and nonsymbolic number processing are correlated with activation of the right IPS. However, the right IPS is also activated by nonnumerical magnitudes. Therefore, perhaps the right IPS supports a general magnitude system used to process both numerical and nonnumerical magnitudes rather than an abstract ANS. In other words, activation that is similar across formats in the right IPS may reflect a general magnitude processing system. Unlike the right IPS, the left IPS seems to respond to numerical stimuli in a format-specific manner. Specifically, the left IPS is often, though not always, activated during symbolic but not nonsymbolic numerical processing. The finding that activation in the left IPS increases across development (Vogel et al., 2014), as children acquire symbolic number knowledge further supports the idea that the left IPS is used to process number symbols. This hemispheric asymmetry in the neural underpinnings of magnitude processing needs to be explored further. Quantitative coordinate based metaanalytic techniques are an important avenue that should be utilized to synthesize data from empirical studies examining magnitude processing. These metaanalytic investigations should explore the consistency of format-dependent hemispheric asymmetry for the processing of numerical magnitudes.

Although research has begun to illuminate which brain regions support number processing, the critical issue of how numerical symbols acquire meaning and what brain mechanisms are engaged during this process remains unanswered. The most prominent theory is that symbols are mapped onto nonsymbolic number representations. This theory, while elegant, has not been confirmed by the available literature. In particular, across many recent studies, the link between the neural correlates of symbolic and nonsymbolic processing has been found to be, at best, weak. The inherent confound between numerical magnitude and nonnumerical variables such as area and contour length in nonsymbolic numerical processing tasks preclude the ability to conclude that symbols map onto a pure quantity estimation. In other words, perhaps continuous visual properties associated with dot arrays (such as area or density of dots) rather than solely the numerosities of dot arrays partially drive neuronal responses during nonsymbolic numerosity tasks. If this is true, it is possible that nonsymbolic numerical magnitudes are processed using continuous properties (or through the integration of discrete, numerical and continuous, nonnumerical properties). Therefore, symbols might be mapped onto a general magnitude system rather than an ANS. However, the mechanism that might underpin such mapping processes remains unclear. Future research should explore the extent to which any links between symbolic and nonsymbolic numbers in the brain vary as a function of the relationship between discrete and continuous variables in nonsymbolic stimuli.

References

Ansari, D. (2007). Does the parietal cortex distinguish between "10," "ten," and ten dots? *Neuron, 53*(2), 165–167.

Ansari, D. (2008). Effects of development and enculturation on number representation in the brain. *Nature Reviews Neuroscience, 9*(4), 278–291.

Ansari, D., & Dhital, B. (2006). Age-related changes in the activation of the intraparietal sulcus during nonsymbolic magnitude processing: an event-related functional magnetic resonance imaging study. *Journal of Cognitive Neuroscience, 18*(11), 1820–1828.

Ansari, D., Garcia, N., Lucas, E., Hamon, K., & Dhital, B. (2005). Neural correlates of symbolic number processing in children and adults. *Neuroreport, 16*(16), 1769–1773. Retrieved from http://www.ncbi.nlm.nih.gov/pubmed/16237324.

Ansari, D., Dhital, B., & Siong, S. C. (2006a). Parametric effects of numerical distance on the intraparietal sulcus during passive viewing of rapid numerosity changes. *Brain Research, 1067*(1), 181–188.

Ansari, D., Fugelsang, J. A., Dhital, B., & Venkatraman, V. (2006b). Dissociating response conflict from numerical magnitude processing in the brain: an event-related fMRI study. *NeuroImage, 32*(2), 799–805.

Brannon, E. M. (2006). The representation of numerical magnitude. *Current Opinion in Neurobiology, 16*(2), 222–229.

Buckley, P. B., & Gillman, C. B. (1974). Comparisons of digits and dot patterns. *Journal of Experimental Psychology, 103*(6), 1131–1136.

Bugden, S., Price, G. R., McLean, D. A., & Ansari, D. (2012). The role of the left intraparietal sulcus in the relationship between symbolic number processing and children's arithmetic competence. *Developmental Cognitive Neuroscience, 2*(4), 448–457.

Bulthé, J., De Smedt, B., & Op de Beeck, H. P. (2014). Format-dependent representations of symbolic and non-symbolic numbers in the human cortex as revealed by multi-voxel pattern analyses. *NeuroImage, 87*, 311–322.

Cantlon, J. F. (2012). Math, monkeys, and the developing brain. *Proceedings of the National Academy of Sciences of the United States of America, 109*(Suppl. 1), 10725–10732.

Cantlon, J. F., Brannon, E. M., Carter, E. J., & Pelphrey, K. A. (2006). Functional imaging of numerical processing in adults and 4-y-old children. *PLoS Biology, 4*(5), e125.

Cantlon, J. F., Libertus, M. E., Pinel, P., Dehaene, S., Brannon, E. M., & Pelphrey, K. A. (2009a). The neural development of an abstract concept of number. *Journal of Cognitive Neuroscience, 21*(11), 2217–2229.

Cantlon, J. F., Platt, M. L., & Brannon, E. M. (2009b). Beyond the number domain. *Trends in Cognitive Sciences, 13*(2), 83–91.

Castelli, F., Glaser, D. E., & Butterworth, B. (2006). Discrete and analogue quantity processing in the parietal lobe: a functional MRI study. *Proceedings of the National Academy of Sciences of the United States of America, 103*(12), 4693–4698.

Chassy, P., & Grodd, W. (2012). Comparison of quantities: core and format-dependent regions as revealed by fMRI. *Cerebral Cortex, 22*(6), 1420–1430.

Cipolotti, L., Butterworth, B., & Denes, G. (1991). A specific deficit for numbers in a case of dense acalculia. *Brain, 114*(6), 2619–2637.

Cohen Kadosh, R. (2008). Numerical representation: abstract or nonabstract? *Quarterly Journal of Experimental Psychology (2006), 61*(8), 1160–1168.

Cohen Kadosh, R., & Walsh, V. (2009). Numerical representation in the parietal lobes: abstract or not abstract? *Behavioral and Brain Sciences, 32*, 313–373.

Cohen Kadosh, R., Henik, A., Rubinsten, O., Mohr, H., Dori, H., van de Ven, V., ... Linden, D.E. J. (2005). Are numbers special? The comparison systems of the human brain investigated by fMRI. *Neuropsychologia, 43*(9), 1238–1248.

Cohen Kadosh, R., Cohen Kadosh, K., Kaas, A., Henik, A., & Goebel, R. (2007). Notation-dependent and -independent representations of numbers in the parietal lobes. *Neuron, 53*(2), 307–314.

Cohen Kadosh, R., Lammertyn, J., & Izard, V. (2008). Are numbers special? An overview of chronometric, neuroimaging, developmental and comparative studies of magnitude representation. *Progress in Neurobiology, 84*(2), 132–147.

Cohen Kadosh, R., Bahrami, B., Walsh, V., Butterworth, B., Popescu, T., & Price, C. J. (2011). Specialization in the human brain: the case of numbers. *Frontiers in Human Neuroscience, 5*, 62.

Coolidge, F. L., & Overmann, K. A. (2012). Numerosity, abstraction, and the emergence of symbolic thinking. *Current Anthropology, 53*(2), 204–225.

Crollen, V., Grade, S., Pesenti, M., & Dormal, V. (2013). A common metric magnitude system for the perception and production of numerosity, length, and duration. *Frontiers in Psychology, 4*(July), 449.

Culham, J. C., & Kanwisher, N. G. (2001). Neuroimaging of cognitive functions in human parietal cortex. *Current Opinion in Neurobiology, 11*(2), 157–163.

Damarla, S. R., & Just, M. A. (2013). Decoding the representation of numerical values from brain activation patterns. *Human Brain Mapping, 34*(10), 2624–2634.

Dehaene, S. (1996). The organization of brain activations in number comparison: event-related potentials and the additive-factors method. *Journal of Cognitive Neuroscience, 8*(1), 47–68.

Dehaene, S. (1997). *The number sense: how the mind creates mathematics*. Oxford University Press, New York, USA. Available from http://www.amazon.com/The-Number-Sense-Creates-Mathematics/dp/0199753873/ref=pd_sim_sbs_b_1?ie=UTF8&refRID=0D2G1RM5K1933GCCSV5G

Dehaene, S., & Cohen, L. (1997). Cerebral pathways for calculation: double dissociation between rote verbal and quantitative knowledge of arithmetic. *Cortex; a Journal Devoted to the Study of the Nervous System and Behavior, 33*(2), 219–250.

Dehaene, S., Dehaene-Lambertz, G., & Cohen, L. (1998). Abstract representations of numbers in the animal and human brain. *Trends in Neurosciences, 21*(8), 355–361. Retrieved from http://www.ncbi.nlm.nih.gov/pubmed/9720604.

Dehaene, S., Piazza, M., Pinel, P., & Cohen, L. (2003). Three parietal circuits for number processing. *Cognitive Neuropsychology, 20*(3), 487–506.

Demeyere, N., Rotshtein, P., & Humphreys, G. W. (2014). Common and dissociated mechanisms for estimating large and small dot arrays: value-specific fMRI adaptation. *Human Brain Mapping, 35*(8), 3988–4001.

Dolores, M., Hevia, D., Izard, V., Coubart, A., Spelke, E. S., & Streri, A. (2014). Representations of space, time, and number in neonates. *Proceedings of the National Academy of Sciences, 111*(13), 4809–4813.

Dormal, V., & Pesenti, M. (2009). Common and specific contributions of the intraparietal sulci to numerosity and length processing. *Human Brain Mapping, 30*(8), 2466–2476.

Dormal, V., Andres, M., & Pesenti, M. (2012a). Contribution of the right intraparietal sulcus to numerosity and length processing: an fMRI-guided TMS study. *Cortex; a Journal Devoted to the Study of the Nervous System and Behavior, 48*(5), 623–629.

Dormal, V., Dormal, G., Joassin, F., & Pesenti, M. (2012b). A common right fronto-parietal network for numerosity and duration processing: an fMRI study. *Human Brain Mapping, 33*(6), 1490–1501.

Eger, E., Sterzer, P., Russ, M. O., Giraud, A. -L., & Kleinschmidt, A. (2003). A supramodal number representation in human intraparietal cortex. *Neuron, 37*(4), 719–725. Retrieved from http://www.ncbi.nlm.nih.gov/pubmed/12597867.

Eger, E., Michel, V., Thirion, B., Amadon, A., Dehaene, S., & Kleinschmidt, A. (2009). Deciphering cortical number coding from human brain activity patterns. *Current Biology, 19*(19), 1608–1615.

Feigenson, L., Dehaene, S., & Spelke, E. S. (2004). Core systems of number. *Trends in Cognitive Sciences, 8*(7), 307–314.

Fias, W., Lammertyn, J., Reynvoet, B., Dupont, P., & Orban, G. A. (2003). Parietal representation of symbolic and nonsymbolic magnitude. *Journal of Cognitive Neuroscience, 15*(1), 47–56.

Franklin, M. S., & Jonides, J. (2008). Order and magnitude share a common representation in parietal cortex. *Journal of Cognitive Neuroscience, 21*(11), 2114–2120.

Gebuis, T., & Reynvoet, B. (2012). Continuous visual properties explain neural responses to nonsymbolic number. *Psychophysiology, 49*(11), 1481–1491.

Gebuis, T., & Reynvoet, B. (2014). The neural mechanism underlying ordinal numerosity processing. *Journal of Cognitive Neuroscience, 26*(5), 1013–1020.

Gerstmann, J. (1940). Syndrome of finger agnosia, disorientation for right and left, agraphia, and acalculia. *Archives of Neurology and Psychiatry, 44*(2), 398–408.

Göbel, S. M., & Rushworth, M. F. (2004). Cognitive neuroscience: acting on numbers. *Current Biology, 14*(13), R517–R519.

Göbel, S. M., Johansen-Berg, H., Behrens, T., & Rushworth, M. F. S. (2004). Response-selection-related parietal activation during number comparison. *Journal of Cognitive Neuroscience, 16*(9), 1536–1551.

Grill-Spector, K., Henson, R., & Martin, A. (2006). Repetition and the brain: neural models of stimulus-specific effects. *Trends in Cognitive Science, 10*(1), 14–23.

Haist, F., Wazny, J. H., Toomarian, E., & Adamo, M. (2014). Development of brain systems for nonsymbolic numerosity and the relationship to formal math academic achievement. *Human Brain Mapping, 36*(2), 804–826.

Halberda, J., & Feigenson, L. (2008). Developmental change in the acuity of the "number sense": the approximate number system in 3-, 4-, 5-, and 6-year-olds and adults. *Developmental Psychology, 44*(5), 1457–1465.

He, L., Zuo, Z., Chen, L., & Humphreys, G. (2013). Effects of number magnitude and notation at 7T: separating the neural response to small and large, symbolic and nonsymbolic number. *Cerebral Cortex, 24*(8), 2199–2209.

Henschen, S. E. (1919). Über Sprach-, Musik- Und Rechenmechanismen Und Ihre Lokalisationen Im Großhirn. *Zeitschrift Für Die Gesamte Neurologie Und Psychiatrie, 52*(1), 273–298.

Holloway, I. D., & Ansari, D. (2010). Developmental specialization in the right intraparietal sulcus for the abstract representation of numerical magnitude. *Journal of Cognitive Neuroscience, 22*(11), 2627–2637.

Holloway, I. D., Price, G. R., & Ansari, D. (2010). Common and segregated neural pathways for the processing of symbolic and nonsymbolic numerical magnitude: an fMRI study. *NeuroImage, 49*(1), 1006–1017.

Holloway, I. D., Battista, C., Vogel, S. E., & Ansari, D. (2013). Semantic and perceptual processing of number symbols: evidence from a cross-linguistic fMRI adaptation study. *Journal of Cognitive Neuroscience, 25*(3), 388–400.

Izard, V., Dehaene-Lambertz, G., & Dehaene, S. (2008). Distinct cerebral pathways for object identity and number in human infants. *PLoS Biology, 6*(2), e11.

Jacob, S. N., & Nieder, A. (2009). Tuning to non-symbolic proportions in the human frontoparietal cortex. *The European Journal of Neuroscience, 30*(7), 1432–1442.

Kaufmann, L., Koppelstaetter, F., Delazer, M., Siedentopf, C., Rhomberg, P., Golaszewski, S., ... Ischebeck, A. (2005). Neural correlates of distance and congruity effects in a numerical Stroop task: an event-related fMRI study. *NeuroImage, 25*(3), 888–898.

Kaufmann, L., Koppelstaetter, F., Siedentopf, C., Haala, I., Haberlandt, E., Zimmerhackl, L.-B., ... Ischebeck, A. (2006). Neural correlates of the number-size interference task in children. *Neuroreport, 17*(6), 587–591.

Kaufmann, L., Vogel, S. E., Wood, G., Kremser, C., Schocke, M., Zimmerhackl, L. -B., & Koten, J. W. (2008). A developmental fMRI study of nonsymbolic numerical and spatial processing. *Cortex; a Journal Devoted to the Study of the Nervous System and Behavior, 44*(4), 376–385.

Kaufmann, L., Wood, G., Rubinsten, O., & Henik, A. (2011). Meta-analyses of developmental fMRI studies investigating typical and atypical trajectories of number processing and calculation. *Developmental Neuropsychology, 36*(6), 763–787.

Le Corre, M., & Carey, S. (2007). One, two, three, four, nothing more: an investigation of the conceptual sources of the verbal counting principles. *Cognition, 105*(2), 395–438.

Leibovich, T., & Henik, A. (2013). Magnitude processing in non-symbolic stimuli. *Frontiers in Psychology, 4*, 375.

Lyons, I. M., & Ansari, D. (2009). The cerebral basis of mapping nonsymbolic numerical quantities onto abstract symbols: an fMRI training study. *Journal of Cognitive Neuroscience, 21*(9), 1720–1735.

Lyons, I. M., & Beilock, S. L. (2013). Ordinality and the nature of symbolic numbers. *The Journal of Neuroscience, 33*(43), 17052–17061.

Lyons, I. M., Ansari, D., & Beilock, S. L. (2014). Qualitatively different coding of symbolic and nonsymbolic numbers in the human brain. *Human Brain Mapping, 36*(2), 475–488.

Moyer, R. S., & Landauer, T. K. (1967). Time required for judgements of numerical inequality. *Nature, 215*(2), 1519–1520. Retrieved from http://www.nature.com/nature/journal/v215/n5109/abs/2151519a0.html.

Nieder, A., & Dehaene, S. (2009). Representation of number in the brain. *Annual Review of Neuroscience, 32*, 185–208.

Notebaert, K., Pesenti, M., & Reynvoet, B. (2010). The neural origin of the priming distance effect: distance-dependent recovery of parietal activation using symbolic magnitudes. *Human Brain Mapping, 31*(5), 669–677.

Notebaert, K., Nelis, S., & Reynvoet, B. (2011). The magnitude representation of small and large symbolic numbers in the left and right hemisphere: an event-related fMRI study. *Journal of Cognitive Neuroscience, 23*(3), 622–630.

Piazza, M., Mechelli, A., Butterworth, B., & Price, C. J. (2002). Are subitizing and counting implemented as separate or functionally overlapping processes? *NeuroImage, 15*(2), 435–446.

Piazza, M., Izard, V., Pinel, P., Le Bihan, D., & Dehaene, S. (2004). Tuning curves for approximate numerosity in the human intraparietal sulcus. *Neuron, 44*(3), 547–555.

Piazza, M., Mechelli, A., Price, C. J., & Butterworth, B. (2006). Exact and approximate judgements of visual and auditory numerosity: an fMRI study. *Brain Research, 1106*(1), 177–188.

Piazza, M., Pinel, P., Le Bihan, D., & Dehaene, S. (2007). A magnitude code common to numerosities and number symbols in human intraparietal cortex. *Neuron, 53*(2), 293–305.

Pica, P., Lemer, C., Izard, V., & Dehaene, S. (2004). Exact and approximate arithmetic in an Amazonian indigene group. *Science, 306*(5695), 499–503.

Pinel, P., Le Clec, H. G., van de Moortele, P. F., Naccache, L., Le Bihan, D., & Dehaene, S. (1999). Event-related fMRI analysis of the cerebral circuit for number comparison. *Neuroreport, 10*(7), 1473–1479.

Pinel, P., Dehaene, S., Rivière, D., & LeBihan, D. (2001). Modulation of parietal activation by semantic distance in a number comparison task. *NeuroImage, 14*(5), 1013–1026.

Pinel, P., Piazza, M., Le Bihan, D., & Dehaene, S. (2004). Distributed and overlapping cerebral representations of number, size, and luminance during comparative judgments. *Neuron, 41*(6), 983–993. Retrieved from http://www.ncbi.nlm.nih.gov/pubmed/15046729.

Raizada, R. D. S., & Poldrack, R. A. (2007). Selective amplification of stimulus differences during categorical processing of speech. *Neuron, 56*(4), 726–740.

Roggeman, C., Santens, S., Fias, W., & Verguts, T. (2011). Stages of nonsymbolic number processing in occipitoparietal cortex disentangled by fMRI adaptation. *The Journal of Neuroscience: The Official Journal of the Society for Neuroscience, 31*(19), 7168–7173.

Shuman, M., & Kanwisher, N. (2004). Numerical magnitude in the human parietal lobe; tests of representational generality and domain specificity. *Neuron, 44*(3), 557–569.

Simon, T. (1999). The foundations of numerical thinking in a brain without numbers. *Trends in Cognitive Sciences, 3*(10), 363–365. Retrieved from http://www.ncbi.nlm.nih.gov/pubmed/10498924.

Temple, E., & Posner, M. I. (1998). Brain mechanisms of quantity are similar in 5-year-old children and adults. *Proceedings of the National Academy of Sciences of the United States of America, 95*(13), 7836–7841. Retrieved from http://www.pubmedcentral.nih.gov/articlerender.fcgi?artid=22775&tool=pmcentrez&rendertype=abstract.

Venkatraman, V., Ansari, D., & Chee, M. W. L. (2005). Neural correlates of symbolic and nonsymbolic arithmetic. *Neuropsychologia, 43*(5), 744–753.

Vogel, S. E., Grabner, R. H., Schneider, M., Siegler, R. S., & Ansari, D. (2013). Overlapping and distinct brain regions involved in estimating the spatial position of numerical and nonnumerical magnitudes: an fMRI study. *Neuropsychologia, 51*(5), 979–989.

Vogel, S. E., Goffin, C., & Ansari, D. (2014). Developmental specialization of the left parietal cortex for the semantic representation of Arabic numerals: an fMR-adaptation study. *Developmental Cognitive Neuroscience, 12C*, 61–73.

Walsh, V. (2003). A theory of magnitude: common cortical metrics of time, space and quantity. Trends in Cognitive Sciences, 7(11), 483–488.

CHAPTER

16

What Do We Measure When We Measure Magnitudes?

Tali Leibovich[*,†,§], *Arava Y. Kallai*[**,†,‡], *Shai Itamar*[**,†]

[*]Department of Cognitive and Brain Sciences, Ben-Gurion University of the Negev, Beer-Sheva, Israel; [**]Department of Psychology, Ben-Gurion University of the Negev, Beer-Sheva, Israel; [†]The Zlotowski Center for Neuroscience, Ben-Gurion University of the Negev, Beer-Sheva, Israel; [‡]Department of Psychology, Max Stern Yezreel Valley College, Emek Yezreel, Israel; [§]Department of Psychology, the University of Western Ontario, Canada

OUTLINE

Continuous Issues in Numerical Cognition. http://dx.doi.org/10.1016/B978-0-12-801637-4.00016-0

16.1 INTRODUCTION

Every object or set of objects we encounter can be described by their magnitude. For example, waiting in line in the supermarket, one can assess the number of people standing in line, their average height, how crowded each cart is, and so forth. Humans are able to represent such magnitudes using an elaborate symbolic system (eg, Arabic numerals, Roman numerals, number words, etc.). That kind of man-made representation depends upon culture, education, and an experience with a specific symbolic system and its rules (Pica, Lemer, Izard, & Dehaene, 2004).

However, processing magnitudes is a basic skill that was found across species, and is not specific to humans. For example, lions can tell if they are outnumbered by an opponent group by listening to the sound of their roars (McComb, Packer, & Pusey, 1994) and fish can estimate the size of a shoal and join the larger shoal in order to increase their chances of survival (Agrillo & Piffer, 2012). In order to investigate a more universal and basic abilities of magnitude processing, nonsymbolic representation of magnitudes are being used, and similarly to the theme of this book, these will be at the focus of the current chapter.

16.1.1 Nonsymbolic Magnitudes

Any item can be represented by magnitude such as its size and circumference. A group of items can be represented by its numerosity (ie, how many items the group contains?). The same group can also be represented by its continuous properties; the average size of the items, their density, their total surface area, and so on. Numerosity is a discrete magnitude: given unlimited time, one can be exact in stating the numerosity of a set, without using any measuring tool. In contrast, without any measuring tool and even without time limitation, one can only estimate continuous magnitudes such as density and average size (Leibovich & Henik, 2013). This is the main difference that we will use here to define discrete and continuous magnitudes. Using nonsymbolic representations of magnitudes (of a group of items) always involves both discrete and continuous magnitudes. Moreover, the correlation between discrete and continuous magnitudes is very obvious; usually, more items will be denser, occupy more space, have greater total surface area, and so forth than fewer items.

In many everyday situations, sets of objects are to be processed as a proportion of a subset from the total set of items or as a part of a whole object (both are nonsymbolic notations of rational numbers—fractions). In these cases, the to-be-judged magnitude is, in fact, relative. Namely, the magnitude of the part is judged relatively to the whole before it is judged relatively to anything else (eg, to another part). For example, the

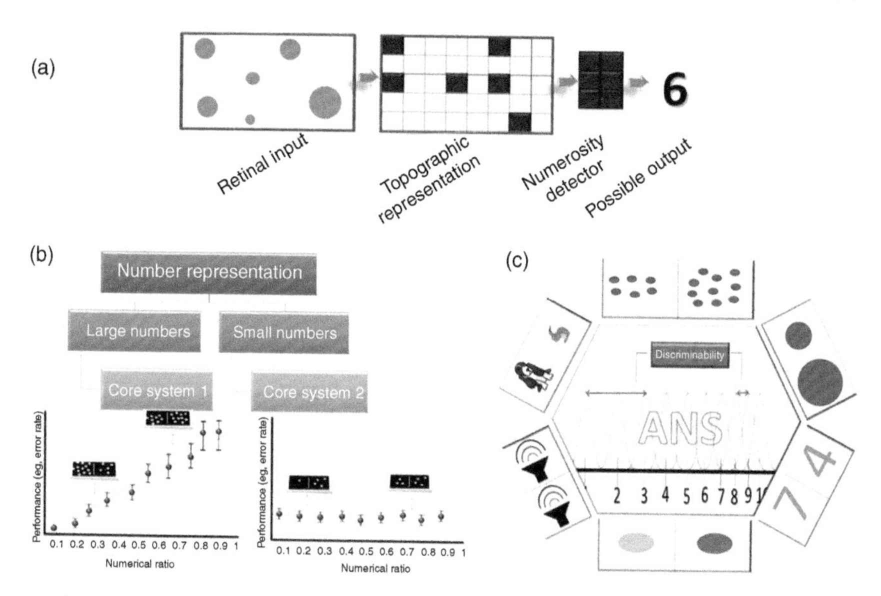

FIGURE 16.1 **Main theories in numerical cognition.** (a) The model suggested by Dehaene and Changeux (1993), emphasizing the existence of an innate "numerosity detector" that sum numerosities regardless of continuous magnitudes. (b) The "two core systems" theory (Feigenson, Dehaene, & Spelke, 2004) separating processing of small and large numbers; the main difference is ratio-dependent performance in core system 1, and violation of this dependency in core system 2. (c) The ANS theory suggesting that different magnitudes (see examples) are being processed by an approximate analog representation of number known as the Approximate Number System, or ANS.

magnitude of a quarter of glass of water is being judged relatively to the whole glass (ie, quarter) before it can be compared to the magnitude of half a glass of water.

Similarly to magnitudes of whole objects, relative magnitudes can also be categorized as discrete or as continuous. The proportion of a subset of objects from the total number of objects (eg, the proportion of girls in a class) is a discrete relative magnitude and so is the ratio between two subsets (eg, the ratio of girls to boys in a class). On the other hand, a part of a continuous object (eg, partially filled glass of water) is a continuous relative magnitude, and so is the ratio between the full and the empty parts of the glass.

While symbolic fractions (eg, 2/3)—a man-made representation that is dependent on culture and education—did not show automatic processing (Bonato, Fabbri, Umiltà, & Zorzi, 2007; Kallai & Tzelgov, 2009; Meert, Grégoire, & Noël, 2009, 2010), nonsymbolic fractions and ratios have shown some evidence for being innately mentally represented. For instance, recent studies with primates and humans showed some evidence that the brain can process ratios. Vallentin and Nieder (2008) found that monkeys can process the ratio of two line lengths similarly to humans' performance (ie, both had distance effect). The authors further found that different populations of neurons were sensitive to different ratios. Other species have also shown evidence for the processing of ratios. For instance, pigeons have shown to successfully learn rules based on either discrete (Honig & Stewart, 1989) or continuous (Emmerton, 2001) ratios. Human infants also show sensitivity for relative magnitude. Duffy, Huttenlocher, and Levine (2005) had demonstrated that 6.5-month-olds were able to identify differences in a relative but not an absolute magnitude. Thus, processing of relative magnitude might be considered as a basic skill, similarly to processing of whole numbers.

16.1.2 Main Theories in Numerical Cognition

In spite of the intimate correlation between discrete and continuous magnitudes, some of the most dominant theories in the field of numerical cognition suggest that humans and other animals process only (or mostly) numerosities (the discrete property of the set). These theories are summarize in Fig. 16.1. For example, Dehaene and Changeux (1993) suggested a neural model for the development of rudimentary numerical abilities. This model was the basis for a later computational model (Verguts & Fias, 2004). These models assume three steps: First, there is an input module which represents the location of every item in a set; in the second step, the location of the objects is mapped onto an intermediate topographic map. Critically, this map registers only locations, and ignores all other magnitudes and properties of the item. In the third step, the numerosity of the items is being summed by a "numerosity detector" that is assumed to be innate in both humans and other animals. This was at the core of the "number sense" (Dehaene, 1997).

A later theory—the two core systems (Feigenson et al., 2004)—addressed both continuous and discrete magnitudes, and tried to find the differences between the processing of each, as will be discussed in detail later. Core system 1 (a descendant of the "number sense") is considered an *innate* system that gives an approximate and noisy representation of *discrete* magnitudes of any given number. This theory assumes that numbers are represented on a "mental number line" (Dehaene, 1997). Numbers' representation on the mental number line is thought to be noisy. Put differently, numerosities are represented as approximate analog magnitudes, with a Gaussian distribution to represent their magnitudes. The distributions of different numerosities overlap; this overlap increases as the numbers increase. This explains why, when comparing 2 and 4, the comparison is very simple: The mental representations of 2 and 4 are minimally overlap in comparison to 6 and 8 whose representations overlap more considerably. When comparing two numerosities (eg, which number is larger, 4 or 8?), the ratio between the numerosities (smaller divided by larger numerosity) affects performance. This ratio-dependent performance is a hallmark characteristic of core system 1. This characteristic feature is based on findings from studies done with infants and adults. It is believed that infants are able to discriminate between numerosities and that their discriminability is modulated by the ratio between the compared numerosities. For example, it was found that when the ratio reaches 0.75, infants are not able to discriminate numerosities (Xu & Spelke, 2000). Halberda, Mazzocco, and Feigenson (2008) performed similar studies of numerosity discrimination with children and adults and found that the ability to discriminate numerosities (acuity) improves with age.

Similar to core system 1, core system 2 is also an innate numerosity processing system. However, this system is used for precisely keeping track of small numerosities (ie, up to 3 to 4 items). Unlike core system 1, which does not take into account any continuous magnitudes of the given object, core system 2 was shown to be able to process both continuous magnitudes and numerosity. In tasks in which infant had to choose between two sets of food items, infants responded in a manner which reflects a preference guided by the stimuli's continuous magnitude (eg, total area of biscuits) and not their numerosity (Feigenson, Carey, & Hauser, 2002). This was taken as an indication for processing continuous magnitudes. On the other hand, when infants were shown a number of food items and then these food items were hidden, infants stop searching when they matched the numerosity of the food items hidden, not their total surface area. Namely, the search process was based on numerosity and not continuous magnitudes. Importantly, this pattern held only with up to three food items (Feigenson & Carey, 2003). Core system 2 is considered to be able to process both numerosity and continuous magnitudes of objects according to the task at hand and the behavior suiting it.

The approximate number system (ANS) (Cantlon, Brannon, Carter, & Pelphrey, 2006; Cantlon et al., 2009a) is a theoretical term that was closely formulated upon the definition of core system 1 (Feigenson et al., 2004). Similarly, it is defined as an innate system that processes and represents discrete magnitudes (eg, numerosities of objects or events) on a mental analog scale. The result is an approximate coding of discrete magnitudes. The ratio-dependent performance in numerosity comparison which characterizes core system 1 also characterizes the ANS. This was found for both monkeys and adult humans, revealing that the ANS is a system that is shared across species (Cantlon et al., 2006). The ANS (as well as its predecessor theories) is considered the foundation for the symbolic number system used for arithmetic (Cantlon et al., 2009a; Feigenson et al., 2004) and a correlation between ANS acuity and mathematic achievements was reported (Mazzocco, Feigenson, & Halberda, 2008).

The only theory that acknowledges continuous magnitude is the accumulator model (Gallistel & Gelman, 1992; Gallistel, Gelman, &

Cordes, 2006; Gibbon, 1977; Meck & Church, 1983). According to this theory, the nonverbal representatives of number are mental magnitudes (ie, real numbers) with scalar variability that represent both discrete (countable) and continuous (uncountable) magnitudes. The accumulator model suggests that magnitude is estimated by the accumulation of neuronal impulses. At the end of the estimation process, the accumulated activity is compared to memory (Gibbon, 1977; Meck & Church, 1983). Meck and Church demonstrated that estimation of duration (continuous magnitude) and of numerosity (discrete magnitude) behave in a similar way, and thus, serve as evidence for a joined mechanism. The development of language singled out the whole numbers from the real numbers, and by that turned them to the foundation of cultural number systems. Notably, none of the theories reviewed previously take relative magnitudes into account and try to explain how such magnitudes are represented. However, there are references to processing of proportions by researchers using the accumulator model (Gallistel & Gelman, 2000).

To summarize, although we discuss these theories separately, they have many points of convergence: core system 1 and the ANS are used interchangeably; the accumulator model is an outcome of the approximate representation suggested in the ANS and core system 1, and core system 2 corresponds with the object file system (Carey, 2001).

16.2 RETHINKING THE PREMISES

The theories mentioned earlier share two basic premises: first, that humans are born with the ability to detect and discriminate numerosities and second, that numerosities are being processed preattentively and without conscious control. However, another line of theoretical arguments seems to undermine these fundamental premises. Mix, Huttenlocher, and Levine (2002a) demonstrated that the results of developmental studies providing evidence that numerosity is processed very early in life can be explained by the discrimination and processing of continuous magnitudes. For example, both Feigenson et al. (2004) and Cantlon, Platt, and Brannon (2009) referred to the work of Xu and Spelke (2000) in order to establish the fact that infants are able to process numerosities automatically. Xu and Spelke conducted a study in which infants were presented with habituation display of dots in different sizes but in the same range of numerosity. After completing the habituation stage, in the test stage, they were shown two displays—one with the same range as the habituation stage and the other with a novel range. It was found that infants looked longer toward the novel range when the ratio between the habituated set and the novel set was 0.5 (eg, 8 vs. 16 dots) and failed to do so when it was 0.66 (eg, 8 vs. 12 dots). The authors attempted to control for overall total area of the test displays so that the overall total area for both familiar and novel ranges was equally different from the habituation display. However, Mix et al. (2002a) pointed out that while area was controlled for in this design, contour was not. When the ratio between the habituated set and the novel set was 0.5, the difference between habituation and test contour length was always small when the test trial had a familiar numerosity (eg, 0.82). When the test trial had a novel numerosity, these differences in counter length became much more distinct (eg, 0.59–0.61). The reported finding regarding the ability of infants to discriminate the sets of dots could have been the result of the infants' ability to discriminate counter length (eg, continuous magnitude) and not numerosity. Further, Mix et al. were able to show that the infants' difficulty in discriminating sets when the ratio between the habituated set and the novel set were 0.66 (eg, 8 vs. 12 dots) could be explained by the fact that for those comparisons the confounded contour length was less strong. Later studies suffered from similar confounds when attempting to control continuous magnitudes (for a review see Leibovich & Henik, 2013). In addition to the important theoretical criticism (Leibovich & Henik, 2013; Mix et al., 2002a; Newcombe, 2002; Simon, 1997), there is also empirical evidence suggesting that first, continuous magnitudes are processed even when irrelevant to the task, and second, that numerosities are not being processed automatically. In support of the first, Leibovich and Henik (2014) had participants compare numerosities of pairs of dot arrays. Continuous magnitudes were minimally or not correlated with numerosities. In a regression analysis, numerosity ratio and the ratio between five different continuous magnitudes (eg, total surface area, density, etc.) were used as predictors of performance (ie, response time and accuracy). The results of this regression demonstrated that about half of the explained variance of response time was accounted for by continuous magnitudes. This suggests that continuous magnitudes are being processed even when irrelevant to the task and are not a reliable predictor of numerosity. In support of the second, Gebuis and Reynvoet (2013) employed a passive viewing paradigm while measuring event-related potential (ERP) activity and demonstrated that while changing continuous magnitudes always registered brain response, changing numerosities did not result in brain response even when participants were told that numerosities will change. Accordingly, the authors concluded that there was no automatic extraction of numerosities, but there was an automatic extraction of continuous magnitude.

Besides overlooking continuous magnitudes, it is hard to infer from the theories described earlier about the representation of parts of the whole and ratios. This is in spite that, as mentioned previously, nonsymbolic notations of rational numbers were found to be processed by nonhuman animals as well as by human infants. Interestingly, just as evidence for automatic continuous magnitude processing is growing, a similar pattern emerges in the nonsymbolic rational numbers front.

Early developmental literature presents conflicting evidence regarding the ability of young children to understand nonsymbolic ratios. Mix et al. (2002a) pointed out that a primary difference between studies reporting early versus later success on proportional reasoning tasks is the kind of stimuli involved. In particular, tasks reporting early success involved relations between continuous magnitudes, like partially filled two-dimensional shapes (Acredolo, O'Connor, Banks, & Horobin, 1989; Goswami, 1989; Huttenlocher, 1999; Sophian, 2000; Spinillo & Bryant, 1991), whereas tasks reporting later success involved relations between discrete magnitudes, like the proportion of red marbles from a set of red and white marbles (Karplus, Pulos, & Stage, 1983; Meadow et al., 1977; Noelting, 1980; Siegler & Vago, 1978). Jeong, Levine, and Huttenlocher (2007) showed that 6-, 8-, and 10-year-olds performed poorly when proportions involved discrete magnitudes, but did better when proportions involved continuous magnitudes. They suggested that young children rely on an early emerging ability to perceptually code relative amounts in reasoning about the proportional relation of continuous magnitudes. In contrast, in the context of discrete magnitudes, young children instead use erroneous counting strategies and ignore the perceptual relation. Thus as is the case for whole magnitudes, ratios seems to be more natural to understand when they are presented as continuous parts of the whole.

16.3 FROM "APPROXIMATE NUMBER SYSTEM" TO "APPROXIMATE MAGNITUDE SYSTEM"

To summarize the previous section, the answer to the question "what do we process when we process magnitudes?" is not quite simple. Until recently, it was clear that the most preliminary property guiding comparative judgment was numerosity, and only numerosity (Dehaene & Changeux, 1993; Dehaene, 1997; Ross & Burr, 2010). However, both theoretical and empirical evidence from the last decade suggest that numerosity is only one dimension of magnitude that plays a part in this process. There are even suggestions that numerosity is not as automatic and basic property as previously suggested (eg, Gebuis & Reynvoet, 2013; Leibovich, Henik, & Salti, 2015; Leibovich, Vogel, Ansari, & Henik, 2016).

In light of such findings, we would like to suggest using the term "approximate magnitude system" (AMS) that takes these complexities into account. Namely, we suggest that both discrete (ie, numerosity) and continuous magnitudes are being processed when comparing arrays of items. Due to the inherently high correlation between different continuous magnitudes, and between discrete and continuous magnitudes, it is reasonable to assume that one system processes all these magnitudes—hence the AMS. This suggestion is also reasonable given the contribution (and

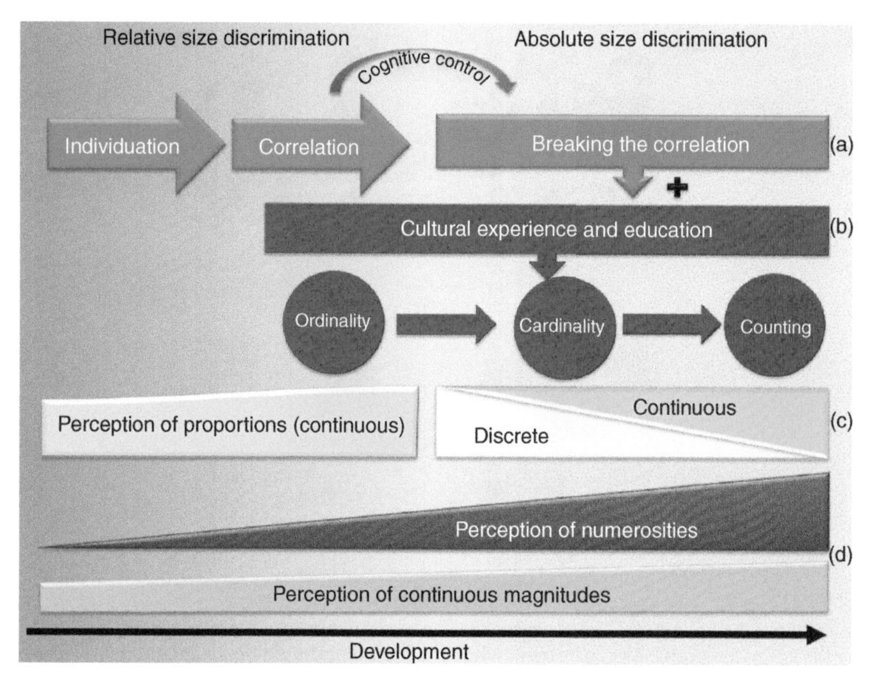

FIGURE 16.2 **Theoretical integrative developmental model of the AMS and its relation to basic mathematical abilities.** The upper *blue* segment (a) describes the major landmarks in the development of the approximate magnitude system (AMS). The other parts describe how the AMS affects different aspects of numerical cognition. The first landmark to be reached in the AMS is the ability to individuate items. The second landmark is reached once the natural correlation between numerosity and continuous magnitudes is learned. At this time of development, size discrimination is always relative to a standard. The third landmark depends on the development of cognitive control, when a child understands that the correlation between numerosities and continuous magnitudes is not perfect, and can be violated. When this landmark is reached, size discrimination can be made according to absolute size and not just relatively (see background shifts from *blue* to *gray*). The red segment (b) describes processes that are related to culture and education that can take place once the third landmark has been reached. The ability to perceive numerosities (in *purple*) does not exist from birth; it improves with age. In contrast, the ability to perceive continuous magnitude (in *green*) is innate; it improves with age but reaches a plateau sooner than perception of numerosities. The orange segment (c) describes the perception of nonsymbolic proportions; in early stages, the ability to differentiate between different proportions relies on continuous magnitudes and therefore the level of performance in such a task is high. However, when learning to count, there is a decrease in performance (in *yellow*) because children employ counting strategies instead of relying on continuous magnitudes. The ability to perceive numerosities and continuous magnitudes (d) is modulated throughout development.

interference) of continuous magnitudes to comparative judgments even when they are irrelevant to the task and do not highly correlate with numerosities (Gebuis & Reynvoet, 2012, 2013; Henik, Leibovich, Naparstek, Diesendruck, & Rubinsten, 2012; Leibovich & Henik, 2013, 2014). Since processing of proportions shares many similarities with processing of

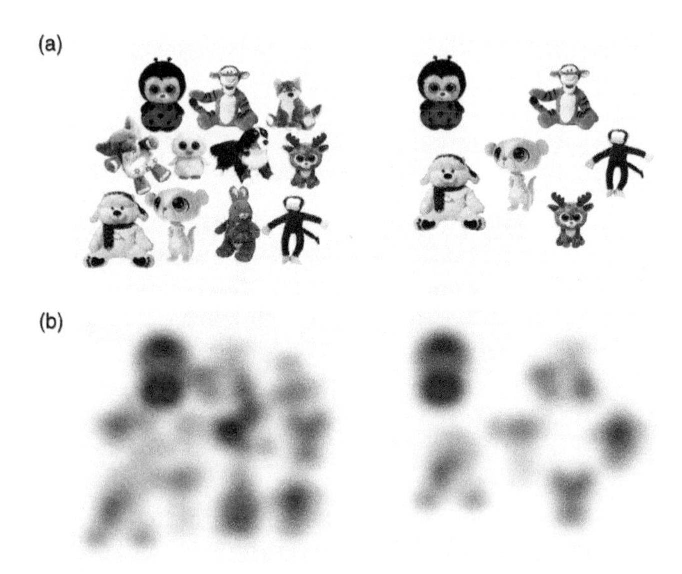

FIGURE 16.3 **Visual acuity develops over time.** (a) Array of items as they are perceived by individuals with normal visual acuity. (b) The same picture as it would look to infants under the age of 5 months. As can be seen, while it is hard to individuate the items and perceive their numerosity, it is still possible to use continuous magnitudes such as density and total surface area to discriminate the magnitudes of these two arrays.

numerosities, at least in the nonsymbolic form, we suggest that the AMS is also used to process proportions.

16.4 DEVELOPMENTAL MODEL OF THE AMS

As was already suggested (Cantrell & Smith, 2013; Mix et al., 2002a), it is important to study the involvement of continuous magnitudes in the development of mathematical abilities. Accordingly we will introduce an integrative theoretical model for the development of the AMS. This model extends an already suggested model (Leibovich & Henik, 2013) and includes processing of continuous magnitudes, numerosities, and proportions. In general, the model describes a developmental continuum, while highlighting several central landmarks in which important changes occur (Fig. 16.2a).

At birth, newborns can only discriminate between continuous magnitudes. During the first months of life, babies learn to individuate items. When this ability develops, the first landmark is reached. The second landmark is reached when the child learns the natural correlation between continuous magnitudes and numerosities (eg, usually, more toys will take more space, be denser, etc., compared with fewer toys). With the improvement of cognitive control abilities, and specifically inhibition, the third significant landmark is reached; now, the child is able to use both numerosities and continuous magnitudes to compare between two arrays, even if the correlation is violated (eg, even though two big toy trucks takes more space than four tennis balls, there are more tennis balls than trucks). We will now elaborate on each landmark. Next, we will explain how the development of the AMS affects, and is affected by, formal and cultural learning of basic concepts of mathematics. We will also discuss different aspects of numerical cognition in light of the suggested model.

16.4.1 Major Landmarks of the AMS Development

In order to distinguish between individual items in an array, several visual-motor mechanisms must be engaged. Specifically, maturation of muscles (smooth muscles in the middle layer of the eye and muscles in the eye orbit), the fovea, and the brain areas related to visual processing needs to take place to allow the fovea to curve and focus on items, to project clear images to the retina and to visually process this image. It has been shown that newborns and babies until the age of 4 weeks are unable to focus their vision in such a way (Fig. 16.3). This ability develops only later; at the age of 8 weeks babies can focus their vision on objects at a distance of 25–100 cm. This ability improves dramatically in the age of 5–6 months (Banks, 1980). It is interesting to mention that these findings regarding the development of visual acuity were studied via methods of preferential looking time, and verified with electrophysiological studies (Dobson & Teller, 1978). The looking time paradigm is abundantly useful in studies aiming to evaluate the abilities of newborns and infant to discriminate numerosities, as demonstrated in the next paragraph.

In spite of such findings, there are studies showing that newborns are able to discriminate numerosities. For example, Coubart, Izard, Spelke, Marie, and Streri (2014) tested the ability of 54-h-old newborns to discriminate numerosities. They used arrays of different animal figures in a habituation paradigm. The newborns saw the stimuli from a viewing distance of 80 cm. The authors reported controlling continuous magnitudes by keeping total luminance and the total occupied area constant. However, newborns at that age are not able to focus their eyes to sharply discriminate the different items in the array. It is more likely that what the newborns were able to discriminate were very crude continuous magnitudes that were not controlled, like the density of the items. Accordingly, we suggest that in the first weeks of life, babies process continuous magnitudes when comparing arrays of items. Note that even if humans are born with the ability to discriminate numerosities (ie, "numerosity detector"), there are physiological systems that need to develop (eg, sharper visual acuity) before we can use such a "numerosity detector."

Another important ability that needs to develop before babies can discriminate numerosities is individuation: the identification of object's boundaries (Carey, 2001; Mix, Huttenlocher, & Levine, 2002b). Before this ability develops, the concept of "numerosity" is inapplicable. Thus, in addition to the crucial maturation of the visual-motor mechanisms, the concept of individuation (not only visual individuation) should develop. This occurs at about the age of 5 months (Bower, 1974). Thus, individuation is the first important developmental landmark of the AMS. It is interesting to note that both visual acuity and individuation abilities mature at the same age of 5–6 months. It might suggest that there is at least an indirect relationship between these abilities. Namely, visual acuity is a prerequisite for individuation ability.

With the development of visual acuity, babies can distinguish individual items and arrays of items. At this time, babies are also exposed to the natural correlation between numerosities and continuous magnitudes; usually the denser array that occupies more space has more items. Understanding this correlation signifies the second landmark. The correlation at this point of development is so solid that children rely completely on continuous magnitudes when asked about numerosities. Evidence for this phenomenon is described in the seminal works of Jean Piaget on the logico-mathematical development in children (Piaget, 1952), in the number conservation task. In this task, children were presented with two rows of coins. The two rows had the same number of coins. However, in one row the coins were more spaced than in the other row, making the total length and density of the rows different. The children were asked to indicate if the number of coins in the two rows was the same or different. Children under the age of 7 indicated that the "stretched" row had more coins. Importantly, when the coins in the two rows were equally spaced, all children successfully indicated that the number of coins in the two rows is the same. Accordingly, Piaget suggested that young children failed to understand that the manipulation done to the rows' length did not change the number of coins in each row. These failures to "conserve number" disappeared after the age of 7 years. This finding is robust and was replicated in many studies (Bjorklund & Harnishfeger, 1995; Houdé, 1997, 2000). Rousselle, Palmers, and Noël (2004) showed that, when comparing the magnitudes of two sets, preschoolers are unable to use numerosities. Instead, they rely on continuous magnitudes. Additional study of this group (Rousselle & Noël, 2008) further showed that sensitivity to irrelevant numerosities increases with age (from age 3 to 6 years), whereas sensitivity to irrelevant continuous magnitudes remains stable.

If we try to understand the findings described so far in this chapter using our suggested perspective and model, it is possible that both numerosities (ie, number of coins) and continuous magnitudes (ie, length and density of row) are processed, but not in a symmetrical way. Namely, the comparative judgment was biased toward continuous magnitudes. We suggest that in order to overcome the bias of continuous magnitudes, sufficient development of the cognitive control mechanism should take place. When cognitive control abilities are sufficiently developed, children can overcome the bias toward continuous magnitudes. This is when the third landmark is being reached. Now, children are able to rely on numerosities when asked to compare them, even in cases when the correlation of numerosities and continuous magnitudes is violated.

Some studies have demonstrated that even adults are biased toward continuous magnitude processing. The difference is that adults are able to recruit cognitive control and inhibit continuous magnitudes processing when irrelevant. For example, Houdé et al. (2011) conducted an imaging study in which children and adults were presented with a task similar to Piaget's number conservation task (eg, same or different number of items). Behavioral results revealed that only after the age of 6, children were able to succeed in the task. This indicates that only at this age, children have the ability to ignore continuous magnitudes when they are not correlated with numerosity. Furthermore, imaging results suggested that the differences in brain activity between those who succeeded or failed in the task (across ages) were in a frontoparietal network activity. This network activation was interpreted by the researchers as a necessary inhibitory process that enables subjects to successfully ignore continuous magnitudes.

As an outcome of the described landmarks, the ability to perceive numerosities and continuous magnitudes changes throughout development. The ability to perceive (ie, to differentiate) continuous magnitudes exists at birth and reaches a plateau relatively early in life (around the age of 4; Odic, Libertus, Feigenson, & Halberda, 2013). On the other hand, the ability to perceive numerosities starts when individuation matures and gradually improves (Fig. 16.2d).

16.4.2 Cultural Experience and Education

The progress which leads to the ability to inhibit the bias toward continuous magnitudes occurs alongside an informal learning process of counting. These two processes reciprocally affect each other. The process of informal learning of counting takes place in most cultures as soon as basic language skills appear. According to Carey's developmental model (Carey, 2001, 2004), the ability to recite basic counting list (eg, "one, two, three, …") is learned as soon as the age of 2 years. At this point the child understands the basic concept of ordinality, namely, that the number words are sequential, so that "two" appears after "one" and before "three." Two-year-olds contrast between one and all other number words. This enables them to recognize the quantity of one, but with no distinction between two and five. The ability to recognize the meaning of two, three, four, five, and so on (ie, cardinality) is achieved slowly and gradually, taking several months to become a "three knower" and then a "four knower," and so forth. At each of these points the child is able to recognize quantities up to that specific number and contrast them to all larger quantities, which cannot be distinguished. This process refines the understanding of cardinality. Carey considers this slow and difficult change as a marker for a qualitative change in which the child creates a discrete representation (ie, positive integers) where none existed before. If we consider the interaction between the developments of the AMS to the process of learning to count, there are several connections worth exploring (Fig. 16.3b). First, the basic ability of individuation is fundamental for the understanding of ordinality. Without the ability to recognize items as distinct, it will be impossible to perceive a sequence with a clear order. Second, the slow and gradual process in which the cardinal value understanding is extended from one number word to the next happens in parallel to the learning of the correlation between numerosity and continuous magnitudes. It is reasonable to assume that both processes influence each other. Furthermore, our suggestion that the continuous magnitudes are more dominantly processed than numerosities in early childhood is supported by Carey's insight, namely, that discrete representations are created throughout development. They do not simply exist at birth. If indeed there are no discrete representations of magnitudes, it is no wonder that continuous magnitudes are the main dimension of magnitude that infants and young children process.

16.4.3 Perception of Proportions

Young infants have shown an ability to perceive proportions. To the best of our knowledge, all studies exploring such abilities in infants used discrete stimuli (eg, proportions of arrays of items). As shown by Mix et al. (2002a), here too the results can be explained by processing of the proportion between continuous magnitudes (Boyer, Levine, & Huttenlocher, 2008) (Fig. 16.2c, left panel). Six-month-old infants have shown to discriminate ratios of two types of objects, as long as the two ratios were different by a factor of 2 (McCrink & Wynn, 2007). The authors used a similar paradigm to that used in order to test habituation to specific numerosity, and showed that just as infants of this age can tell 20 from 40 (but not 20 from 30), they can also tell 2:1 from 4:1 (but not 2:1 from 3:1). They concluded that infants' representations of ratios are parallel to those of sets of objects.

In correspondence with perception of sets of objects, perception of continuous proportions precedes the perception of discrete proportions (Henik et al., 2012; Leibovich & Henik, 2013; Mix et al., 2002a; Simon, 1997). As mentioned earlier, Jeong et al. (2007) demonstrated that children (up to about 10 years of age) understand ratios better when they are presented as continuous stimuli than as discrete stimuli. The authors used partially colored donut-shaped stimuli where the colored part could be either continuous or separated into segments. When the number of discrete segments was incongruent with the proportion (ie, when the donut with more colored segments had overall smaller proportion of colored parts), children failed to recognize the larger proportion. The difference between congruent and incongruent conditions suggested that children were misled by the numerosity of the segments. Thus, before the development of cognitive control, proportions, like sets of objects, are processed based on continuous magnitudes. The age at which proportions are being processed based on discrete magnitudes is yet to be determined.

The poor performance of older children (Jeong et al., 2007) in comparison with the success of younger infants (McCrink & Wynn, 2007) in discriminating discrete proportions (Fig. 16.2c, right panel) is attributed to the acquisition of counting skills (Fig. 16.2b). As is the case in other domains (like language), after acquiring a new rule, children overgeneralize it (Clark, 1995). Thus, after learning that counting the items in a set determines the quantity of a set, children use the counting schema even when it is not appropriate. Boyer et al. (2008) used a proportional choice task. In the task, two choice stimuli were presented following a target stimulus. The target and the choice alternatives could be either continuous (partially colored bar) or discrete (the same bar but with lines that demarcated each 1-cm unit). The authors found that children erroneously rely on the number of parts only when comparison by counting the units was possible (ie, in the discrete–discrete condition). This erroneous strategy was most evident in kindergarteners and first graders and was lessened only in third and fourth grades (ie, at the ages of 9–10 years). Interestingly, the erroneous counting strategy is still evident in adults, when testing their automatic processing of discrete proportions (Fabbri, Caviola, Tang, Zorzi, & Butterworth, 2012). In contrast, Kallai and Henik (in preparation) showed that, unlike the case of discrete ratios, when using continuous ratios (ie, part of the whole), both absolute size and relative size were found to be processed automatically. Thus, the same pattern that is found in intentional tasks in early childhood (ie, success in judging continuous ratios and failure when judging discrete proportions) is found under automatic processing conditions in adults. This similarity suggests once again that while processing of continuous magnitude (in this case, proportion) is more automatic, the processing of discrete magnitude (proportion) requires cognitive control.

16.4.4 Relative Versus Absolute Size Discrimination

In the background of size perception development (more specifically, of continuous size perception) stands the development of the ability to judge the size of an object without external reference (see the graded blue background color of the model in Fig. 16.2). Studies of size perception demonstrated that infants and young children show sensitivity to relative size of an object but not to absolute size (Duffy, Huttenlocher, Levine, & Duffy, 2005b; Duffy et al., 2005a; Gao, Levine, & Huttenlocher, 2000). For instance, Duffy and coworkers showed that infants and young children up to about 4 years of age process the size of a wooden dowel relative to the size of a container but not as absolute size. When placing the same dowel in a different-sized container, infants and children did not recognize it as the same object. In contrast, placing a larger dowel in a larger container, keeping the ratio between them constant, infants and young children perceived it as the same object (Duffy et al., 2005a; Huttenlocher, Duffy, & Levine, 2002). Eight-year-olds no longer rely solely on the relation provided by an available standard and can ignore misleading relative information provided by the standard (Duffy et al., 2005a). We placed this developmental course at the background of all other numerical developmental processes because it refers to the question of whether the magnitude of an object is processed in and by itself or only in reference to the magnitude of another object. To link this to our discussions of proportions, it is easy to see how the early ability to process relative size enables processing of part of the whole (where the part can be assessed in relation to the whole), whereas the processing of discrete proportions develops much later.

16.5 SUMMARY

In this chapter we reviewed the known theoretical frameworks regarding the developmental trajectory of magnitude processing. We presented evidence suggesting that all studies discussing the ability of babies and infants to discriminate numerosities can be better explained by discrimination of continuous magnitudes. Accordingly, we have suggested an integrative developmental model addressing the development of the AMS, and shown how this model is relevant to the understanding of numerosities, continuous magnitudes, and proportions throughout development. We also discussed how learning and cultural context interact with our model. This model is theoretical. Studies aimed to further explore this model are encouraged since we believe they will shed new light on the foundations of numerical cognition.

References

Acredolo, C., O'Connor, J., Banks, L., & Horobin, K. (1989). Children's ability to make probability estimates: skills revealed through application of Anderson's functional measurement methodology. *Child Development, 60*, 933–945.

Agrillo, C., & Piffer, L. (2012). Musicians outperform nonmusicians in magnitude estimation: evidence of a common processing mechanism for time, space and numbers. *Quarterly Journal of Experimental Psychology, 65*(12), 2321–2332.

Banks, M. S. (1980). The development of visual accommodation during early infancy. *Child Development*, *51*(3), 646–666.

Bjorklund, D. F., & Harnishfeger, K. K. (1995). *Interference and inhibition in cognition*. San Diego, CA: Academic Press.

Bonato, M., Fabbri, S., Umiltà, C., & Zorzi, M. (2007). The mental representation of numerical fractions: real or integer? *Journal of Experimental Psychology. Human Perception and Performance*, *33*(6), 1410–1419.

Bower, T. G. (1974). In T. G. Bower (Ed.), *Development in infancy*. San Francisco: W. H. Freeman.

Boyer, T., Levine, S., & Huttenlocher, J. (2008). Development of proportional reasoning: where young children go wrong. *Developmental Psychology*, *44*(5), 1478–1490.

Cantlon, J. F., Brannon, E. M., Carter, E. J., & Pelphrey, K. A. (2006). Functional imaging of numerical processing in adults and 4-y-old children. *PLoS Biology*, *4*(5), e125.

Cantlon, J. F., Libertus, M. E., Pinel, P., Dehaene, S., Brannon, E. M., & Pelphrey, K. A. (2009a). The neural development of an abstract concept of number. *Journal of Cognitive Neuroscience*, *21*(11), 2217–2229.

Cantlon, J. F., Platt, M. L., & Brannon, E. M. (2009b). Beyond the number domain. *Trends in Cognitive Sciences*, *13*(2), 83–91.

Cantrell, L., & Smith, L. B. (2013). Open questions and a proposal: a critical review of the evidence on infant numerical abilities. *Cognition*, *128*(3), 331–352.

Carey, S. (2001). Cognitive foundations of arithmetic: evolution and ontogenisis. *Mind & Language*, *16*(1), 37–55.

Carey, S. (2004). Bootstrapping & the origin of concepts. *Daedalus*, *133*(1), 59–68.

Clark, E. V. (1995). *The lexicon in acquisition*. Cambridge: Cambridge University Press.

Coubart, A., Izard, V., Spelke, E. S., Marie, J., & Streri, A. (2014). Dissociation between small and large numerosities in newborn infants. *Developmental Science*, *17*(1), 11–22.

Dehaene, S. (1997). *The number sense: how the mind creates mathematics*. New York: Oxford University Press.

Dehaene, S., & Changeux, J. P. (1993). Development of elementary numerical abilities: a neuronal model. *Journal of Cognitive Neuroscience*, *5*(4), 390–407.

Dobson, V., & Teller, D. Y. (1978). Visual acuity in human infants: a review and comparison of behavioral and electrophysiological studies. *Vision Research*, *18*(11), 1469–1483.

Duffy, S., Huttenlocher, J., & Levine, S. (2005a). It is all relative: how young children encode extent. *Journal of Cognition and Development*, *6*(1), 51–63.

Duffy, S., Huttenlocher, J., Levine, S., & Duffy, R. (2005b). How infants encode spatial extent. *Infancy*, *8*(1), 81–90.

Emmerton, J. (2001). Pigeons' discrimination of color proportion in computer-generated visual displays. *Animal Learning & Behavior*, *29*(1), 21–35.

Fabbri, S., Caviola, S., Tang, J., Zorzi, M., & Butterworth, B. (2012). The role of numerosity in processing nonsymbolic proportions. *Quarterly Journal of Experimental Psychology*, *65*(12), 2435–2446.

Feigenson, L., & Carey, S. (2003). Tracking individuals via object-files: evidence from infants' manual search. *Developmental Science*, *6*(5), 568–584.

Feigenson, L., Carey, S., & Hauser, M. (2002). The representations underlying infants' choice of more: object files versus analog magnitudes. *Psychological Science*, *13*(2), 150–156.

Feigenson, L., Dehaene, S., & Spelke, E. (2004). Core systems of number. *Trends in Cognitive Sciences*, *8*(7), 307–314.

Gallistel, C. R., & Gelman, R. (1992). Preverbal and verbal counting and computation. *Cognition*, *44*(1), 43–74.

Gallistel, C., & Gelman, R. (2000). Non-verbal numerical cognition: from reals to integers. *Trends in Cognitive Sciences*, *4*, 59–65.

Gallistel, C. R., Gelman, R., & Cordes, S. (2006). The cultural and evolutionary history of the real numbers. In *Evolution and culture: a Fyssen Foundation symposium* (p. 296).

Gao, F., Levine, S., & Huttenlocher, J. (2000). What do infants know about continuous quantity? *Journal of Experimental Child Psychology*, *77*(1), 20–29.

Gebuis, T., & Reynvoet, B. (2012). Continuous visual properties explain neural responses to nonsymbolic number. *Psychophysiology*, *49*(11), 1481–1491.

Gebuis, T., & Reynvoet, B. (2013). The neural mechanisms underlying passive and active processing of numerosity. *NeuroImage*, *70*, 301–307.

Gibbon, J. (1977). Scalar expectancy theory and Weber's law in animal timing. *Psychological Review*, *84*, 279–325.

Goswami, U. (1989). Relational complexity and the development of analogical reasoning. *Cognitive Development*, *4*(3), 251–268.

Halberda, J., Mazzocco, M., & Feigenson, L. (2008). Individual differences in non-verbal number acuity correlate with maths achievement. *Nature*, *455*(7213), 665–668.

Henik, A., Leibovich, T., Naparstek, S., Diesendruck, L., & Rubinsten, O. (2012). Quantities, amounts, and the numerical core system. *Frontiers in Human Neuroscience*, *5*(186), 1–4.

Honig, W., & Stewart, K. (1989). Discrimination of relative numerosity by pigeons. *Animal Learning & Behavior*, *17*(2), 134–146.

Houdé, O. (1997). Numerical development: from the infant to the child. Wynn's (1992) paradigm in 2- and 3-year olds. *Cognitive Development*, *12*(3), 373–391.

Houdé, O. (2000). Inhibition and cognitive development: object, number, categorization, and reasoning. *Cognitive Development*, *15*(1), 63–73.

Houdé, O., Pineau, A., Leroux, G., Poirel, N., Perchey, G., Lanoë, C., & Mazoyer, B. (2011). Functional magnetic resonance imaging study of Piaget's conservation-of-number task in preschool and school-age children: a neo-Piagetian approach. *Journal of Experimental Child Psychology*, *110*(3), 332–346.

Huttenlocher, J. (1999). Spatial scaling in young children. *Psychological Science*, *10*(5), 393–398.

Huttenlocher, J., Duffy, S., & Levine, S. (2002). Infants and toddlers discriminate amount: are they measuring? *Psychological Science*, *13*(3), 244–249.

Jeong, Y., Levine, S., & Huttenlocher, J. (2007). The development of proportional reasoning: effect of continuous versus discrete quantities. *Journal of Cognition and Development*, *8*(2), 237–256.

Kallai, A. Y., & Henik, A. (in preparation). Absolute or relative size: what do we perceive when we look at the full half of the glass?

Kallai, A., & Tzelgov, J. (2009). A generalized fraction: an entity smaller than one on the mental number line. *Journal of Experimental Psychology: Human Perception Performance*, *35*(6), 1845–1864.

Karplus, R., Pulos, S., & Stage, E. (1983). Early adolescents' proportional reasoning on "rate" problems. *Educational Studies in Mathematics*, *14*(3), 219–233.

Leibovich, T., & Henik, A. (2013). Magnitude processing in non-symbolic stimuli. *Frontiers in Psychology*, *4*, 375.

Leibovich, T., & Henik, A. (2014). Comparing performance in discrete and continuous comparison tasks. *Quarterly Journal of Experimental Psychology*, *67*(5), 899–917.

Leibovich, T., Henik, A., & Salti, M. (2015). Numerosity processing is context driven even in the subitizing range: an fMRI study. *Neuropsychologia*, *77*, 137–147.

Leibovich, T., Vogel, S. E., Henik, A., & Ansari, D. (2016). Asymmetric processing of numerical and nonnumerical magnitudes in the brain: an fMRI study. *Journal of Cognitive Neuroscience*, *28*(1), 166–176.

Mazzocco, M. M. M., Feigenson, L., & Halberda, J. (2008). Impaired acuity of the approximate number system underlies mathematical learning disability (dyscalculia). *Child Development*, *82*(4), 1224–1237.

McComb, K., Packer, C., & Pusey, A. (1994). Roaring and numerical assessment in contests between groups of female lions *Panthera leo*. *Animal Behaviour*, *47*(2), 379–387.

McCrink, K., & Wynn, K. (2007). Ratio abstraction by 6-month-old infants. *Psychological Science*, *18*(8), 740–745.

Meadow, K. P., Droz, R., Rahmy, M., Diamanti, J., Piaget, J., Inhelder, B., & Fishbein, H. D. (1977). The origin of the idea of chance in children. In P. B. Trans, L. Leake, & H. D. Fishbein (Eds.), *Contemporary sociology* (Vol. 6). New York: WW Norton.

Meck, W. H., & Church, R. M. (1983). A mode control model of counting and timing processes. *Journal of Experimental Psychology: Animal Behavior Processes, 9*(3), 320–334.

Meert, G., Grégoire, J., & Noël, M. (2009). Rational numbers: componential versus holistic representation of fractions in a magnitude comparison task. *The Quarterly Journal of Experimental Psychology, 62*(8), 1598–1616.

Meert, G., Grégoire, J., & Noël, M. (2010). Comparing 5/7 and 2/9: adults can do it by accessing the magnitude of the whole fractions. *Acta Psychologica, 135*(3), 284–292.

Mix, K. S., Huttenlocher, J., & Levine, S. C. (2002a). Multiple cues for quantification in infancy: is number one of them? *Psychological Bulletin, 128*(2), 278–294.

Mix, K. S., Huttenlocher, J., & Levine, S. C. (2002b). In K. S. Mix, J. Huttenlocher, & S. C. Levine (Eds.), *Quantitative development in infancy and early childhood.* New York, NY: Oxford University Press.

Newcombe, N. S. (2002). The nativist-empiricist controversy in the context of recent research on spatial and quantitative development. *Psychological Science, 13*(5), 395–401.

Noelting, G. (1980). The development of proportional reasoning and the ratio concept part I—differentiation of stages. *Educational Studies in Mathematics, 11*(2), 217–253.

Odic, D., Libertus, M. E., Feigenson, L., & Halberda, J. (2013). Developmental change in the acuity of approximate number and area representations. *Developmental Psychology, 49*(6), 1103–1112.

Piaget, J. (1952). *The child's conception of number.* Hove: Psychology Press.

Pica, P., Lemer, C., Izard, V., & Dehaene, S. (2004). Exact and approximate arithmetic in an Amazonian indigene group. *Science, 306*(5695), 499–503.

Ross, J., & Burr, D. C. (2010). Vision senses number directly. *Journal of Vision, 10*(2), 1–8.

Rousselle, L., & Noël, M. (2008). The development of automatic numerosity processing in preschoolers: evidence for numerosity-perceptual interference. *Developmental Psychology, 44*, 544–560.

Rousselle, L., Palmers, E., & Noël, M. P. (2004). Magnitude comparison in preschoolers: what counts? Influence of perceptual variables. *Journal of Experimental Child Psychology, 87*(1), 57–84.

Siegler, R., & Vago, S. (1978). The development of a proportionality concept: judging relative fullness. *Journal of Experimental Child Psychology, 25*(3), 371–395.

Simon, T. (1997). Reconceptualizing the origins of number knowledge: a "non-numerical" account. *Cognitive Development, 12*(3), 349–372.

Sophian, C. (2000). Perceptions of proportionality in young children: matching spatial ratios. *Cognition, 75*(2), 145–170.

Spinillo, A., & Bryant, P. (1991). Children's proportional judgments: the importance of "half". *Child Development, 62*(3), 427–440.

Vallentin, D., & Nieder, A. (2008). Behavioral and prefrontal representation of spatial proportions in the monkey. *Current Biology, 18*(8), 1420–1425.

Verguts, T., & Fias, W. (2004). Representation of number in animals and humans: a neural model. *Journal of Cognitive Neuroscience, 16*(9), 1493–1504.

Xu, F., & Spelke, E. S. (2000). Large number discrimination in 6-month-old infants. *Cognition, 74*(1), B1–B11.

CHAPTER

17

How Do Humans Represent Numerical and Nonnumerical Magnitudes? Evidence for an Integrated System of Magnitude Representation Across Development

Stella F. Lourenco

Emory University, Atlanta, Georgia, United States of America

OUTLINE

Continuous Issues in Numerical Cognition. http://dx.doi.org/10.1016/B978-0-12-801637-4.00017-2

17.1 INTRODUCTION

It is difficult to imagine a world without information related to magnitude in it. This information pervades much of our daily decision making. The tip we leave at the restaurant is based on an estimate, or precise computation, of numerical magnitude. Selecting the shortest checkout line at the grocery store may reflect a comparison of spatial magnitude, namely, the length of people comprising the lines. An approximation of the time it will take to drive from one side of town to the other during rush hour is rooted in representations of duration. We regularly recruit representations of magnitude to solve a variety of problems, and accumulating evidence suggests that our reliance on magnitude emerges early in our development (for review, see Dehaene, 2011). Many findings also suggest that the ability to represent magnitude is present throughout the animal kingdom. Like humans, other animals rely on magnitude information for everyday activities, including those critical to survival such as foraging and the avoidance of predators (Dehaene, 2011; Gallistel & Gelman, 2000; Geary, Berch, & Mann Koepke, 2015). Humans, however, also have access to a system of symbols and formal rules for mathematical computations that they enact on magnitude information. We acquire a system of symbols that we use to represent numbers precisely and that allow us to engage in computations related to space and time (Dehaene, 2011; Devlin, 2000; Lakoff & Núñez, 2000).

Within a given context, we generally have access to multiple sources of magnitude information. A visual display of objects provides at least two such sources. One can make reference to the number of items within the display. This type of information is often considered "discrete" because the items may be precisely enumerated, at least in a symbolic context such as when items are counted (eg, "one," "two," "three," and so on). One can also estimate the so-called "continuous" properties of the display such as the cumulative surface area occupied by the set of items. The display in this case conflates number and cumulative area. In the physical world, such conflations generally reflect positive correlations; that is, increasing the number of a set is accompanied by an increase in cumulative area for the set. Although many questions exist about how different magnitudes are represented, especially when simultaneously available, one common perspective is that it makes much adaptive sense to encode and rely on multiple magnitudes for decision making.

Magnitudes come in different forms and though we may ultimately be able to make reference to different magnitudes such as "number," "cumulative area," and so on, they share analog format. In analog format, magnitudes are represented approximately, such that activation for a given value is accompanied by activation for adjacent values (eg, the presentation of 7 items will activate representations of 7, but also of 6 and 8), albeit to a lesser extent. Moreover, the larger the magnitude, the greater the "noise" in the underlying representation, that is, the wider the spread of activation of nearby values. A behavioral consequence of representations that are analog in format is that discrimination follows Weber's law. For example, discriminating 8 versus 4 items, a 2:1 ratio, is easier than discriminating 12 versus 8 items, a 3:2 ratio, even though the absolute difference between the items (4 items) is the same in both cases (eg, Buckley & Gillman, 1974). Ease of discrimination in human adults is often indexed by decision speed (ie, faster reaction times) or accuracy. Weber's law is similarly obeyed when discriminating objects that vary in physical size, or interobject relations varying in distance, or items varying in duration (eg, Droit-Volet, Clément, & Fayol, 2008; Fias, Lammertyn, Reynvoet, Dupont, & Orban, 2003; Henmon, 1906).

Despite widespread acknowledgment that cues to magnitudes are often conflated in the physical environment and that representations of different magnitudes share analog format, claims of domain specificity are prevalent, particularly in the case of number. One proposal is that representations of numerical magnitude are distinct from those of other magnitudes (eg, Barth, Kanwisher, & Spelke, 2003; Dehaene & Changeux, 1993; Ross & Burr, 2010; Stoianov & Zorzi, 2012). For instance, researchers who argue for an approximate number system (ANS) hold that number is represented independently of other magnitudes, even when other magnitude cues are conflated with number in visual displays. Visual models of the ANS involve normalizing over nonnumerical magnitudes such as the individual sizes of objects or the cumulative area of objects within a set, as well as other perceptual features such as the shape and position of individual objects (eg, Chen & Verguts, 2010; Dehaene & Changeux, 1993; Stoianov & Zorzi, 2012; Verguts & Fias, 2004). More extreme perspectives of domain specificity hold that representations of numerical magnitude (and arithmetic processes) are modular, that is, isolated in their cognitive and neural processes from other magnitudes. For instance, Butterworth (1999, 2005) has proposed that dyscalculia, a mathematical learning disability, is the result of a defective number module.

17.2 THEORY OF INTEGRATION ACROSS MAGNITUDES: THE GENERAL MAGNITUDE SYSTEM

In contrast to the ANS, a system of generalized magnitude representation, or the so-called *general magnitude system*, emphasizes interactions across different magnitudes, such as number, area, and duration (eg, Bueti & Walsh, 2009; Lourenco & Longo, 2011; Lourenco, 2015a; Walsh, 2003). These interactions are supported not only by common analog structure but, critically, also by shared neural mechanisms, particularly in and around the intraparietal sulcus (IPS; see Section 17.3). Models from visual perception that are consistent with this proposal hold that representations of numerical magnitude rely on the processing of other magnitudes, such as average element size, cumulative surface area, and/or density (eg, Allik & Tuulmets, 1991; Dakin, Tibber, Greenwood, Kingdom, & Morgan, 2011; Gebuis & Reynvoet, 2012). Number may from the outset be tied to other magnitudes. According to these models, number is never independent of other magnitudes, but rather, it emerges from them early in visual processing.

There are also perspectives that emphasize the ontogenetic dependence between number and other magnitudes. One proposal is that representations of numerical magnitude build on continuous magnitudes such as spatial extent over ontogenetic time (Henik, Leibovich, Naparstek, Diesendruck, & Rubinsten, 2011; see also, Geary et al., 2015). Studies using computational models have even demonstrated that number processing is better characterized as growing out of size discrimination than emerging de novo (Katz, Benbassat, Diesendruck, Sipper, & Henik, 2013). On this view, continuous dimensions serve a primary role in quantification. Another possibility, originally proposed by Bueti and Walsh (2009; Walsh, 2003), is that number representations coopted a system for action within parietal cortex because it was already equipped with analog coding for spatial and temporal information. On this view, representations of numerical magnitude came to be integrated with other magnitudes because of evolutionary pressures related to action planning and the need for analog format available within parietal cortex.

The notion that number and other magnitudes might be associated with one another on structural and neural grounds has historical roots. The theory of a general magnitude system revisits and extends previous claims about the mental connections between different magnitudes. In "An Essay Concerning Human Understanding," the philosopher John Locke speculated about the association between space and time, arguing that: "expansion and duration do mutually embrace and comprehend each other; every part of space being in every part of duration, and every part of duration in every part of expansion" [Locke, 1689 (1995), pp. 140]. In reviewing neuropsychological studies, the neurologist Macdonald Critchley noted the cooccurrence of time, space, and number deficits, highlighting neurological disorders that showed comorbidity of deficits (Critchley, 1953). With respect to the comorbidity between time and space, he wrote: "pure temporal disorientation... occurring independently of spatial disorders, is a rarer phenomenon, for more often, the two are combined" (Critchley, 1953, pp. 352).

There is also the classic work of developmental psychologist Jean Piaget who showed that nonnumerical cues interfered with children's numerical judgments (Piaget, 1965). In his so-called "number conservation task," Piaget asked children to judge the relative numerosity of two rows of objects that differed in length (eg, two rows of coins). Children between 3 and 6 years of age frequently judged longer rows as greater in number, even if they had fewer objects. Children also claimed that the number of objects in a single row increased if the experimenter spread the objects apart. Piaget concluded that, for children, number was mentally conflated with space. More recently, these data have been taken as evidence for a general magnitude system that processes information related to different magnitudes—in this case number and spatial extent. More specifically, on this view, children's errors are the result of partially integrated or overlapping representations of number and spatial extent. Importantly, though, children's errors were consistently asymmetric—that is, children erred because they favored information about spatial extent over number, raising separate questions about how numerical and nonnumerical magnitudes are weighted in a system where representations of number and spatial extent may overlap. One possibility is that the weighting of magnitudes, or the reliance of one magnitude over the other, may depend on the relative salience of associated cues available within a given context (Bonn & Cantlon, 2012; Lourenco & Longo, 2011). Another possibility is that differential automaticity of processing different magnitudes (cf., Arend, Cappelletti, & Henik, 2014), or the ease with which different magnitudes are inhibited, affects what magnitude one relies on in a given task. These possibilities assume some differentiation in the representation of magnitudes, and although it is unclear at what stage of processing differentiation might be critical, it is important to highlight that a model of generalized magnitude representation does not imply a complete lack of differentiation (eg, Bueti & Walsh, 2009; Cappelletti, Freeman, & Butterworth, 2011; Dormal & Pesenti, 2009; Lourenco, 2015a). It has been suggested that within such a system magnitudes also dissociate because of distinct, along with shared, patterns of neural activation. Later I will discuss evidence from single cell recording with monkeys demonstrating a portion of neurons that code for both number and line length (eg, Tudusciuc & Nieder, 2007). Critically, this research also highlights individual neurons that code separately for number and line length, suggesting a mechanism by which dissociation is supported.

Drawing on a variety of paradigms, other investigators have demonstrated that children's magnitude judgments involve conflation across different dimensions of magnitude. For instance, Levin (1977) asked kindergarteners to judge which of two lights was presented for the longer duration. The lights differed in size (and brightness). Children consistently judged the larger (and brighter) stimuli to have lasted longer in time (see also Siegler & Richards, 1979). More recently, Casasanto, Fotakopoulou, & Boroditsky (2010) examined distance and duration judgments by having children indicate which of two snails in a movie traveled farther in space or longer in time. Again, children showed congruity effects, with farther in space mapping to longer in time.

On the one hand, children's reliance on the availability of multiple magnitude cues makes good sense since different magnitudes are generally positively correlated. For instance, greater in numerical magnitude is often associated with greater spatial extent such as overall area. Other examples include positive correlations between distance and time so that walking farther in distance generally translates to longer in time, as well as between number and time so that putting more coins into a coin sorter takes longer than when there are fewer coins. Relying on multiple magnitudes in these cases may be adaptive both in the short and long term, since the accuracy of one's estimates will increase when based on more than one source of information. However, in cases in which the available magnitudes are inversely correlated, as in tasks in which they are pitted against each other, the better strategy would be to focus exclusively on the magnitude that is relevant to the judgment. This may be difficult (perhaps especially in children), however, if different magnitudes form part of a general magnitude system with overlapping representations.

Stavy and Tirosh (2000) proposed that children's "more X—more Y" mappings were detrimental to their acquisition of mathematical and scientific concepts because many of these concepts require that magnitudes be dissociated. For instance, the concept of *five* when used to refer to "five mice" involves information about both the quantity of five and the spatial magnitudes (ie, the physical size of each mouse or the cumulative area occupied by all mice). In another example, "five elephants" is still about the numerical concept of *five*, but the spatial magnitudes are now vastly different. Learning "five," and engaging in numerical abstraction more generally, requires dissociating numerical and spatial magnitudes, even if the magnitudes overlap to some extent as proposed by a general magnitude system. Much recent research suggests that representations of numerical magnitude, particularly their associated precision, are related to mathematical competence such as arithmetic fluency (Bonny & Lourenco, 2013; Halberda, Mazzocco, & Feigenson, 2008; Libertus, Feigenson, & Halberda, 2011; Lourenco, Bonny, Fernandez, & Rao, 2012). Gilmore et al. (2013) have argued that this relationship may be mediated by individual differences in inhibitory control such that it is children's ability to override irrelevant nonnumerical cues that predicts mathematical competence (see also Fuhs & McNeil, 2013). This argument holds that dissociating magnitudes may be better for supporting the acquisition of mathematical concepts or that inhibitory control more generally, when multiple magnitudes are activated, is associated with higher mathematical aptitude.

Additional paradigms with children and adults have provided converging evidence for cross-magnitude interactions. In a classic study, Henik and Tzelgov (1982) adapted the well-known Stroop paradigm for use with magnitude information, namely number and physical size. It was found that adults' judgments of magnitude were either interfered with (as in Piaget's number conservation task) or facilitated depending on the pairing of magnitudes. For instance, adults judged the numerical values or physical sizes of Arabic digits faster when number and physical size were congruent than when they were incongruent. More specifically, when participants were asked to indicate which number was greater in value, their response times were affected by the physical sizes of the digits (which were irrelevant to the judgment of number). There was facilitation when size was congruent with number (eg, 2 7) but interference when it was incongruent (eg, 2 7). This paradigm has been used to show that congruity effects also exist between Arabic digits and the duration of the stimuli (Oliveri et al., 2008), and with nonsymbolic displays of number (Dormal, Seron, & Pesenti, 2006; Hurewitz, Gelman, & Schnitzer, 2006; Tokita & Ishiguchi, 2011; Xuan, Zhang, He, & Chen, 2007). When visual displays are arrays of dots, there is evidence of both interference and facilitation in participants as young as 3 years of age, at least from spatial extent and duration onto judgments of number (Gebuis, Kadosh, de Haan, & Henik, 2009; Lourenco, Levine, & Degner, 2014; Piaget, 1965; Rousselle, Palmers, & Noël, 2004). By school age, such effects are generally bidirectional in that number also either interferes with or facilitates judgments of other magnitudes (Rousselle & Noël, 2008; Stavy & Tirosh, 2000).

On the one hand, cross-magnitude effects such as facilitation and interference pose a challenge for models such as the ANS that posit independence between numerical and nonnumerical magnitudes (eg, Feigenson, Dehaene, & Spelke, 2004; Harvey, Klein, Petridou, & Dumoulin, 2013; Ross & Burr, 2010). Indeed, as noted earlier, these interactions have been taken as evidence for an alternative model, namely the general magnitude system, which holds that number is not fully dissociable from nonnumerical magnitudes. In line with a general magnitude system, these interactions reflect at least some overlap among representations of different magnitudes. On the other hand, the cross-magnitude interactions would not be incompatible with differentiated, nonoverlapping magnitude representations, as posited by the ANS, if they were *indirectly* mediated by another system. One possibility is that language facilitates the connection between magnitudes, with common magnitude words such as "larger" and "greater" mediating the interactions. A related proposal is that there is a metaphorical relationship between magnitudes, with interactions supported by specific conceptual metaphors (for review, see Winter, Marghetis, & Matlock, 2015). Yet another possibility, still consistent with differentiated magnitude representations, is that cross-magnitude interactions reflect postrepresentational processing, specifically the decision stage of a task in which the judgment elicits comparisons across distinct magnitudes (eg, Gabay, Leibovich, Henik, & Gronau, 2013; Yates, Loetscher, & Nicholls, 2012).

In this chapter, I argue for integration across numerical and nonnumerical magnitudes, particularly "prothetic" dimensions such as spatial extent and duration in which the more-versus-less ordering of the dimensions is unambiguous (Stevens, 1957, 1975; see also, Lourenco & Longo, 2011). This integration is consistent with a general magnitude system, in which representations of number and other magnitudes are at least partially overlapping. The notion of *overlap* highlights two important components. The first is that the cognitive interactions observed across different tasks are *direct*—in other words, they do not involve mediation by other systems such as language (Lourenco, 2015a). The second, related, component is shared neural mechanisms (eg, Walsh, 2003). That number and other magnitudes rely on activation in and around the IPS has been reviewed and discussed elsewhere (eg, Cantlon, Platt, & Brannon, 2009; Cohen Kadosh, Lammertyn, & Izard, 2008; Walsh, 2003). Though I will provide a brief overview of this evidence (Section 17.3), I focus primarily on the first component—the direct interactions between magnitudes.

In the sections that follow, I argue for a general magnitude system that exists beginning early in human life and that it is continuous throughout human development. I will present two types of evidence in support of these claims. One comes from work with preverbal children (infants). As described later, the findings with infants not only shed light on the developmental origins of the cross-magnitude interactions in humans but they also serve to rule out linguistic mediation as a mechanism for these interactions (Section 17.4). Another line of evidence comes from work with adults using training and subliminal priming procedures (Section 17.5). Both types of procedures demonstrate transfer across number and spatial extent, and both rule out mediation by language or postrepresentational strategies as potential accounts of cross-magnitude interactions. These data additionally point to the developmental continuity of a general magnitude system. I will also describe data from a recent longitudinal study that directly support the claim of developmental continuity by demonstrating that individual differences in the strength of magnitude associations assessed in infancy predict interindividual variability in these associations at preschool age, even when accounting for the effects of other cognitive factors (Section 17.5).

17.3 SHARED NEURAL CODING FOR NUMBER AND OTHER MAGNITUDES

Several publications have now reviewed the neural evidence in support of shared neural coding for number and other magnitudes. For a treatment of this evidence, I direct the reader to references such as Cantlon et al. (2009); Cohen Kadosh et al. (2008); and Walsh (2003). Here I provide a brief overview of this evidence in order to ground the later behavioral evidence I describe in support of nonmediated interactions between representations of different magnitudes.

Neuropsychological studies with patients as well as neuroimaging and neural stimulation studies with healthy participants point to a shared neural substrate for numerical and nonnumerical magnitudes. Patients with deficits in representing magnitude (numerical and/or nonnumerical) often present with parietal damage (eg, Walsh, 2003) and healthy participants show parietal activation when responding to one or more magnitude (eg, Dormal, Andres, & Pesenti, 2008; Dormal, Andres, & Pesenti, 2012; Hayashi et al., 2013; Pinel, Piazza, Le Bihan, & Dehaene, 2004). Nevertheless, concerns about the size of brain lesions as well as the spatial and temporal resolution associated with techniques such as fMRI leave open the possibility that different magnitudes may not be processed by the same neural mechanisms. Instead, the processing of different magnitudes could implicate completely distinct neural regions that are not adequately assessed with these procedures (cf., Haxby, Gobbini, Furey, Ishai, Schouten, & Pietrini, 2001).

Studies using single cell recording with monkeys are not subject to the same limitations of resolution. Indeed, they have provided evidence of shared coding for different magnitudes at the cellular level (Tudusciuc & Nieder, 2007, 2009). Nieder (2012) found that a small proportion of neurons in monkey ventral intraparietal area (VIP) are tuned to both number and line length. Importantly, despite the overlap, these neurons do not appear to align in their coding of relative magnitude. That is, the evidence points to preferred numerical values and line lengths that do not match in their more-versus-less relations. More specifically, individual neurons tuned to larger numerosities were not more likely to be tuned to longer line lengths compared to shorter line lengths (see Tudusciuc & Nieder, 2007). The lack of alignment with respect to relative magnitude raises obvious questions about how cross-magnitude interactions shown behaviorally might be supported by individual neurons where facilitation and interference across magnitudes reflect matching or mismatching of more-versus-less relations.

As I discussed in Lourenco (2015a), one possible explanation for the lack of alignment within individual neurons is that the alignment of relative magnitude takes place at the population level. This possibility follows from research showing that even within a single magnitude—namely number—there is a lack of alignment that pertains to cardinal values when they involve different sensory modalities. In monkey VIP, neurons tuned to specific cardinal values (eg, 1, 2, 3, or 4) do not show reliable cross-modal matching, except in the case of 1 (Nieder, 2012). That is, a neuron that responds selectively to numerosity 3 for visual stimuli may respond to auditory sequences of numerosity 2 or 4. And yet, behavioral paradigms show that nonhuman animals (Jordan, Brannon, Logothetis, & Ghazanfar, 2005) and preverbal children (Izard, Sann, Spelke, & Streri, 2009; Starkey, Spelke,

& Gelman, 1983) match numerosities across sensory modalities (vision and audition). It is thus possible that alignment both within and across magnitudes occurs at the population level. In the case of number, this may apply to specific cardinalities that cut across sensory modality. In the case of different magnitudes, this may apply to the common more-versus-less ordering that cuts across magnitudes such as number and spatial extent.

Another possibility is that alignment of relative magnitude within individual neurons takes place in the prefrontal cortex (PFC), rather than the parietal cortex. Although much discussion concerning the neural instantiation of a general magnitude system has focused on the parietal cortex, particularly the IPS, there is clear evidence from humans and nonhuman animals that frontal regions such as the dorsolateral PFC and the frontal gyrus play a central role in processing number and nonnumerical magnitudes (Ansari & Dhital, 2006; Coull, Vidal, Nazarian, & Macar, 2004; Fulbright, Manson, Skudlarski, Lacadie, & Gore, 2003; Genovesio, Tsujimoto, & Wise, 2011; Hayashi et al., 2013; Nieder, 2012; Onoe et al., 2001). Indeed, recent findings suggest that a proportion of neurons in the PFC is tuned abstractly to relative magnitude, that is, responding indiscriminately to numerosity and line length when making "greater than/less than" judgments (Eiselt & Nieder, 2013).

In summary, research with human and nonhuman animals suggests a common neural substrate for the processing of number and other magnitudes. This work, however, leaves open questions about how shared neural coding within portions of the IPS and/or PFC may support cross-magnitude behavioral effects such as facilitation and interference. Moreover, even in the face of evidence of shared neural activation, controversy remains about whether cross-magnitude interactions observed in behavioral paradigms reflect this shared activation. Could the interactions instead engage specialized, nonshared coding within these regions, which are mediated by other systems? Or could the shared coding be a consequence of this mediation? It is thus critical to consider cross-magnitude interactions in the context of different paradigms and different age groups. By using different paradigms, researchers are in a better position to hone in on the mechanism or mechanisms that underlie the interactions. By comparing different age groups and even comparing individual differences in the strength of the interactions, researchers can also begin to shed light on the stability of the putative general magnitude system. These approaches are discussed next.

17.4 DEVELOPMENTAL ORIGINS OF A GENERAL MAGNITUDE SYSTEM

Accumulating evidence with human infants suggests that the cross-magnitude interactions observed in older children and adults are also present in preverbal children (de Hevia & Spelke, 2010; de Hevia, Izard,

Coubart, Spelke, & Streri, 2014; Lourenco & Longo, 2010; Lourenco, 2015b; Srinivasan & Carey, 2010). Demonstrating that these interactions are possible among preverbal children goes a long way to ruling out language as a mechanism for supporting the associations between numerical and nonnumerical magnitudes. As already noted, one possibility is that the interactions observed in verbal participants are mediated by language, and shared neural activation could thus be a consequence of this mediation. Here I present research with preverbal children. Across studies, it has been shown that infants expect the more-versus-less ordering of one magnitude (eg, number) to generalize to other magnitudes (eg, spatial extent or duration). It is thus not the case that language, and more specifically magnitude words, are necessary for cross-magnitude interactions to exist. The effects in early development have been taken as evidence for a general magnitude system in which the representations of different magnitudes interface directly (eg, Lourenco & Longo, 2011; Lourenco, 2015a).

In one study with infants, Lourenco and Longo (2010) found bidirectional effects for various combinations of number, spatial extent, and duration. In the initial (habituation) phase of this study, 9-month-olds were taught that one magnitude (eg, physical size) mapped systematically to a specific pattern (eg, stripes or dots). The infants were then tested on whether the learning of this arbitrary mapping generalized to another magnitude (eg, number or duration). During habituation, infants might be shown, for example, that the larger-sized rectangles were striped and the smaller-sized rectangles were dotted (Fig. 17.1). When subsequently tested with arrays that varied in number, trials that maintained the mapping to pattern (ie, congruent test trials) featured a larger numerical array with striped rectangles and a smaller numerical array with dotted rectangles; trials that violated the mapping to pattern (ie, incongruent test trials) featured a larger numerical array with dotted rectangles and a smaller numerical array with striped rectangles (Fig. 17.1). The same logic was applied to duration (eg, congruent test trials: longer-lasting objects as striped and shorter-lasting objects as dotted; incongruent test trials: longer-lasting objects as dotted and shorter-lasting objects as striped). Across all combinations, infants looked longer to incongruent than congruent test trials, providing evidence of transfer across different magnitudes. An interpretation of these findings is that the cross-magnitude interactions reflect an abstract coding of the more-versus-less relations shared by the different magnitudes. Such coding could be supported by a general magnitude system that includes at least some neural overlap in the coding of number, spatial extent, and duration.

More recently, de Hevia et al. (2014) found that the interaction between number and spatial extent (ie, line length) occurs from the first days of life (1–3 days of age). Infants who were habituated to an ascending or a descending sequence of magnitude subsequently looked longer to trials

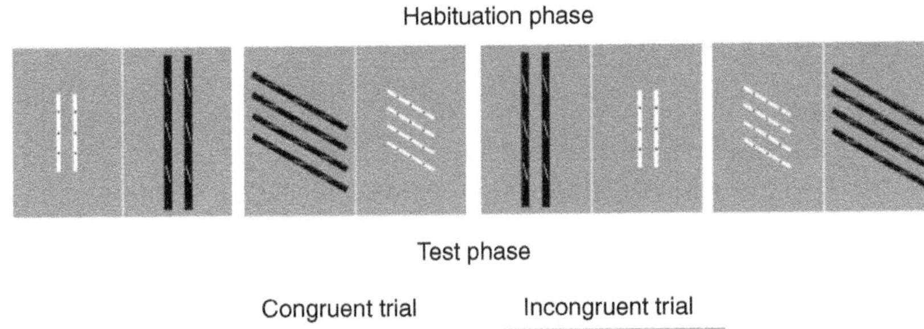

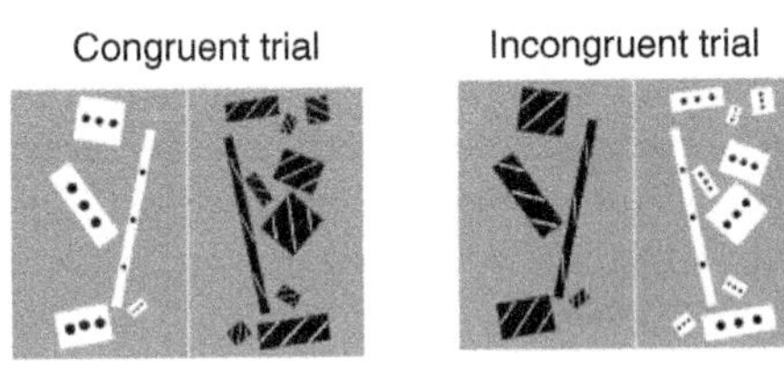

FIGURE 17.1 **An example of the stimuli used in the study of Lourenco & Longo (2010).** In all conditions, infants were habituated to an arbitrary mapping between relative magnitude and pattern (striped vs. dotted). In this example, the habituation phase (top row) involved trials in which smaller rectangles were dotted and larger rectangles were striped. On each trial, infants saw two sets of rectangles against a gray background and separated by a blue divider. Across these trials, factors such as the orientation and number of rectangles were varied; what remained constant was the mapping between relative magnitude and pattern, which, in this example, was that small rectangles were always dotted and large rectangles were always striped. During the test phase (bottom row), infants were presented with two types of trials: congruent and incongruent. In the congruent trials, the smaller number of rectangles was dotted and the larger number of rectangles was striped. In the incongruent trials, the smaller number was striped and the larger number was dotted. Within trials, the overall spatial properties for small and large numerical arrays were held approximately equally.

in which another magnitude showed the opposite sequence (that is, descending following ascending or ascending following descending). This cross-over based on the monotonicity of the sequence had previously been shown for number, such that infants who were familiarized to a sequence of ascending or descending numerical values subsequently looked longer to the reverse monotonicity of numerical values (eg, Brannon, 2002). That infants behave similarly even when comparing the monotonicity of different magnitudes suggests that their behaviors are supported by a general magnitude system with overlapping representations of numerical and nonnumerical magnitude.

Other work with infants has tested whether cross-magnitude interactions are possible when number and other magnitudes are presented in different sensory modalities (Lourenco, 2015b). This work follows cross-modal matching paradigms with number in which human infants and nonhuman animals have been shown to look at visual displays that match auditory sequences in numerical value. For instance, if four sounds are presented over a loud speaker, human infants look longer to a visual display with four items than one with eight items (Izard et al., 2009). Research using a modified version of this paradigm has found that matching cuts across magnitude as well. By at least 7 months of age, infants matched stimuli that corresponded to different magnitudes. For instance, when infants heard a short duration, they looked longer at a visual display with a smaller numerical array than at a larger numerical array. In contrast, when they heard a long duration, they looked longer at the visual display with the larger than smaller numerical array. Taken together, the research demonstrates that not only are infants capable of matching approximate number across sensory modality (ie, vision and audition) but they also match numerical and nonnumerical magnitudes (ie, number, duration, and area) according to their common more-versus-less ordering. This pattern of findings is what one would predict if number forms part of a general magnitude system that also processes nonnumerical magnitudes. Because infants in some of these experiments were as young as a few days old, and because similar effects have been observed with nonhuman animals (Meck & Church, 1983; Merritt, Casasanto, & Brannon, 2010), these findings are consistent with a general magnitude system that is rooted in nonverbal representations and that may be innate in human ontogeny as well as phylogenetically quite ancient.

The effects displayed among infants bear a striking similarity to those observed across a variety of tasks with older children and adults. In particular, infants show congruity effects across different magnitudes that are based on the common more-versus-less organization for number and other magnitudes (eg, more number = more area). One interpretation is that the findings in infants, older children, and adults point to a general magnitude system that is continuous throughout development. An alternative possibility is that different mechanisms support congruity effects at different developmental time points. As noted previously, verbal organisms could rely on language, specifically common magnitude words, to mediate the interactions between distinct, nonoverlapping representations. Thus, even if infants' cross-magnitude interactions were supported by a general magnitude system with overlapping representations, there is the possibility that older children and adults could rely on other mechanisms. Indeed, it has been suggested that initially overlapping representations of magnitude become differentiated over development (Lourenco & Longo, 2011). If true, then mediation by other systems (eg, language) would be necessary for supporting later cross-magnitude interactions.

Among infants, it has been suggested that cross-magnitude interactions could be supported by the broader phenomenon of synesthesia (Cantlon, 2012; Cohen Kadosh & Gertner, 2011), rather than a general magnitude system. Unlike a general magnitude system, the view from synesthesia is that it supports a variety of automatic mappings, not just those based on common more-versus-less organization. Importantly, across typical development, synesthetic connections are thought to be pruned or to become inhibited, such that the intersensory connections responsible for infants' cross- and intramodal percepts would eventually subside (eg, Spector & Maurer, 2009). If cross-magnitude interactions are the result of synesthesia, then the prediction is that they should become weaker over development. Although it is difficult to make comparisons across age because of the large differences in task demands, it is clear that cross-magnitude interactions are present throughout development, suggesting that they are not supported by the mechanism of synesthesia.

As discussed further, we recently completed a longitudinal investigation in which we examined whether cross-magnitude associations in infants were predictive of cross-magnitude associations at preschool age. This work demonstrates stability in the strength of magnitude interactions during early development and thus provides direct support for the continuity of a general magnitude system, at least from infancy to childhood. We have also examined the performance of adults under conditions in which language is not possible or in which decision stage effects can be ruled out as an explanation for cross-magnitude interactions. These data suggest that similar mechanisms support magnitude associations at different ages, including adulthood. I turn to the work with adults next.

17.5 INTEGRATION ACROSS NUMERICAL AND NONNUMERICAL MAGNITUDES IN HUMAN ADULTS

Rather than a general magnitude system, an explanation of the congruity effects described in adults and older children is that they involve mediation by language. When people are asked to make ordinal judgments following prompts such as "larger," "greater," or "more," cross-magnitude interactions could be the result of a match or mismatch in verbal labels, not direct exchanges, or overlap, between the magnitudes. For instance, if asked to identify the larger of two items, then "larger" denotes the more end of the continuum, whether the judgment concerns number or physical size. "Larger" for number and "larger" for size constitute a match ("larger" for number and "smaller" for size constitute a mismatch). Although congruity effects in prelinguistic children (eg, Lourenco & Longo, 2010; described in Section 17.4) and nonverbal animals (eg, Merritt et al., 2010) provide reason to doubt mediation by language, it remains a possibility for human adults and verbal children. Linguistic coding is thus a likely mechanism for cross-magnitude interactions observed on tasks in which it is possible to code different magnitudes with common words.

As already noted, there is also another account that emphasizes decision stage effects. This account is that although the preparation of a response may recruit different magnitudes, perhaps because each has organizational structure that is relevant to the task, the representations of the magnitudes themselves do not interface with one another. On this account, cross-magnitude interactions are not the result of direct connections or overlapping representations, but rather they may reflect a mapping of each magnitude to the task-relevant decision.

17.5.1 Transfer of Subliminally Primed Magnitude Information

Using a subliminal priming paradigm with adult participants, we recently took on these alternative accounts. Across two experiments, participants made judgments of cumulative area following subliminally primed numbers (Lourenco, Ayzenberg, & Lyu, in press). Under conditions of subliminal priming, stimuli that are below the threshold of conscious detection are nevertheless processed, as evidenced by their effects on other tasks. Subliminal priming has been previously used to show that number representations are flexible enough to accommodate different symbolic notations (words and digits; Naccache & Dehaene, 2001; Reynvoet, Brysbaert, & Fias, 2002), different sensory modalities (vision and audition; Kouider & Dehaene, 2009), and that they incorporate the distribution (mean) of a set of numerical values (Van Opstal, de Lange, & Dehaene, 2011). In our work, we asked whether subliminally primed Arabic numerals would affect subsequent judgments of cumulative area, even though the number primes could not be labeled according to magnitude relevant words such as "more"/"less" or "greater than"/"less than." We tested for transfer from number to area across two contexts. In one experiment, participants judged which array was *larger* in area. In another experiment, they judged whether the two arrays were the *same* or *different* in area. The first experiment was a first test of whether transfer was possible in the absence of conscious detection and thus linguistic mediation (since stimuli that cannot be detected are not labeled). The latter experiment pitted the account based on a general magnitude system against the one based on decision stage effects, allowing us to test directly the basis for cross-magnitude transfer under conditions of subliminal priming.

The subliminal primes in these experiments were pairs of black and white Arabic digits (eg, white 4 and black 8). To ensure that these stimuli were not consciously detected, they were presented briefly (43 ms) and sandwiched between two masks. (Subsequent tests of prime awareness confirmed that primes were not detected by participants.) Following subliminal priming, participants judged the cumulative area of black and white arrays, which were presented rapidly (200 ms) but were clearly visible. When participants were tasked with judging relative cumulative area, we found clear congruity effects such that participants were both more accurate and faster when the mapping between relative number and color was congruent with the mapping between relative area and color compared to when it was incongruent. In this experiment, the primes always differed in value, with one number presented in white and the other presented in black (Fig. 17.2). Congruent trials were ones in which the mapping between relative magnitude and color was shared. That is, if the smaller number was white and the larger number was black, then the congruent test trial was one in which the array with the larger surface area was black and the array with the smaller surface area was white. Incongruent trials were ones in which the mapping was reversed (Fig. 17.2). In this example, incongruent trials were ones in which the arrays with the larger surface area was white and the smaller surface area was black. That cross-magnitude effects were possible under conditions of subliminal priming suggests that interactions between number and another magnitude (in this case, cumulative area) are not strictly mediated by an explicit verbal strategy, but instead that there may be integration, at least partially, across these magnitudes.

An alternative explanation, raised previously, is that cross-magnitude transfer could be an effect that occurs at the decision stage of processing

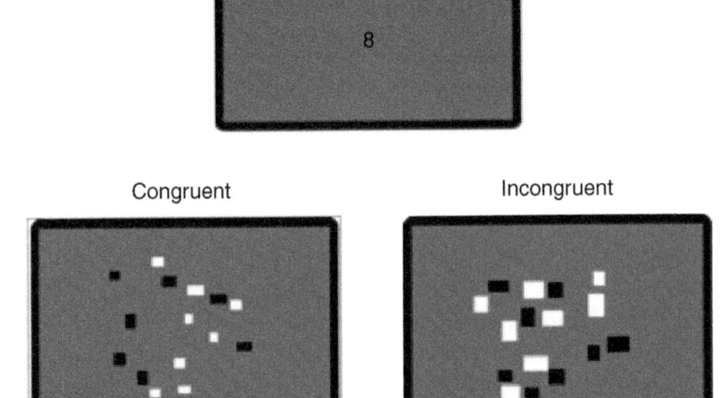

FIGURE 17.2 **Examples of the stimuli used in the study of Lourenco et al. (in press).** The top image is an example of a prime display and the two bottom images are examples of congruent and incongruent trials when participants were asked to judge which array (black or white) was larger in cumulative surface area. On a congruent trial, the mapping between Arabic digits and color (black/white) matched that between cumulative surface area and color—for example, a prime display with a white 4 and black 8 followed by a target display with smaller white surface area and larger black surface area. On an incongruent trial, there was a mismatch across prime and target displays—for example, a white 4 and black 8 followed by smaller black surface area and larger white surface area.

rather than between the magnitude representations themselves (Cohen Kadosh, Gevers, & Notebaert, 2011; Santens & Verguts, 2011; Yates et al., 2012). Contrary to a general magnitude system, behavioral interactions could be due to participants preparing to respond "larger" to cumulative area (the judgment in the first experiment), which, although only technically relevant to the task of judging cumulative area, also concerns the structure of numerical magnitude and thus could involve recruiting each magnitude (independently) during a decision stage. To address this possibility, the second experiment in our study required that participants judge whether arrays were the *same* or *different* in cumulative area. Subliminal primes that were identical in magnitude were also included (eg, a pair of 5s, one white and one black) and their effects on subsequent cumulative area judgments were compared to those that were preceded by subliminal primes that were different in numerical magnitude (eg, a white 4 and a black 8). Critically, the two accounts—general magnitude system and decision stage—make different predictions about performance when judging cumulative area. For instance, if congruity effects occur at the decision stage, then priming with a small white number and a large black number would result in facilitation when the judgments are "different area" compared to "same area." In contrast, if congruity effects are supported by a general magnitude system, then facilitation (and interference) should depend on the mapping between relative magnitude and color, even though this mapping is irrelevant to deciding whether the pairs are the same or different. In this case, priming with a small white number and a large black number would result in facilitation of area judgments if the array with the larger area was black and the array with the smaller area was white. There would be interference if the array with the larger area was white and the array with the smaller area was black. This contrasts with the prediction based on a decision stage account in which both types of trials are congruent because the arrays and primes are both different in magnitude. In this example, the incongruent trials in the case of a decision stage account are only those in which the arrays are the same in area. Our data were consistent with an account based on a general magnitude system and are bolstered by findings from ERP studies that suggest that cross-magnitude interactions occur prior to the decision stage of a task (eg, Schwarz & Heinze, 1998).

But how exactly does a general magnitude system support such interactions, especially given that one source of information (magnitude) was subliminally primed? Consistent with the neural evidence reviewed earlier, we have proposed that congruity effects are possible in the context of subliminal priming because of shared neural codes for numerical and nonnumerical magnitudes. Accumulating evidence from humans and nonhuman animals suggests that the posterior parietal cortex is recruited when processing number in various formats (symbolic and nonsymbolic;

Piazza, Pinel, Le Bihan, & Dehaene, 2007) and sensory modalities (eg, vision and audition; Eger, Sterzer, Russ, Giraud, & Kleinschmidt, 2003; Nieder, 2012). There is also evidence from neuroimaging with humans and electrophysiology with monkeys that nonnumerical magnitudes recruit neural codes in and around the IPS and, critically, a proportion of these codes is shared with number (Bueti & Walsh, 2009; Cappelletti et al., 2013; Dormal et al., 2012). It is possible, then, that the activation in parietal cortex associated with the number primes served to facilitate or interfere with subsequent judgments of area (cf., Schacter & Buckner, 1998; Schnyer, Ryan, Trouard, & Forster, 2002).

Another possible account of cross-magnitude interactions is that a process of analogy allows for "cross-talk" between distinct magnitude representations. Because different magnitudes share analog format and because at least the ones described here are unambiguously organized in their more-versus-less ordering, comparisons and alignment could be possible across them (eg, Cantlon et al., 2009; Vicario & Martino, 2010). In other words, an analogical strategy could support the interactions across distinct representations because of the structural similarity between numerical and nonnumerical magnitudes (Kotovsky & Gentner, 1996; Srinivasan & Carey, 2010). This account seems unlikely in the context of subliminal priming, however. Evidence from neuroimaging studies suggests that comparisons by way of analogy recruit the PFC (Bunge, Wendelken, Badre, & Wagner, 2005; Geake & Hansen, 2005; Speed, 2010; Vicario & Martino, 2010), and because subliminally processed information is not known to reach frontal cortical regions (for review, Dehaene, Changeux, Naccache, Sackur, & Sergent, 2006), it is unlikely that the effects observed in our study were mediated by an analogical strategy within this region (Dehaene et al., 1998). Nevertheless, this possibility remains an alternative in other contexts. I return to this issue in the next section where I discuss the use of training procedures to test for the transfer of improvement in representational precision from one magnitude to another.

17.5.2 Transfer of Representational Precision Across Magnitudes

Another approach that has been used to assess the legitimacy of a general magnitude system involves training the representational precision of one magnitude and then testing for transfer to another magnitude. Next I discuss two behavioral studies that recently used this approach. This type of approach provides a strong test of a general magnitude system because it is not possible to counter the transfer of representational precision with alternative mechanisms such as linguistic mediation or an analogical strategy.

In one of these studies, DeWind and Brannon (2012) first showed that it was possible to improve the precision of adult participants' number representations. Following multiple sessions of computerized training with (immediate) feedback on their performance (ie, correct/incorrect on each trial), participants showed significant improvement in judging which of two arrays was larger or smaller in number. Importantly, however, DeWind and Brannon found that the improved precision for number was not accompanied by improved precision for spatial extent (line lengths). That is, representations of line length did not become more precise following training with numerical comparisons. An interpretation of the lack of transfer is that representations of number and spatial extent do not form part of a general magnitude system with overlap in these representations. However, a concern with this interpretation is that participants' representations of line length were highly precise, which may have left little room for improvement and, thus, could have made it difficult to capture transfer from number to line length. Another concern is that line length, unlike number, is a property of a single element. This difference could affect the processes implicated during comparison judgments or, more specifically, the mechanisms recruited in each task.

These concerns motivated research in my laboratory where we similarly asked about the transfer of representational precision across numerical and nonnumerical magnitudes. Unlike DeWind and Brannon (2012), however, we used the cumulative area of arrays of dots rather than the lengths of individual lines as the nonnumerical magnitude of interest Lourenco et al. (in press). Like number, cumulative area is a property of a set, which likely makes the attentional demands related to visual input more comparable across magnitudes. Moreover, judgments of cumulative area are less accurate than those of line length but comparable to those of number (Lourenco et al., 2012). Despite this, we found that participants trained on numerical comparisons did not generalize greater representational precision to cumulative area, as in the study of DeWind and Brannon. Critically, however, a separate group of participants was trained on cumulative area comparisons and these participants *did* show transfer of representational precision. That is, participants who demonstrated greater improvement on the area task were more accurate on the number task following training. On the one hand, transfer from area to number provides clear support for a general magnitude system, with overlapping representations of numerical and nonnumerical magnitude. On the other hand, it is unclear how such a system would accommodate asymmetrical transfer; that is, transfer from area to number, but not vice versa.

Data from an additional color-word Stroop task in our study (where participants judged the font color of a printed word while ignoring the color denoted by the word, which, on some trials, was incongruent with the font; eg, the word "blue" printed in red) suggests an important effect of the stimuli used on the number task. Any task that assesses the precision of numerical magnitude using nonsymbolic arrays of number must incorporate trials in which the spatial properties such as cumulative area are incongruent with the number. This procedure is used to ensure that participants are indeed judging numerical, rather than nonnumerical, magnitude. In contrast, our area task involved no incongruent trials; that is, within a given trial, the two arrays being compared were always equal in numerical value, such that there was never any irrelevant information related to a different magnitude (see also, Lourenco et al., 2012). This led us to hypothesize that training with numerical comparisons might have served to dissociate numerical and nonnumerical magnitudes by promoting inhibition of the irrelevant spatial information. Consistent with this possibility, participants who showed greater improvement in numerical precision following training with comparisons of number arrays also showed less interference on the color-word Stroop task, suggesting improvement in inhibitory control more generally. There was no such change on the color-word Stroop task following area training. However, a separate study employing transcranial random noise stimulation (tRNS) of parietal cortex, which temporarily potentiates neural activity, found that improved precision associated with representations of numerical magnitude can in fact transfer to nonnumerical magnitudes, namely, spatial extent and duration (Cappelletti et al., 2013). Although this research provides additional support for a general magnitude system, likely with overlapping representations of numerical and nonnumerical magnitudes, it leaves open questions about how tRNS affects representational precision, an important area for continued research.

17.6 DEVELOPMENTAL CONTINUITY OF THE GENERAL MAGNITUDE SYSTEM

We recently conducted a longitudinal investigation in my laboratory where we examined whether the cross-magnitude interactions documented in infancy (ie, 9-month-olds) were similarly present among the same children at preschool age (ie, 3.5-year-olds). By assessing the extent to which individual differences in the strength of magnitude associations are correlated over development, this study offers a strong test of the general magnitude system, which holds that the interactions across numerical and nonnumerical magnitudes are not mediated by other mechanisms such as language available to verbal organisms and thus should show continuity over human ontogeny. Infants were tested at 9 months of age in the study of Lourenco and Longo (2010) described earlier (Section 17.4) and then a proportion of them returned to the lab at 3.5 years of age to complete a child-friendly Stroop-like task that involved the pairing of number and cumulative area. Other measures (described later) were also given to the 3- to 5-year-olds.

One important finding in the study of Lourenco and Longo (2010) was that infants showed cross-magnitude interactions regardless of the magnitudes presented during the habituation and test phases of the procedure. That is, for every combination of number, area, and duration, infants expected the mapping between magnitude and pattern to generalize from one magnitude to the other. The task of primary interest given to the 3.5-year-olds when they returned to the lab was designed to assess the association between nonsymbolic number and area. In this task, children were asked to judge which of two arrays was larger in number. We focused specifically on the potential association between number and area at preschool age because interactions between these two magnitudes have been previously documented among children in this age range on this type of task (eg, Rousselle & Noël, 2008). Moreover, we asked the children to judge only number because the effects from spatial extent to number are more robust than the reverse, at least at this age (eg, Rousselle et al., 2004). Importantly, we also gave the children tests of inhibitory control (ie, day-night Stroop task; Gerstadt, Hong, & Diamond, 1994), analogical reasoning (Kotovsky & Gentner, 1996), and verbal competence (Peabody Picture Vocabulary Test or PPVT; Dunn & Dunn, 2007). These additional measures were included in order to allow us to isolate interindividual variability due to a general magnitude system as opposed to other cognitive factors.

We found that the strength of infants' cross-magnitude associations predicted the strength of preschoolers' cross-magnitude associations. The strength of the magnitude associations in infancy was measured as the difference between looking time to incongruent and congruent trials. Recall that the incongruent trials involved a mismatch in the mapping between relative magnitude and pattern whereas the congruent trials involved a match in this mapping (Section 17.4). The strength of the cross-magnitude associations at preschool age was the difference in children's errors on trials in which the number-area relations were incongruent compared to when they were congruent. Incongruent trials involved conflicting information between number and area; that is, a larger number array that was smaller in cumulative area than a smaller number array (Fig. 17.3). Congruent trials involved the opposite—a larger number array that was larger in area than a smaller number array. Critically, the positive correlation between infancy and preschool age held even when accounting for variability in inhibitory control, analogical reasoning, and verbal competence (assessed at preschool age). Our findings thus demonstrate consistency in interindividual variability of cross-magnitude associations from infancy to preschool age that cannot be accounted for by individual differences in other cognitive factors such as inhibitory control or analogical reasoning. Our findings also suggest that even when individuals have access to language, cross-magnitude associations are not exclusively mediated by language. The correlation between infancy and preschool age suggests that preverbal magnitude associations may be important precursors to those observed later in life.

That inhibitory control did not account for the continuity in magnitude associations is important because individual differences in inhibition could be one reason for the consistency in the strength of these associations. The role of inhibition in numerosity judgments has been shown

FIGURE 17.3 **The task used to assess cross-magnitude interactions at 3.5 years of age in the longitudinal study of Lourenco and Aulet (2016).** Children were always asked "Who had more things—Bert or Ernie?" after a brief presentation of the visual displays. The top row represents a congruent trial because number and spatial magnitudes are congruent (Bert has a greater number of squares and the squares are both larger in individual size and their total surface area compared to those of Ernie). The bottom row represents an incongruent trial because number and spatial magnitudes are incongruent (Bert has a greater number of dots but the dots are smaller both in individual size and in their total surface area compared to those of Ernie). The side of the larger numerical value was counterbalanced across trials.

under a variety of conditions, particularly when individuals judge number for spatially incongruent displays (eg, Cappelletti, Didino, Stoianov, & Zorzi, 2014; Gebuis & Reynvoet, 2012). The measure of inhibitory control in our study required that children say "day" when shown a moon and "night" when shown a sun. These trials were more difficult than those in which responses were semantically congruent, that is, children were required to say "day" for sun and "night" for moon. As with measures of cross-magnitude associations, this task requires the coordination of two types of information, which may be either congruent or incongruent in their relation. One way to think about the potential role of inhibition in accounting for cross-magnitude interactions is that inhibitory control may be required to dissociate between magnitudes when one is relevant to the task but the other is not. Thus, the children who displayed stronger magnitude associations across development could be the ones with poorer inhibitory control across this period. Individual differences in inhibitory control, however, were not responsible for the correlation in magnitude associations between infancy and preschool age.

That analogical reasoning also did not account for the continuity in magnitude associations is important for ruling out accounts that posit analogy as the basis for cross-magnitude interactions. The measure of analogical reasoning in this study required that children select which of two choices was an analogical match to a target. For instance, if the target showed a blue circle in between two red circles, then the match was a (black) triangle in between two (black) squares rather than a (black) triangle to the right of two (black) squares. It has been suggested that comparisons based on analogy mediate the interactions between distinct, nonoverlapping magnitudes (Section 17.5). That is, analogy may serve to support "cross-talk" between distinct magnitudes that do not directly interact themselves. On this view, children with stronger magnitude associations could be the ones with better analogical reasoning skills. One potential concern, however, is that children's performance on the analogical reasoning task in this study was generally low (many performed at chance), which could have affected the predictive value of this measure. Thus, the question of whether analogical processes can account for cross-magnitude interactions and, in particular, their continuity over development remains an issue for future investigation.

17.7 CONCLUSIONS

In this chapter, I have presented evidence for a system of magnitude representation that includes integration across number and other magnitudes. I focused primarily on direct exchanges between representations of number and spatial extent, such as the physical sizes of individual objects or the cumulative area of a set of objects. I also noted conditions in which interactions between magnitudes included the timing of a stimulus presentation such as the duration of an individual object or the total sequence of objects. Unlike models that may posit complete differentiation (or domain specificity) between magnitude representations, the strong version of a general magnitude system holds that there is overlap in the representations. I described this overlap as involving direct exchanges that are not mediated by other systems such as language or analogical comparisons. The consequence may be an abstract coding of the more-versus-less relations for number and other prothetic magnitudes such as spatial extent and duration. Importantly, these relations may be grounded in analog format that is common to many continuous dimensions, but, critically, analog format alone (which can account for the ratio effects that characterize discrimination of different magnitudes) is not sufficient for characterizing the cognitive interactions observed across various tasks and across development. That is, common format does not guarantee communication across representations, which could exist within distinct (perhaps even modular) systems. I, and others, have thus proposed that shared neural codes may be necessary for supporting exchanges between magnitudes.

From the perspective of a general magnitude system, cognitive interactions are not mediated indirectly by other systems. I have argued against mediation under different conditions by suggesting that language is not needed to support cross-magnitude interactions, nor is analogical processing, and these interactions cannot be explained by decision stage effects. The nature of the neural codes and processes that underlie the direct connections between magnitudes are admittedly vague and much work is needed to provide the needed detail. One possibility is that populations of neurons, some of which include shared coding for different magnitudes, establish abstractions in terms of more-versus-less relations. Another possibility, which I did not consider but which should be acknowledged, is that there may be different types of cognitive interactions. In this chapter, I focused exclusively on the ones that I believe demonstrate the existence of a general magnitude system. There may be others, however, that reflect distinct, nonoverlapping representations. These would have to be mediated by other systems such as language (or conceptual metaphor) or broader analogical processes. Humans may experience numerous cognitive interactions when processing different cues to magnitude in the physical world and these interactions extend beyond the prothetic dimensions highlighted in this chapter, namely, magnitudes such as number, spatial extent, and duration. It will be critical for future research to understand the mechanisms that support the many possible cognitive interactions that we are likely to experience from context to context.

References

Allik, J., & Tuulmets, T. (1991). Occupancy model of perceived numerosity. *Perception & Psychophysics, 49*, 303–314.

Ansari, D., & Dhital, B. (2006). Age-related changes in the activation of the parietal sulcus during nonsymbolic magnitude processing: an event-related functional magnetic resonance imaging study. *Journal of Cognitive Neuroscience, 18*, 1820–1828.

Arend, I., Cappelletti, M., & Henik, A. (2014). Time counts: bidirectional interaction between time and numbers in human adults. *Consciousness and Cognition, 26*, 3–12.

Barth, H., Kanwisher, N., & Spelke, E. (2003). The construction of large number representations in adults. *Cognition, 86*, 201–221.

Bonn, C. D., & Cantlon, J. F. (2012). The origins and structure of quantitative concepts. *Cognitive Neuropsychology, 29*, 149–173.

Bonny, J. W., & Lourenco, S. F. (2013). The approximate number system and its relation to early math achievement: evidence from the preschool years. *Journal of Experimental Child Psychology, 114*, 375–388.

Brannon, E. M. (2002). The development of ordinal numerical knowledge in infancy. *Cognition, 83*, 223–240.

Buckley, P. B., & Gillman, C. B. (1974). Comparisons of digits and dot patterns. *Journal of Experimental Psychology, 103*, 1131–1136.

Bueti, D., & Walsh, V. (2009). The parietal cortex and the representation of time, space, number and other magnitudes. *Philosophical Transactions of the Royal Society B, 364*, 1831–1840.

Bunge, S. A., Wendelken, C., Badre, D., & Wagner, A. D. (2005). Analogical reasoning and prefrontal cortex: evidence for separable retrieval and integration mechanisms. *Cerebral Cortex, 15*, 239–249.

Butterworth, B. (1999). *The mathematical brain*. London, England: Macmillan.

Butterworth, B. (2005). Developmental dyscalculia. In J. I. D. Campbell (Ed.), *Handbook of mathematical cognition* (pp. 455–467). Hove, England: Psychology Press.

Cantlon, J. F. (2012). Math, monkeys, and the developing brain. *Proceedings of the National Academy of Sciences, 109*, 10725–10732.

Cantlon, J. F., Platt, M. L., & Brannon, E. M. (2009). Beyond the number domain. *Trends in Cognitive Sciences, 13*, 83–91.

Cappelletti, M., Freeman, E., & Butterworth, B. (2011). Time processing in dyscalculia. *Frontiers in Psychology, 2*, 364.

Cappelletti, M., Gessaroli, E., Hithersay, R., Mitolo, M., Didino, D., Kanai, R., & Walsh, V. (2013). Transfer of cognitive training across magnitude dimensions achieved with concurrent brain stimulation of the parietal lobe. *The Journal of Neuroscience, 33*, 14899–14907.

Cappelletti, M., Didino, D., Stoianov, I., & Zorzi, M. (2014). Number skills are maintained in healthy aging. *Cognitive Psychology, 69*, 25–45.

Casasanto, D., Fotakopoulou, O., & Boroditsky, L. (2010). Space and time in the child's mind: evidence for a cross-dimensional asymmetry. *Cognitive Science, 34*, 387–405.

Chen, Q., & Verguts, T. (2010). Beyond the mental number line: a neural network model of number–space interactions. *Cognitive Psychology, 60*, 218–240.

Cohen Kadosh, R., & Gertner, L. (2011). Synesthesia: gluing together time, number and space. In S. Dehaene, & E. Brannon (Eds.), *Space, time and number in the brain: searching for the foundations of mathematical thought* (pp. 123–132). London, UK: Elsevier.

Cohen Kadosh, R., Lammertyn, J., & Izard, V. (2008). Are numbers special? An overview of chronometric, neuroimaging, developmental and comparative studies of magnitude representation. *Progress in Neurobiology, 84*, 132–147.

Cohen Kadosh, R., Gevers, W., & Notebaert, W. (2011). Sequential analysis of the numerical Stroop effect reveals response suppression. *Journal of Experimental Psychology, 37*, 1243–1249.

Coull, J. T., Vidal, F., Nazarian, B., & Macar, F. (2004). Functional anatomy of the attentional modulation of time estimation. *Science, 303*, 1506–1508.

Critchley, M. (1953). *The parietal lobes*. Oxford, England: Williams and Wilkins.

Dakin, S. C., Tibber, M. S., Greenwood, J. A., Kingdom, F. A., & Morgan, M. J. (2011). A common visual metric for approximate number and density. *Proceedings of the National Academy of Sciences USA, 108*, 19552–19557.

de Hevia, M. D., & Spelke, E. S. (2010). Number-space mapping in human infants. *Psychological Science, 21*, 653–660.

de Hevia, M. D., Izard, V., Coubart, A., Spelke, E. S., & Streri, A. (2014). Representations of space, time, and number in neonates. *Proceedings of the National Academy of Sciences, 111*, 4809–4813.

Dehaene, S. (2011). *The number sense: how the mind creates mathematics*. New York, NY: Oxford University Press.

Dehaene, S., & Changeux, J. (1993). Development of elementary numerical abilities: a neuronal model. *Journal of Cognitive Neuroscience, 5*, 390–407.

Dehaene, S., Naccache, L., Le Clec'H, G., Koechlin, E., Mueller, M., Dehaene-Lambertz, G., & Le Bihan, D. (1998). Imaging unconscious semantic priming. *Nature, 395*, 597–600.

Dehaene, S., Changeux, J. P., Naccache, L., Sackur, J., & Sergent, C. (2006). Conscious, preconscious, and subliminal processing: a testable taxonomy. *Trends in Cognitive Sciences, 10*, 204–211.

Devlin, K. J. (2000). *The math gene: how mathematical thinking evolved and why numbers are like gossip*. New York: Basic Books.

DeWind, N. K., & Brannon, E. M. (2012). Malleability of the approximate number system: effects of feedback and training. *Frontiers in Human Neuroscience, 6*, 68.

Dormal, V., & Pesenti, M. (2009). Common and specific contributions of the intraparietal sulci to numerosity and length processing. *Human Brain Mapping, 30*, 2466–2476.

Dormal, V., Seron, X., & Pesenti, M. (2006). Numerosity-duration interference: a Stroop experiment. *Acta Psychologica, 121*, 109–124.

Dormal, V., Andres, M., & Pesenti, M. (2008). Dissociation of numerosity and duration processing in the left intraparietal sulcus: a transcranial magnetic stimulation study. *Cortex, 44*, 462–469.

Dormal, V., Andres, M., & Pesenti, M. (2012). Contribution of the right intraparietal sulcus to numerosity and length processing: an fMRI-guided TMS study. *Cortex, 48*, 623–629.

Droit-Volet, S., Clément, A., & Fayol, M. (2008). Time, number and length: similarities and differences in discrimination in adults and children. *The Quarterly Journal of Experimental Psychology, 61*, 1827–1846.

Dunn, L. M., & Dunn, D. M. (2007). *Peabody picture vocabulary test* (4th ed.). Minneapolis, MN: Pearson.

Eger, E., Sterzer, P., Russ, M. O., Giraud, A. -L., & Kleinschmidt, A. (2003). A supramodal number representation in human intraparietal cortex. *Neuron, 37*, 719–725.

Eiselt, A. K., & Nieder, A. (2013). Representation of abstract quantitative rules applied to spatial and numerical magnitudes in primate prefrontal cortex. *The Journal of Neuroscience, 33*, 7526–7534.

Feigenson, L., Dehaene, S., & Spelke, E. (2004). Core systems of number. *Trends in Cognitive Sciences, 8*, 307–314.

Fias, W., Lammertyn, J., Reynvoet, B., Dupont, P., & Orban, G. A. (2003). Parietal representation of symbolic and non-symbolic magnitude. *Journal of Cognitive Neuroscience, 15*, 47–56.

Fuhs, M. W., & McNeil, N. M. (2013). ANS acuity and mathematics ability in preschoolers from low-income homes: contributions of inhibitory control. *Developmental Science, 16*, 136–148.

Fulbright, R., Manson, C., Skudlarski, P., Lacadie, C. M., & Gore, C. L. J. (2003). Quantity determination and the distance effect with letters, numbers and shapes: a functional MR imaging study of number processing. *American Journal of Neuroradiology, 23*, 197–200.

Gabay, S., Leibovich, T., Henik, A., & Gronau, N. (2013). Size before numbers: conceptual size primes numerical value. *Cognition, 129*, 18–23.

Gallistel, C., & Gelman, R. (2000). Non-verbal numerical cognition: from reals to integers. *Trends in Cognitive Sciences, 4*, 59–65.

Geake, J. G., & Hansen, P. C. (2005). Neural correlates of intelligence as revealed by fMRI of fluid analogies. *NeuroImage, 26*, 555–564.

Geary, D. C., Berch, D. B., & Mann Koepke, K. (2015). The evolution of number systems. In D. C. Geary, D. B. Berch, & K. Mann Koepke (Eds.), *Evolutionary origins and early development of number processing (Vol. 1, Mathematical Cognition and Learning)* (pp. 335–353). San Diego, CA: Elsevier Academic Press.

Gebuis, T., & Reynvoet, B. (2012). The interplay between nonsymbolic number and its continuous visual properties. *Journal of Experimental Psychology, 141*, 642–648.

Gebuis, T., Kadosh, R. C., de Haan, E., & Henik, A. (2009). Automatic quantity processing in 5-year olds and adults. *Cognitive Processing, 10*, 133–142.

Genovesio, A., Tsujimoto, S., & Wise, S. P. (2011). Prefrontal cortex activity during the discrimination of relative distance. *The Journal of Neuroscience, 31*, 3968–3980.

Gerstadt, C. L., Hong, Y. J., & Diamond, A. (1994). The relationship between cognition and action: performance of children 3½–7 years old on a Stroop-like day-night test. *Cognition, 53*, 129–153.

Gilmore, C., Attridge, N., Clayton, S., Cragg, L., Johnson, S., Marlow, N., & Inglis, M. (2013). Individual differences in inhibitory control, not non-verbal number acuity, correlate with mathematics achievement. *PLoS One, 8*, e67374.

Halberda, J., Mazzocco, M. M. M., & Feigenson, L. (2008). Individual differences in non-verbal number acuity correlate with maths achievement. *Nature, 455*, 665–668.

Harvey, B. M., Klein, B. P., Petridou, N., & Dumoulin, S. O. (2013). Topographic representation of numerosity in the human parietal cortex. *Science, 341*, 1123–1126.

Haxby, J. V., Gobbini, M. I., Furey, M. L., Ishai, A., Schouten, J. L., & Pietrini, P. (2001). Distributed and overlapping representations of faces and objects in ventral temporal cortex. *Science, 293*, 2425–2430.

Hayashi, M. J., Kanai, R., Tanabe, H. C., Yoshida, Y., Carlson, S., Walsh, V., & Sadato, N. (2013). Interaction of numerosity and time in prefrontal and parietal cortex. *The Journal of Neuroscience, 33*, 883–893.

Henik, A., & Tzelgov, J. (1982). Is three greater than five: the relation between physical and semantic size in comparison tasks. *Memory & Cognition, 10*, 389–395.

Henik, A., Leibovich, T., Naparstek, S., Diesendruck, L., & Rubinsten, O. (2011). Quantities, amounts, and the numerical core system. *Frontiers in Human Neuroscience, 5*, 186.

Henmon, V. A. C. (1906). The time of perception as a measure of differences in sensation. *Archives of Philosophical, Psychological and Science Method, 8*, 5–75.

Hurewitz, F., Gelman, R., & Schnitzer, B. (2006). Sometimes area counts more than number. *Proceedings of the National Academy of Sciences, 103*, 19599–19604.

Izard, V., Sann, C., Spelke, E. S., & Streri, A. (2009). Newborn infants perceive abstract numbers. *Proceedings of the National Academy of Sciences, 106*, 10382–10385.

Jordan, K. E., Brannon, E. M., Logothetis, N. K., & Ghazanfar, A. A. (2005). Monkeys match the number of voices they hear to the number of faces they see. *Current Biology, 15*, 1034–1038.

Katz, G., Benbassat, A., Diesendruck, L., Sipper, M., & Henik, A. (2013). From size perception to counting: an evolutionary computation point of view. In *Proceedings of the fifteenth annual conference companion on genetic and evolutionary computation* (pp. 1675–1678). ACM.

Kotovsky, L., & Gentner, D. (1996). Comparison and categorization in the development of relational similarity. *Child Development, 67*, 2797–2822.

Kouider, S., & Dehaene, S. (2009). Subliminal number priming within and across the visual and auditory modalities. *Experimental Psychology, 56*, 418–433.

Lakoff, G., & Núñez, R. E. (2000). *Where mathematics comes from: how the embodied mind brings mathematics into being*. New York, NY: Basic Books.

Levin, I. (1977). The development of time concepts in young children: reasoning about duration. *Child Development, 48*, 435–444.
Libertus, M. E., Feigenson, L., & Halberda, J. (2011). Preschool acuity of the approximate number system correlates with school math ability. *Developmental Science, 14*, 1292–1300.
Locke, J. (1689). An essay concerning human understanding. Amherst: Promethius Books, 1995.
Lourenco, S. F. (2015a). On the relation between numerical and non-numerical magnitudes: evidence for a general magnitude system. In D. C. Geary, D. B. Berch, & K. Mann Koepke (Eds.), *Evolutionary origins and early development of number processing (Vol. 1, Mathematical Cognition and Learning)* (pp. 145–174). San Diego, CA: Elsevier Academic Press.
Lourenco, S. F. (2015b). Cross-dimensional and cross-modal matching of magnitude in infancy. Manuscript under review.
Lourenco, S. F., & Aulet, L. S. (2016). Cross-magnitude interactions across development: longitudinal evidence for a general magnitude system. (Under review.)
Lourenco, S. F., & Longo, M. R. (2010). General magnitude representation in human infants. *Psychological Science, 21*, 873–881.
Lourenco, S. F., & Longo, M. R. (2011). Origins and development of generalized magnitude representation. In S. Dehaene, & E. Brannon (Eds.), *Space, time and number in the brain: searching for the foundations of mathematical thought* (pp. 225–244). London, UK: Elsevier.
Lourenco, S. F., Bonny, J. W., Fernandez, E. P., & Rao, S. (2012). Nonsymbolic number and cumulative area representations contribute shared and unique variance to symbolic math competence. *Proceedings of the National Academy of Sciences USA, 109*, 18737–18742.
Lourenco, S. F., Levine, S. C., & Degner, H. (2014). The influence of temporal cues on preschoolers' numerical judgments. Manuscript in preparation for submission.
Lourenco, S. F., Ayzenberg, V., & Lyu, J. (in press). A general magnitude system in human adults: evidence from a subliminal priming paradigm. *Cortex*.
Meck, W. H., & Church, R. M. (1983). A mode control model of counting and timing processes. *Journal of Experimental Psychology, 9*, 320–334.
Merritt, D. J., Casasanto, D., & Brannon, E. M. (2010). Do monkeys think in metaphors? Representations of space and time in monkeys and humans. *Cognition, 117*, 191–202.
Naccache, L., & Dehaene, S. (2001). Unconscious semantic priming extends to novel unseen stimuli. *Cognition, 80*, 215–229.
Nieder, A. (2012). Supramodal numerosity selectivity of neurons in primate prefrontal and posterior parietal cortices. *Proceedings of the National Academy of Sciences, 109*, 11860–11865.
Oliveri, M., Vicario, C. M., Salerno, S., Koch, G., Turriziani, P., Mangano, R., Chillemi, G., & Caltagirone, C. (2008). Perceiving numbers alters temporal perception. *Neuroscience Letters, 432*, 308–311.
Onoe, H., Komori, M., Onoe, K., Takechi, H., Tsukada, H., & Watanabe, Y. (2001). Cortical networks recruited for time perception: a monkey positron emission tomography (PET) study. *NeuroImage, 13*, 37–45.
Piaget, J. (1965). *The child's conception of number*. Oxford, England: W.W. Norton.
Piazza, M., Pinel, P., Le Bihan, D., & Dehaene, S. (2007). A magnitude code common to numerosities and number symbols in human intraparietal cortex. *Neuron, 53*, 293–305.
Pinel, P., Piazza, M., Le Bihan, D., & Dehaene, S. (2004). Distributed and overlapping cerebral representations of number, size, and luminance during comparative judgments. *Neuron, 41*, 983–993.
Reynvoet, B., Brysbaert, M., & Fias, W. (2002). Semantic priming in number naming. *The Quarterly Journal of Experimental Psychology, 55*, 1127–1139.
Ross, J., & Burr, D. C. (2010). Vision senses number directly. *Journal of Vision, 10*, 1–8.
Rousselle, L., & Noël, M. P. (2008). The development of automatic numerosity processing in preschoolers: evidence for numerosity-perceptual interference. *Developmental Psychology, 44*, 544–560.

Rousselle, L., Palmers, E., & Noël, M. P. (2004). Magnitude comparison in preschoolers: what counts? Influence of perceptual variables. Journal of Experimental Child Psychology, 87, 57–84.

Santens, S., & Verguts, T. (2011). The size congruity effect: is bigger always more? *Cognition, 118*, 94–110.

Schacter, D. L., & Buckner, R. L. (1998). Priming and the brain. *Neuron, 20*, 185–195.

Schnyer, D. M., Ryan, L., Trouard, T., & Forster, K. (2002). Masked word repetition results in increased fMRI signal: a framework for understanding signal changes in priming. *NeuroReport, 13*, 281–284.

Schwarz, W., & Heinze, H. J. (1998). On the interaction of numerical and size information in digit comparison: a behavioral and event-related potential study. *Neuropsychologia, 36*, 1167–1179.

Siegler, R. S., & Richards, D. D. (1979). Development of time, speed, and distance concepts. *Developmental Psychology, 15*, 288–298.

Spector, F., & Maurer, D. (2009). Synesthesia: a new approach to understanding the development of perception. *Developmental Psychology, 45*, 175–189.

Speed, A. (2010). Abstract relational categories, graded persistence, and pre-frontal cortical representation. *Cognitive Neuroscience, 1*, 126–137.

Srinivasan, M., & Carey, S. (2010). The long and the short of it: on the nature and origin of functional overlap between representations of space and time. *Cognition, 116*, 217–241.

Starkey, P., Spelke, E. S., & Gelman, R. (1983). Detection of intermodal numerical correspondences by human infants. *Science, 222*, 179–181.

Stavy, R., & Tirosh, D. (2000). *How students (mis-)understand science, mathematics: intuitive rules*. New York, London, UK: Teachers College Press, Columbia University.

Stevens, S. S. (1957). On the psychophysical law. *Psychological Review, 64*, 153–181.

Stevens, S. S. (1975). *Psychophysics: introduction to its perceptual, neural, and social prospects*. New York: Wiley.

Stoianov, I., & Zorzi, M. (2012). Emergence of a "visual number sense" in hierarchical generative models. *Nature Neuroscience, 15*, 194–196.

Tokita, M., & Ishiguchi, A. (2011). Temporal information affects the performance of numerosity discrimination: behavioral evidence for a shared system for numerosity and temporal processing. *Psychonomic Bulletin & Review, 18*, 550–556.

Tudusciuc, O., & Nieder, A. (2007). Neuronal population coding of continuous and discrete quantity in the primate posterior parietal cortex. *Proceedings of the National Academy of Sciences, 104*, 14513–14518.

Tudusciuc, O., & Nieder, A. (2009). Contributions of primate prefrontal and posterior parietal cortices to length and numerosity representation. *Journal of Neurophysiology, 101*, 2984–2994.

Van Opstal, F., de Lange, F. P., & Dehaene, S. (2011). Rapid parallel semantic processing of numbers without awareness. *Cognition, 120*, 136–147.

Verguts, T., & Fias, W. (2004). Representation of number in animals and humans: a neural model. *Journal of Cognitive Neuroscience, 16*, 1493–1504.

Vicario, C. M., & Martino, D. (2010). The neurophysiology of magnitude: one example of extraction analogies. *Cognitive Neuroscience, 1*, 144–145.

Walsh, V. (2003). A theory of magnitude: common cortical metrics of time, space and quantity. *Trends in Cognitive Sciences, 7*, 483–488.

Winter, B., Marghetis, T., & Matlock, T. (2015). Of magnitudes and metaphors: explaining cognitive interactions between space, time, and number. *Cortex, 64*, 209–224.

Xuan, B., Zhang, D., He, S., & Chen, X. (2007). Larger stimuli are judged to last longer. *Journal of Vision, 7*, 1–5.

Yates, M. J., Loetscher, T., & Nicholls, M. E. (2012). A generalized magnitude system for space, time, and quantity? A cautionary note. Journal of Vision, 12, 1–7.

CHAPTER

18

Sensory Integration Theory: An Alternative to the Approximate Number System

Wim Gevers, Roi Cohen Kadosh**, Titia Gebuis*[†]

*Center for Research in Cognition and Neurosciences (CRCN), Université Libre de Bruxelles and UNI—ULB Neurosciences Institute, Brussels, Belgium; **Department of Experimental Psychology, Oxford, United Kingdom; [†]Department of Molecular and Cellular Neurobiology, Center for Neurogenomics and Cognitive Research, Neuroscience Campus Amsterdam, VU Amsterdam, Amsterdam, The Netherlands

OUTLINE

Continuous Issues in Numerical Cognition. http://dx.doi.org/10.1016/B978-0-12-801637-4.00018-4

18.1 INTRODUCTION

In everyday life we use symbolic numbers to understand and communicate numerical information. From computers to shopping, from sports events to banking, symbolic numbers play a fundamental role for these and many other activities. While adults can learn about the magnitude of new symbolic numbers without associating it with numerosity (Tzelgov, Yehene, Kotler, & Alon, 2000), infants and children rely on an innate *nonsymbolic* number system (Cantlon, Brannon, Carter, & Pelphrey 2006). Such a nonsymbolic system is composed of two subsystems: (1) a system that represents large numerosities (larger than four or five items) and (2) a system that represents a smaller number of objects (Cantlon, Platt, & Brannon, 2009; Feigenson, Dehaene, & Spelke, 2004). In the first subsystem, numerosities are approximated, while in the second subsystem a rapid and more accurate estimate is performed that has been termed subitizing (Kaufman, Lord, Reese, & Volkmann, 1949; Trick & Pylyshyn, 1994). In this chapter, we focus on the first subsystem, which has also been termed the approximate number system (ANS) (Halberda, Mazzocco, & Feigenson, 2008; Park & Brannon, 2013). The ANS is believed to extract large numerosities independent from the visual input like the total surface area or density of the display (Burr & Ross, 2008; Piazza, 2010). This means that the estimate or the comparison of numerosities will not be biased by the size, density, surface, or other sensory cues present in the image. Furthermore, the ANS is also considered to be the foundation of more complex mathematical skills. Studies have shown that participants who perform better in estimating or comparing numerosities tasks are also better in more advanced mathematics (Halberda et al., 2008; Libertus, Feigenson, & Halberda, 2011; Park & Brannon, 2013; Piazza, 2010, but see De Smedt & Gilmore, 2011; Gilmore et al., 2013; Holloway & Ansari, 2009; Rousselle & Noel, 2007; Sasanguie, De Smedt, Defever, & Reynvoet, 2012; Soltesz, Szucs, & Szucs, 2010).

As we will argue, the processing of numerosity depends on a variety of cognitive processes that are required for successful task performance. Central to our argument is that nonsymbolic number stimuli (eg, arrays of dots) are by definition confounded with sensory cues. However, instead of treating this confound as a problem to study pure numerosity processes, we would like to stress its potential role in numerosity processing. Our hypothesis is that large numerosities can be estimated or compared by integrating their different sensory cues. This idea that the judgment of larger numerosities relies on sensory information has been proposed before (eg, Allik & Tuulmets, 1991; Dakin, Tibber, Greenwood, Kingdom, & Morgan, 2011). Although this view has been acknowledged as a plausible alternative for the ANS theory (Barth, Kanwisher, & Spelke, 2003; Dehaene, 1992; Izard & Dehaene, 2008; Piazza, Izard, Pinel, Le Bihan, &

Dehaene, 2004; Stoianov & Zorzi, 2012), it has not yet received large support. We additionally present the assumption that nonnumerical abilities such as conservation (Piaget, 1965) can explain the variability in performance on numerosity tasks (ie, the integration of the sensory cues) and might be a powerful tool to improve numerosity performance and hence more complex mathematical abilities.

18.2 CONCEPT

When one needs to compare the numerical quantity of two sets of dots, large differences in the accuracy of different observers is evident, especially when numerosity increases (Gebuis & Reynvoet, 2012c; Izard & Dehaene, 2008). This ability to compare or estimate larger numerosities is suggested to be supported by a mechanism referred to as the approximate number system (ANS). The theory suggests that the number of items would be estimated or compared independent of the sensory properties such as the size or density of the dots that are present in the visual scene (Dehaene & Changeux, 1993; Stoianov & Zorzi, 2012; Verguts & Fias, 2004). On the basis of arguments outlined in detail elsewhere (Gebuis, Cohen Kadosh, & Gevers, submitted), we reasoned that the numerosity of a visual display of dots can be estimated, but this would not be done independently from sensory cues such as size, brightness, circumference, loudness, pitch, or frequency (Gebuis, Gevers, & Cohen Kadosh, 2014).

It is important to examine whether what is assumed to reflect numerosity processing is indeed numerosity, rather than sensory cues that may be processed by similar cognitive and neural mechanisms (Bueti & Walsh, 2009; Cantlon et al., 2009; Cohen Kadosh, Lammertyn, & Izard, 2008; Lourenco, 2015; Walsh, 2003). This similarity at the mechanistic level is not surprising because, as previously mentioned, visual or auditory properties are confounded with numerosity in everyday life. Take, for instance, the situation of two cues at the airport passport control. Here, you want to quickly estimate the number of people in each line to pick the one with the least people to save as much time as possible. Usually, the longer line holds the larger number of people, and it is therefore not unlikely that this information is used to guide our behavior.

Such a confound between numerical quantity and other visual features is not new and was already described by Piaget on the basis of the conservation paradigm (Piaget, 1965). In the conservation of number task, a child is typically shown two rows of buttons, each consisting of the same number of buttons. When one row of buttons is distributed more widely, the child at a certain developmental age is likely to say that the longer row now contains more buttons. Piaget claims that the childrens' thoughts in the preoperational stage (roughly age 2 until 5–6 years of age) are prelogical, as the children are only able to focus on one feature of a problem at a time and are dominated by their immediate perception of things. For instance, during a conservation task with liquids, a child will typically say that one glass contains *more* water because the water in that glass is *higher*. Or, in a number conservation task, a child will say that the row of more widely distributed buttons contains *more* buttons than the more densely packed row of buttons, simply because it is *longer*. In other words, at this stage a child is seemingly not able to differentiate between the visual features of a stimulus (eg, length, height, or density) and the numerosity associated with that stimulus (more or less items). Only later in development will a child become able to differentiate between the abstract notion of numerosity and visual features such as length, density, and size.

Once a person is able to differentiate between the perceptual cues and the abstract notion of numerosity, the important question is how a person derives the numerosity estimate. One possibility is that the person normalizes all visual cues, filtering out the numerosity information per se. This normalization process is one of the necessary steps in models that aim to explain how we represent numerosity and that forms the basis for the ANS theory (Dehaene & Changeux, 1993; Stoianov & Zorzi, 2012; Verguts & Fias, 2004). According to this theory then, pure numerosity is derived independently from or in parallel with the perceptual information. Another possibility is that the person will actually *use* the sensory information to perform some sort of integration across the different visual cues to derive a numerosity estimate. In recent work (Gebuis, Cohen Kadosh, & Gevers, submitted), we outlined in detail why we believe that the second option is more likely to be the case. Therefore, in the following section only a rapid overview of this argumentation is provided. Subsequently, assuming that integration of perceptual cues is the process resulting in a numerosity estimate or comparison, we will make a proposal of how this integration process develops and how it could be studied. Finally, we will provide some ideas on how the integration process could be related to arithmetic performance.

18.3 SENSORY CUES REMAIN TO INFLUENCE NUMEROSITY PROCESSES EVEN WHEN THEY ARE SEEMINGLY CONTROLLED

Studies that manipulate sensory cues show that changes in visual properties affect numerosity estimation (Allik & Tuulmets, 1991; Dakin, Tibber, Greenwood, Kingdom, & Morgan, 2011; Frith & Frith, 1972; Gebuis & Reynvoet, 2012a; Ginsburg, 1991; Ginsburg & Nicholls, 1988; Sophian, 2007; Sophian & Chu, 2008; Szucs & Soltesz, 2008; Szucs, Soltesz, Jarmi, & Csepe, 2007), while studies that tried to control the different sensory cues still found effects of the sensory cues such as congruency effects (Gebuis & Reynvoet, 2012b; Gilmore, Attridge, & Inglis, 2011; Halberda & Feigenson, 2008). This implies that the sensory cues influence numerosity processing even when sensory controls are applied. This can be explained as follows: studies that controlled sensory cues to investigate numerosity processing neglected the most important fact, namely that a perfect control for sensory cues is practically impossible. Two sets of items that differ in numerosity *always* differ in one or more sensory cues. Consequently, changing the sensory cues, that is, by making for instance dots larger or denser, etc. does not prevent reliance on these sensory cues. Instead, numerosity estimates could become less accurate, possibly because the integration of the different sensory cues becomes more difficult (Gebuis & Reynvoet, 2012b).

A few studies have investigated the effects of sensory cues. These studies showed that congruency effects change when the sensory cues change (Gebuis & Reynvoet, 2012b; Gebuis & Van der Smagt, 2011; Hurewitz, Gelman, & Schnitzer, 2006). Some of these changes cause the cancellation of a congruency effect, or even the reversal. A likely explanation for these results is that when multiple sensory cues exist, they tend to compete with each other in their weight given to the numerosity estimate. For instance, if there is a stimulus that is incongruent in surface but congruent in diameter, more weight could be given to diameter if this cue is more salient (eg, differs to a larger extent between the two stimuli that need to be compared). In this case, better performance for surface incongruent trials can be observed. According to this explanation two or more sensory cues can also cancel each other's effect (Gebuis & Reynvoet, 2012b). This interesting interplay between the different sensory cues to reach a numerosity estimate shows striking resemblances with the process of conservation, as previously explained.

18.4 SENSORY INTEGRATION AND ANS TASKS

As discussed earlier in this chapter, integrating the sensory cues when performing an ANS task like the numerosity comparison task seems to hold strong similarities with the well-known process of conservation (Piaget, 1952). Piaget divided children in different stages on the basis of their performance. The stage termed preoperational representation (beginning around 18 months of age and lasting until 6 or 7 years of age) is related to our argument. This is the period during which a child makes the famous conservation errors. A classic example is the liquid conservation problem where two identical glasses of water (eg, tall and narrow glasses, filled up to the same level) are presented to a child. Subsequently, while the child is watching, the water is poured from a tall and narrow glass into a lower and wider glass. If a child is not able yet to conserve, he or she will typically indicate that the narrow tall glass contains more water. Another conservation example relates to numerosity; here the same number of buttons is placed in two parallel lines of equal length. Then the experimenter moves the buttons of one line such that this line becomes longer than the other and again asks the child which line contains most buttons. A child that cannot conserve yet will typically say that now the longer line contains more buttons (Piaget, 1965). These are clear examples of situations or tasks where a single perceptual characteristic interferes with the correct numerical answer. An important factor to explain the difficulties observed in a conservation task is the notion that children have to understand that an increase in one dimension is compensated for by a decrease in another dimension (Piaget, 1952). This means that in the liquid conservation task children should understand that an increase in height is compensated for by a decrease in width, whereas in a numerosity task, children have to understand that a group of large dots scattered over a large area generally contains fewer dots than a group of small dots scattered over a smaller area at a higher density. The understanding that all of these sensory variables are related to each other and all together give information about the numerosity presented is the basis for what we termed the "integration procedure. " In this view, one individual cue cannot inform about numerosity; instead an integration of different visual cues is required. In other words, the integration is not derived from a calculation on the parts but a system on its own that receives input from the different sensory streams (for a similar position, see Anobile, Cicchini, & Burr, 2014; Arrighi, Togoli, & Burr, 2014). The integration of visual cues seems to relate to the concept of an inverse relationship between dimensions as observed in the conservation task. Indeed, in conservation tasks, children are tested on their understanding that number can remain the same even when physical appearances change. One could interpret this such that representations of number exist as well as its physical properties. In our sensory-based theory, we do not dispute that we have a concept of numerosity. What we do dispute is that this concept is derived independently from the different sensory properties via an ANS system. We believe that the children's inability to perform the Piaget task at an early age could indicate that they think of numerosity in terms of sensory cues and are not equipped with an innate number system that judges number separate from the visual cues. It is only at a later stage that they start to understand the concept of conservation. The children now understand that even though one row of buttons is longer, it does contain more space in between the individual buttons, and therefore might consist of an equal number of buttons as the more compressed row. In other words, understanding conservation would be similar to understanding that numerosity is comprised of several sensory cues that together give an insight in the number presented.

Conservation abilities are acquired throughout development. Therefore, if a parallel can be drawn between performance in numerosity comparison tasks and conservation tasks, it could be expected that children become better with age in comparison tasks. In other words, one would expect that the size of the congruency effect observed in numerosity comparison tasks decreases with increasing age. Such a decrease in the size of the congruency effect with age has indeed been observed (Szűcs, Nobes, Devine, Gabriel, & Gebuis, 2013; Gebuis, Cohen Kadosh, & Gevers, submitted). Furthermore, Szűcs et al. (2013) showed that performance of adults and children was the same for congruent trials but largely differed for incongruent trials. The authors concluded that the difference in the size of the congruency effect was related to inhibition abilities, arguing that younger children have more difficulty to inhibit irrelevant information compared to adults. This notion fits nicely with the theory of Piaget (1952) that children have to learn to inhibit the false heuristic of choosing the longer line or the taller glass, and instead base their judgment on numerosity. Indeed, very bad performance in numerosity tasks is visible at ages where inhibition is not yet fully developed. For instance, it has indeed been observed that children of 3 (Rousselle & Noël, 2008; Rousselle et al., 2004) or 4 (Soltezs et al., 2010) years of age were unable to perform the numerosity comparison task if visual cues were manipulated inconsistently. Performance is above chance starting from the age of 5 (Gebuis, Cohen Kadosh, de Haan, & Henik, 2009).

However, the ability to judge numerosity does not only depend on the general ability to inhibit false heuristics (Houdé et al., 2011) but also on the ability to weigh the different perceptual dimensions (Defever, Reynvoet, & Gebuis, 2013). Defever et al. (2013) compared the size of the congruency effect in a numerosity comparison task across different ages (first, second, third, and sixth grade of primary school). Surprisingly, the congruency effect increased with age. Closer inspection of the data demonstrated that visual cues were important for all age groups, but that younger children relied on a subset of the sensory cues. Not all of these children relied on the same subset of sensory cues [for similar heterogeneity in congruency effects, see Halberda & Feigenson (2008)]. In the youngest age group, about half of the participants associated a large visual cue with a larger numerosity, while the other half associated the larger visual cues with smaller numerosities. This response pattern resulted in opposite congruency effects, which led to the cancellation of the overall congruency effect. These opposite congruency effects diminished with increasing age, most likely because the older children stopped responding to a single subset of sensory cues but instead adopted a more diverse strategy of taking into account the full range of sensory cues. That this was a more effective strategy was visible in the overall performance as it increased with increasing age.

New research could investigate the direct link between conservation abilities, knowledge of the inverse rule, and numerosity comparison performance across different ages by disentangling the congruency effect with respect to different sensory manipulations. More work is needed on this topic. For instance, it would be interesting to know whether the developmental trajectory for a given child on conservation tasks is exactly the same as his trajectory for the congruency effect in typical ANS tasks like comparing larger numerosities.

18.5 SENSORY INTEGRATION AND ARITHMETIC

Previous studies suggested that performance on a numerosity comparison task is related to different mathematics abilities (Gilmore et al., 2011; Halberda et al., 2008; Libertus et al., 2011; Libertus, Feigenson, & Halberda, 2013; Lourenco, Bonny, Fernandez, & Rao, 2012; Mundy & Gilmore, 2009). Three reasons for this association have been put forth: (1) the acuity of the ANS system—the higher the acuity, the better the performance on ANS-like tasks and consequently mathematics (eg, Halberda, Libertus, etc.); (2) inhibition—the better the ability to inhibit responses, the smaller the congruency effects in number-size congruency tasks, and the better math performance (eg, Szűcs et al., 2013, Gilmore et al., 2013); and (3) conservation ability or the integration of different sensory cues—the better both abilities, the better math performance (Defever et al., 2013).

In our view, both inhibitory processes and conservation or integration abilities are required for numerosity tasks and consequently could both relate to math ability. More specifically inhibition is required to suppress a direct response to the most prominent sensory cues and in a subsequent step the participant has to be able to integrate the various sensory cues to make a numerosity judgment. Gilmore et al. (2013) suggested a different role for inhibition ability. Instead of suppressing the most prominent sensory cues to allow integration of the various sensory cues, they suggested that inhibition is necessary to inhibit all sensory cues to allow an ANS-based numerosity judgment independent of the sensory cues. In their first experiment, children performed a dot comparison task indicating which of two stimuli contained the largest number of dots. Children showed a congruency effect with worse performance on incongruent compared to congruent trials. The size of this congruency effect was taken as a marker of inhibitory skills and correlated with math ability. In the view of Gilmore et al. (2013), inhibition is required to suppress a response to the visual cues so that in the end participants can respond on the basis of their ANS. Freely translated, inhibition can be taken as the functional processing mechanism to accomplish normalization of the visual cues. The better the inhibition, the less influence of the visual cues, and the smaller the congruency effect. The idea now would be that better inhibitory functions (ie, the ability to suppress visual information) would relate to math achievement. Interestingly however, Gilmore also observed that the correlation between math achievement and inhibitory skills resulted from the incongruent trials only. This is not in line with the idea that inhibition is applied to all trials. Rather, inhibition would be applied to incongruent trials only, and it would be exactly this inhibition, which is related to math ability. One can ask the question—how is the child able to selectively apply inhibition to incongruent trials only? In other words, how does the child know that a trial is incongruent if he did not apply inhibition on the visual cues yet? In our proposal, inhibition is an important feature, but it would be needed to suppress the initial tendency to respond to the most salient visual feature, and do this regardless of whether a trial is congruent or incongruent. The difference in performance between congruent and incongruent trials would result simply because weighing (ie, the integration) of the different sensory variables is more difficult on incongruent compared to congruent trials.

If conservation abilities underlie performance on numerosity comparison tasks, a logical consequence is to expect a relationship between sensory integration or conservation ability and math achievement as well. This is indeed the case: a large number of studies observed strong correlations between sensory processes (Lourenco et al., 2012; Tibber et al., 2013) and conservation ability and arithmetic achievement at the end of first grade (Dimitrovsky & Almy, 1975; Dodwell, 1961; Kaufman & Kaufman, 1972). The correlation between conservation abilities and math achievement persist even when IQ was controlled for (Taloumis, 1979).

Further research was performed trying to establish the value and the meaning of this correlation. One line of research focused on intervention studies. Researchers asked the question whether training children the conservation method would result in better arithmetic performance (Bearison, 1975). The answer was negative. Children who spontaneously achieved conservation earlier benefited more from instructions in arithmetic but this benefit could not be induced by training the children on conservation-like tasks (Bearison, 1975). Can a similar pattern of observations be made in studies relating numerosity processing abilities to math achievement? To establish the association between the age onset of numerosity processing abilities and math achievement later on, a developmental individual differences approach is needed, which to our knowledge has not been conducted yet. However, a few intervention studies have been conducted that included some form of numerosity training and measures of math achievement (Räsänen, Wilson, Aunio, & Dehaene, 2009; Wilson et al., 2006a; Wilson, Revkin, Cohen, Cohen, & Dehaene, 2006). However, none included control conditions and therefore do not allow for strong conclusions. One recent study did use language processing as a control condition (Obersteiner, Reiss, & Ufer, 2013). In line with the results observed in studies linking conservation to math achievement, no main effect of numerosity training on math achievement was observed. Caution is needed as this parallel is based on a null finding (eg, no effect of training numerosity on math achievement). Clearly, more work is needed here.

On the basis of an extensive review of the literature, Hiebert and Carpenter (1982) used a different approach. They investigated which task characteristic in conservation tasks would be responsible for the observed association with certain aspects of math achievement.

A first relevant observation was the strong relation (beyond IQ) between conservation abilities and math achievement. Importantly, however, success on the conservation task was not a prerequisite for success on a number of basic mathematics tasks. In other words, even when children had difficulties with some aspects of the conservation task, mathematic skills could (at least in some children) still be acquired (Hiebert, Carpenter, & Moser, 1982; Steffe, 1976). The ability that seemed most directly related to math achievement was again the notion of an inverse relationship between unit number and unit size. Maybe this observation could provide a clue to answer the important remaining question as to why the ability to compare more accurately larger numerosities, which requires the integration of the different sensory cues, is related to math achievement. In our view, such a link, if existent, might be mediated by inhibitory mechanisms (Houdé, 2009; Pina, Moreno, Cohen Kadosh, & Fuentes, 2015). The ability to combine different sensory cues to make an estimate of numerosity seems to be key to strong performance on numerosity studies. However, a parallel reasoning would also have a more pessimistic implication, namely that intervention studies improving numerosity judgments would not easily result in improvements on math achievement. However, to properly address this relationship, more intervention studies focusing on the specific strategies used to judge numerosities and relate it to math achievement are needed.

18.6 CONCLUSIONS

In this chapter, we have described a number of important caveats in the theory of the ANS, which describes our ability to estimate and or compare large numerosities. Empirical evidence is not consistent with, or even contradicts, the existence of an ANS. We therefore proposed an alternative sensory integration mechanism that, without invoking a number sense, seems to be able to explain our performance in numerosity judgment tasks. By making a theoretical parallel between the integration mechanism and conservation abilities, some testable hypotheses could be formulated. These hypotheses concern the processing of numerosities itself as well as the observed link between numerosity processing and math achievement. We hope that such insights will open new venues for studies that will be able to take into account the effect of sensory cues, and will allow better theoretical progress with impact not only on basic science but also on numerical deficiencies (eg, dyscalculia) and intervention studies.

References

Allik, J., & Tuulmets, T. (1991). Occupancy model of perceived numerosity. *Attention Perception Psychophysics*, *49*(4), 303–314.

Anobile, G., Cicchini, G. M., & Burr, D. C. (2014). Separate mechanisms for perception of numerosity and density. *Psychological Science*, *1*(25), 265–270.

Arrighi, R., Togoli, I., & Burr, D. (2014). A generalized sense of number. *Proceedings of the Royal Society of London B*, *281*:20141791.

Barth, H., Kanwisher, N., & Spelke, E. (2003). The construction of large number representations in adults. *Cognition*, *86*(3), 201–221.

Bearison, D. J. (1975). Induced versus spontaneous attainment of concrete operations and their relationship to school achievement. *Journal of Educational Psychology*, *67*, 576–580.

Bueti, D., & Walsh, V. (2009). The parietal cortex and the representation of time, space, number and other magnitudes. *Philosophical Transactions of the Royal Society B*, *364*, 2369–2380.

Burr, D., & Ross, J. (2008). A visual sense of number. *Current Biology*, *18*(6), 425–428.

Cantlon, J. F., Brannon, E. M., Carter, E. J., & Pelphrey, K. A. (2006). Functional imaging of numerical processing in adults and 4-y-old children. *PLoS Biology*, *4*(5), e125.

Cantlon, J. F., Platt, M. L., & Brannon, E. M. (2009). Beyond the number domain. *Trends in Cognitive Sciences*, *13*, 83–91.

Cohen Kadosh, R., Lammertyn, J., & Izard, V. (2008). Are numbers special? An overview of chronometric, neuroimaging, developmental and comparative studies of magnitude representation. *Progress in Neurobiology*, *84*, 132–147.

Dakin, S. C., Tibber, M. S., Greenwood, J. A., Kingdom, F. A., & Morgan, M. J. (2011). A common visual metric for approximate number and density. *Proceedings of the National Academy of Sciences USA*, *108*(49), 19552–19557.

De Smedt, B., & Gilmore, C. K. (2011). Defective number module or impaired access? Numerical magnitude processing in first graders with mathematical difficulties. *Journal of Experimental Child Psychology*, *108*(2), 278–292.

Defever, E., Reynvoet, B., & Gebuis, T. (2013). Task- and age-dependent effects of visual stimulus properties on children's explicit numerosity judgments. *Journal of Experimental Child Psychology*, *116*, 216–233.

Dehaene, S. (1992). Varieties of numerical abilities. *Cognition*, *44*(1–2), 1–42.

Dehaene, S., & Changeux, J. P. (1993). Development of elementary numerical abilities: A neuronal model. *Journal of Cognitive Neuroscience*, *5*(4), 390–407.

Dimitrovsky, L., & Almy, M. (1975). Early conservation as a predictor of arithmetic achievement. *The Journal of Psychology*, *91*, 65–70.

Dodwell, P. C. (1961). Children s understanding of number concepts: characteristics of an individual and of a group test. *Canadian Journal of Psychology*, *15*, 29–36.

Feigenson, L., Dehaene, S., & Spelke, E. C. (2004). Core systems of number. *Trends in Cognitive Science*, *8*(7), 307–314.

Frith, C. D., & Frith, U. (1972). The solitaire illusion: an illusion of numerosity. *Perception and Psychophysics*, *11*(6), 409–410.

Gebuis, T., & Reynvoet, B. (2012a). Continuous visual properties explain neural responses to nonsymbolic number. *Psychophysiology*, *49*(11), 1649–1659.

Gebuis, T., & Reynvoet, B. (2012b). The interplay between nonsymbolic number and its continuous visual properties. *Journal of Experimental Psychology*, *141*, 642–648.

Gebuis, T., & Reynvoet, B. (2012c). The role of visual information in numerosity estimation. *PLoS One*, *7*(5), 37426.

Gebuis, T., & Van der Smagt, M. J. (2011). False approximations of the approximate number system? *PLoS One*, *6*(10), 25405.

Gebuis, T., Cohen Kadosh, R., de Haan, E., & Henik, A. (2009). Automatic quantity processing in 5-year olds and adults. *Cognitive Processing*, *10*(2), 133–142.

Gebuis, T., Gevers, W., & Cohen Kadosh, R. (2014). Topographic representation of high-level cognition: numerosity or sensory processing? *Trends in Cognitive Sciences*, *18*, 1–3.

Gilmore, C., Attridge, N., & Inglis, M. (2011). Measuring the approximate number system. *Quarterly Journal of Experiment Psychology*, *64*(11), 2099–2109.

Gilmore, C., Attridge, N., Clayton, S., Cragg, L., Johnson, S., Marlow, N., & Inglis, M. (2013). Individual differences in inhibitory control, not non-verbal number acuity, correlate with mathematics achievement. *PLoS One*, *8*(6), e67374.

Ginsburg, N. (1991). Numerosity estimation as a function of stimulus organization. *Perception*, *20*(5), 681–686.

Ginsburg, N., & Nicholls, A. (1988). Perceived numerosity as a function of item size. *Perceptual and Motor Skills*, *67*(2), 656–658.

Halberda, J., & Feigenson, L. (2008). Developmental change in the acuity of the "number sense": The approximate number system in 3-, 4-, 5-, and 6-year-olds and adults. *Developmental Psychology*, *44*(5), 1457–1465.

Halberda, J., Mazzocco, M. M., & Feigenson, L. (2008). Individual differences in non-verbal number acuity correlate with maths achievement. *Nature*, *455*(7213), 665–668.

Hiebert, J., & Carpenter, T. P. (1982). Piagetian tasks as readiness measures in mathematics instruction: a critical review. *Educational Studies in Mathematics*, *13*(3), 329–345.

Hiebert, J., Carpenter, T. P., & Moser, J. M. (1982). Cognitive development and children's solutions to verbal arithmetic problems. *Journal of Research in Mathematics Education*, *13*, 83–98.

Holloway, I. D., & Ansari, D. (2009). Mapping numerical magnitudes onto symbols: the numerical distance effect and individual differences in children's mathematics achievement. *Journal of Experimental Child Psychology*, *103*(1), 17–29.

Houdé, O. (2009). Abstract after all? Abstraction through inhibition in children and adults. *Behavioral Brain Sciences*, *32*, 339–340.

Houdé, O., Pineau, A., Leroux, G., Poirel, N., Perchey, G., Lanoë, C., Lubin, A., Turbelin, M.-R., Rossi, S., Simon, G., Delcroix, N., Lamberton, F., Vigneau, M., Wisniewski, G., Vicet, J.-R., & Mazoyer, B. (2011). Functional MRI study of Piaget's conservation-of-number task in preschool and school-age children: a neo-Piagetian approach. *Journal of Experimental Child Psychology*, *110*, 332–346.

Hurewitz, F., Gelman, R., & Schnitzer, B. (2006). Sometimes area counts more than number. *Proceedings of the National Academy of Sciences USA*, *103*(51), 19599–19604.

Izard, V., & Dehaene, S. (2008). Calibrating the mental number line. *Cognition*, *106*(3), 1221–1247.

Kaufman, A. S., & Kaufman, N. L. (1972). Tests build from Piaget's and Gesell's tasks as predictors of first grade achievement. *Child Development*, *43*, 521–535.

Kaufman, E. L., Lord, M. W., Reese, T. W., & Volkmann, J. (1949). The discrimination of visual number. *American Journal of Psychology*, *62*(4), 498–525.

Libertus, M. E., Feigenson, L., & Halberda, J. (2011). Preschool acuity of the approximate number system correlates with school math ability. *Developmental Science*, *14*(6), 1292–1300.

Libertus, M., Feigenson, L., & Halberda, J. (2013). Is approximate number precision a stable predictor of math ability? *Learning and Individual Differences*, *25*, 126–133.

Lourenco, S. F. (2015). On the relation between numerical and non-numerical magnitudes: evidence for a general magnitude system. In D. C. Geary, D. B. Berch, & K. Mann-Koepke (Eds.), *Mathematical cognition and learning: evolutionary origins and early development of basic number processing* (pp. 145–174). (Vol. 1). New York: Elsevier.

Lourenco, S. F., Bonny, J. W., Fernandez, E. P., & Rao, S. (2012). Nonsymbolic number and cumulative area representations contribute shared and unique variance to symbolic math competence. *Proceedings of the National Academy of Sciences USA, 109*(46), 18737–18742.

Mundy, E., & Gilmore, C. K. (2009). Children's mapping between symbolic and nonsymbolic representations of number. *Journal of Experimental Child Psychology, 103*(4), 490–502.

Obersteiner, A., Reiss, K., & Ufer, S. (2013). How training on exact or approximate mental representations of number can enhance first-grade students' basic number processing and arithmetic skills. *Learning and Instruction, 23*, 125–135.

Park, J., & Brannon, E. M. (2013). Training the approximate number system improves math proficiency. *Psychological Science, 24*(10), 2013–2019.

Piaget, J. (1952). *The child's conception of number*. London: Routledge & KeganPaul.

Piaget, J. (1965). *The child's conception of number*. New York: W. Norton Company & Inc.

Piazza, M. (2010). Neurocognitive start-up tools for symbolic number representations. *Trends in Cognitive Sciences, 14*(12), 542–551.

Piazza, M., Izard, V., Pinel, P., Le Bihan, D., & Dehaene, S. (2004). Tuning curves for approximate numerosity in the human intraparietal sulcus. *Neuron, 44*(3), 547–555.

Pina, V., Moreno, A. C., Cohen Kadosh, R., & Fuentes, L. J. (2015). Intentional and automatic numerical processing as predictors of mathematical abilities in primary school children. *Frontiers in Psychology, 6*, 375.

Räsänen, P. J., Wilson, A. J., Aunio, P., & Dehaene, S. (2009). Computer-assisted intervention for children with low numeracy skills. *Cognitive Development, 24*(4), 450–472.

Rousselle, L., & Noël, M. P. (2007). Basic numerical skills in children with mathematics learning disabilities: a comparison of symbolic vs non-symbolic number magnitude processing. *Cognition, 102*(3), 361–395.

Rousselle, L., & Noël, M.-P. (2008). The development of automatic numerosity processing in preschoolers: evidence for numerosity-perceptual interference. *Developmental Psychology, 44*, 544–560.

Rousselle, L., Palmers, E., & Noël, M.-P. (2004). Magnitude comparison in preschoolers: what counts? Influence of perceptual variables. *Journal of Experimental Child Psychology, 87*, 57–84.

Sasanguie, D., De Smedt, B., Defever, E., & Reynvoet, B. (2012). Association between basic numerical abilities and mathematics achievement. *British Journal of Developmental Psychology, 30*(Pt. 2), 344–357.

Soltesz, F., Szucs, D., & Szucs, L. (2010). Relationships between magnitude representation, counting and memory in 4- to 7-year-old children: a developmental study. *Behavioral and Brain Functions, 6*, 13.

Sophian, C. (2007). Measuring spatial factors in comparative judgments about large numerosities. In D. Schmorrow, & L. Reeves (Eds.), *Foundations of Augmented Cognition: Third International Conference* (pp. 157–165). Secaucus, NJ: Springer.

Sophian, C., & Chu, Y. (2008). How do people apprehend large numerosities? *Cognition, 107*(2), 460–478.

Steffe, L. P. (1976). On a model for teaching young children mathematics. In A. R. Osborne (Ed.), *Models for Learning Mathematics, ERIC/SMEAC*. Ohio: Columbus.

Stoianov, I., & Zorzi, M. (2012). Emergence of a 'visual number sense' in hierarchical generative models. *Nature Neuroscience, 15*, 194–196.

Szucs, D., & Soltesz, F. (2008). The interaction of task-relevant and task-irrelevant stimulus features in the number/size congruency paradigm: an ERP study. *Brain Research, 1190*, 143–158.

Szucs, D., Soltesz, F., Jarmi, E., & Csepe, V. (2007). The speed of magnitude processing and executive functions in controlled and automatic number comparison in children: an electroencephalography study. *Behavioral and Brain Functions, 3*, 23.

Szűcs, D., Nobes, A., Devine, A., Gabriel, F. C., & Gebuis, T. (2013). Visual stimulus parameters seriously compromise the measurement of approximate number system acuity and comparative effects between adults and children. *Frontiers in Psychology, 4*, 444.

Taloumis, T. (1979). Scores on Piagetian area tasks as predictors of achievement in mathematics over a four-year period. *Journal of Researcher in Mathematics Education, 10*, 120–134.

Tibber, M. S., Manasseh, G. S. L., Clarke, R. C., Gagin, G., Swanbeck, S. N., Butterworth, B., Lotto, R. B., & Dakin, S. C. (2013). Sensitivity to numerosity is not a unique visuospatial psychophysical predictor of mathematical ability. *Vision Research, 89*, 1–9.

Trick, L. M., & Pylyshyn, Z. W. (1994). Why are small and large numbers enumerated differently? A limited-capacity preattentive stage in vision. *Psychological Review, 101*(1), 80–102.

Tzelgov, J., Yehene, V., Kotler, L., & Alon, A. (2000). Automatic comparisons of artificial digits never compared: learning linear ordering relations. *Journal of Experimental Psychology, 26*, 103–120.

Verguts, T., & Fias, W. (2004). Representation of number in animals and humans: a neural model. *Journal of Cognitive Neuroscience, 16*(9), 1493–1504.

Walsh, V. (2003). A theory of magnitude: common cortical metrics of time, space and quantity. *Trends in Cognitive Sciences, 7*, 483–488.

Wilson, A. J., Dehaene, S., Pinel, P., Revkin, S. K., Cohen, L., & Cohen, D. (2006a). Principles underlying the design of "The Number Race," an adaptive computer game for remediation of dyscalculia. *Behavioral and Brain Function, 2*, 19.

Wilson, A. J., Revkin, S. K., Cohen, D., Cohen, L., & Dehaene, S. (2006b). An open trial assessment of "The Number Race," an adaptive computer game for remediation of dyscalculia. *Behavioral and Brain Function, 2*, 20.

Subject Index

O

W

X

Y

Printed and bound by CPI Group (UK) Ltd, Croydon, CR0 4YY
08/06/2025
01896870-0008